Attention all "Enzyme Handbook" users:

A file with the complete volume indexes Vols. 1 through 9 in delimited ASCII format is available for downloading at no charge from the Springer EARN mailbox. Delimited ASCII format can be imported into most databanks.

The file has been compressed using the popular shareware program "PKZIP" (Trademark of PKware Inc., PKZIP is available from most BBS and shareware distributors).

This file is distributed without any expressed or implied warranty.

To receive this file send an e-mail message to:
SVSERV@DHDSPRI6.BITNET

The message must be:
GET /CHEMISTRY/ENZ_HB.ZIP

SVSERV is an automatic data distribution system. It responds to your message. The following commands are available:

HELP	returns a detailed instruction set for the use of SVSERV,
DIR *(name)*	returns a list of files available in the directory "name",
INDEX *(name)*	same as "DIR"
CD *<name>*	changes to directory "name",
SEND *<filename>*	invokes a message with the file "filename",
GET *<filename>*	same as "SEND".

D. Schomburg · D. Stephan (Eds.)
GBF – Gesellschaft für Biotechnologische Forschung

Enzyme Handbook

9

Class 1.1: Oxidoreductases

EC 1.1.1.1 – EC 1.1.1.149

*for EC 1.1.1.150 – EC 1.1.99.26
see Vol. 10*

Springer-Verlag Berlin
Heidelberg GmbH

Professor Dr. Dietmar Schomburg
Dr. Dörte Stephan

GBF – Gesellschaft für Biotechnologische Forschung mbH
Mascheroder Weg 1

38124 Braunschweig
FRG

This collection of datasheets was generated from the database „BRENDA"

ISBN 978-3-642-48987-7 ISBN 978-3-642-85200-8 (eBook)
DOI10.1007/978-3-642-85200-8

Library of Congress Cataloging-in-Publication Data. (Revised for vol. 9). Enzyme Handbook. Vols. 6–7 edited by D. Schomburg, M. Salzmann, D. Stephan, vols. 8–9 edited by Schomburg and Stephan. Loose-leaf. Includes bibliographical references and indexes. Contents: 1. Class 4: Lyases – 2. Class 5: Isomerases. Class 6: Ligases – [etc.] – 9. Class 1.1: Oxidoreductases. 1. Enzymes – Handbooks, manuals, etc. I. Schomburg, D. (Dietmar). II. Salzmann, M. (Margit). III. Stephan, D. (Dörte).
QP601.E5158 1990 660'.634 91-145566
ISBN 978-3-642-48987-7

Media conversion, printing and bookbinding: Brühlsche Universitätsdruckerei, Giessen
Production of the plasticfiles: Lux-Plastik oHG, Murnau
SPIN: 10056100 51/3020 - 5 4 3 2 1 0 - Printed on acid-free paper

Preface

Recent progress on enzyme immobilisation, enzyme production, coenzyme regeneration and enzyme engineering has opened up fascinating new fields for the potential application of enzymes in a large range of different areas. As more progress in research and application of enzymes has been made the lack of an up-to-date overview of enzyme molecular properties has become more apparent. Therefore, we started the development of an enzyme data information system as part of protein-design activities at GBF. The present book "Enzyme Handbook" represents the printed version of this data bank. In future a computer searchable version will be also available.

The enzymes in this Handbook are arranged according to the Enzyme Commission list of enzymes. Some 3000 "different" enzymes will be covered. Frequently enzymes with very different properties are included under the same EC number. Although we intend to give a representative overview on the characteristics and variability of each enzyme the Handbook is not a compendium. The reader will have to go to the primary literature for more detailed information. Naturally it is not possible to cover all the numerous literature references for each enzyme (for special enzymes up to 40000) if the data representation is to be concise as is intended.

It should be mentioned here that the literature data are extracted from literature and critically evaluated by qualified scientists. On the other hand the original authors' nomenclature for enzyme forms and subunits is retained as is their nomenclature for organisms and strains even if the organism is reclassified in the meantime. The cross references to the protein sequence data bank and to the Brookhaven protein 3D structure data bank are taken directly from their data files without further verification by the authors. In order to keep the tables concise redundant information is avoided as far as possible (e.g. if K_m values are measured in the presence of an obvious cosubstrate, only the name of the cosubstrate is given in parentheses as a commentary without reference to its specific role).

The authors are grateful to the following biologists and chemists for invaluable help in the compilation of data: Margit Salzmann, Cornelia Munaretto, Dr. Ida Schomburg, Dr. Astrid Beermann and Astrid Haberz. In addition we would like to thank Mrs. C. Munaretto and Dr. I. Schomburg for the correction of the final manuscript.

Braunschweig, Spring 1995

Dörte Stephan
Dietmar Schomburg

BRENDA – Compilation of Enzyme Data

To collect basic characteristics of enzymes – is that not a kind of archaic activity in the times of molecular biology and computer-aided data banks providing sequences of nucleic acids and proteins with little more delay than a few days as well as their three-dimensional structures? What should be the purpose of compiling turnover numbers, Michaelis constants, substrate specificities, sources, synonyms etc. of enzymes from sometimes remote publications? The answer sounds as simple as surprising: The aim of the compilation of data is to make use of the overwhelming abundance of structural knoweldge we owe to the new techniques of molecular biology.

Admittedly, it was not primarily enzymology which caused the explosion of knowledge in biology during the last decade. This was due to the advance of molecular biology which enabled us to isolate genes, to amplify them *ad libidum* and to elucidate their primary structure within days only. Also, the optimization and automatization of techniques for the analysis of macromolecules has provided detailed insights into a large variety of complex biomolecules nobody would have anticipated in the early seventies. Due to powerful computers it has now become feasible to propose fairly realistic models of macromolecules based solely on primary structures and homology considerations.

Nevertheless – or therefore – it appears as mandatory as rewarding to know the brave world of enzymology in which one had and often still has to come along without any detailed structural knowledge. We should not ignore that nature has not generated the multiplicity of structures, because it simply felt obliged to the principle of diversification or because it wanted to test our computing capacity to handle sequence data. It had to create new structures to cope with the steadily changing demands of a variable environment. Thus, amino acid sequences, folding of peptide chains and conformational details are only the technical tools of nature to catalyse specific biological functions. In consequence, *it is the functional profile of an enzyme which enables a biologist or physician to analyze a metabolic pathway and its disturbance; it is the substrate specificity of an enzyme which tells an analytical biochemist how to design an assay; it is the stability, specificity and efficiency of an enzyme which determines its usefulness in the biotechnical transformation of a molecule.* And the sum of all these functional data will have to be considered when the designer of artificial biocatalysts has to choose the optimum prototype to start with.

Unfortunately, it is by no means as simple to design (organize) a meaningful and systematic compilation of functional enzymological data as to enter sequences of amino acids or nucleotides into a data base. Functional data are less well defined, are never devoid of a trace of ambiguity, their selection remains inevitably subjective, and their complexity requires simplification. The present compilation of enzymological data, therefore, can and will not be a substitute for original publications but rather offer a key to the literature. But I do think that the Enzyme Handbook is indeed an excellent key to open or reopen the mysterious world of

enzyme to all those who there have to find the solutions of their problems: to biologists, physicians, structural biochemists, biochemical analysts, biotechnologists and also to the molecular biologists.

Braunschweig, Spring 1993

Leopold Flohé
GBF, Scientific Director

List of Abbreviations

A	adenosine
Ac	acetyl
ACP	acyl-carrier-protein
ADP	adenosine 5'-diphosphate
Ala	alanine
All	allose
Alt	altrose
AMP	adenosine 5'-monophosphate
Ara	arabinose
Arg	arginine
Asn	asparagine
Asp	aspartic acid
ATP	adenosine 5'-triphosphate
Bicine	N,N'-bis(2-hydroxyethyl) glycine
C	cytidine
cal	calorie
CDP	cytidine 5'-diphosphate
CDTA	trans-1,2-diaminocyclo-hexane-N,N,N,N-tetra-aceticacid
CHAPS	3-[(3-cholamidopropyl)-dimethylammonio]-1-propanesulfonate
CHAPSO	3-[(3-cholamidopropyl)-dimethylammonio]-2-hydroxy-1-propane-sulfonate
CMP	cytidine 5'-monophosphate
CoA	coenzyme A
CTP	cytidine 5'-triphosphate
Cys	cysteine
d	deoxy-
D- and L-	prefixes indicating configuration
DFP	diisopropylfluorophosphate
DNA	deoxyribonucleic acid
DPN	diphosphopyridinium nucleotide (now NAD)
DTNB	5,5'-dithiobis(2-nitrobenzoate)
DTT	dithiothreitol (i.e. Cleland's reagent)
e	electron
EC	number of enzyme in Enzyme Commission's system
E. coli	Escherichia coli
EDTA	ethylene diaminetetraacetate
EGTA	ethylene glycol bis (β-amino-ethylether) tetraacetate
EPR	electron paramagnetic resonance
ER	endoplasmic reticulum
Et	ethyl
EXAFS	extended X-ray absorption fine structure
FAD	flavin-adenine dinucleotide
FMN	flavin mononucleotide (riboflavin 5'-monophosphate)
FPLC	fast protein liquid chromatography
Fru	fructose
Fuc	fucose
G	guanosine
Gal	galactose
GDP	guanosine 5'-diphosphate
Glc	glucose
GlcN	glucosamine
GlcNAc	N-acetylglucosamine
Gln	glutamine
Glu	glutamic acid
Gly	glycine
Glygly	glycylglycine
GMP	guanosine 5'-monophosphate
GSH	glutathione
GSSG	oxidized glutathione
GTP	guanosine 5'-triphosphate
Gul	gulose
h	hour
H_4	tetrahydro
HEPES	4-(2-hydroxyethyl)-1-piperazineethane sulfonic acid
His	histidine
HPLC	high performance liquid chromatography
Hyl	hydroxylysine
Hyp	hydroxyproline
IAA	iodoacetamide
Ig	immunoglobulin
Ile	isoleucine
Ido	idose
IDP	inosine 5'-diphosphate
IMP	inosine 5'-monophosphate

ir	irreversible
ITP	inosine 5'-triphosphate
K_m	Michaelis constant
L-	see D-
Leu	leucine
Lys	lysine
Lyx	lyxose
M	mol/l
m-	meta-
Man	mannose
MES	2-(N-morpholino)ethane sulfonate
Met	methionine
min	minute
MOPS	3-(N-morpholino) propane sulfonate
Mur	muramic acid
MW	molecular weight
NAD	nicotinamide-adenine dinucleotide
NADH	reduced NAD
NADP	NAD phosphate
NADPH	reduced NADP
NAD(P)H	indicates either NADH or NADPH
NDP	nucleoside 5'-diphosphate
NEM	N-ethylmaleimide
Neu	neuraminic acid
NMN	nicotinamide mononucleotide
NMP	nucleoside 5'-monophosphate
NTP	nucleoside 5'-triphosphate
o-	ortho-
Orn	ornithine
p-	para-
PCMB	p-chloro-mercuribenzoate
PEG	polyethylene glycol
PEP	phosphoenolpyruvate
pH	$-\log_{10} [H^+]$
Ph	phenyl
Phe	phenylalanine
PIXE	proton-induced X-ray emission
PMSF	phenylmethane-sulfonylfluoride
Pro	proline
Q_{10}	factor for the change in reaction rate for a 10° temperature increase
r	reversible
Rha	rhamnose
Rib	ribose
RNA	ribonucleic acid
mRNA	messenger RNA
rRNA	ribosomal RNA
tRNA	transfer RNA
Sar	N-methylglycine (sarcosine)
SDS-PAGE	sodium dodecyl sulphate polyacrylamide gel electrophoresis
Ser	serine
SFK-525A	2-diethylaminoethyl-2,2-diphenylvalerate
sp.	species
T	ribosylthymine
$t^{1/2}$	time for half-completion of reaction
Tal	talose
TDP	ribosylthymine 5'-diphosphate
TEA	triethanolamine
THF	tetrahydrofolate
Thr	threonine
TMP	ribosylthymine 5'-monophosphate
Tos-	tosyl-(p-toluenesulfonyl-)
TPN	triphosphopyridinium nucleotide (now NADP)
Tris	tris(hydroxymethyl)-aminomethane
Trp	tryptophan
TTP	ribosylthymine 5'-triphosphate
Tyr	tyrosine
U	uridine
U/mg	µmol/(mg·min)
UDP	uridine 5'-diphosphate
UMP	uridine 5'-monophosphate
UTP	uridine 5'-triphosphate
UV	ultraviolet
Val	valine
Xaa	symbol for an amino acid of unknown constitution in peptide formula
XAS	X-ray absorption spectroscopy
XTP	xanthosine 5'-triphosphate
Xyl	xylose

Index

(Alphabetical order of Enzyme names)

For EC 1.1.1.1 – EC 1.1.1.149 see Vol. 9

For EC 1.1.1.150 – EC 1.1.99.26 see Vol. 10

EC-No.	Name
1.1.1.194	Coniferyl-alcohol dehydrogenase [1]
1.1.1.174	Cyclohexane-1,2-diol dehydrogenase
1.1.1.245	Cyclohexanol dehydrogenase
1.1.1.163	Cyclopentanol dehydrogenase
1.1.1.127	2-Dehydro-3-deoxy-D-gluconate 5-dehydrogenase
1.1.1.126	2-Dehydro-3-deoxy-D-gluconate 6-dehydrogenase
1.1.99.4	Dehydrogluconate dehydrogenase
1.1.1.130	3-Dehydro-L-gulonate 2-dehydrogenase
1.1.1.169	2-Dehydropantoate 2-reductase
1.1.1.168	2-Dehydropantoyl-lactone reductase (A-specific)
1.1.1.214	2-Dehydropantoyl-lactone reductase (B-specific)
1.1.1.102	3-Dehydrosphinganine reductase
1.1.1.125	2-Deoxy-D-gluconate 3-dehydrogenase
1.1.1.229	Diethyl 2-methyl-3-oxosuccinate reductase
1.1.1.160	Dihydrobunolol dehydrogenase
1.1.1.219	Dihydrokaempferol 4-reductase
1.1.1.96	Diiodophenylpyruvate reductase
1.1.1.84	Dimethylmalate dehydrogenase
1.1.1.133	dTDP-4-dehydrorhamnose reductase
1.1.1.134	dTDP-6-deoxy-L-talose 4-dehydrogenase
1.1.1.186	dTDPgalactose 6-dehydrogenase
1.1.3.16	Ecdysone oxidase
1.1.1.162	Erythrulose reductase
1.1.1.148	Estradiol 17alpha-dehydrogenase

EC-No.	Name
1.1.1.62	Estradiol 17beta-dehydrogenase
1.1.1.216	Farnesol dehydrogenase
1.1.1.182	Fenchol dehydrogenase
1.1.1.234	Flavanone 4-reductase
1.1.99.11	Fructose 5-dehydrogenase
1.1.1.124	Fructose 5-dehydrogenase ($NADP^+$)
1.1.1.57	Fructuronate reductase
1.1.1.16	Galactitol 2-dehydrogenase
1.1.3.24	L-Galactonolactone oxidase
1.1.1.48	Galactose 1-dehydrogenase
1.1.1.120	Galactose 1-dehydrogenase ($NADP^+$)
1.1.3.9	Galactose oxidase
1.1.1.187	GDP-4-dehydro-D-rhamnose reductase
1.1.1.135	GDP-6-deoxy-D-talose 4-dehydrogenase
1.1.1.132	GDPmannose 6-dehydrogenase
1.1.1.183	Geraniol dehydrogenase
1.1.1.215	Gluconate 2-dehydrogenase
1.1.99.3	Gluconate 2-dehydrogenase
1.1.1.69	Gluconate 5-dehydrogenase
1.1.99.10	Glucose dehydrogenase (acceptor)
1.1.99.17	Glucose dehydrogenase (pyrroloquinoline-quinone)
1.1.1.47	Glucose 1-dehydrogenase
1.1.1.118	Glucose 1-dehydrogenase (NAD^+)
1.1.1.119	Glucose 1-dehydrogenase ($NADP^+$)
1.1.3.4	Glucose oxidase
1.1.1.49	Glucose-6-phosphate 1-dehydrogenase
1.1.99.13	Glucoside 3-dehydrogenase
1.1.1.19	Glucuronate reductase
1.1.1.20	Glucuronolactone reductase
1.1.1.29	Glycerate dehydrogenase
1.1.1.6	Glycerol dehydrogenase
1.1.99.22	Glycerol dehydrogenase (acceptor)

EC-No.	Name
1.1.1.72	Glycerol dehydrogenase ($NADP^+$)
1.1.1.156	Glycerol 2-dehydrogenase ($NADP^+$)
1.1.99.5	Glycerol-3-phosphate dehydrogenase
1.1.1.8	Glycerol-3-phosphate dehydrogenase (NAD^+)
1.1.1.94	Glycerol-3-phosphate dehydrogenase ($NAD(P)^+$)
1.1.1.177	Glycerol-3-phosphate 1-dehydrogenase ($NADP^+$)
1.1.3.21	Glycerol-3-phosphate oxidase
1.1.99.14	Glycolate dehydrogenase
1.1.1.185	L-Glycol dehydrogenase
1.1.1.26	Glyoxylate reductase
1.1.1.79	Glyoxylate reductase($NADP^+$)
1.1.1.45	L-Gulonate 3-dehydrogenase
1.1.3.8	L-Gulonolactone oxidase
1.1.1.164	Hexadecanol dehydrogenase
1.1.3.5	Hexose oxidase
1.1.1.23	Histidinol dehydrogenase
1.1.1.155	Homoisocitrate dehydrogenase
1.1.1.3	Homoserine dehydrogenase
1.1.99.6	D-2-Hydroxy-acid dehydrogenase
1.1.3.15	(S)-2-Hydroxy-acid oxidase
1.1.99.24	Hydroxyacid-oxoacid transhydrogenase
1.1.1.35	3-Hydroxyacyl-CoA dehydrogenase
1.1.1.152	3alpha-Hydroxy-5beta-androstan-17-one 3alpha-dehydrogenase
1.1.1.97	3-Hydroxybenzyl-alcohol dehydrogenase
1.1.1.30	3-Hydroxybutyrate dehydrogenase
1.1.1.61	4-Hydroxybutyrate dehydrogenase
1.1.1.157	3-Hydroxybutyryl-CoA dehydrogenase

EC-No.	Name
1.1.1.52	3alpha-Hydroxycholanate dehydrogenase
1.1.1.241	6-endo-Hydroxycineole dehydrogenase
1.1.1.166	Hydroxycyclohexanecarboxylate dehydrogenase
1.1.1.226	4-Hydroxycyclohexanecarboxylate dehydrogenase
1.1.99.26	3-Hydroxycyclohexanone dehydrogenase
1.1.1.66	omega-Hydroxydecanoate 1-dehydrogenase
1.1.1.232	5-Hydroxyeicosatetraenoate dehydrogenase
1.1.1.98	(R)-2-Hydroxy-fatty-acid dehydrogenase
1.1.1.99	(S)-2-Hydroxy-fatty-acid dehydrogenase
1.1.99.2	2-Hydroxyglutarate dehydrogenase
1.1.1.230	3alpha-Hydroxyglycyrrhetinate dehydrogenase
1.1.1.31	3-Hydroxyisobutyrate dehydrogenase
1.1.1.167	Hydroxymalonate dehydrogenase
1.1.3.19	4-Hydroxymandelate oxidase
1.1.1.178	3-Hydroxy-2-methylbutyryl-CoA dehydrogenase
1.1.1.170	3beta-Hydroxy-4beta-methylcholestenecarboxylate 3-dehydrogenase (decarboxylating)
1.1.1.88	Hydroxymethylglutaryl-CoA reductase
1.1.1.34	Hydroxymethylglutaryl-CoA reductase (NADPH)
1.1.1.60	2-Hydroxy-3-oxopropionate reductase
1.1.1.222	(R)-4-Hydroxyphenyllactate dehydrogenase
1.1.1.237	Hydroxyphenylpyruvate reductase
1.1.3.27	Hydroxyphytanate oxidase
1.1.1.59	3-Hydroxypropionate dehydrogenase

EC-No.	Name
1.1.1.141	15-Hydroxyprostaglandin dehydrogenase (NAD$^+$)
1.1.1.197	15-Hydroxyprostaglandin dehydrogenase (NADP$^+$)
1.1.1.196	15-Hydroxyprostaglandin-D dehydrogenase (NADP$^+$)
1.1.1.231	15-Hydroxyprostaglandin-I dehydrogenase (NADP$^+$)
1.1.1.81	Hydroxypyruvate reductase
1.1.1.209	3(or 17)alpha-Hydroxysteroid dehydrogenase
1.1.1.51	3(or 17)beta-Hydroxysteroid dehydrogenase
1.1.1.53	3alpha(or 20beta)-Hydroxysteroid dehydrogenase
1.1.1.210	3beta(or 20alpha)-Hydroxysteroid dehydrogenase
1.1.1.145	3beta-Hydroxy-DELTA^5-steroid dehydrogenase
1.1.1.159	7alpha-Hydroxysteroid dehydrogenase
1.1.1.146	11beta-Hydroxysteroid dehydrogenase
1.1.1.176	12alpha-Hydroxysteroid dehydrogenase
1.1.1.238	12beta-Hydroxysteroid dehydrogenase
1.1.1.147	16alpha-Hydroxysteroid dehydrogenase
1.1.1.149	20alpha-Hydroxysteroid dehydrogenase
1.1.1.213	3alpha-Hydroxysteroid dehydrogenase (A-specific)
1.1.1.50	3alpha-Hydroxysteroid dehydrogenase (B-specific)
1.1.1.239	3alpha(17beta)-Hydroxy steroid dehydrogenase (NAD$^+$)
1.1.1.201	7beta-Hydroxysteroid dehydrogenase (NADP$^+$)
1.1.1.150	21-Hydroxysteroid dehydrogenase (NAD$^+$)
1.1.1.151	21-Hydroxysteroid dehydrogenase (NADP$^+$)
1.1.1.15	D-Iditol 2-dehydrogenase
1.1.1.14	L-Iditol 2-dehydrogenase
1.1.1.128	L-Idonate 2-dehydrogenase
1.1.1.111	3-(Imidazol-5-yl)lactate dehydrogenase
1.1.1.205	IMP dehydrogenase
1.1.1.112	Indanol dehydrogenase
1.1.1.190	Indole-3-acetaldehyde reductase (NADH)
1.1.1.191	Indole-3-acetaldehyde reductase (NADPH)
1.1.1.110	Indolelactate dehydrogenase
1.1.1.18	myo-Inositol 2-dehydrogenase
1.1.1.41	Isocitrate dehydrogenase (NAD$^+$)
1.1.1.42	Isocitrate dehydrogenase (NADP$^+$)
1.1.1.223	Isopiperitenol dehydrogenase
1.1.1.80	Isopropanol dehydrogenase (NADP$^+$)
1.1.1.85	3-Isopropylmalate dehydrogenase
1.1.1.86	Ketol-acid reductoisomerase
1.1.1.78	D-Lactaldehyde dehydrogenase
1.1.1.77	Lactaldehyde reductase
1.1.1.55	Lactaldehyde reductase (NADPH)
1.1.1.28	D-Lactate dehydrogenase
1.1.1.27	L-Lactate dehydrogenase
1.1.2.5	D-Lactate dehydrogenase (cytochrome c-553)
1.1.2.4	D-Lactate dehydrogenase (cytochrome)
1.1.2.3	L-Lactate dehydrogenase (cytochrome)
1.1.99.7	Lactate-malate transhydrogenase
1.1.1.192	Long-chain-alcohol dehydrogenase
1.1.3.20	Long-chain-alcohol oxidase
1.1.1.211	Long-chain-3-hydroxyacyl-CoA dehydrogenase
1.1.1.37	Malate dehydrogenase

EC-No.	Name
1.1.99.16	Malate dehydrogenase (acceptor)
1.1.1.39	Malate dehydrogenase (decarboxylating)
1.1.1.83	D-Malate dehydrogenase (decarboxylating)
1.1.1.82	Malate dehydrogenase ($NADP^+$)
1.1.1.38	Malate dehydrogenase (oxaloacetate-decarb-oxylating)
1.1.1.40	Malate dehydrogenase (oxaloacetate-decarb-oxylating, $NADP^+$)
1.1.3.3	Malate oxidase
1.1.1.67	Mannitol 2-dehydrogenase
1.1.1.138	Mannitol 2-dehydrogenase ($NADP^+$)
1.1.2.2	Mannitol dehydrogenase (cytochrome)
1.1.1.17	Mannitol 1-phosphate 5-dehydrogenase
1.1.1.224	Mannose-6-phosphate 6-reductase
1.1.1.131	Mannuronate reductase
1.1.1.207	(-)-Menthol dehydrogenase
1.1.1.244	Methanol dehydrogenase
1.1.3.31	Methanol oxidase
1.1.1.32	Mevaldate reductase
1.1.1.33	Mevaldate reductase (NADPH)
1.1.1.218	Morphine 6-dehydrogenase
1.1.1.208	(+)-Neomenthol dehydrogenase
1.1.3.28	Nucleoside oxidase
1.1.1.73	Octanol dehydrogenase
1.1.1.92	Oxaloglycolate reductase (decarboxylating)
1.1.1.100	3-Oxoacyl-[acyl-carrier-protein] reductase
1.1.1.212	3-Oxoacyl-[acyl-carrier-protein] reductase (NADH)
1.1.1.172	2-Oxoadipate reductase
1.1.1.235	8-Oxocoformycin reductase
1.1.1.104	4-Oxoproline reductase
1.1.1.106	Pantoate 4-dehydrogenase
1.1.1.144	Perillyl-alcohol dehydrogenase
1.1.1.43	Phosphogluconate 2-dehydrogenase
1.1.1.44	Phosphogluconate dehydrogenase (decarboxylating)
1.1.1.95	Phosphoglycerate dehydrogenase
1.1.1.142	D-Pinitol dehydrogenase
1.1.99.23	Polyvinyl-alcohol dehydro-genase (acceptor)
1.1.3.30	Polyvinyl-alcohol oxidase
1.1.1.202	1,3-Propanediol dehydro-genase
1.1.1.7	Propanediol-phosphate dehydrogenase
1.1.1.189	Prostaglandin-E_2 9-reductase
1.1.1.188	Prostaglandin-F synthase
1.1.1.246	Pterocarpin synthase
1.1.3.10	Pyranose oxidase
1.1.1.107	Pyridoxal 4-dehydrogenase
1.1.1.65	Pyridoxine 4-dehydrogenase
1.1.99.9	Pyridoxine 5-dehydrogenase
1.1.3.12	Pyridoxine 4-oxidase
1.1.1.220	6-Pyruvoyltetrahydropterin 2'-reductase
1.1.1.24	Quinate dehydrogenase
1.1.99.25	Quinate dehydrogenase (pyrroloquinoline-quinone)
1.1.1.105	Retinol dehydrogenase
1.1.1.173	L-Rhamnose 1-dehydrogenase
1.1.1.56	Ribitol 2-dehydrogenase
1.1.1.137	Ribitol-5-phosphate 2-dehydrogenase
1.1.1.115	Ribose 1-dehydrogenase ($NADP^+$)
1.1.1.228	(+)-cis-Sabinol dehydrogenase
1.1.3.18	Secondary-alcohol oxidase
1.1.1.153	Sepiapterin reductase
1.1.1.143	Sequoyitol dehydrogenase
1.1.1.25	Shikimate 5-dehydrogenase

EC-No.	Name
1.1.99.21	D-Sorbitol dehydrogenase (acceptor)
1.1.1.140	Sorbitol-6-phosphate 2-dehydrogenase
1.1.99.12	Sorbose dehydrogenase
1.1.1.123	Sorbose 5-dehydrogenase ($NADP^+$)
1.1.3.11	L-Sorbose oxidase
1.1.1.58	Tagaturonate reductase
1.1.1.93	Tartrate dehydrogenase
1.1.1.63	Testosterone 17beta-dehydrogenase
1.1.1.64	Testosterone 17beta-dehydrogenase ($NADP^+$)
1.1.3.23	Thiamine oxidase
1.1.1.129	L-Threonate 3-dehydrogenase
1.1.1.103	L-Threonine 3-dehydrogenase
1.1.1.206	Tropine dehydrogenase
1.1.1.236	Tropinone reductase
1.1.1.136	UDP-N-acetylglucosamine 6-dehydrogenase
1.1.1.158	UDP-N-acetylmuramate dehydrogenase
1.1.1.22	UDPglucose 6-dehydrogenase
1.1.99.19	Uracil dehydrogenase
1.1.1.154	Ureidoglycolate dehydrogenase
1.1.1.203	Uronate dehydrogenase
1.1.1.199	(S)-Usnate reductase
1.1.4.1	Vitamin-K-epoxide reductase (warfarin sensitive)
1.1.4.2	Vitamin-K-epoxide reductase (warfarin-insensitive)
1.1.1.221	Vomifoliol 4-dehydrogenase
1.1.1.204	Xanthine dehydrogenase
1.1.3.22	Xanthine oxidase
1.1.1.175	D-Xylose 1-dehydrogenase
1.1.1.113	L-Xylose 1-dehydrogenase
1.1.1.179	D-Xylose 1-dehydrogenase ($NADP^+$)
1.1.1.9	D-Xylulose reductase
1.1.1.10	L-Xylulose reductase
1.1.1.242	Zeatin reductase

1 NOMENCLATURE

EC number
1.1.1.1

Systematic name
Alcohol:NAD^+ oxidoreductase

Recommended name
Alcohol dehydrogenase

Synonymes
Aldehyde reductase
Dehydrogenase, alcohol
ADH
Alcohol dehydrogenase (NAD)
Aliphatic alcohol dehydrogenase
Ethanol dehydrogenase
NAD-dependent alcohol dehydrogenase
NAD-specific aromatic alcohol dehydrogenase
NADH-alcohol dehydrogenase
NADH-aldehyde dehydrogenase
Primary alcohol dehydrogenase
Yeast alcohol dehydrogenase

CAS Reg. No.
9031-72-5

---

2 REACTION AND SPECIFICITY

Catalysed reaction
An alcohol + NAD^+ →
→ an aldehyde or ketone + NADH (mechanism [15, 26, 30–33, 35, 41, 54, 56, 57, 88, 90–92], ordered bi-bi mechanism [23, 69, 74, 91], structure and mechanism [32–34, 91], esterase mechanism [39], peroxidase mechanism [40])

Reaction type
Redox reaction

Natural substrates

Ethanol + NAD^+ (principal and rate-limiting enzyme of ethanol metabolism [22], metabolic role not limited to ethanol oxidation [25, 39, 40], involved in development of male hamsters' reproductive system [47], the distinction of isozymes/different enzymes/intermediate forms in one organism suggests multiple functions [57], inducible [81], involved in metabolism of dietary wax esters in salmonid fish [59], overview over enzyme family [57]) [22, 25, 39, 40, 47, 57, 59, 81]

Substrate spectrum

1 Ethanol + NAD^+ (broad specificity (isozymes AA, CC [48]) [18, 48], high enantiomeric stereospecificity, r (pyrazole-sensitive isozyme [25], not isozyme BB [48], reduction preferred [82]) [3, 4, 6, 15, 24, 25, 27, 35, 41, 42, 48, 49, 51, 54, 67, 69, 71, 72, 74, 75, 78, 79, 81–83, 86, 90, 91], best substrate (not isozyme set III [79]) [45, 59, 67–69, 79, 83, 87], class I isozyme [46], oxidation only at concentrations above 0.5 M (isozyme BB) [48], benzoylpyridine adenine dinucleotide can replace NAD^+ [33], poor substrate (isozyme ADH-1 [49]) [11, 13, 14, 16, 21, 38], no saturation of anodic pyrazole-insensitive isozyme with up to 2 M ethanol [25], no saturation of isozyme ADH-2 [49]) [2–4, 6, 10–16, 18, 21–25, 27, 33, 35, 38, 41, 42, 45–49, 51, 54, 56, 57, 59, 60, 67–69, 71, 72, 74–79, 81–92]

2 Methanol + NAD^+ (r (isozyme AA [48]) [48, 81], class I isozyme [11], isozymes $beta_2beta_2$ and $beta_1beta_1$ [19], poor substrate (pyrazole-sensitive isozyme [24], isozymes AA [48], ADH-3 [49]) [24, 45, 48, 49, 59, 69, 81], no substrate (class II isozyme [14, 16, 18, 21], class III isozyme [11], $beta_3beta_3$-isozyme [19], pyrazole-insensitive [24, 25] cathodic isozyme [25]) [11, 14, 16, 18, 21, 24, 25, 67, 68, 71, 75, 82, 84, 85], no substrate at pH 7.5 [45]) [11–13, 19, 24, 45, 48, 49, 59, 66, 69, 81]

3 Monohalosubstituted ethanol + NAD^+ [12]

4 Primary alcohol + NAD^+ (r, short chain length: preferred substrate [7, 85, 90], medium-chain length (isozyme BB) [48], reaction rates decrease with increasing chain length [4], long-chain preferred (anodic pyrazole-insensitive isozyme [25], isozyme ADH-2 [49]) [25, 49], C_2/C_3 (isozyme ADHII [68]) [68, 72], C_2-C_4 [67], C_2-C_5 (isozyme ADHI [68]) [66–68, 77–79], C_2-C_6 [77, 78, 91], aliphatic linear and branched [66, 84], n- and iso-aliphatic C_2-C_8-alcohols [71, 75], no saturation with short-chain alcohols or aldehydes [25]) [4, 7, 25, 47–49, 66–68, 71, 72, 75, 77–79, 84, 85, 90, 91]

5 1-Propanol + NAD^+ (r [42, 79, 81], best substrate (isozyme set III) [79], isozyme ADHII [68], benzoylpyridine adenine dinucleotide can replace NAD^+ [33]) [18, 33, 42, 45, 59, 67, 68, 77–79, 81, 83–85, 87, 90, 91]

6 1-Butanol + NAD^+ (r [42, 49, 79, 86, 90], isozyme ADH-1 and 3, no saturation of ADH-2 [49], benzoylpyridine adenine dinucleotide can replace NAD^+ [33], poor substrate [87]) [16, 18, 19, 24, 33, 42, 45, 47, 49, 59, 67, 77–79, 81–87, 90, 91]

7 1-Pentanol + NAD+ (isozyme AA, not BB [48], isozymes ADH-3 and ADH-An [60], poor substrate [16, 87], benzoylpyridine adenine dinucleotide can replace NAD+ [33]) [11, 14, 16, 18, 19, 21, 24, 25, 33, 45, 48, 49, 60, 77–79, 84, 85, 87, 91]
8 1-Hexanol + NAD+ (r [42]) [18, 42, 47, 48, 77, 78, 84, 85, 91]
9 1-Heptanol + NAD+ [84]
10 1-Octanol + NAD+ (r [49, 60], isozyme ADH-3 [60]) [11, 13, 14, 16, 19, 25, 48, 49, 60, 84, 85]
11 Nonanol + NAD+ (r, isozyme set III) [79]
12 Allyl alcohol + NAD+ (i.e. 2-propen-1-ol, r [68], best substrate [4, 78, 83], isozymes ADH-I/air, ADH-II [83]) [4, 68, 71, 73, 75, 78, 81, 83, 85]
13 Hexadecanol + NAD+ (poor substrate) [59]
14 2-Methyl-1-propen-1-ol + NAD+ [75]
15 2-Methylpropanol + NAD+ (poor substrate [78]) [77, 78, 84]
16 2-Methylpropan-2-ol + NAD+ [81, 84, 85]
17 3-Methylpropanol + NAD+ [77, 78]
18 2-Propanol + NAD+ (r [31], benzoylpyridine adenine dinucleotide cannot replace NAD+ [33], poor substrate [67, 69, 90], not [79]) [14, 30, 31, 33, 45, 65, 67, 69, 81, 84, 85, 90]
19 2-Butanol + NAD+ (r, enantioselective reaction [31], poor substrate [47, 59, 75], not [79]) [19, 31, 45, 47, 59, 75, 81, 83–85]
20 2-Pentanol + NAD+ (r, enantioselective reaction [31], benzoylpyridine adenine dinucleotide cannot replace NAD+ [33]) [31, 33, 84, 85]
21 2-Octanol + NAD+ (r, enantioselective reaction) [31]
22 Secondary alcohol + NAD+ (r [39], poor substrates [47], linear and cyclic [66], not [79]) [39, 47, 66, 84]
23 3-Pentanol + NAD+ (r [31], benzoylpyridine adenine dinucleotide cannot replace NAD+ [33]) [31, 33]
24 Isobutanol + NAD+ [19]
25 Isopentanol + NAD+ (not [76]) [19]
26 Ethylene glycol + NAD+ (poor substrate [14], isozyme AA, not BB or CC [48], not [76, 79]) [12–14, 48]
27 Benzyl alcohol + NAD+ (r (isozymes AA, CC, not BB [48]) [42, 48], class II and III isozymes [46], isozyme ADH-3 [60], no saturation of isozyme ADH-2 [49], poor substrate [16]) [13, 14, 16, 30, 42, 46, 48, 49, 60, 70, 81]
28 Cyclopentanol + NAD+ [77, 78]
29 Cyclohexanol + NAD+ (r [28, 42], isozyme $beta_1beta_1$ [19], isozyme AA, not BB or CC [48], pyrazole-sensitive isozyme [25], poor substrate [14, 16, 47], no substrate for pyrazole-insensitive cathodic isozyme [25]) [13, 14, 16, 19, 25, 28, 42, 47, 48, 60, 77, 78]
30 3-Methylcyclohexanol + NAD+ (best substrate) [66]
31 2-Buten-1-ol + NAD+ [49, 75]
32 3-Buten-1-ol + NAD+ [75]

33 trans-2-Hexen-1-ol + NAD^+ [48]
34 3-Pyridylcarbinol + NAD^+ [18]
35 all-trans-Retinol + NAD^+ (i.e. vitamin A) [12, 47]
36 Leaf alcohols + NAD^+ (i.e. E-prop-2-en-ol, Z-/E-hex-3-en-1-ol) [73]
37 2-Ethylhexan-1-ol + NAD^+ [84]
38 3-Methylpentan-1,5-diol + NAD^+ [84]
39 Cinnamic alcohol + NAD^+ (no activity with benzoylpyridine adenine dinucleotide, best substrate [83], not [76]) [33, 81, 83, 85]
40 2-Phenylethanol + NAD^+ (i.e. phenethyl alcohol, poor substrate) [81]
41 Vanillyl alcohol + NAD^+ [14, 16]
42 Tryptophol + NAD^+ (poor substrate [16]) [14, 16]
43 3-Phenyl-1-propanol + NAD^+ [14]
44 Formaldehyde + NADH [81, 82]
45 Acetone + NADH (poor substrate) [81]
46 Ketones (linear and cyclic) + NADH [28, 42, 66]
47 p-Anisaldehyde + NADH (i.e. p-methoxybenzaldehyde) [66, 70]
48 Methylglyoxal + NADH [72]
49 D-Glucitol + NAD^+ [77, 78]
50 12-Hydroxydodecanoic acid + NAD^+ (best substrate [16], not at pH 7.5 [47], isozyme ADH-3 [60]) [11, 14, 16, 25, 47, 49, 60]
51 16-Hydroxyhexadecanoic acid + NAD^+ [13, 14, 25]
52 3-Keto-beta-cholanoic acid + NADH [35]
53 4-Nitrobenzaldehyde + NADH (isozyme AA, CC, not BB) [48]
54 3-Nitrobenzaldehyde + NADH [49]
55 5beta-Androstan-3beta-ol-17-one + NAD^+ (oxidation at 44–56% the rate of ethanol oxidation [45]) [45, 51]
56 5beta-Pregnan-21-ol-3,20-dione hemisuccinate + NADH [42]
57 3beta-Hydroxysteroids + NAD^+ (substrates for SS- and ES-, not EE-isozyme) [27]
58 3beta-Hydroxy bile acid epimers + NAD^+ [28, 51]
59 5alpha-Androstan-17beta-ol-3-one + NADH [51]
60 3beta-Dihydrotestosterone + NADH [27]
61 3-Oxo-5beta-androstan-17beta-ol + NADH [28, 42]
62 3beta-Hydroxy-5beta-androstan-17-one + NADH [28, 42]
63 2-Deoxy-D-ribose + NAD^+ [14]
64 Phenyloctanoate + H_2O (esterase activity) [39]
65 Thiophenyloctanoate + H_2O (esterase activity) [39]
66 p-Chlorophenyloctanoate + H_2O (esterase activity) [39]
67 p-Nitrophenyloctanoate + H_2O (esterase activity) [39]
68 H_2O_2 + NAD^+ (peroxidase activity) [40]
69 More (enantioselectivity not changed by substitution of Co(II) for Zn(II) [31], at neutral pH the isozymes are more effective in aldehyde reduction than in alcohol oxidation [49], poor substrates: furfuryl alcohol, 2-octanol, phenylalaninol, digitose, DL-synephrine, 17beta-hydroxyethiocho-

lan-3-one [16], 2,3-pentanedione, phenylglyoxal, benzaldehyde, hydroxyacetone [72], tert-butanol, citronellol (r, isozyme set II and III) [79], no substrates are methanol (class II isozyme [14, 16, 18, 21], class III isozyme [11], beta$_3$beta$_3$-isozyme [19], pyrazole-insensitive [24, 25] cathodic isozyme [25], isozymes BB,CC [48], ADH-1 and –2 [49]) [11, 14, 16, 18, 19, 21, 24, 25, 48, 49, 60, 67, 68, 71, 75, 76, 82, 84, 85], primary alcohols and aldehydes (except C_2/C_3 or allyl alcohol and acrolein), isozyme ADHII [68], butanol [76], pentanol (isozyme BB) [48], hexanol [76], nonanol [81], allyl alcohol [87], secondary alcohols (ADH-I [68]) [68, 75, 79], tertiary alcohols [75], cyclohexanol (beta$_3$beta$_3$-isozyme [16], isozyme BB and CC [48]) [16, 19, 48, 75, 81], aromatic alcohols [71, 75], glycerol [18, 67, 79, 85, 87], ethylene glycol (isozyme BB and CC [48]) [48, 60, 75, 79], propanediol [87], acetone [85], diethylene glycol [76], diols [75, 76], isopropanol [67, 76, 87], isopentanol, isooctanol [76, 87], 2-mercaptoethanol, malate [75], lactate [75, 81], pyruvate, acetoacetate, serine, threonine, 3-hydroxybutan-2-one (oxidation), 3-hydroxybutyrate, butan-2-one, pentan-2-one, cyclohexanone [81], fructose, glucose, diacetyl, dihydroxyacetone, 3-deoxyglucosane, glyceraldehyde [72], methoxyethanol and ethoxyethanol [76], formaldehyde [68], cinnamyl alcohol [76], 2-methyl-1-propenol [75], colamine (aminoethanol) [76], 3alpha-hydroxy-bile acids [51], digitalis glycosides, norepinephrine, 3beta-hydroxyethiocholan-17-one [16], exhibits esterase activity: hydrolyzes phenyloctanoate, thiophenyloctanoate, p-chlorophenyloctanoate, p-nitrophenyloctanoate, not p-methylphenyloctanoate [39], exhibits peroxidase activity with H_2O_2 and beta-NAD$^+$ [40]) [11, 14, 16, 18, 19, 21, 24, 25, 31, 39, 40, 48, 49, 51, 60, 67, 68, 71, 72, 75, 76, 79, 81, 82, 84, 85, 87]

Product spectrum

1 Acetaldehyde + NADH [2–4, 6, 10, 12–14, 16, 18, 21–23]
2 Formaldehyde + NADH [48]
3 ?
4 Primary aldehydes + NADH [4, 7, 25, 47–49, 66, 68, 71, 75, 77–79, 84, 85]
5 Propionaldehyde + NADH
6 Butyraldehyde + NADH
7 ?
8 ?
9 ?
10 Octanal (i.e. caprylaldehyde) + NADH [16, 48, 60]
11 Nonanal + NADH [79]
12 ?
13 ?
14 ?
15 ?

16 ?
17 ?
18 ?
19 2-Butanone + NADH [31]
20 2-Pentanone + NADH [31]
21 2-Octanone + NADH [31]
22 Ketone + NADH [39, 47, 66]
23 3-Pentanone + NADH [31]
24 ?
25 ?
26 ?
27 Benzaldehyde + NADH [42, 48]
28 ?
29 ?
30 ?
31 ?
32 ?
33 ?
34 ?
35 Retinal + NADH [12, 47]
36 ?
37 ?
38 ?
39 ?
40 ?
41 ?
42 ?
43 ?
44 ?
45 ?
46 ?
47 ?
48 ?
49 ?
50 ?
51 ?
52 ?
53 ?
54 ?
55 ?
56 ?
57 ?
58 ?
59 ?

60 ?
61 ?
62 ?
63 ?
64 Phenoate + octanol [39]
65 Thiophenoate + octanol [39]
66 ?
67 ?
68 ?
69 ?

Inhibitor(s)

Pyrazole (specific, strong, formation of abortive ternary complex with NAD^+-enzyme [29, 66], competitive to ethanol [61], human: partial, rat: complete [10], not [23], not isozyme BB [48]) [1, 5, 10, 12, 29, 45, 48, 49, 61, 65, 66, 71, 88]; 4-Methylpyrazole (weak, competitive to acetaldehyde [23, 47], not [11, 21], 3 isozyme classes: I and II more or less sensitive, not: III [46], not class II isozyme [14], not isozyme ADH-An [60]) [2, 23–25, 45–47, 51, 60, 91]; Substituted pyrazoles (e.g. 4-cyano-, 4-propyl- [23], 4-pentyl- (strong, competitive to ethanol [23]) [14, 23], 4-octyl- [14], 4-methoxy- (weak, competitive to acetaldehyde) [23], 4-bromo- (competitive to ethanol [23]) [23, 25], 4-iodo- [23, 34] or 4-nitro-substituted (competitive to ethanol) [23]) [12, 14, 23, 25, 34]; PCMB (not [76]) [45, 68, 72, 75, 85]; p-Hydroxymercuribenzoate [75]; o-Phenanthroline (non-competitive to NADH [69], mixed-type [47], fully reversible by dilution or addition of Zn^{2+} [14, 16, 25, 68] or Co^{2+} [68], not isozyme BB [48]) [2, 14, 16, 21, 24, 25, 45, 47, 48, 68, 69, 75]; 2-Mercaptopropionyl glycine (weak) [45]; Iodoacetamide [69, 75, 76]; PMSF (slight) [75]; Iodoacetate (inactivation, NAD^+, AMP, adenosine, nicotinamide protect [4]) [4, 45, 85]; NEM [45, 75]; Imidazole (competitive [56]) [56, 61]; Pyridoxal phosphate [4]; Isobutyramide (formation of fluoroscent ternary complex with NADH and enzyme [12]) [12, 61]; Butyramide [91]; 8-Amino-6-methoxyquinoline (and 8-amino-substituted derivatives) [12]; 2-Chloroethanol [61]; 2,2,2-Trifluoroethanol [30, 61, 91]; Heptafluorbutanol [91]; EDTA (irreversible [24], reactivation with Zn^{2+}, not Fe^{2+} [87], weak [45], not [68, 72, 75]) [14, 24, 25, 45, 66, 69, 76, 86, 87]; Zn^{2+} (denaturation at high concentration [36]) [36, 45, 76, 82]; Fe^{2+} [87]; Co^{2+} (denaturation at high concentration) [36]; Ag^{2+} [45]; Cu^{2+} (not [76]) [45]; Hg^{2+} (not [76]) [45]; Borate [76, 79]; NaN_3 [69]; 2,2'-Bipyridyl (pyrazole-sensitive isozyme [25]) [14, 25, 45]; 8-Hydroxyquinoline-5-sulfonic acid (irreversible [24]) [21, 24]; Dipicolinic acid (inactivation, irreversible [24], 2 Zn^{2+}-ions remain enzyme-bound, reactivation by addition of Zn^{2+} or Co^{2+} [36]) [21, 24, 36]; Benzoylpyridine adenine dinucleotide (competitive to NAD^+) [33]; NAD^+/acetone (inactivation, irreversible formation of enzyme-NADH-acetone-complex) [62]; NAD^+ [91]; NADH (competitive to

NAD+) [15, 17, 23, 35, 61, 63, 91]; ADP-ribose [86]; Cibachron Blue (competitive) [61]; GSH (strong inactivation [78]) [45, 78]; Acetaldehyde (non-competitive [23, 35]) [23, 35, 91]; Ethanol (mixed non-competitive inhibitor to acetaldehyde [15, 17], substrate inhibition above 75 mM [84], 130 mM [18], 300 mM [12], 500 mM (beta$_3$beta$_3$ isozyme), competitive inhibitor of peroxidase reaction [40]) [12, 15, 17–19, 40, 84, 91]; 12-Hydroxydodecanoic acid (competitive) [47]; 3-alpha-Hydroxysteroids (strong) [27]; S-5-Chloro-3-(imidazol-5-yl)propionate (inactivation, not (R)-enantiomer, mechanism) [29]; 2-Bromo-3-(imidazolyl)propionate [29]; p-Nitrophenyl octanoate (substrate inhibition, esterase reaction) [39]; p-Nitrophenol (esterase reaction) [39]; Octanoic acid (esterase reaction) [39]; H_2O_2 (progressive inactivation during catalytic peroxidase reaction, NAD^+ protects) [40]; 1-Dodecane sulfonic acid sodium salt (mixed-type) [53]; Bis(2-ethylhexyl)sodium sulfosuccinate [53]; SDS (competitive) [53]; Membrane-associated inactivator (depressed by DTT, GSH, mercaptoethanol) [80]; More (inhibitor design studies based on 3-dimensional enzyme structure [32], no inhibition by citrate [75], 4,7-phenanthroline [14], 4-methylpyrazole [11, 14, 21], 3-methylpyrazole [23], 1-hexane sulfonic acid sodium salt [53], 2-methylpenten-2,4-diol, 2,3-dimethylbutan-2-ol [61], EDTA [68, 72, 75], metal ions, metal binding agents, reducing agents [85]) [11, 14, 21, 23, 32, 53, 61, 68, 72, 75, 85]

Cofactor(s)/prosthetic group(s)/activating agents

NAD^+ (specific requirement, 2 mol NAD^+ per dimer [14, 25, 61], kinetics of NAD^+/NADH-binding [26, 35], biphasic kinetics with NAD^+ [71], not $NADP^+$ (isozyme ADH-2 [68]) [14, 16, 18, 21, 45–49, 66, 68, 73, 81] at concentrations up to 25 mM [14]) [1–43, 45–49, 54, 56, 57, 59–79, 81–92]; beta-NAD^+ [40, 53]; NADH (requirement, NADH-oxidation is the favoured reaction [69], binding of NADH is accompanied by conformational change of protein from an open to a closed structure [19], enzyme-NADH-complex is optically active [25], tight NADH-SS-isozyme complex in pH-range 10–12 [27] whose fluorescence is increased by isobutyramide [28]) [3, 4, 6, 15, 19, 25, 27, 28, 48, 49, 54, 56, 57, 59, 60, 65, 66, 68–72, 78, 79, 81–91]; NADPH (SS-isozyme [27], 1.8% as effective as NADH [72], not [14, 16, 18, 21, 45–49, 66, 68, 73, 81] at concentrations up to 25 mM [14]) [27, 72]; NAD^+-analogues (activation, monomeric and polymeric analogues [88]) [54, 88]; EDTA (activation) [82]; 2-Mercaptoethanol (slight activation) [76]; More (NAD^+ induces an active form of the enzyme by a large conformational change [34], enzyme-surfactant interaction [53]) [34, 53]

Metal compounds/salts

Zn^{2+} (requirement, structure-stabilizing [36], zinc protein [12], tightly bound [75], 4 mol Zn^{2+} per dimer [14, 16, 18, 19, 27, 42, 45, 49, 60, 70], 1 mol Zn^{2+} per subunit [67, 68, 81], 3.6–4.2 gatom/mol enzyme, atomic absorption spectroscopy [21, 23, 24], each catalytic domain binds 2 Zn^{2+} [34], 1 Zn^{2+} at each catalytic site [53], denaturation at high concentrations [36]) [12, 14, 16, 18, 19, 21, 23–27, 29–31, 34, 36, 42, 43, 45, 48, 49, 52, 53, 55–57, 60, 66–70, 75, 81, 87]; Co^{2+} (activation, isozyme ADH-2 [68], can replace Zn^{2+}: 1.7 gatom per dimer [30], specific active-site-substituted Co(II)-alcohol dehydrogenase [31, 36], can substitute Zn^{2+} in Zn^{2+}-depleted $beta_1beta_1$-isozyme [19], binds specifically at catalytic site [36], denaturation at high concentration [36]) [19, 30, 31, 36, 68]; Mg^{2+} (activation) [82]; Fe^{2+} (activation [82], not [87]) [82]; L-Cysteine (slight activation) [76]; More (metalloprotein [69], no metal requirement, plasma emission spectroscopy [61]) [61, 69]

Turnover number (min^{-1})

More (kinetic data for various alcohols at pH 10.5 and 7.5 [49], for ethanol with various surfactants [53], for modified enzyme or coenzyme [54], for different isozymes [24, 25]) [24, 25, 30, 31, 42, 49, 53, 54]; 6.12 (methanol) [12]; 41.9 (propionaldehyde, reduction) [42]; 69.6 (propanol, oxidation) [42]; 78 (retinol) [47]; 82 (benzyl alcohol) [47]; 90 (ethanol, highly purified preparation) [12]; 122 (cyclohexanol) [47]; 123 (3-oxo-5beta-androstan-17beta-ol) [42]; 150 (ethanol, crystalline enzyme) [12]; 200 (acetaldehyde, pyrazole-insensitive isozyme) [24]; 285 (2-butanol) [47]; 306 (ethanol) [53]; 354 (cyclohexanone, reduction) [42]; 450 (1-butanol) [47]; 463 (3-methylcyclohexanol) [66]; 465 (NAD^+) [47]; 480 (ethanol) [47]; 1000 (acetaldehyde, pyrazole-sensitive isozyme) [24]

Specific activity (U/mg)

More (modifications of amino-groups of the isozymes increase activity [35], activity of cells grown in presence or absence of Zn^{2+} [86]) [14–16, 18, 21, 27, 35, 42, 44, 46–49, 51, 69, 72, 73, 75, 81–83, 86–88, 91]; 1.4 (caecum) [59]; 1.6 (liver) [59]; 2.5 (isozyme ADH-2) [60]; 2.86 [19]; 3.3 (isozyme ADH-3) [12, 60]; 3.4 (pH 10) [42]; 3.92 [66]; 5.31 (activity at 65°C, 3 times higher at 95°C) [70]; 19.0 ($beta_2beta_2$ isozyme) [15]; 32.5 (isozyme ADH-1, stomach) [49]; 38.1 [71]; 55.0 [65]; 64.0 (isozyme CC) [48]; 79.1 [85]; 93.0 (isozyme ADHI) [68]; 120 (dry seeds) [79]; 121.6 (isozyme ZADH-I) [67]; 196.3 [84]; 470 (isozyme ADHII) [68]; 729 (isozyme I, shoot tissue) [79]; 17167 [72]; 28700 (isozyme I) [77]; 31900 (isozyme II) [77]; 36000 (isozyme III) [77]; 55900 [78]

K_m-value (mM)

More (kinetic constants for mouse enzyme for a range of alcohol and aldehyde substrates [48], for native and isonicotinimidylated isozymes [35], for modified enzyme or coenzyme [54], with benzoylpyridine adenine dinucleotide as coenzyme [33], for EE- [41] and SS-isozyme [41, 42], kinetic constants for specific-active-site substituted Co(II)-enzyme [31], kinetic constants for oxidation and reduction [68] at different pH-values [74], for different isozymes [51, 75], esterase reaction [39], for peroxidase reaction [40], for ethanol oxidation in presence of surfactants [53], kinetic data with mono- and polymeric NAD^+-analogues [88], pH-dependence of kinetic data of free or immobilized enzyme [89]) [6, 10, 11, 13, 14, 16, 17, 31, 33, 35, 39–42, 48, 51, 53, 54, 68, 74, 75, 77, 78, 81, 83–86, 88–91]; 0.0004 (below, 3beta-hydroxy-5alpha-cholanoate) [28]; 0.004–120 (benzyl alcohol, different isozymes [13]) [13, 14]; 0.005–0.019 (1-octanol, different isozymes [13]) [13, 14]; 0.005–23.0 (cyclohexanol, different isozymes) [13]; 0.006–0.5 (octanol, different isozymes) [25]; 0.0069–0.06 (NAD^+ (pH 7.5 [45]) [24, 45], 16-hydroxyhexadecanoic acid [13, 14, 25], different isozymes [13, 24, 25]) [13, 14, 24, 25, 45]; 0.0074–0.013 (NAD^+, homodimeric isozymes of subunit $beta_1$, alpha, $gamma_1$ and 2 [15, 17], butanol [24], pentanol, pyrazole-sensitive isozyme [24, 25]) [15, 17, 24, 25]; 0:012–19.0 (pentanol, different isozymes) [25]; 0.02 (retinol) [47]; 0.0311 (NADH) [6]; 0.052–0.08 (NAD^+, pH 10.8 [45], different isozymes [25]) [25, 45, 47, 59, 77, 78]; 0.10–0.13 (acetaldehyde, pyrazole-sensitive isozyme [24], pentanol, class I isozyme [11]) [11, 24]; 0.15–0.18 (ethanol, pyrazole-sensitive isozyme [24], isozyme AA [48]) [24, 48]; 0.2–0.24 (tryptophol [14], 12-hydroxydodecanoic acid [25]) [14, 25]; 0.44–0.8 (1-butanol [47], 1-octanol, class III isozyme [11], ethanol (rat [10], pH 10.8 [45]) [10, 45, 59]) [10, 11, 45, 47, 59]; 0.91–4.0 (acetaldehyde, pyrazole-sensitive isozyme [25], ethanol, human [10], pyrazole-insensitive isozyme [24]) [10, 13, 24, 25, 42, 47]; 1.35 ((S)-2-butanol) [31]; 2.2–400 (ethanol, different isozymes) [25]; 6.0–150.0 (methanol, different isozymes [13]) [11–13, 24, 25]; 7.05–11.4 (acetaldehyde [6, 24], pyrazole-insensitive isozyme [24]) [6, 24]; 7.5 ((R)-2-butanol) [31]; 13.0–220.0 (ethylene glycol, different isozymes) [13]; 27.0–31.0 (pentanol, class III-isozyme [11], cyclohexanol [47]) [11, 47]; 40.0–47.3 (ethanol) [6, 42]; 120 (ethanol) [14]; 250 (2-butanol) [47]; 400 (ethanol, pyrazole-insensitive cathodic isozyme) [25]

pH-optimum

More (pI: 5.1 (ADH-1), 5.95–6.3 (ADH-2), 8.25–8.4 (ADH-3) [49], pI: 5.2 [66], pI: 5.4–6.1 [71], pI: 5.66–5.76 [77], pI: 6.4 (class III isozyme) [16], pI: 6.6 [84], pI: 6.9 (isozyme 4), 8.0 (3), 8.7 (2), 8.8 (1) [51], pI: 7.5–7.95 [61], isoelectric-focusing yields 5 isozymes: pI: 7.4 (E2E2), pI: 7.7 (E1E2), pI: 8.0 (E1E1), pI: 8.4 (ES'), pI: 8.6 (ES) [37], pI: 8.6 (class II isozyme) [14], effect of pH on kinetic constants [47]) [14, 16, 37, 47, 49, 51, 61, 66, 71, 77, 84, 91]; 4.5 (reduction, ADHI) [68]; 5.5 (acetaldehyde reduction, isozyme $beta_1 beta_1$) [15]; 5.5–7.5 (isozyme 1) [75]; 5.5–9.0 (isozyme 2) [75]; 5.6–6.5 (oxidation) [84]; 5.7 (reduction) [72]; 6.0–9.0 (isozyme 3) [75]; 6.5 (reduction, ADH-2 [68], acetaldehyde [6], octanal [60]) [6, 60, 68]; 6.7 (reduction) [71]; 6.8 (reduction) [69]; 7.0 (ethanol, isozyme $beta_3 beta_3$) [19]; 7.2–7.5 (acetaldehyde reduction) [15]; 7.5 (reduction [3, 4, 74], anisaldehyde [66], two maxima pH 7.5 and 10.8 [45], variable NAD^+-concentration [89]) [3, 4, 45, 66, 74, 89]; 8.0 (variable ethanol-concentration) [89]; 8.0–8.4 (reduction) [84]; 8.0–8.5 [85]; 8.0–10.0 (plateau, reduction, isozyme sets II and III) [79]; 8.2 (pyrazole-sensitive isozyme) [24]; 8.3 [87]; 8.5 (ethanol, $beta_2 beta_2$-isozyme [19], benzyl alcohol [66], esterase reaction [39]) [4, 19, 39, 66]; 8.5–8.8 (atypical form [10]) [10, 12, 15]; 8.6–8.8 (reduction, isozyme set I) [79]; 8.7 [76]; 8.9 (oxidation) [72]; 9.0–10.0 (oxidation) [71]; 9.5 (ethanol oxidation [6], isozyme 1 [51], ZADH-I [67], ADHI [68], isozyme set II [79]) [6, 51, 67, 68, 79]; 10.0 (above [18], oxidation, steady increase of activity up to, in 5% (v/v) aqueous dioxane, isozyme sets I and III [79], esterase reaction [39], isozyme ZADH-II [67]) [18, 39, 67, 75, 79]; 10.4 (white Americans and Europeans) [15]; 10.5 ($beta_1 beta_1$-isozyme [19]) [2, 10, 19]; 10.6 (pyrazole-sensitive isozyme, two maxima [24], pyrazole-insensitive isozyme, 1-octanol [25]) [24, 25]; 10.7 (isozymes 2–4) [51]; 10.8 (two maxima, pH 7.5 and 10.8 [45]) [12, 45]; 11.0 (ethanol [60], 1-octanol [16], pyrazole-insensitive isozyme [24]) [16, 24, 60]; 11.2 (pyrazole-sensitive, ethanol) [25]

pH-range

5.0 (no activity below: pyrazole-insensitive isozyme, pyrazole-sensitive isozyme: about 42% of maximal activity) [24]; 5.0–9.5 (about half-maximal activity at pH 5.0 and 9.5) [75]; 5.2–7.0 (about half-maximal activity at pH 5.2 and 7.0, oxidation) [84]; 6.0 (no activity below: pyrazole-sensitive isozyme [25]) [25, 45]; 6.5–10.0 (about 60% of activity at pH 6.5 and 10.0, isozyme set II, about half-maximal activity at pH 6.5 and 75% of activity at pH 10.0, isozyme set III) [79]; 6.5–11.0 (about half-maximal activity at pH 6.5 and about 65% at 11.0, isozyme ZADH-I) [67]; 6.9–7.6 (about half-maximal activity at pH 6.9 and 7.6) [45]; 7.2–9.2 (about half-maximal activity at pH 7.2 and about 60% at pH 9.2, reduction) [84]; 7.4–10.2 (about half-maximal activity at pH 7.4 and 10.2, isozyme I) [83]; 7.5–9.8 (about half-maximal activity at pH 7.5 and 9.8, isozyme set I) [79]; 7.8–8.6 (about half-maximal activity at pH 7.8 and 8.6, isozyme II) [83]; 8.3–9.3 (about half-maximal activity at pH 8.3 and 9.3) [87]; 8.5–11.0 (about half-maximal activity at pH 8.5 and 65% at

11.0, isozyme ZADH-II) [67]; 9.5–11.6 (pyrazole-sensitive isozyme: about half-maximal activity at pH 9.5 and 80% of maximal activity at pH 11.6) [25]; 9.6–11.0 (pyrazole-insensitive isozyme: about half-maximal activity at pH 9.6 and 70% of maximal activity at pH 11.0) [25]; 9.8–11.0 (about half-maximal activity at pH 9.8 and 11.0) [45]; 9.8–11.2 (about half-maximal activity at pH 9.8 and 11.2) [16]

Temperature optimum (°C)

More (Q_{10}: 2.07) [84]; 20 (assay at) [33]; 22 (assay at) [71]; 25 (assay at) [13, 24, 25, 31, 37, 42, 46, 47, 49, 53, 54, 60, 65, 69, 81, 86, 89, 90]; 30 (assay at [39, 48, 87, 88]) [84]; 35 (Saccharomyces cerevisiae) [3]; 37 (assay at) [48]; 45 (isozyme ADH-2) [68]; 45–50 (Kluyveromyces marxianus) [3]; 50–55 [85]; 55 (isozyme ADHI) [68]; 65 (assay at) [66]; 70 (above) [82]; 95 (instrumental limitation) [66, 70]

Temperature range (°C)

40–95 (continuous increase of activity) [66]

3 ENZYME STRUCTURE

Molecular weight

60000 (Vicia faba, gel filtration) [4]
60000–80000 (Saccharomyces cerevisiae, gel filtration) [87]
68000 (parsley, gel filtration) [72]
70000 (Zea mays, gel filtration, sedimentation equilibrium centrifugation) [75]
71000 (Sulfolobus solfataricus, sucrose gradient centrifugation, gel filtration) [66]
71500 (rabbit, amino acid analysis) [45]
72000 (Saccharomyces cerevisiae/Zn^{2+}-deficient, gel filtration) [86]
74000 (Glycine max, gel filtration [71], Sulfolobus solfataricus, sucrose gradient centrifugation, gel filtration [70]) [70, 71]
74500 (rabbit, class I isozyme, gel filtration) [46]
76000–77000 (Macaca mulatta, pyrazole-sensitive isozyme, ultracentrifugation) [25]
76230 (Saimiri sciureus boliviensis, pyrazole-sensitive isozyme, low-speed sedimentation equilibrium method) [24]
78000 (human, gel filtration) [18, 23]
78430 (Saimiri sciureus boliviensis, pyrazole-insensitive isozyme, sedimentation equilibrium method) [24]
79000 (mouse, isozyme BB, gel filtration) [48]
79000–84000 (human, gel filtration) [21]
80000 (horse [27, 28], Misocricetus auratus [47], rat [49], Salmo gairdneri [59], Zymomonas mobilis isozyme ADHI [68], gel filtration [27, 28, 47, 49,

59, 68], Saccharomyces cerevisiae, sucrose density gradient centrifugation [86]) [27, 28, 47, 49, 59, 68, 86]
82700 (human [16], Coturnix japonica, isozyme ADH-3 [60], sedimentation equilibrium method) [16, 60]
82800 (Macaca mulatta, pyrazole-insensitive isozyme, ultracentrifugation) [25]
83000 (mouse, isozyme AA, gel filtration) [48]
85000 (human, amino acid analysis [18], mouse, isozyme CC, gel filtration [48]) [18, 48]
86000 (Oryza sativa, gel filtration) [6]
87000 (human) [12]
90000 (Hordeum vulgare, gel filtration) [79]
94000 (Chlamydomonas moewusii, gel filtration) [69]
95000 (Zymomonas mobilis, isozyme ADH-2, gel filtration) [68]
116000 (Triticum, gel filtration) [77, 78]
140000 (Saccharomyces cerevisiae, gel filtration) [86]
150000 (tea, gel filtration [73], Astasia longa, isozymes ADH-I/air and ADH-II, PAGE [83]) [73, 83]
152000 (Candida guilliermondii, gel filtration) [85]
176000 (Sporotrichum pulverulentum, PAGE) [84]
180000 (Candida sp. N-16, gel filtration) [82]
290000 (Aspergillus nidulans, gel filtration in the presence of DTT, PAGE [81], Astasia longa, isozyme ADH-I/O_2, PAGE [83]) [81, 83]
More (primary structure [38, 43], amino acid composition [44, 81], peptide mapping [91], peptide and gene analysis [70], structural similarities: in-register comparison of yeast and horse liver enzyme [52]) [38, 43, 44, 52, 70, 81, 91]

Subunits

Dimer (2 × 30000, Vicia faba, SDS-PAGE [4], 2 × 31100, Zymomonas mobilis, isozyme ZADH-II, SDS-PAGE [67], 2 × 35000, Zea mays, SDS-PAGE [75], 2 × 36000, rabbit, SDS-PAGE [45], 2 × 37000, Sulfolobus solfataricus [66, 70], Glycine max [71], parsley [72], SDS-PAGE [66, 70–72], 2 × 37588, Sulfolobus solfataricus, amino acid analysis [70], 2 × 38000, Zymomonas mobilis, isozyme ADH-II, SDS-PAGE [68], 2 × 39000, human, amino acid analysis [23], mouse, isozyme BB [48], rat, isozyme ADH-2 [49], SDS-PAGE [48, 49], 2 × 40000, human testis [11], human liver [12, 21], Saimiri sciureus boliviensis [24], horse [28, 37], chicken [44], rat, isozyme ADH-3 [49], Salmo gairdneri [59], Zymomonas mobilis, isozyme ADHI [68], Saccharomyces cerevisiae, isozyme ADH-IV [87], SDS-PAGE [11, 12, 21, 24, 28, 37, 44, 49, 59, 68, 87], 2 × 41000, human [16], Misocricetus auratus [47], SDS-PAGE [16, 47], 2 × 41700, rabbit, class I isozymes, SDS-PAGE [46], 2 × 42000, human (class III isozyme [14]) [14, 18], Coturnix japonica [60], Chlamydomonas moewusii [69], SDS-PAGE [14, 18, 60, 69], 2 × 43000, mouse isozyme AA [48], rat, isozyme ADH-1 [49], SDS-PAGE [48, 49], 2 × 45000, Hordeum vul-

gare, isozyme I, SDS-PAGE [79], 2 × 46000, Oryza sativa, SDS-PAGE [6], 2 × 47000, mouse, isozyme CC, SDS-PAGE [48], 2 × 58000, Triticum, SDS-PAGE [77, 78]) [4, 6, 11, 12, 14, 16, 18, 21, 23, 24, 28, 37, 44–49, 59, 60, 66–72, 75, 77–79, 87]
Tetramer (4 × 34700, Zymomonas mobilis, isozyme ZADH-I, SDS-PAGE [67], 4 × 38000, Candida guilliermondii, SDS-PAGE [85], 4 × 42000, Sporotrichum pulverulentum, SDS-PAGE [84], 4 × 45000, Candida sp. N-16, SDS-PAGE [82]) [67, 82, 84, 85]
Octamer (8 × 37500, Aspergillus nidulans, SDS-PAGE) [81]
Oligomer (x × 28000, Ctenopharyngodon idellus, SDS-PAGE [2], x × 40000, human [15], human, $beta_3beta_3$ isozyme [19], SDS-PAGE) [2, 15, 19]
More (each subunit is divided by a long cleft into two domains: a coenzyme binding and a catalytic domain) [34]

Glycoprotein/Lipoprotein

–

4 ISOLATION/PREPARATION

Source organism

Human (adults and premature infants [21], white American, European and Japanese adults [15]) [11–23, 58]; Saimiri sciureus boliviensis (squirrel monkey) [24]; Macaca mulatta (rhesus monkey) [25]; Horse [26–42, 52, 54, 56, 57, 92]; Rabbit (New Zealand White) [45, 46]; Misocricetus auratus (gold hamster) [47]; Chinese Hamster [1]; Mouse (strains C57BL/LiA (isozyme AA), ddN (isozyme BB) CBA/H (isozyme CC) [48]) [9, 48]; Rat (strain Wistar [10], Sprague-Dawley (castrated [51]) [49–51]) [10, 49–51, 92]; Suncus murinus (shrew) [5]; Coturnix japonica (japanese quail, different isozyme pattern depending on sex and age [60]) [43, 60]; Chicken [44]; Salmo gairdneri (rainbow trout) [59]; Ctenopharyngodon idellus (grass carp) [2]; Drosophila melanogaster (strain Samarkand [65], 3 alloenzymes [63]) [8, 62–65]; Drosophila simulans [8, 63, 64]; Drosophila virilis [8, 64]; Drosophila funebris [8]; Drosophila immigrans [8]; Drosophila lebanonensis [8, 61]; Vicia faba (bean) [4]; Glycine max (soybean, var. Prize) [71]; Brassica napus (rape) [74]; Parsley [72]; Tea [73]; Oryza sativa (rice, cv S-6) [6, 80]; Zea mays (maize, inbred line W64 [75]) [75, 76]; Triticum (wheat, species turgidum strain Rollette (tetraploid) [77], monococcum (diploid) [78]) [77, 78]; Najas marina [7]; Hordeum vulgare (barley, cv Sundance) [79]; Chlamydomonas moewusii (strain 11–5/10, green alga) [69]; Yeast [52, 56, 57, 88, 89, 92]; Kluyveromyces marxianus [3]; Saccharomyces cerevisiae (grown in presence and absence of Zn^{2+} [86]) [3, 86, 87, 90, 91]; Astasia longa (Jahn strain) [83]; Sporotrichum pulverulentum [84]; Candida guilliermondii (Y4, mutant strain B10-O5) [85]; Candida sp. N-16 [82]; Aspergillus nidulans (wild-type strain O51) [81]; Schizosaccharomyces pombe [91]; Zymomonas mobilis [57, 67, 68]; Sulfolobus solfataricus [66, 70]; More (overview [55], commercial preparation [53]) [53, 55]

Source tissue

Liver (isozymes AA and BB [48], not ADH-1 [49]) [1, 2, 5, 10, 12–46, 48, 49, 51, 52, 54, 56–60]; Intestine (intestinal caecum [59]) [49, 59]; Kidney (not isozyme ADH-1) [49]; Adrenal (not isozyme ADH-3) [49]; Stomach (isozyme CC [48], not ADH-3 [49]) [48–50]; Lung [49]; Cornea [49]; Ovary (not isozyme ADH-3) [49]; Uterus [49]; Epididymis [49]; Testis [11, 47, 49]; Spleen (isozyme ADH-2) [49]; Thymus (isozyme ADH-2) [49]; Heart (isozyme ADH-2) [49]; Brain (isozyme ADH-2) [49]; Pancreas (isozyme ADH-2) [49]; Muscle (isozyme ADH-2) [49]; Fly [61–65]; Cell (suspension culture [79]) [3, 66–70, 79, 81–87, 91]; Seed (germinating [4, 73–75, 77, 80], dry [79], scutellum [75]) [4, 6, 7, 73–75, 77, 79, 80]; Embryo tissue [74, 77]; Leaf [72]; Shoot [6, 79]; Root [79]; Tissue [7]

Localisation in source

Cytoplasm [45, 46, 66, 73, 79, 80, 83, 87, 91]; Mitochondria [91]

Purification

Human (2 isozymes [11], class I isozymes [13], class II isozyme: pi-ADH [14, 23], 4 of six isozymes [15], class III isozyme: chi-ADH, anodic protein [16, 18], chi_1-ADH and chi_2-ADH [21], ten isozymes [17], $beta_3beta_3$ isozyme [20], affinity chromatography on (4-[3-N-6-aminocapropyl]aminopropyl)-pyrazole-Sepharose (alpha-, beta- and gamma-subunit separation [15]) [11, 13, 15], Agarose-hexane-AMP ($beta_2beta_2$, $beta_2beta_1$, $beta_1beta_1$ isozyme separation [15]) [11, 14, 15]) [11–23]; Saimiri sciureus boliviensis (agarose-hexane-AMP and (4-[3-N-6-aminocapropyl]aminopropyl)-pyrazole-Sepharose chromatography, 2 isozymes) [24]; Macaca mulatta (3 electrophoretically distinct classes of isozymes) [25]; Horse (affinity chromatography on magnetic 5'-AMP-Sepharose, 3 isozymic forms: EE, ES, SS, different methods (review) [27], starch gel or ion agar electrophoresis [28], phosphocellulose chromatography [35], affinity chromatography on 5'-AMP-Sepharose [42]) [27, 28, 35, 42]; Rabbit (5'-AMP-Sepharose affinity chromatography, 3 electrophoretically distinct isozyme classes I–III) [45]; Chinese Hamster (2 major classes of isozymes of different electrophoretic mobility encoded by different structural gene loci derived from a common ancestor) [1]; Misocricetus auratus (affinity chromatography on (4-[3-N-6-aminocapropyl]aminopropyl)-pyrazole-Sepharose) [47]; Mouse (triazine-dye affinity chromatography) [48]; Rat (3 electrophoretically distinct isozymes: ADH-1, ADH-2, ADH-3 [49], class IV isozyme, peptide structure [50], affinity chromatography on N^6-(6-aminohexyl)-5'-AMP-Sepharose, 4 isozymes [51]) [49–51]; Coturnix japonica (isozyme of ADH-3-type [43], 4 isozymes [60]) [43, 60]; Chicken (5'-AMP-Sepharose affinity chromatography) [44]; Salmo gairdneri (affinity chromatography on Blue-Sepharose CL-6B) [59]; Ctenopharyngodon idellus [2]; Drosophila melanogaster (affinity chromatography on 8-(6-aminohexyl)amino-ATP/AMP) [65]; Chlamydomonas moewusii [69]; Glycine max (4 isozymes by isoelectric focusing) [71]; Parsley [72]; Tea [73]; Brassica napus [74]; Zea mays (3 isozymes) [75]; Triticum (3 isozymes, streptomycin precipitation/gel filtration chromatography) [77, 78]; Hordeum vulgare [79]; Aspergillus nidulans [81]; Candida sp. N-16 [82]; Candida guilliermondii (isozyme ADH-1) [85]; Saccharomyces cerevisiae (partially [86], isozyme ADH-IV, dye-ligand chromatography [87]) [86, 87]; Astasia longa (partial, several isozyme fractions depending on growth conditions) [83]; Sporotrichum pulverulentum [84]; Sulfolobus solfataricus [66]; Zymomonas mobilis (2 electrophoretically separable isozymes of different amino acid composition [67], affinity chromatography) [67, 68]; More (streptomycin precipitation and gel filtration are essential for obtaining a stable enzyme preparation [77, 78]) [77, 78]

Crystallization

(human [12], horse (3-dimensional structure by X-ray crystallography [27], enzyme structure in crystal forms with NAD^+ differs from unliganded enzyme [34], EE-isozyme [57]) [27, 34, 35, 57], Saccharomyces cerevisiae [86]) [12, 27, 34, 35, 57, 86]

Cloned

(mouse, cDNA for chi-like class III ADH (ADH-B2) cloned and sequenced [9], human, class I isozyme, beta-ADH cDNA clone pADH12 [58], Drosophila melanogaster, sequencing and partial characterization [65], Saccharomyces cerevisiae, Ty-mediated cloning and overexpression of ADH4-gene, multicopy vector [87]) [9, 58, 65, 87]

Renaturated

–

5 STABILITY

pH

More (at neutral pH stable [6], denaturation in acidic media [74]) [6, 74]; 5.5–11.2 (pyrazole-insensitive isozyme stable) [25]; 6.0 (below, inactivation) [85]; 6.0–9.0 (isozymes 2 and 3, stable) [75]; 6.5–7.0 (most stable) [74]; 7.0–10.6 (stable if the ionic strength is above 0.05) [12]; 8.0 (optimal stability) [12]; 8.5–10.0 (more stable in diphosphate than in sodium phosphate buffer) [85]; 9.0 (isozyme 1, 20% loss of activity) [75]; 10 (unstable) [75]; 11.0 (above, rapid inactivation, pyrazole-sensitive isozyme) [24]

Temperature (°C)

More (acetimidylation or methylation do not affect thermal stability [37], thermostability is lost after dialysis against chelating agents, or incubation in 6 M urea or 0.1% SDS at 30°C [70], resistant to heat inactivation [82, 83]) [37, 70, 82, 83]; 20 (and above unstable) [12]; 40 (1 h stable [6], stable below [85]) [6, 85]; 50 (70% of initial activity after 24 h) [66]; 55 (remaining activities after 1 h: 27%, 90% and 79%, isozymes 1–3, respectively) [75]; 60 ($t_{1/2}$: about 100 min (isozyme E1E1), $t_{1/2}$: 45 min (soluble or bound to octyl-Sepharose), about 70% of maximal activity after 100 min (bound to CNBr-Sepharose) [37], $t_{1/2}$: 20 h [66], inactivation above [74]) [37, 66, 74]; 65 (inactivation after 30 min [68], at least 16 h stable at 2.5 mg/ml enzyme protein, $t_{1/2}$: 8 h at 0.5 mg/ml [70]) [68, 70]; 70 ($t_{1/2}$: 5 h) [66]; 85 ($t_{1/2}$: 3 h at 0.5 mg/ml enzyme protein) [70]

Oxidation

Ascorbate or dithioerythritol prevents oxidation during purification [16]

Organic solvent

Aqueous dioxane, 5% v/v, stable at neutral, inactivation at high pH-values [39]

General stability information

Freezing inactivates [12, 85]; Freeze-drying inactivates [83]; Crystalline, unstable in crystallization medium [12]; Chromatography on Sephadex C-50 stabilizes class II isozyme [14]; Ethanol, 0.01 M, stabilizes [18]; Glycerol, 50% v/v stabilizes [19]; Glycerol, 10% v/v stabilizes [79]; NAD^+ stabilizes [19]; N_2 stabilizes during dialysis [86]; DTT stabilizes [19, 45, 46, 49, 60, 73, 77–79]; Dithioerythritol stabilizes [69]; 2-Mercaptoethanol stabilizes (10 mM [73]) [46, 73, 77–79]; Sucrose, 20% (w/v), does not stabilize [86]; Dialysis inactivates, DTT stabilizes, GSH or 2-mercaptopropionyl glycine partly reverses inactivation [45]; Precipitation with ammonium sulfate, acetone or alcohol inactivates [71]

Storage

Unstable to storage [83]; –20°C, isozyme 2 at least 3 months in 50% v/v ethylene glycol, isozyme 1 and 3: $t_{1/2}$ is less than 2 weeks in 50% v/v ethylene glycol [75]; 4°C, several months [70]; 4°C, 2–3 weeks [14]; 4°C, under nitrogen plus DTT, 6–8 weeks [14]; 4°C, plus DTT up to 10 days [16]; 4°C, $t_{1/2}$: 24 h [18, 23], glycerol or NAD^+ partially protects [23]; 4°C, plus 10 mM ethanol, up to 2 weeks [23]; 4°C, up to 7 days (isozymes 2 and 3 [51]) [21, 45, 51]; 4°C, with 1 mM DTT, 76% activity retained after 18 days [45]; 4°C, with 2 mM 2-mercaptoethanol, $t_{1/2}$: 66 days [46]; 4°C, $t_{1/2}$: around 7 days (isozyme 4) [51]; 4°C, 83% loss of activity after 7 days (isozyme 1) [51]; 4°C, $t_{1/2}$: 6–7 weeks in 20 mM Tris/HCl pH 8.6 [61]; 4°C, $t_{1/2}$: less than 16 h [86]

6 CROSSREFERENCES TO STRUCTURE DATABANKS

PIR/MIPS code

PIR3:S14270 (Acetobacter polyoxogenes); PIR3:S14271 (Acetobacter polyoxogenes); PIR2:A30196 (Alcaligenes eutrophus); PIR3:S35669 (American alligator); PIR1:DEMUAM (Arabidopsis thaliana); PIR3:S42883 (Arabidopsis thaliana); PIR2:S07218 (Bacillus stearothermophilus (fragment)); PIR2:A10207 (Bacillus stearothermophilus (fragment)); PIR2:S11074 (Baltic cod (fragments)); PIR3:A40731 (Drosophila guanche); PIR2:A29054 (Emericella nidulans); PIR3:S41377 (Entamoeba histolytica); PIR1:DEEC (Escherichia coli); PIR3:S14809 (Escherichia coli); PIR3:S23871 (Escherichia coli); PIR1:DEZPA (fission yeast (Schizosaccharomyces pombe)); PIR2:S16453 (fruit fly (Drosophila ambigua)); PIR2:E23724 (fruit fly (Drosophila differens)); PIR3:S20713 (fruit fly (Drosophila erecta)); PIR2:A23724 (fruit fly (Drosophila heteroneura)); PIR1:DEFFRL (fruit fly (Drosophila lebanonensis)); PIR3:JN0564 (Fruit fly (Drosophila lebanonensis)); PIR2:S09633 (fruit fly (Drosophila mauritiana)); PIR1:DEFFA (fruit fly (Drosophila melanogaster)); PIR3:S17083 (fruit fly (Drosophila melanogaster)); PIR3:S17084 (fruit fly (Drosophila melanogaster)); PIR3:S17085 (fruit fly (Drosophila melanogaster)); PIR2:A40553 (fruit fly (Drosophila mettleri)); PIR3:S26785 (fruit fly (Drosophila miranda)); PIR2:S09634 (fruit fly (Drosophila orena)); PIR3:S26784 (fruit fly (Drosophila persimilis)); PIR2:B23724 (fruit fly (Drosophila picticornis)); PIR2:C23724 (fruit fly (Drosophila planitibia)); PIR2:S06292 (fruit fly (Drosophila pseudoobscura)); PIR2:S25453 (fruit fly (Drosophila pseudoobscura)); PIR2:S25452 (fruit fly (Drosophila pseudoobscura)); PIR2:S25451 (fruit fly (Drosophila pseudoobscura)); PIR2:S25450 (fruit fly (Drosophila pseudoobscura)); PIR2:S25449 (fruit fly (Drosophila pseudoobscura)); PIR2:S25448 (fruit fly (Drosophila pseudoobscura)); PIR2:S25447 (fruit fly (Drosophila pseudoobscura)); PIR2:S25446 (fruit fly (Drosophila pseudoobscura)); PIR3:S25918 (fruit fly (Drosophila pseudoobscura)); PIR3:S26766 (fruit fly (Drosophila pseudoobscura)); PIR3:S26767 (fruit fly (Drosophila pseudoobscura)); PIR3:S26768 (fruit fly (Drosophila pseudoobscura)); PIR3:S26769 (fruit fly (Drosophila pseudoobscura)); PIR3:S26770 (fruit fly (Drosophila pseudoobscura)); PIR3:S26771 (fruit fly (Drosophila pseudoobscura)); PIR3:S26772 (fruit fly (Drosophila pseudoobscura)); PIR3:S26773 (fruit fly (Drosophila pseudoobscura)); PIR3:S26774 (fruit fly (Drosophila pseudoobscura)); PIR3:S26775 (fruit fly (Drosophila pseudoobscura)); PIR3:S26776 (fruit fly (Drosophila pseudoobscura)); PIR3:S26777 (fruit fly (Drosophila pseudoobscura)); PIR3:S26778 (fruit fly (Drosophila pseudoobscura)); PIR3:S26779 (fruit fly (Drosophila pseudoobscura)); PIR3:S26780 (fruit fly (Drosophila pseudoobscura)); PIR3:S26781 (fruit fly (Drosophila pseudoobscura)); PIR3:S26782 (fruit fly (Drosophila pseudoobscura)); PIR3:S26783 (fruit fly (Drosophila pseudoobscura)); PIR2:S07439 (fruit fly (Drosophila sechellia)); PIR2:D23724 (fruit fly (Drosophila silvestris)); PIR2:S07333 (fruit fly (Drosophila simulans)); PIR3:S18273 (fruit fly (Drosophila simulans)); PIR3:S30595 (fruit fly (Drosophila subobscura)); PIR3:S20715 (fruit fly (Drosophila teissieri)); PIR3:S18277 (fruit fly (Drosophila yakuba)); PIR3:S18278 (fruit fly (Drosophila yakuba)); PIR3:S18279 (fruit fly (Drosophila yakuba)); PIR3:S18280 (fruit fly (Drosophila yakuba)); PIR3:S18281 (fruit fly (Drosophila yakuba)); PIR3:S18282 (fruit fly (Drosophila yakuba)); PIR3:S18284 (fruit fly (Drosophila yakuba)); PIR3:S18287 (fruit fly (Drosophila yakuba)); PIR3:S20718 (fruit fly (Drosophila yakuba)); PIR2:S10019 (garden strawberry); PIR2:A35837 (Japanese quail); PIR2:A05102 (maize); PIR3:S20593 (mallard); PIR2:JC1376 (Mycobacterium bovis); PIR2:S11075 (Perez's frog (fragment)); PIR3:S27651 (Pseudomonas sp.); PIR3:S29343 (rabbit); PIR3:S30353 (rabbit); PIR2:A26468 (rat); PIR2:JN0629 (rat); PIR2:A26211 (rat (fragments)); PIR2:A32973 (Thermoanaerobacter brockii); PIR3:S16038 (Tomato (fragment)); PIR3:S20613 (Tomato (fragment)); PIR2:A61024 (wheat (cv. Millewa)); PIR3:S32521 (yeast (Kluyveromyces marxianus)); PIR2:A25978 (Zymomonas mobilis); PIR2:S00912 (1 (clone lambda-PG8) garden pea); PIR2:S01893 (1 barley); PIR2:S15712 (1 fruit fly (Drosophila hydei)); PIR2:S01902 (1 fruit fly (Drosophila mojavensis)); PIR2:A24268 (1 fruit fly (Drosophila mulleri)); PIR2:S06001 (1 fruit fly (Drosophila navojoa)); PIR1:DEPJA1 (1 garden petunia); PIR2:S04571 (1 maize); PIR2:A00342 (1 maize (fragment)); PIR2:A38405 (1 Perez's frog); PIR2:S11853 (1 potato); PIR2:JQ0474 (1 rice); PIR1:DEJYAW (1 white clover); PIR2:S09475 (1 yeast (Kluyveromyces marxianus var. lactis)); PIR1:DEILSP (1-S pearl millet); PIR2:S04039 (2 barley); PIR2:S15711 (2 fruit fly (Drosophila hydei)); PIR2:JN0565 (2 fruit fly (Drosophila lebanonensis)); PIR2:S01901 (2 fruit fly (Drosophila mojavensis)); PIR3:A45962 (2 fruit fly (Drosophila mojavensis) (fragment)); PIR2:B24268 (2 fruit fly (Drosophila mulleri)); PIR2:A23084 (2 maize); PIR1:DEPOA2 (2 potato); PIR1:DERTA (2 rat); PIR1:DERZA2 (2 rice); PIR2:JT0618 (2 tomato (fragment)); PIR2:S07584 (2 Zymomonas mobilis); PIR2:B24801 (2 Zymomonas mobilis (fragment)); PIR2:S04040 (3 barley); PIR1:DEPOA3 (3 potato); PIR2:S09293 (3 precursor yeast (Kluyveromyces marxianus var. lactis)); PIR2:S17252 (3 precursor yeast (Kluyveromyces marxianus var. lactis)); PIR2:S17253 (4 precursor yeast (Kluyveromyces marxianus var. lactis)); PIR2:JH0789 (5 human); PIR2:A41274 (6 human); PIR2:JS0326 (72K chain precursor Acetobacter aceti); PIR3:S20429 (A Indian spiny-tailed lizard); PIR1:DEMSAA (A mouse); PIR2:A27322 (A mouse); PIR2:A29628 (A mouse); PIR2:A37092 (allozyme ADH-71K fruit fly (Drosophila melanogaster)); PIR1:DEHUAA (alpha human); PIR3:S20430 (B Indian spiny-tailed lizard); PIR2:A60269 (B-2 mouse (fragment)); PIR2:A26281 (beta human); PIR2:B25428 (beta human); PIR2:A05182 (beta human (fragments)); PIR2:A25849 (beta mouse (fragment)); PIR1:DEHUAB (beta-1 human); PIR2:A23607 (beta-1 human); PIR2:S05202 (beta-2 (allele 2) human); PIR3:S10621 (beta-2 human); PIR2:A26826 (beta-3 human (fragment)); PIR2:S02617 (chi horse (fragment)); PIR1:DEHUC2 (chi-2 precursor human);

PIR2:A33372 (class I rat); PIR3:B40731 (dup Drosophila guanche); PIR3:S33434 (E Clostridium acetobutylicum); PIR1:DEHOAL (E horse); PIR2:A39872 (E horse); PIR2:C25428 (gamma human); PIR1:DEHUAG (gamma-1 human); PIR2:A24798 (gamma-2 human); PIR1:DECHA1 (I chicken); PIR1:DEBYA (I yeast (Saccharomyces cerevisiae)); PIR2:S38795 (I yeast (Saccharomyces cerevisiae)); PIR2:A35260 (I Zymomonas mobilis); PIR2:E40649 (I Zymomonas mobilis (fragment)); PIR2:A33909 (I hepatic baboon); PIR2:S20911 (II yeast (Kluyveromyces marxianus var. lactis)); PIR1:DEBYA2 (II yeast (Saccharomyces cerevisiae)); PIR2:S38796 (II yeast (Saccharomyces cerevisiae)); PIR2:A24648 (III Emericella nidulans); PIR2:S22244 (III yeast (Saccharomyces cerevisiae)); PIR2:A33419 (III chi horse); PIR1:DEBY4 (IV yeast (Saccharomyces cerevisiae)); PIR2:A42343 (major form Baltic cod); PIR2:S18820 (pi human); PIR2:A27109 (pi human); PIR2:A49340 (precursor Acetobacter pasteurianus (strain NCI1380)); PIR1:DEHOAS (S horse); PIR2:B39872 (S horse); PIR2:S31140 (SFA yeast (Saccharomyces cerevisiae)); PIR3:S21170 (sigma human); PIR2:A44245 (NAD(+)-dependent Sulfolobus solfataricus); PIR2:A42654 (thermostable Bacillus stearothermophilus)

Brookhaven code

7ADH (Horse (Equus caballus) liver); 3HUD (Human (Homo sapiens) expressed and purified from)

7 LITERATURE REFERENCES

[1] Talbot, B.G., Qureshi, A.A., Cohen, R., Thirion, J.P.: Biochem. Genet.,19,813–829 (1981)
[2] Fong, W.P.: Comp. Biochem. Physiol. B Comp. Biochem.,98B,297–302 (1991)
[3] Pessione, E., Pergola, L., Cavaletto, M., Giunta, C., Trotta, A., Vanni, A.: Ital. J. Biochem.,39,71–82 (1990)
[4] Leblova, S., El Ahmad, M.: Collect. Czech. Chem. Commun.,54,2519–2527 (1989)
[5] Keung, W.M., Ho, Y.W., Fong, W.P., Lee, C.Y.: Comp. Biochem. Physiol. B Comp. Biochem.,93B,169–173 (1989)
[6] Tong, W.F., Lin, S.W.: Bot. Bull. Acad. Sin.,29,245–253 (1988)
[7] Van Geyt, J., Jacobs, M., Triest, L.: Aquat. Bot.,28,129–141 (1987)
[8] Vilageliu, L., Juan, E., Gonzalez-Duarte, R. in "Adv. Genet., Dev., Evol. Drosophila, [Proc. Eur. Drosophila Res. Conf.]" (Lakovaara, S., ed.) 7,237–250, Plenum, New York, N.Y. (1982)
[9] Edenberg, H.J., Brown, C.J., Carr, L.G., Ho, W.H., Hu, M.W.: Adv. Exp. Med. Biol.,284 (Enzymol. Mol. Biol. Carbonyl Metab.3) ,253–262 (1991)
[10] Herrera, E., Zorzano, A., Fresneda, V.: Biochem. Soc. Trans.,11,729–730 (1983) (Review)
[11] Dafeldecker, W.P., Vallee, B.L.: Biochem. Biophys. Res. Commun.,134,1056–1063 (1986)
[12] Woronick, C.L.: Methods Enzymol.,41B,369–374 (1975) (Review)
[13] Wagner, F.W., Burger, A.R., Vallee, B.L.: Biochemistry,22,1857–1863 (1983)

[14] Ditlow, C.C., Holmquist, B., Morelock, M.M., Vallee, B.L.: Biochemistry,23, 6363–6368 (1984)
[15] Yin, S.-J., Bosron, W.F., Magnes, L.J., Li, T.-K.: Biochemistry,23,5847–5853 (1984)
[16] Wagner, F.W., Parés, X., Holmquist, B., Vallee, B.L.: Biochemistry,23,2193–2199 (1984)
[17] Bosron, W.F., Magnes, L.J., Li, T.-K.: Biochemistry,22,1852–1857 (1983)
[18] Bosron, W.F., Li, T.-K.: Biochem. Biophys. Res. Commun.,74,85–91 (1977)
[19] Schneider-Bernlöhr, H., Formicka-Kozlowska, G., Bühler, R., Wartburg, J.-P., Zeppezauer, M.: Eur. J. Biochem.,173,275–280 (1988)
[20] Burnell, J.C., Li, T.-K., Bosron, W.F.: Biochemistry,28,6810–6815 (1989)
[21] Parés, X., Vallee, B.L.: Biochem. Biophys. Res. Commun.,98,122–130 (1981)
[22] Bosron, W.F., Li, T.-K., Vallee, B.L.: Proc. Natl. Acad. Sci. USA,77,5784–5788 (1980)
[23] Bosron, W.F., Li, T.-K., Dafeldecker, W.P., Vallee, B.L.: Biochemistry,18,1101–1105 (1979)
[24] Dafeldecker, W.P., Parés, X., Vallee, B.L., Bosron, W.F., Li, T.-K.: Biochemistry,20, 856–861 (1981)
[25] Dafeldecker, W.P., Meadow, P.E., Parés, X., Vallee, B.L.: Biochemistry,20,6729–6734 (1981)
[26] Kvassman, J., Pettersson, G.: Eur. J. Biochem.,166,167–172 (1987)
[27] Andersson, L., Mosbach, K.: Methods Enzymol.,89,435–445 (1982) (Review)
[28] Pietruszko, R.: Methods Enzymol.,89,429–435 (1982) (Review)
[29] Dahl, K.H., Eklund, H., McKinley-McKee, J.S.: Biochem. J.,211,391–396 (1983)
[30] Maret, W., Makinen, M.W.: J. Biol. Chem.,266,20636–20644 (1991)
[31] Adolph, H.W., Maurer, P., Schneider-Bernlöhr, H., Sartorius, C., Zeppezauer, M.: Eur. J. Biochem.,201,615–625 (1991)
[32] Freudenreich, C., Samama, J.-P., Biellmann, J.-F.: J. Am. Chem. Soc.,106,3344–3353 (1984)
[33] Samama, J.-P., Hirsch, D., Goulas, P., Biellmann, J.-F.: Eur. J. Biochem.,159, 375–380 (1986)
[34] Eklund, H.: Biochem. Soc. Trans.,17,293–296 (1989)
[35] Dworschack, R.T., Plapp, B.V.: Biochemistry,16,111–116 (1977)
[36] Maret, W., Andersson, I., Dietrich, H., Schneider-Bernlöhr, H., Einarsson, R., Zeppezauer, M.: Eur. J. Biochem.,98,501–512 (1979)
[37] Skerker, P.S., Clark, D.S.: Biotechnol. Bioeng.,33,62–71 (1989)
[38] Kaiser, R., Holmquist, B., Vallee, B.L., Jörnvall, H.: Biochemistry,28,8432–8438 (1989)
[39] Tsai, C.S.: Arch. Biochem. Biophys.,213,635–642 (1982)
[40] Favilla, R., Cavatorta, P., Mazzini, A., Fava, A.: Eur. J. Biochem.,104,223–227 (1980)
[41] Ryzewski, C.N., Pietruszko, R.: Biochemistry,19,4843–4848 (1980)
[42] Ryzewski, C.N., Pietruszko, R.: Arch. Biochem. Biophys.,183,73–82 (1977)
[43] Kaiser, R., Nussrallah, B., Dam, R., Wagner, F.W., Jörnvall, H.: Biochemistry,29, 8365–8371 (1990)
[44] Von Bahr-Lindström, H., Andersson, L., Mosbach, K., Jörnvall, H.: FEBS Lett.,89,293–297 (1978)
[45] Hoshino, T., Ishigura, I., Ohta, Y.: J. Biochem.,97,1163–1172 (1985)
[46] Keung, W.M., Yip, P.K.: Biochem. Biophys. Res. Commun.,158,445–453 (1989)
[47] Keung, W.M.: Biochem. Biophys. Res. Commun.,156,38–45 (1988)
[48] Algar, E.M., Seeley, T.-L., Holmes, R.S.: Eur. J. Biochem.,137,139–147 (1983)
[49] Julià, P., Farrés, J., Parés, X.: Eur. J. Biochem.,162,179–189 (1987)

[50] Parés, X., Moreno, A., Cederlund, E., Höög, J.-O., Jörnvall, H.: FEBS Lett.,277,115–118 (1990)
[51] Mezey, E., Potter, J.J.: Arch. Biochem. Biophys.,225,787–794 (1983)
[52] Jörnvall, H.: Eur. J. Biochem.,72,443–452 (1977)
[53] Creagh, A.L., Prausnitz, J.M., Blanch, H.W.: Biotechnol. Bioeng.,41,156–161 (1993)
[54] Plapp, B.V., Sogin, D.C., Dworschack, R.T., Bohlken, D.P., Woenckhaus, C.: Biochemistry,25,5396–5402 (1986)
[55] Jörnvall, H., Persson, B., Krook, M., Kaiser, R.: Biochem. Soc. Trans.,18,169–171 (1988) (Review)
[56] Silverstein, E.: Curr. Top. Cell. Regul.,24,209–218 (1984) (Review)
[57] Jörnvall, H., Höög, J.-O., Von Bahr-Lindström, H., Johansson, J., Kaiser, R., Persson, B.: Biochem. Soc. Trans.,16,223–227 (1988) (Review)
[58] Retzios, A., Thatcher, D.R.: Biochem. Soc. Trans.,9,298–299 (1981)
[59] Bauermeister, A., Sargent, J.: Biochem. Soc. Trans.,6,222–224 (1978)
[60] Nussrallah, B.A., Dam, R., Wagner, F.W.: Biochemistry,28,6245–6251 (1989)
[61] Winberg, J.-O., Hovik, R., McKinley-McKee, J.S., Juan, E., Gonzalez-Duarte, R.: Biochem. J.,235,481–490 (1986)
[62] Winberg, J.-O., McKinley-McKee, J.S.: Biochem. J.,251,223–227 (1988)
[63] Heinstra, P.W.H., Thörig, G.E.W., Scharloo, W., Drenth, W., Nolte, R.J.M.: Biochim. Biophys. Acta,967,224–233 (1988)
[64] Juan, E., Gonzalez-Duarte, R.: Biochem. J.,195,61–69 (1981)
[65] Lee, C.-Y.: Methods Enzymol.,89,445–450 (1982) (Review)
[66] Rella, R., Raia, C.A., Pensa, M., Pisani, F.M., Gambacorta, A., De Rosa, M., Rossi, M.: Eur. J. Biochem.,167,475–479 (1987)
[67] Wills, C., Kratofil, P., Londo, D., Martin, T.: Arch. Biochem. Biophys.,210,775–785 (1981)
[68] Kinoshita, S., Kakizono, T., Kadota, K., Das, K., Taguchi, H.: Appl. Microbiol. Biotechnol.,22,249–254 (1985)
[69] Grondal, E.J.M., Betz, A., Kreuzberg, K.: Phytochemistry,22,1695–1699 (1983)
[70] Ammendola, S., Raia, C.A., Caruso, C., Camardella, L., D'Auria, S., De Rosa, M., Rossi, M.: Biochemistry,31,12514–12523 (1992)
[71] Tihanyi, K., Talbot, B., Brzezinski, R., Thirion, J.-P.: Phytochemistry,28,1335–1338 (1989)
[72] Liang, Z.Q., Hayase, F., Nishimura, T., Kato, H.: Agric. Biol. Chem.,54,1717-1719 (1990)
[73] Hatanaka, A., Harada, T.: Agric. Biol. Chem.,36,2033–2035 (1972)
[74] Stiborová, M., Leblová, S.: Phytochemistry,18,23–24 (1979)
[75] Lai, Y.-K., Chandlee, J.M., Scandalios, J.G.: Biochim. Biophys. Acta,706,9–18 (1982)
[76] Leblová, S., Ehlichová, D.: Phytochemistry,11,1345–1346 (1972)
[77] Langston, P.J., Pace, C.N., Hart, G.E.: Plant Physiol.,65,518–522 (1980)
[78] Langston, P.J., Hart, G.E., Pace, C.N.: Arch. Biochem. Biophys.,196,611–618 (1979)
[79] Mayne, M.G., Lea, P.J.: Phytochemistry,24,1433–1438 (1985)
[80] Shimomura, S., Beevers, H.: Plant Physiol.,71,742–746 (1983)
[81] Creaser, E.H., Porter, R.L., Britt, K.A., Pateman, J.A., Doy, C.H.: Biochem. J.,225,449–454 (1985)
[82] Yabe, M., Shitara, K., Kawashima, J., Shinoyama, H., Ando, A., Fujii, T.: Biosci. Biotechnol. Biochem.,56,338–339 (1992)

[83] Morosoli, R., Bégin-Heick, N.: Biochem. J.,141,469–475 (1974)
[84] Rudge, J., Bickerstaff, G.F.: Enzyme Microb. Technol.,8,120–124 (1986)
[85] Indrati, R., Ohita, Y.: Can. J. Microbiol.,38,953–957 (1992)
[86] Dickenson, C.J., Dickinson, F.M.: Biochem. J.,153,309–319 (1976)
[87] Drewke, C., Ciriacy, M.: Biochim. Biophys. Acta,950,54–60 (1988)
[88] Yamazaki, Y., Maeda, H., Satoh, A., Hiromi, K.: J. Biochem.,95,109–115 (1984)
[89] Mazid, M.A., Laidler, K.J.: Can. J. Microbiol.,60,100–107 (1982)
[90] Dickinson, F.M., Monger, G.P.: Biochem. J.,131,261–270 (1973)
[91] Ganzhorn, A.J., Green, D.W., Hershey, A.D., Gould, R.M., Plapp, B.V.: J. Biol. Chem.,262,3754–3761 (1987)
[92] Feraudi, M., Kohlmeier, M., Schmolz, G.: Biochem. Soc. Trans.,3,1063–1066 (1975) (Review)

1 NOMENCLATURE

EC number
1.1.1.2

Systematic name
Alcohol:$NADP^+$ oxidoreductase

Recommended name
Alcohol dehydrogenase ($NADP^+$)

Synonymes
Aldehyde reductase (NADPH)
Dehydrogenase, alcohol (nicotinamide adenine dinucleotide phosphate)
NADP-alcohol dehydrogenase
NADP-aldehyde reductase
NADP-dependent aldehyde reductase
NADPH-aldehyde reductase
NADPH-dependent aldehyde reductase
Nonspecific succinic semialdehyde reductase (EC 1.1.1.2)
ALR 1 [23, 25]
Low-K_m aldehyde reductase [31, 40]
High-K_m aldehyde reductase [31, 33, 40]
More (may be identical with EC 1.1.1.19, EC 1.1.1.33 and EC 1.1.1.55, cf. EC 1.1.1.184, ALR 1 may be identical with hexonate dehydrogenase or daunorubicin reductase [25])

CAS Reg. No.
9028-12-0

2 REACTION AND SPECIFICITY

Catalysed reaction
An alcohol + $NADP^+$ →
→ an aldehyde or ketone + NADPH (mechanism [5, 11, 17, 22, 25, 29, 34])

Reaction type
Redox reaction

Natural substrates

Aldehyde + NADPH (physiological role not clearly defined [3, 5], reduction of various aromatic, medium [2] or long [31] chain aliphatic aldehydes [2], involved in catabolism of biogenic amines [24, 30], detoxification processes [38]) [2, 3, 5, 24, 30, 31, 38]
Glucuronate + NADPH (biosynthesis of ascorbic acid) [28, 36]
Primary alcohol + $NADP^+$ (physiological role not clear) [42]

Substrate spectrum

1 Aliphatic aldehydes + NADPH (non-polyhydroxylated [4], long-chain: e.g. hexanal [27, 31] or octadecanal [31], laurylaldehyde (poor substrate) [36], short-chain: r [38], e.g. acetaldehyde (not [2, 18, 33]) [38], propionaldehyde (ir, poor substrate for reductase II [1]) [1, 3, 4, 38], n-butyraldehyde (best substrate for reductase I [1] and reductases AR1 and 3 [36]) [1, 2, 18, 21, 36], heptanal [21], n-octanal [31], short-chain aldehydes: poor substrates [2], no substrates [24, 37], aldehyde reduction preferred to alcohol oxidation [5, 21, 22]) [1–5, 18, 21, 22, 24, 27, 31, 36, 38–40]
2 Aromatic aldehydes + NADPH (preferred substrates [2], e.g. benzaldehyde [1, 2, 18, 21, 24, 30, 38], aromatic aldehydes with an electron-withdrawing group in para-position are the best substrates [21], oxidation of 4-nitrophenol less than 2% the rate of alcohol formation from p-nitrobenzaldehyde [2], aldehyde reduction is favoured reaction [21, 22]) [1, 2, 18, 21, 22, 24, 30, 38–40]
3 Substituted benzaldehydes + NADPH (ir [37], best substrates [2], e.g. p-carboxybenzaldehyde (best substrate [18, 21], poor substrate [36]) [2, 16, 18, 21, 22, 24, 28, 30, 36], p-nitrobenzaldehyde [1, 2, 16, 18, 19, 21, 24, 25, 27, 28, 30, 31, 33, 36, 37], m- and o-nitrobenzaldehyde (poor substrates for reductase AR1 and 3 [36]) [1, 2, 16, 18, 19, 21, 22, 24, 25, 27, 28, 30, 31, 33, 36, 37], chlorobenzaldehyde (p- (poor substrate [37]) [2, 18, 37] and o- [21]) [2, 18, 21, 37], cyanobenzaldehyde (p- (best substrate [18]) [16, 18, 24], m- [18]) [16, 18, 24], p-fluorobenzaldehyde [18], anisaldehyde (r) [18, 21, 31], no substrate: o-carboxybenzaldehyde [33]) [1, 2, 16, 18, 19, 21, 22, 24, 25, 27, 28, 30, 31, 33, 36, 37, 39, 40]
4 3-Pyridinecarboxaldehyde + NADPH (i.e. 3-pyridinealdehyde or nicotinaldehyde, r [18], ir [3], best substrate (reductase I [4]) [4, 19], reduction at 320% the rate of D-glyceraldehyde reduction [19], poor substrate (reductase I) [1], not [27]) [1, 3, 4, 16, 18, 19, 21, 22, 25, 28, 30, 36, 37, 39]
5 4-Pyridinecarboxaldehyde + NADPH (i.e. 4-pyridinealdehyde, reduction at 121% the rate of 3-pyridinecarboxaldehyde reduction [18], not [27]) [18, 34, 36]
6 p-Tolualdehyde + NADPH [24]
7 p-Nitroacetophenone + NADPH (reductase I: poor substrate [21], no substrate [18, 31]) [21, 27]
8 4-Carboxyacetophenone + NADPH (reductase I, poor substrate) [21]

9 Glyceraldehyde + NADPH (ir, poor, reductase I [1, 3], reduction at 41% the rate of p-nitrobenzaldehyde reduction [33], D-enantiomer (reduction at 25% [18], 42% (reductase II), 62% (reductase I) [21] the rate of 3-pyridinecarboxaldehyde reduction) [18, 19, 21, 36, 37], L-enantiomer (r, reduction at 5% the rate of 3-pyridinecarboxaldehyde reduction, oxidation of glycerol at 64% the rate of 3-pyridinemethanol oxidation [18]) [18, 36]) [1–5, 18, 19, 21, 26, 30, 32, 33, 36, 37, 39]
10 D-Lactaldehyde + NADPH (reduction at 28% the rate of 3-pyridinecarboxaldehyde reduction) [18]
11 Methylglyoxal + NADPH (i.e. pyruvaldehyde, reduction at 24% the rate of 3-pyridinecarboxaldehyde reduction [18]) [2, 18, 36, 37]
12 Phenylglyoxal + NADPH (best substrate [36], reduction at 120% (reductase I), 132% (reductase II) the rate of 3-pyridinecarboxaldehyde reduction [21]) [21, 36]
13 4-Carboxyphenylglyoxal + NADPH (reductase I, reduction at 56% the rate of 3-pyridinecarboxaldehyde reduction) [21]
14 Glycolaldehyde + NADPH (reductase AR 2 [36], not [37]) [36]
15 Benzalacetone + NADPH [27]
16 Camphorquinone + NADPH ((+)-enantiomer (reduction at 52% the rate of 3-pyridinecarboxaldehyde reduction [18]) [2, 18], (-)-enantiomer (reduction at 2% the rate of 3-pyridinecarboxaldehyde reduction) [18]) [2, 18]
17 3,4-Dihydroxyphenylglycolaldehyde + NADPH [16]
18 3,4-Dihydroxyphenylacetaldehyde + NADPH [16]
19 p-Hydroxyphenylacetaldehyde + NADPH (r [24]) [24, 30]
20 p-Hydroxyphenylglycolaldehyde + NADPH [24, 30]
21 Phenylacetaldehyde + NADPH [24]
22 Indole-3-acetaldehyde + NADPH (reductase I, reduction at 25% the rate of 3-pyridinecarboxaldehyde reduction [21], not [18, 24]) [21, 36]
23 5-Hydroxyindoleacetaldehyde + NADPH [16]
24 Succinic semialdehyde + NADPH (reductase I, reduction at 80% the rate of 3-pyridinecarboxaldehyde reduction [21], reduction at 83% the rate of p-nitrobenzaldehyde reduction [33], not (reductase II) [21]) [21, 28, 33]
25 2,3-Butanedione + NADPH (i.e. diacetyl, reductase I [21], reduction at 13% [18], 83% [21] the rate of 3-pyridinecarboxaldehyde reduction) [18, 21, 36, 37]
26 2,3-Pentanedione + NADPH (reduction at 12% the rate of 3-pyridinecarboxaldehyde reduction) [18]
27 1,2-Cyclohexanedione + NADPH [36]
28 Ethyl acetoacetate + NADPH [31, 36]
29 Chloralhydrate + NADPH [30]
30 D-Aldoses + NADPH (e.g. D-galactose (poor substrate, reductase II [1], not (reductase I [1]) [1, 4, 18, 28, 37]) [1], D-xylose (reductase II [1],

poor substrate [26], not (reductase I [1, 37]) [1, 18, 37]) [1, 26, 30, 36], D-ribose (reductase AR 2, not Ar 1 and 3 [36]) [18, 36], D-erythrose (reduction at 9% the rate of 3-pyridinecarboxaldehyde reduction [18], reduction at 6% the rate of D-glyceraldehyde reduction [19]) [18, 19], no substrate: D-arabinose [19], aldoses [3, 32]) [1, 18, 19, 26, 30, 36, 39, 40]

31 Glucose + NADPH (reductase I [4], not (reductase II [4]) [1–4, 37]) [4]

32 D-Glucuronate + NADPH (r [29], reductase II [1], 3-acetylpyridine adenine dinucleotide can replace NADPH [5], reduction at 54% the rate of D-glyceraldehyde reduction [19], reduction at 14% [18], 10% (reductase II), 100% (reductase I) [21] the rate of 3-pyridinecarboxaldehyde reduction, reduction at 87% the rate of p-nitrobenzaldehyde reduction [33], not reductase AR 2 [36], reductase I [1, 4, 37]) [1, 3–5, 18, 19, 21, 25, 26, 28–30, 32, 33, 36]

33 D-Glucuronolactone + NADPH (reduction at 5% the rate of 3-pyridinecarboxaldehyde reduction [18]) [18, 36]

34 D-Galacturonate + NADPH (not reductase AR 2) [36]

35 Norepinephrine + NADPH [16]

36 Serotonin + NADPH [16]

37 Dopamine + NADPH [16]

38 Octopamine + NADPH [16]

39 4-Halo-3-oxobutanoate esters + NADPH (ir, asymmetric reduction) [37]

40 Primary alcohols + NADP+ (r, ethanol (not [2, 18, 31, 33]) [38], C-2 to C-7 [41]) [38, 41, 42]

41 Secondary alcohols + NADP+ (overview) [41]

42 Benzyl alcohol + NADP+ (oxidation at 244% the rate of 3-pyridinemethanol oxidation) [18]

43 p-Methoxybenzyl alcohol + NADP+ (oxidation rate at pH 10 less than 0.5% of the reduction rate of 3-pyridinecarboxaldehyde at pH 6.2) [21]

44 Sugar alcohols + NADP+ (e.g. L-arabitol (oxidation at 126% the rate of 3-pyridinemethanol oxidation), D-sorbitol (oxidation at 80% the rate of 3-pyridinemethanol oxidation, not [1]), erythritol (oxidation at 126% the rate of 3-pyridinemethanol oxidation), adonitol (oxidation at 55% the rate of 3-pyridinemethanol oxidation), D-arabitol (oxidation at 18% the rate of 3-pyridinemethanol oxidation)) [18]

45 Tris + NADP+ (oxidation at 56% the rate of 3-pyridinemethanol oxidation) [18]

46 L-Sorbose + NADP+ (oxidation at 53% the rate of 3-pyridinemethanol oxidation) [18]

47 D-Aldoses + NADP+ (e.g. ribose (oxidation at 36% the rate of 3-pyridinemethanol oxidation), D-arabinose (oxidation at 91% the rate of 3-pyridinemethanol oxidation)) [18]

48 L-Gulonate + NADP+ (3-acetyl-pyridine adenine dinucleotide can replace NADP+ [5]) [5]

49 gamma-Hydroxybutyrate + $NADP^+$ (oxidation at about 10% the rate of succinic semialdehyde reduction) [33]
50 Propane-1,2-diol + $NADP^+$ (oxidation at about 10% the rate of succinic semialdehyde reduction) [33]
51 More (overlapping substrate specificity with aldose reductase [39, 40], no substrates for reduction are D-hexoses [26], D-fructose [2, 18, 33], pyridoxal [2, 18], acetone, dihydroxyacetone, acetol, acetoine [37], pyridoxal-5-phosphate, glyceraldehyde phosphate [2], daunorubicin [21], cyclohexanone [18, 28], menadione [28], dehydro-L-ascorbate, propiophenone, pyruvate, aldosterone, progesterone [18], phenylpyruvic acid, sodium bisulfite [24], androstane-3,17-dione and 5alpha-dihydroxytestosterone (reductase AR 1 and 3) [36], no oxidation of methanol [41], octanol [31], methoxyethanol, 2-aminoethanol, tert-butanol, 1,4-butanediol, pyruvate, formaldehyde [41], cyclohexanol, phenylpropyl alcohol, glycerol phosphate [2], galactitol, xylitol, glycerol [1], inositol, perseitol, DL-lactate [18, 41], glycolate [18], acetoacetate, acetoacetyl-CoA, palmitoyl-CoA [31]) [1, 2, 18, 21, 24, 26, 28, 31, 33, 36, 37, 39–41]

Product spectrum

1 Corresponding aliphatic alcohols + $NADP^+$ [4, 27, 38]
2 Corresponding aromatic alcohols + $NADP^+$ [22, 38]
3 Corresponding substituted benzyl alcohols + $NADP^+$ [27]
4 Pyridine-3-methanol + $NADP^+$ [25]
5 Pyridine-4-methanol + $NADP^+$ [34]
6 ?
7 ?
8 ?
9 Glycerol + $NADP^+$ [18, 36]
10 ?
11 ?
12 ?
13 ?
14 ?
15 ?
16 ?
17 ?
18 ?
19 ?
20 ?
21 ?
22 ?
23 ?
24 ?
25 ?
26 ?

27 ?
28 Ethyl L-(+)-3-hydroxybutyrate [31]
29 ?
30 ?
31 ?
32 L-Gulonate + NADP+ [25, 29]
33 ?
34 ?
35 ?
36 ?
37 ?
38 ?
39 (R)-4-Halo-3-hydroxybutanoate esters + NADP+ [37]
40 Primary aldehyde + NADPH [41]
41 Secondary aldehyde + NADPH [41]
42 Benzaldehyde + NADPH [18, 38]
43 ?
44 ?
45 ?
46 ?
47 ?
48 D-Glucuronate + NADPH [5]
49 ?
50 ?
51 ?

Inhibitor(s)

Phenobarbital (not [27]) [1–4, 21, 24, 36]; Amobarbital [18, 24]; Barbiturates (strong) [16, 18, 21, 23, 30, 38]; Sodium barbitone (kinetics [11], strong [20]) [11, 20, 28, 33]; Phenobarbitone [25, 30]; Ethacrynic acid (strong) [20, 36]; Valproate (not [27]) [8, 25, 26, 29, 30]; Chlorpromazine (weak [30], not [33]) [16, 24, 30, 36]; Alrestatin (i.e. 1,3-dioxo-1H-benz[de]isoquinoline-2(3-hydroxy acetic acid)) [1, 3, 25]; Chromone [1]; Sorbinil (i.e. (4S)-6-fluorospiro[chroman-4,4'-imidazoline]-2',5'-dione, strong [4], weak [9]) [1, 3, 4, 9, 25, 26, 32]; 3,3'-Tetramethyleneglutaric acid (weak [9]) [1, 3, 4, 9, 30, 36]; Alconil (pyridoxal-5-phosphate protects [32]) [26, 32]; Tolrestat [26, 32]; M79175 [26]; AY22284A [30]; Epalrestat (pyridoxal-5-phosphate protects [32]) [26, 32]; Ponalrestat [26]; Quercetrin [3, 30, 36]; Ethylphenylhydantoin [2]; Diphenylhydantoin [16, 30, 33]; 5,5'-Dithiobis(2-nitrobenzoate) (reversible) [20]; Homovanillic acid [30, 36]; Indomethacin [36]; 4,4'-Dithiodipyridine (NADPH protects) [25]; PCMB [36, 38]; p-Hydroxymercuribenzoate (not [18]) [1, 3, 4]; p-Substituted mercuribenzoate (partially reversible by 2-mercaptoethanol [19], not [2]) [19]; NEM (strong [2, 3], not [18]) [2, 3, 4, 10]; Iodoacetic acid (not [20]) [18]; IAA (not [2, 18]) [38]; Hydroxylamine [18]; 1,10-Phenanthroline (not [21]) [38]; $HgCl_2$ (strong, not reductase AR 2) [36]; $CuSO_4$ [36];

Li_2SO_4 (0.4 M, complete [3], reductase AR 1 and 3 [36]) [1, 3, 4, 36]; LiCl [3]; $MgSO_4$ (reductase AR 1 and 3) [36]; NaCl [1, 3, 4, 21, 26, 32]; NaF [18]; NH_4Cl [26, 32]; $(NH_4)_2SO_4$ (reductase AR 1 and 3 [36]) [3, 36]; Ammonium molybdate [36]; SO_4^{2-} [10]; $NADP^+$ (competitive) [5, 22, 30]; 5-Hydroxyindoleacetic acid [33]; 3-Methoxy-4-hydroxymandelic acid [33]; 1,4,5,6-Tetrahydronicotinamide adenine dinucleotide phosphate (reduced form, competitive to NADPH) [21]; 2'-AMP [30]; 5'-AMP [30]; 2'5'-AMP [30]; Blue Dextran (reversible) [30]; Glutaric acid [30]; Gulonic acid [29]; Benzoic acid [29]; Dimethylglutaric acid [30]; Procion Red HE3B [30]; p-Nitrobenzaldehyde [33]; Succinic semialdehyde [33]; Phenylpyruvic acid [34]; Cibacron Blue (dead-end inhibition [34]) [30, 34]; 4-Carboxybenzaldehyde (uncompetitive substrate inhibition [11], at high concentrations [22]) [11, 22]; 3-Pyridinecarboxaldehyde (at high concentrations) [22]; D-Glucuronate (substrate inhibition, non-competitive, at high concentrations [25]) [5, 25]; ATPribose (competitive) [5]; Bovine serum albumin (inhibits reduction of dodecanal and octanal, but stimulates reduction of n-hexa- and n-octadecanal) [31]; Glycerol (non-competitive to D-glyceraldehyde) [17]; Diphenic acid [6]; Butane-2,3-dione (NADPH and L-gulonate protect) [25]; 2,3-Dimethylsuccinic acid [6]; Pyrazole (slight, no inhibition [16, 18, 21, 23, 24, 31, 33]) [2, 42]; Warfarin (strong) [20]; More (inhibition kinetics [25], product inhibition kinetics [29, 34], no inhibition by clorazepam, flurazepam, nitrazepam [28], 2-mercaptoethanol [1, 18], EDTA [18, 21], disulfiram, diethyldicarbonate (at pH 6, 7, or 8 at 25°C, in the presence or absence of cosubstrate) [20], gamma-aminobutyrate, serotonin, noradrenalin, acetaldehyde, methanol, ethanol, Triton X-100 [33], Na_2SO_4 [36]) [1, 18, 20, 21, 25, 28, 29, 33, 34, 36]

Cofactor(s)/prosthetic group(s)/activating agents

NADPH (requirement, specific for isozyme II [1, 3], A-specific with respect to NADPH [19, 21, 40], mechanism [5, 17, 19, 21, 22, 40]) [1–5, 19, 21–41]; $NADP^+$ (requirement, not NAD^+) [2, 24, 33, 37, 38, 41, 42]; NADH (requirement, isozyme I [1, 3], reductase AR 2: 26% as effective as NADPH (may be identical with aldose reductase [40]) [36], less than 5% as effective as NADPH [2], cannot replace NADPH (isozyme II [1, 3], reductase AR 1 and 3 [36]) [1, 3, 23–26, 32, 33, 36–38]) [1–3, 36]; 3-Acetylpyridine adenine dinucleotide (can replace NADPH) [5]; Bovine serum albumin (stimulates reduction of n-hexa- and n-octadecanal, but inhibits reduction of dodecanal and octanal) [31]

Metal compounds/salts

Zn^{2+} (zinc protein, no other metal requirement) [41]; Phosphate (increase of activity with increasing concentrations from 10 to 500 mM [2]) [2, 18, 21]; Sulfate (slight increase of activity [8], stimulates reductase AR 2 [36]) [8, 18, 36]; More (enzyme contains no Zn^{2+}, Mn^{2+}, Cu^{2+} [20], no Zn^{2+} requirement [21]) [20, 21]

Turnover number (min^{-1})

More (pH-dependence of kinetic data [5], kinetic data for reductases AR 1, 2 and 3 [36]) [5, 32, 36]; 27.6 (propionaldehyde) [2]; 28.2 (benzaldehyde) [2]; 40.2 (butyraldehyde) [2]; 96.0 (p-chlorobenzaldehyde) [2]; 114 (camphorquinone) [2]; 180 (DL-glyceraldehyde) [2]; 282 (m-nitrobenzaldehyde) [2]; 336 ((+)-camphorquinone) [18]; 438 (p-carboxybenzaldehyde, methylglyoxal) [2]; 570 (p-nitrobenzaldehyde) [2]; 693 (D-glucuronate) [18]; 820 (D-lactaldehyde) [18]; 865 (D-glyceraldehyde) [18]; 998 (4-cyanobenzaldehyde) [18]; 1160 (4-pyridinecarboxaldehyde) [18]; 1300 (4-carboxybenzaldehyde) [18]; 1320 (4-nitrobenzaldehyde) [18]; 1610 (3-pyridinecarboxybenzaldehyde) [18]

Specific activity (U/mg)

More [2–4, 18–20, 24, 26, 28, 32, 39, 41]; 1.09 (yeast) [35]; 1.5 (chicken) [35]; 2.25 [38]; 2.47 (reductase I [1], fruit fly [35]) [1, 35]; 5.86 (reductase AR 2) [36]; 5.9–6.0 (liver and brain) [31]; 6.53 (rat) [35]; 9.59 (rat brain) [33]; 11.0 (kidney) [31]; 18.8 [23]; 19.7–19.9 (reductase AR 1 and 3) [36]; 29.62 (reductase II) [1]; 35.0 (reductase II, pig liver) [21]; 40.0 (reductase I, pig liver) [21]; 57.5 (reductase II) [4]; 81.7 [37]

K_m-value (mM)

More (pH-dependence of kinetic data [5], kinetic studies [17], kinetic data for reductases AR 1, 2 and 3 [36]) [1–5, 16–19, 21, 24–27, 31, 36, 37, 41, 42]; 0.00143 (NADPH) [24]; 0.003–0.006 (octanal [31], NADPH) [1, 2, 4, 31–33]; 0.008 (n-hexadecanal) [31]; 0.011–0.012 (propionaldehyde [3], 3,4-dihydroxyphenylglycolaldehyde [16]) [3, 16]; 0.025–0.027 (p-carboxybenzaldehyde [2], 5-hydroxyindole acetaldehyde [16]) [2, 16]; 0.039 (3-pyridinecarboxaldehyde) [3]; 0.15–0.31 (p-nitrobenzaldehyde [2, 27, 31], octanal [27], succinic semialdehyde [33]) [2, 27, 31, 33]; 0.08 (p-nitrobenzaldehyde) [33]; 0.22 (propionaldehyde) [38]; 0.51–1.5 (DL-glyceraldehyde [2, 3], benzaldehyde [38]) [2, 3, 38]; 0.94 (p-chlorobenzaldehyde) [2]; 1.2–1.3 (pyridinecarboxaldehyde [31], methylglyoxal, butyraldehyde [2]) [2, 31]; 1.4–1.5 (camphorquinone [2], p-nitroacetophenone [27], DL-glyceraldehyde [33], benzyl alcohol (oxidation) [38]) [2, 27, 33, 38]; 2.4 (m-nitrobenzaldehyde) [2]; 2.9–3.3 (propionaldehyde) [2, 31]; 3.4–3.8 (benzaldehyde [2], acetaldehyde [31]) [2, 31]; 4.5 (D-glucuronate) [33]; 25 (ethylacetoacetate) [31]; 43 (D-glucuronate) [31]; 100 (n-propanol (oxidation)) [38]

pH-optimum

More (pI: 4.7 [37], 5.06 [3], 5.2 (reductase II) [4], 5.3 [2], 5.4 (reductase II) [1], 5.7 [18], 5.76 (reductase I) [4], 6.1 [26, 32], 6.25 (reductase I) [1], 6.45 [9], 6.7 [31], 6.8 (reductase AR 3) [36], 6.9 [20], 7.1 (reductase AR 2), 7.4 (reductase AR 1) [36], various pI-values of mammalian tissues [39]) [1–4, 9, 18, 20, 26, 31, 32, 36, 37, 39]; 5.9 (reductase I, pig liver, 3-pyridinecarboxaldehyde or p-nitrobenzaldehyde reduction) [21]; 6.0 (reductase I [1, 4], reductase AR 2 [36]) [1, 4, 31, 36]; 6.0–6.5 (reductase AR1 and 3) [36]; 6.2

(reductase II, pig liver, 3-pyridinecarboxaldehyde or p-nitrobenzaldehyde reduction) [21]; 6.6 (p-nitrobenzaldehyde reduction) [2]; 6.8 (aldehyde reduction, bovine) [24]; 7.0 (reductase II [1, 4], 3-pyridinecarboxaldehyde reduction [18]) [1, 3, 4, 9, 10, 18, 37]; 7.0–8.0 (aldehyde reduction) [38]; 8.5 (alcohol oxidation) [5]; 9.0–9.5 (benzyl alcohol [2], oxidation) [2, 38]; 9.7 (p-hydroxyphenylethanol oxidation, bovine) [24]; 10.0 (broad, 3-pyridinemethanol oxidation) [18]

pH-range

5.0–7.6 (aldehyde reduction: about 55% of maximal activity at pH 5.0 and 7.6, bovine) [24]; 5.4–7.4 (reductase I: about 90% of maximal activity at pH 5.4 and about half-maximal activity at 7.4, reductase II: about 60% of maximal activity at pH 5.4 and 7.4) [21]; 5.8–7.6 (about 65% of maximal activity at pH 5.8 and 90% at 7.6, D-glucuronate) [33]; 6.0–7.8 (about 85% of maximal activity at pH 6.0 and 50% at pH 7.8, p-nitrobenzaldehyde) [33]; 6.0–7.9 (about 90% of maximal activity at pH 6.0 and 50% at pH 7.9, succinic semialdehyde) [33]; 6.2–8.0 (about 65% of maximal activity at pH 6.2 and 8.0) [18]; 9.6–10.2 (p-hydroxyphenylethanol oxidation: about half-maximal activity at pH 9.6 and 10.2, bovine) [24]

Temperature optimum (°C)

25 (assay at) [2, 18–20, 23, 24, 26, 32, 34–36, 38]; 30 (assay at) [6, 21, 22, 25, 27–29, 31, 33]; 35 [3]; 37 (assay at) [27, 37]; 60 [37]; 70 (ethanol, secondary alcohol dehydrogenase) [41]; 80 (2-propanol, secondary alcohol dehydrogenase) [41]; 85 (primary alcohol dehydrogenase) [41]

Temperature range (°C)

3 ENZYME STRUCTURE

Molecular weight

25000 (pig, gel filtration) [18]
30000 (bovine [28], chicken (reductase AR 1 and 3 [36]) [35, 36], yeast [35], gel filtration) [28, 35, 36]
30200 (pig, sedimentation equilibrium centrifugation) [18]
32000 (bovine [9], rat kidney [31], Sporobolomyces salmonicolor [37], chicken reductase AR 2 [36], gel filtration [9, 31, 36, 37], sucrose density centrifugation [31]) [9, 31, 36, 37]
32500–32700 (human reductase II, gel filtration) [1, 3, 4]
33000 (pig [19], rat [35], gel filtration) [19, 35]
34000 (pig liver reductase I [21], fruit fly [35], gel filtration) [21, 35]
35000 (pig liver reductase II, gel filtration) [21, 23]
35500 (human, gel filtration) [2]
36300 (human, meniscus depletion method) [2]
40000 (dog, gel filtration) [26]

43700 (pig, short column sedimentation equilibrium centrifugation) [20]
70000 (monkey, sucrose density centrifugation, gel filtration) [16]
72000–74000 (human reductase I, gel filtration) [1, 4]
75000 (rat brain, PAGE) [27]
80000 (Trypanosoma cruzi, gel filtration) [38]
172000 (Thermoanaerobacter ethanolicus secondary alcohol dehydrogenase, gel filtration) [41]
184000 (Thermoanaerobacter ethanolicus primary alcohol dehydrogenase, gel filtration) [41]
More (pig, amino acid composition [19], liver reductase I seems to be identical to kidney enzyme, immunological properties [21]) [19, 21]

Subunits

Monomer (Sporobolomyces salmonicolor, SDS-PAGE [37], 1 × 32500, human reductase II, SDS-/2-mercaptoethanol-/urea-PAGE [3, 4], 1 × 32600, chicken, SDS-PAGE [35], 1 × 33000, human reductase II, SDS-/2-mercaptoethanol-/urea-PAGE [1], yeast, SDS-PAGE [35], 1 × 33500, bovine, SDS-PAGE [28], 1 × 34000, rat kidney, SDS-PAGE [35], 1 × 35000, pig liver, SDS-PAGE [21], 1 × 36300, rat brain, SDS-PAGE [33], 1 × 36900, pig, SDS-PAGE [19], 1 × 37000, pig liver [23], rat kidney [31], SDS-PAGE [23, 31], 1 × 38000, human [8], pig [18], SDS-PAGE [8, 18], 1 × 38500, fruit fly, SDS-PAGE [35], 1 × 39000, chicken reductase AR 1 and 3, SDS-PAGE [36], 1 × 39400, human, SDS-PAGE [2], 1 × 39500, chicken reductase AR 2, SDS-PAGE [36], 1 × 41000, human [10], rat testis [32], SDS-PAGE [10, 32], 1 × 41700, pig kidney, SDS-PAGE [20], 1 × 75000, rat brain [27]) [1–4, 8, 10, 18–21, 23, 27, 28, 31–33, 35–38]
Dimer (1 × 32500 + 1 × 39000, human reductase I, SDS-/2-mercaptoethanol-/urea-PAGE [4], 1 × 35000 + 1 × 42000, human reductase I, SDS-/2-mercaptoethanol-PAGE [1]) [1, 4]
Tetramer (4 × 42000, Thermoanaerobacter ethanolicus secondary alcohol dehydrogenase, SDS-PAGE, 4 × 44500, Thermoanaerobacter ethanolicus primary alcohol dehydrogenase, SDS-PAGE) [41]

Glycoprotein/Lipoprotein

More (no hexosamines in hydrolysate [19], no glycoprotein [36]) [19, 36]

---

4 ISOLATION/PREPARATION

Source organism

Human [1–5, 8, 10, 14, 39, 40]; Monkey [16, 39]; Pig [6, 7, 12, 17–23, 39, 40]; Cat [39]; Dog [26]; Rabbit [34, 39, 40]; Guinea pig [40]; Rat (Wistar strain [27, 31], the enzyme from rat brain microsomes is presumably identical to NADPH-cytochrome c reductase [27]) [15, 21, 24, 27, 30–33, 35, 39, 40]; Mouse [39]; Horse [21]; Bovine (ox) [9, 11, 24, 28, 29, 33]; Sheep [21, 25, 39]; Chicken [35, 36]; Fruit fly [35]; Entamoeba histolytica [13]; Trypanosoma cruzi (Tulahuén strain, Tul 0 stock) [38]; Sporobolomyces salmonicolor [37]; Phycomyces blakesleeanus (Vanderbilt strain and mating type) [42]; Yeast [35]; Thermoanaerobacter ethanolicus [41]

Source tissue

Liver (rat [33]) [1, 2, 5, 21–23, 25, 30, 31, 33, 34, 39, 40]; Kidney (cortex [18, 20, 26], outer and inner renal medulla [26], rat, pig [21]) [11, 17–21, 26, 28, 29, 31, 35, 36, 39, 40]; Heart [39]; Lung [39]; Lens [6, 9, 10, 39]; Brain (distribution [24], subcellular distribution [15], pons, medulla, midbrain [16], pig [21], the enzyme from rat brain microsomes is presumably identical to NADPH-cytochrome c reductase [27]) [7, 12, 14–16, 21, 24, 27, 31, 33, 39, 40]; Testis [8, 32, 39]; Erythrocytes [3]; Placenta [4]; Uterus [39]; Seminal vesicles [39]; Spleen [39]; Whole organism [35]; Epimastigotes [38]; Cell [13, 35, 37, 41, 42]; More (high- and low-K_m-reductases are widely distributed in chicken tissues [36], distribution in mammalian tissues [39]) [36, 39]

Localisation in source

Cytoplasm (reductase I [21]) [18, 21, 24, 28, 30, 31, 36, 38]; Microsomes [27]

Purification

Human (2 aldehyde reductases: I [1, 4, 14] and II [1, 3, 4, 14], separated by ion exchange, purified by affinity chromatography [4]) [1–4, 14]; Monkey [16]; Dog [26]; Pig [18–21, 23]; Horse (liver, partial) [21]; Bovine [9, 24]; Sheep (liver, partial) [21]; Rat (reductase I and II, two isozymic forms: high-K_m (major) and low-K_m aldehyde reductase [31], affinity chromatography with various ligands [30]) [15, 21, 27, 30, 31]; Trypanosoma cruzi (partial) [38]; Sporobolomyces salmonicolor [37]; Thermoanaerobacter ethanolicus (two enzymes: primary and secondary alcohol dehydrogenase) [41]

Crystallization

(Sporobolomyces salmonicolor) [37]

Cloned

–

Renaturated

–

5 STABILITY

pH

6.0–10.0 (at 55°C 10 min stable) [37]; 7.0 (below 60°C 10 min stable) [37]

Temperature (°C)

35 (rapid loss of activity above, reductase II) [4]; 41 (65% loss of activity after 25 min, NADP(H), phenobarbitone, alrestatin, sorbinil, 2-ethylhexanoa-

te or valproate protects, D-glucuronate or L-gulonate enhances activity loss) [25]; 45 (complete inactivation, reductase II) [4]; 50 (inactivation after 5 min [3], complete inactivation [4]) [3, 4]; 55 (from pH 6.0 to 10.0: 10 min stable) [37]; 60 (below, at pH 7.0: 10 min stable) [37]; 70 (2 h stable in 10 mM Tris-HCl buffer, pH 7.6, secondary alcohol dehydrogenase) [41]

Oxidation

Organic solvent

Glycerol, 50% v/v, stable to [21]

General stability information

Freezing inactivates [1, 3, 4]; Freezing/thawing inactivates [2, 24, 33]; Freeze/thawing leads to precipitation [21]; Dilution inactivates [33]; Glycerol, 10% v/v, stabilizes during early purification [18]; Glycerol, 50% v/v, stabilizes during storage [21]; NaN_3, 1 mM, stable to [21]; $NADP^+$, NADPH, phenobarbitone, alrestatin, sorbinil, 2-ethylhexanoate or valproate enhances heat-stability [25]; 2-Mercaptoethanol stabilizes [38]; 2–3 months stable in 0.01 M sodium phosphate buffer, pH 7.0 [1]

Storage

–80°C, at least 6 months [27]; –20°C, at least 6 months [28]; –20°C, 1 month protein concentration above 0.5 mg/ml, in buffer containing 20% glycerol [36]; Frozen, at least 6 months [33]; Frozen, 3–4 months [42]; 0–2°C, several days [24]; 4°C, complete loss of activity within 1 month [28]; 4°C, 2 days [36]; 4°C, at least a week [3]; 4°C, several weeks at concentrations of about 1 mg protein/ml [33]; 4°C, about 2 weeks in the presence of 2-mercaptoethanol [4]; 4°C, several months [2]

6 CROSSREFERENCES TO STRUCTURE DATABANKS

PIR/MIPS code

PIR2:E37334 (Acinetobacter calcoaceticus (strain NCIB 8250) (fragment)); PIR3:A46409 (Entamoeba histolytica); PIR2:A33851 (human); PIR3:S17208 (Methanobacterium sp. (DSM 3108) (fragments)); PIR2:S17644 (Methanocorpusculum parvum (fragment))

Brookhaven code

7 LITERATURE REFERENCES

[1] Petrash, J.M., Srivastava, S.K.: Biochim. Biophys. Acta,707,105–114 (1982)
[2] Wermuth, B., Münch, J.D.B., Von Wartburg, J.-P.: J. Biol. Chem.,252,3821–3828 (1977)
[3] Das, B., Srivastava, S.K.: Arch. Biochem. Biophys.,238,670–679 (1985)
[4] Das, B., Srivastava, S.K.: Biochim. Biophys. Acta,840,324–333 (1985)

[5] Bhatnagar, A., Das, B., Liu, S.-Q., Srivastava, S.K.: Arch. Biochem. Biophys.,287, 329–336 (1991)
[6] Branlant, G.: Eur. J. Biochem.,129,99–104 (1982)
[7] Turner, A.J., Tipton, K.F.: Biochem. J.,130,765–772 (1972)
[8] Tanimoto, T., Ohta, M., Tanaka, A., Ikemoto, I., Machida, T.: Int. J. Biochem.,23, 421–428 (1991)
[9] Das, B., Hair, G.A., Ansari, N.H., Srivastava, S.K.: Lens Res.,5,233–247 (1988)
[10] Das, B., Song, H.P., Ansari, N.H., Hair, G.A., Srivastava, S.K.: Lens Res.,4,309–335 (1987)
[11] Worrall, D.M., Daly, A.K., Mantle, T.J.: J. Enzyme Inhib.,1,163–168 (1986)
[12] Cromlish, J.A., Yoshimoto, C.K., Flynn, T.G.: J. Neurochem.,44,1477–1484 (1985)
[13] Lo, H.S., Chang, C.J.: J. Parasitol.,68,372–377 (1982)
[14] Cash, C.D., Maître, M., Mandel, P.: J. Neurochem.,33,1169–1175 (1979)
[15] Reyes, E., Erwin, V.G.: Neurochem. Res.,2,87–97 (1977)
[16] Bronaugh, R.L., Erwin, V.G.: J. Neurochem.,21,809–815 (1973)
[17] Davidson, W.S., Flynn, T.G.: Biochem. Soc. Trans.,6,943–945 (1978)
[18] Bosron, W.F., Prairie, R.L.: J. Biol. Chem.,247,4480–4485 (1972)
[19] Flynn, T.G., Shires, J., Walton, D.J.: J. Biol. Chem.,250,2933–2940 (1975)
[20] Morpeth, F.F., Dickinson, F.M.: Biochem. J.,191,619–626 (1980)
[21] Branlant, G., Biellmann, J.-F.: Eur. J. Biochem.,105,611–621 (1980)
[22] Magnien, A., Branlant, G.: Eur. J. Biochem.,131,375–381 (1983)
[23] Kovár, J., Plocek, J.: J. Chromatogr.,351,371–375 (1986)
[24] Tabakoff, B., Erwin, V.G.: J. Biol. Chem.,245,3263–3268 (1970)
[25] De Jongh, K.S., Schofield, P.J., Edwards, M.R.: Biochem. J.,242,143–150 (1987)
[26] Ohta, M., Tanimoto, T., Tanaka, A.: Biochim. Biophys. Acta,1078,395–403 (1991)
[27] Takahashi, N., Saito, T., Goda, Y., Tomita, K.: J. Biochem.,99,513–519 (1986)
[28] Daly, A.K., Mantle, T.J.: Biochem. J.,205,373–380 (1982)
[29] Daly, A.K., Mantle, T.J.: Biochem. J.,205,381–388 (1982)
[30] Turner, A.J., Hryszko, J.: Biochim. Biophys. Acta,613,256–265 (1980)
[31] Takahashi, N., Saito, T., Tomita, K.: Biochim. Biophys. Acta,748,444–452 (1983)
[32] Kawasaki, N., Tanimoto, T., Tanaka, A.: Biochim. Biophys. Acta,996,30–36 (1989)
[33] Rivett, A.J., Smith, I.L., Tipton, K.F.: Biochem. J.,197,473–481 (1981)
[34] Sawada, H., Hara, A., Nakayama, T., Hayashibara, M.: J. Biochem.,92,185–191 (1982)
[35] Davidson, W.S., Weihrauch, L., Flynn, T.G.: Biochem. Soc. Trans.,6,940–942 (1978)
[36] Hara, A., Deyashiki, Y., Nakayama, T., Sawada, H.: Eur. J. Biochem.,133,207–214 (1983)
[37] Yamada, H., Shimizu, S., Kataoka, M., Sakai, H., Miyoshi, T.: FEMS Microbiol. Lett.,70,45–48 (1990)
[38] Arauzo, S., Cazzulo, J.J.: FEMS Microbiol. Lett.,58,283–286 (1989)
[39] Markus, H.B., Raducha, M., Harris, H.: Biochem. Med.,29,31–45 (1982)
[40] Flynn, T.G.: Biochem. Pharmacol.,31,2705–2712 (1982) (Review)
[41] Bryant, F.O., Wiegel, J., Ljungdahl, L.G.: Appl. Environ. Microbiol.,54,460–465 (1988)
[42] Hartz, T.K., Houston, M.R., Lockwood, L.B.: Mycologia,70,586–593 (1978)

1 NOMENCLATURE

EC number
1.1.1.3

Systematic name
L-Homoserine:NAD(P)$^+$ oxidoreductase

Recommended name
Homoserine dehydrogenase

Synonymes
HSDH [1, 10]
HSD [7]

CAS Reg. No.
9028-13-1

2 REACTION AND SPECIFICITY

Catalysed reaction
L-Homoserine + NAD(P)$^+$ →
→ L-aspartate 4-semialdehyde + NAD(P)H

Reaction type
Redox reaction

Natural substrates
Aspartate 4-semialdehyde + NAD(P)H (third reaction in the pathway between aspartate and the amino acids threonine, isoleucine, methionine [1]) [1, 7]

Substrate spectrum
1 L-Homoserine + NAD(P)$^+$ (r) [1, 6, 7, 9–12]

Product spectrum
1 L-Aspartate 4-semialdehyde + NAD(P)H [11]

Inhibitor(s)
L-Threonine (sensitive and insensitive isozymes [7, 10], degree of inhibition depends on age of plant [7] not inhibitory [12]) [1, 3–7, 10, 11]; D-Threonine (slight) [11]; DL-allo-Threonine [11]; L-Cysteine (slight inhibition of chloroplast isozyme, strong inhibition of cytoplasmic isozyme [1], slight [11]) [1, 10–12, 15]; p-Chloromercuribenzoate [8]; Tris [9]; Methionine [10]

Cofactor(s)/prosthetic group(s)/activating agents

NADPH (preferred [1], threonine sensitive isozyme can use NADPH or NADH, threonine insensitive isozyme can use NADPH only [7]) [1, 4, 7, 10]; NADH (threonine sensitive isozyme can use NADPH or NADH, threonine insensitive isozyme can use NADPH only [7]) [1, 7]; $NADP^+$ (preferred [11]) [6, 10, 11]; NAD^+ (low activity [11]) [6, 11]

Metal compounds/salts

K^+ (required [11], activation of threonine sensitive isozyme [7], activation [12, 13]) [7, 11–13]; Na^+ (activation of threonine sensitive form [7], can partially replace K^+ [11], activation [12]) [7, 11, 12]; NH_4^+ (activation of threonine sensitive isozyme [7], can partially replace K^+ [11]) [7, 11]; Li^+ (activation of threonine sensitive isozyme [7], can partially replace K^+ [11]) [7, 11]; Cs^+ (activation of threonine sensitive enzyme) [7]; Rb^+ (can partially replace K^+) [11]

Turnover number (min^{-1})

Specific activity (U/mg)

360 (threonine insensitive isozyme) [4]; 343 (threonine sensitive isozyme) [4]; 6.8 (Ricinus communis) [7]; 2.8 (Pisum sativum) [7]; 0.17 (Triticum aestivum) [7]

K_m-value (mM)

0.006 (NADPH, threonine sensitive isozyme) [7]; 0.027 (NADPH, Pisum sativum, isozyme II) [1]; 0.031 (NADPH, threonine sensitive isozyme) [7]; 0.032–0.036 (NADPH, threonine resistant isozyme) [4, 7]; 0.04 (aspartate 4-semialdehyde, Pisum sativum isozyme II [1], NADPH, threonine sensitive isozyme [4]) [1, 4]; 0.046 (NADPH, Pisum sativum, isozyme I) [1]; 0.066 (aspartate 4-semialdehyde, Pisum sativum, isozyme I) [1]; 0.08 (aspartate 4-semialdehyde, threonine insensitive isozyme) [10]; 0.1 (aspartate 4-semialdehyde, Hordeum vulgare, isozyme I) [1]; 0.13–0.15 (aspartate 4-semialdehyde, threonine resistant isozyme) [4, 7]; 0.17–0.18 ($NADP^+$) [11, 12]; 0.24–0.25 (aspartate 4-semialdehyde, threonine sensitive isozyme) [4, 7]; 0.36–0.4 (aspartate 4-semialdehyde, Hordeum vulgare, isozyme II [1], homoserine [11], NADH, Pisum sativum, isozyme I [1]) [1, 11]; 0.46 (NADH, Pisum sativum, isozyme II) [1]; 0.5 (aspartate 4-semialdehyde, threonine sensitive isozyme) [10]

pH-optimum

9 [12]; 9.8 [11]

pH-range

8–10 [11]

Temperature optimum (°C)

70 [11]; More (above 50°C temperature dependent conformational change) [12]

Temperature range (°C)

20–70 (no activity at 80°C) [11]

3 ENZYME STRUCTURE

Molecular weight

34000 (Pisum sativum, isozyme II, cytoplasm, gel filtration) [1]
69000 (Hordeum vulgare, isozyme II, cytoplasm, gel filtration) [1]
70000 (Zea mays, threonine insensitive isozyme, gel filtration) [4]
75000 (Pisum sativum, Triticum aestivum, threonine insensitive isozyme, gel filtration) [7]
80000 (Ricinus communis, threonine insensitive isozyme, gel electrophoresis) [7]
110000 (Rhodospirillum rubrum, gel filtration, sedimentation equilibrium centrifugation) [8]
168000–174000 (Hordeum vulgare, Pisum sativum, isozyme I, chloroplasts, gel filtration) [1]
190000 (Zea mays, threonine sensitive isozyme, gel filtration) [4]
220000 (Rhodospirillum rubrum, aggregated form, gel filtration) [8]
280000 (Ricinus communis, threonine sensitive isozyme, gel electrophoresis) [7]
290000 (Triticum aestivum, threonine sensitive isozyme, gel filtration) [7]

Subunits

Dimer (2 × 38000, Zea mays, threonine resistant isozyme, SDS-PAGE [4], 2 × 55000, Rhodospirillum rubrum, SDS-PAGE [8]) [4, 8]
Dimer or tetramer (x × 89000 + x × 93000, Zea mays, threonine sensitive isozyme, SDS-PAGE) [4]
? (x × 85000, Zea mays, threonine sensitive isozyme, SDS-PAGE) [2]

Glycoprotein/Lipoprotein

–

4 ISOLATION/PREPARATION

Source organism

Pisum sativum (pea) [1, 7, 10]; Hordeum vulgare (barley) [1, 15]; Zea mays (corn) [2–6, 15, 16]; Ricinus communis (castor bean) [7]; Rhodospirillum rubrum [8, 17]; Triticum aestivum (wheat) [7]; Thermophilic bacterium [11]; Thermus flavus [12]; Glycine max [15]; E. coli [9, 13, 14]

Source tissue

Leaf [1]; Seedling [2, 3, 5–7, 10]; Cell suspension cultures [4]; Mesophyll cells [16]

Localisation in source

Chloroplasts (isozyme I [1]) [1, 16]; Cytoplasm (isozyme II) [1]

Purification

Zea mays (threonine sensitive and threonine insensitive isozymes [4]) [2–4]; Pisum sativum [7, 10]; Ricinus communis [7]; Triticum aestivum [7]; Rhodospirillum rubrum [8, 17]; E. coli [14]; Thermophilic bacterium [11]; Thermus flavus (partial) [12]

Crystallization

–

Cloned

–

Renaturated

–

---

5 STABILITY

pH

Temperature (°C)

70 (up to) [12]; 80 (slow inactivation, protection by K^+, Na^+) [12]; 90 (rapid inactivation) [12]

Oxidation

Organic solvent

General stability information

Storage

–25°C, 50 mM potassium phosphate buffer, pH 7.2, 15% v/v glycerol, 1 mM EDTA, 1 mM threonine, 14 mM 2-mercaptoethanol [10]; –20°C, 0.05 M potassium phosphate buffer, pH 7.5, 1.0 mM EDTA, 2.0 mM dithioerythritol, 5.0 mM L-threonine, 20% v/v glycerol, at least 2 years [3]; –20°C, 30 mM potassium phosphate buffer, pH 7.5, 50% v/v glycerol, threonine resistant isozyme at least 2 months stable, threonine sensitive isozyme at least 4 months stable [4]

6 CROSSREFERENCES TO STRUCTURE DATABANKS

PIR/MIPS code

PIR1:DEECHS (Bacillus subtilis); PIR1:DEFKHG (Corynebacterium glutamicum); PIR1:DEPSHA (Pseudomonas aeruginosa); PIR2:S33317 (yeast (Saccharomyces cerevisiae))

Brookhaven code

7 LITERATURE REFERENCES

[1] Sainis, J.K., Mayne, R.G., Wallsgrove, R.M., Lea, P.J., Miflin, B.J.: Planta,152, 491–496 (1981)
[2] Krishnaswamy, S., Bryan, J.K.: Arch. Biochem. Biophys.,246,250–262 (1986)
[3] Krishnaswamy, S., Bryan, J.K.: Arch. Biochem. Biophys.,222,449–463 (1983)
[4] Walter, T.J., Connelly, J.A., Gengenbach, B.G., Wold, F.: J. Biol. Chem.,254,1349–1355 (1979)
[5] Di Camelli, C.A., Bryan, J.K.: Plant Physiol.,65,176–183 (1980)
[6] Bryan, J.K., Locher, N.R.: Plant Physiol.,68,1400–1405 (1981)
[7] Grego, S., Tricoli, D., Di Marco, G.: Phytochemistry,19,1619–1623 (1980)
[8] Epstein, C.C., Datta, P.: Eur. J. Biochem.,82,453–461 (1978)
[9] Ogilvie, J.W., Whitaker, S.C.: Biochim. Biophys. Acta,445,525–536 (1976)
[10] Aarnes, H., Rognes, S.E.: Phytochemistry,13,2717–2724 (1974)
[11] Cavari, B.Z., Grossowicz: Biochim. Biophys. Acta,302,183–190 (1973)
[12] Saiki, T., Shinshi, H., Arima, K.: J. Biochem.,74,1239–1248 (1973)
[13] Patte, J.C., LeBras, G., Loving, T., Cohen, G.N.: Biochim. Biophys. Acta,67,16–30 (1963)
[14] Cohen, G.N.: Curr. Top. Cell. Regul.,1,183–231 (1969)
[15] Bryan, J.K. in "The Biochemistry of Plants" (Miflin, B.J., ed.) 5,403–452, Academic Press, N.Y. (1980)
[16] Bryan, J.K., Kissik, E., Matthew, B.F.: Plant Physiol.,59,673–679 (1977)
[17] Datta, P.: J. Biol. Chem.,245,5779–5787 (1970)

1 NOMENCLATURE

EC number
1.1.1.4

Systematic name
(R,R)-Butane-2,3-diol:NAD+ oxidoreductase

Recommended name
(R,R)-Butanediol dehydrogenase

Synonymes
EC 1.1.1.74 (formerly)
Butyleneglycol dehydrogenase
Dehydrogenase, D-butanediol
D-(-)-Butanediol dehydrogenase
Butylene glycol dehydrogenase
Diacetyl (acetoin) reductase
Dehydrogenase, D-aminopropanol
D-Aminopropanol dehydrogenase
1-Amino-2-propanol dehydrogenase
2,3-Butanediol dehydrogenase
D-1-Amino-2-propanol dehydrogenase
(R)-Diacetyl reductase
(R)-2,3-Butanediol dehydrogenase
D-1-Amino-2-propanol:NAD+ oxidoreductase (enzyme from E. coli is identical with glycerol dehydrogenase) [6]
1-Amino-2-propanol oxidoreductase [6]
Aminopropanol oxidoreductase [6]

CAS Reg. No.
37250-09-2

2 REACTION AND SPECIFICITY

Catalysed reaction
(R,R)-Butane-2,3-diol + NAD+ →
→ (R)-acetoin + NADH

Reaction type
Redox reaction

Natural substrates

(R,R)-2,3-Butanediol + NAD^+ (second step in pathway wherein L-threonine is converted to D-1-amino-2-propanol via the intermediate formation of aminoacetone) [7]

Substrate spectrum

1 2,3-Butanediol + NAD^+ (r [4], D(-)-2,3-butandiol [1, 4], meso-butanediol [1], D-2,3-butanediol [8], 2,3-butanediol without specification of stereochemistry [2, 3]) [1–4, 8]
2 2,3-Butanedione + NADH (2,3-butanedione is identical with diacetyl and dimethylglyoxal) [2]
3 DL-1,2-Propanediol + NAD^+ [7, 8]
4 D-1-Amino-2-propanol + NAD^+ (DL-1-amino-2-propanol [8]) [7–9]
5 Glycerol + NAD^+ [7, 8]
6 DL-1,2-Butanediol + NAD^+ [8]
7 1,2-Ethanediol + NAD^+ [8]
8 3-Amino-1,2-propanediol + NAD^+ [8]
9 3-Chloro-1,2-propanediol + NAD^+ [8]
10 3-Mercapto-1,2-propanediol + NAD^+ [8]

Product spectrum

1 Acetoin + NADH [2]
2 Acetoin + NAD^+ [2]
3 ?
4 1-Amino-2-propanone + NADH
5 ?
6 ?
7 ?
8 ?
9 ?
10 ?

Inhibitor(s)

Iodoacetate [7]; p-Substituted mercuribenzoate [7]; N-Ethylmaleimide [7]; EDTA [7]; 8-Hydroxyquinoline [7]; o-Phenanthroline [7]; Borate buffer [4]

Cofactor(s)/prosthetic group(s)/activating agents

NAD^+ [1–9]; NADH [4]; More (no activity with $NADP^+$ or 3-acetylpyridine-NAD^+) [7]

Metal compounds/salts

No requirement [4]

Turnover number (min^{-1})

Specific activity (U/mg)

0.56 [3]; 0.21 (Bacillus polymyxa) [4]; 0.6 (Acetobacter suboxydans) [4]; 29.7–30.8 [6]; 2.3 [7]; 16.9 [8]

K_m-value (mM)

3.3 (D-(-)2,3-butanediol) [4]; 6.25 (meso-2,3-butanediol) [4]; 0.53 (acetoin) [4]; 0.2 (NAD^+) [4]; 0.1 (NADH) [4]; 87 (diacetyl) [4]; 38 (1,2-propanediol) [4]; 183.6 (1,2-ethanediol) [8]; 450.0 (3-amino-2-propanol) [8]; 2.8 (DL-1,2-propanediol) [8]; 8.0 (DL-1,2-butanediol) [8]; 14.5 (D-1-amino-2-propanol) [8]; 18.7 (3-chloro-1,2-propanediol) [8]; 19.4 (3-mercapto-1,2-propanediol) [8]; 26.1 (DL-2,3-butanediol) [8]; 30.9 (DL-1-amino-2-propanol) [8]; 54.7 (D-2,3-butanediol) [8]; 157.0 (glycerol) [8]; 1.23 (DL-1-amino-2-propanol) [8]; 33.7 (DL-1-amino-2-propanol) [6]; 107.0 (glycerol) [6]; 1.0 (DL-1,2-propanediol) [6]; 25 (DL-1-amino-2-propanol, enzyme form L) [7]; 1.4 (NAD^+, enzyme form L) [7]; 40 (DL-1,2-propanediol, enzyme forms L and S) [7]; 1.1 (NAD^+, enzyme form S) [7]

pH-optimum

7 (acetoin reduction) [4]; 8.0–8.6 [9]; 8.3–8.6 (2 optima: 8.3–8.6 and 10.0–10.2) [8]; 8.5 [2]; 8.6 [7]; 9 (D(-)-2,3-butanediol oxidation) [4]; 10.0–10.2 (2 optima: 8.3–8.6 and 10.0–10.2) [8]

pH-range

Temperature optimum (°C)

32 [2]; 60–62 (Bacillus polymyxa, Acetobacter suboxydans) [4]

Temperature range (°C)

3 ENZYME STRUCTURE

Molecular weight

83300 (E. coli, enzyme form S, sucrose density gradient centrifugation) [7]
87100 (E. coli, enzyme form S, gel filtration) [7]
100000–107000 (Bacillus polymyxa, gel filtration) [4]
291000 (E. coli, enzyme form L, sucrose density gradient centrifugation) [7]
380000 (E. coli, enzyme form L, gel filtration) [7]
417000 (E. coli, enzyme form L, sedimentation equilibrium centrifugation) [8]

Subunits

Octamer (8 × 47000, E. coli, enzyme form L, SDS-PAGE) [8]

Glycoprotein/Lipoprotein

–

4 ISOLATION/PREPARATION

Source organism

Saccharomyces vini [2]; Paracoccus denitrificans [3]; Acetobacter suboxydans [4]; Bacillus polymyxa [4]; Staphylococcus aureus [5]; Aeromonas hydrophila [4]; Bacillus subtilis [4]; E. coli (enzyme is identical with glycerol dehydrogenase [6]) [6–9]; More (overview Micrococcaceae, Bacillaceae, Pseudomonadaceae, Enterobacteriaceae, Brevibacteriaceae) [1]

Source tissue

Localisation in source

Purification

Bacillus polymyxa [4]; E. coli (enzyme is identical with glycerol dehydrogenase [6], form L [8]) [6–9]

Crystallization

–

Cloned

–

Renaturated

–

5 STABILITY

pH

7–7.6 [4]

Temperature (°C)

60–65 [4]; 74 (1 h, slow inactivation) [6]

Oxidation

Organic solvent

General stability information

Storage

Very sensitive to storage, even at −20°C [3]; Stable at −18°C [4]; 4°C, 6 months, stable [8]

6 CROSSREFERENCES TO STRUCTURE DATABANKS

PIR/MIPS code

Brookhaven code

7 LITERATURE REFERENCES

[1] Sadaharu, U., Masuda, H., Miroyuhi, M.: J. Ferment. Technol.,61,467–471 (1983)
[2] Kavadzo, A.V., Rodopulo, A.K., Shaposhnikov, G.L.: Izv. Akad. Nauk SSSR Ser. Biol.,3,435–441 (1979)
[3] Nokhal, T.-H., Schlegel, H.-G.: Arch. Microbiol.,145,197–201 (1986)
[4] Höhn-Bentz, H., Radler, F.: Arch. Microbiol.,116,197–203 (1978)
[5] Strecker, H.J., Harary, I.: J. Biol. Chem.,211,263–270 (1954)
[6] Kelley, J.J., Dekker, E.E.: J. Bacteriol.,162,170–175 (1985)
[7] Campbell, R.L., Swain, R.R., Dekker, E.E.: J. Biol. Chem.,253,7282–7288 (1978)
[8] Kelley, J.J., Dekker, E.E.: J. Biol. Chem.,259,2124–2129 (1984)
[9] Dekker, E.E., Swain, R.R.: Biochim. Biophys. Acta,158,306–307 (1968)

1 NOMENCLATURE

EC number
1.1.1.5

Systematic name
Acetoin:NAD^+ oxidoreductase

Recommended name
Acetoin dehydrogenase

Synonymes
Diacetyl reductase

CAS Reg. No.
9028-49-3

2 REACTION AND SPECIFICITY

Catalysed reaction
Diacetyl + NADH →
→ acetoin + NAD^+

Reaction type
Redox reaction

Natural substrates

Substrate spectrum
1 Diacetyl + NAD(P)H (ir [3, 4, 8], i.e. 2,3-butanedione) [1, 3–5, 8, 11]
2 Ethylpyruvate + NADPH [8]
3 Ethylacetoacetate + NADPH [8]
4 Methylacetoacetate + NADPH [8]
5 Ethylthioacetoacetate + NADPH [8]
6 Acetoacetyl N-acetylcysteamine + NADPH [8]
7 2,3-Butanediol + NAD^+ (r) [10]
8 More (substrates with uncharged alpha-dicarbonyl structure, beta-keto esters but no free beta-keto acids [8], not: acetone [3, 4], pentane-3-one [3, 4], pentane-2,4-dione [3, 4], hexane-2,5-dione [3, 4], pyruvate [3], oxaloacetate [3], acetylmethylcarbinol [3], acetate [4], 2-oxoglutarate [4]) [3, 4, 8]

Product spectrum

1 Acetoin + $NAD(P)^+$ (i.e. 3-hydroxy-2-butanone) [1, 8, 11]
2 ?
3 ?
4 ?
5 ?
6 ?
7 Acetoin + NADH [10]
8 ?

Inhibitor(s)

Acetoin (product inhibition) [1, 10, 12]; 2,3-Butanediol (product inhibition [10, 11], above 40 mM [8]) [8, 10, 11]); NAD^+ (product inhibition) [1, 10, 11]; $NADP^+$ (product inhibition) [1]; p-Chloromercuriphenylsulfonate [1]; Salicylic acid [3]; Ethanol [5]; Acetate [5, 10]; Zn^{2+} [6]; Ba^{2+} [6]; Cu^{2+} [6]; Hg^{2+} [6]; Mg^{2+} [6]; Mn^{2+} [6]; Na^+ (slight) [6]; K^+ (slight) [6]; Ca^{2+} (slight) [6]; Tris buffer [6, 8]; NADPH (above 0.15 mM) [8]; NADH (above 0.3 mM) [8]; 2-Oxoglutarate [9]; Hexane-2,5-dione [9]; Acetone [9]; Pentane-3-one [9]

Cofactor(s)/prosthetic group(s)/activating agents

NADPH (transfer of pro-4S hydrogen to diacetyl, i.e. B-specific [1], NADPH and NADH equally effective [3–5], highly specific [8]) [1, 3–5, 8]; NADH (NADH and NADPH equally effective) [3–5]; NAD^+ [10]

Metal compounds/salts

More (no activation by monovalent and divalent cations [6], no Zn, Cu, Fe [1]) [1, 6]

Turnover number (min^{-1})

Specific activity (U/mg)

80–98 [1]; 16 (bovine) [5]; 12 (pigeon) [5]; 2.45 (E. coli) [5, 8]

K_m-value (mM)

0.008 (NADPH (+ ethylacetoacetate or acetoacetyl N-acetylcysteamine)) [8]; 0.009–0.011 (NADH) [10]; 0.02 (NADPH (+diacetyl), E. coli) [5, 8]; 0.04 (diacetyl, bovine) [5, 12]; 0.067 (diacetyl, bovine) [4]; 0.1 (NADH, bovine [5], NADH [9]) [5, 9]; 0.14–0.18 (NAD^+ (+ 2,3-butanediol)) [10]; 0.46 (NADH, E. coli) [5, 8]; 0.53–0.8 (acetoin) [10]; 3.1–3.5 (diacetyl, pigeon) [4, 9]; 4.44 (diacetyl) [8]; 7.15 (acetoacetyl N-acetylcysteamine) [8]; 7.4 (acetoin) [1]; 11.3–122 (2,3-butanediol, value depends on buffer) [10]; 15.4 (ethylacetoacetate) [8]; More (dependence on pH) [7]

pH-optimum

4.5–8 (acetoin, no defined optimum) [10, 11]; 5.6–6.2 [4]; 6–7 [8]; 6.1 [3, 5, 9]; 7.5 [6]; 9.5 (2,3-butanediol) [10]

pH-range

5.2–8.4 (50% of maximal activity at pH 5.2, 55% of maximal activity at pH 8.4) [6]; 5.5–7.5 (less than 50% of maximal activity above and below) [3]

Temperature optimum (°C)

37 [6]; 40 [3]

Temperature range (°C)

3.5–40 (increase of activity over range) [3]; 18–45 (less than 50% of maximal activity above and below) [6]

3 ENZYME STRUCTURE

Molecular weight

10000 (E. coli, gel filtration) [8]
76000 (bovine, gel filtration) [3]
92000 (hamster, sucrose density centrifugation) [1]
110000 (pigeon, gel filtration) [4]

Subunits

Tetramer (4 × 23500, hamster, SDS-PAGE, gel filtration in presence of 6 M guanidine HCl) [1]

Glycoprotein/Lipoprotein

Glycoprotein [1]

4 ISOLATION/PREPARATION

Source organism

Hamster [1]; Bovine [3–5, 12]; Pigeon [2, 4, 5, 7, 9]; E. coli [5, 8]; Klebsiella pneumoniae [6]; Aerobacter aerogenes [10, 11, 13]

Source tissue

Liver [1–5, 7, 9, 12]

Localisation in source

Soluble part of cell (pigeon and bovine [5]) [5, 6]; Mitochondria (bovine) [5]

Purification

Hamster [1]; Bovine [3, 5]; Pigeon [4, 5]; E. coli [5, 8]; Aerobacter aerogenes [13]; Klebsiella pneumoniae [6]

Crystallization

–

Cloned

–

Renaturated
[1]

5 STABILITY

pH
5.4–7.6 [7]; 7.0–8.0 [6]

Temperature (°C)
37 (up to) [6]

Oxidation

Organic solvent

General stability information
Diluted solution unstable [10]; Inactivation by dialysis against H_2O, no reactivation by addition of Zn^{2+}, Cu^{2+}, Mg^{2+}, Fe^{2+}, or Fe^{3+} [3]

Storage
–20°C [5]; –15°C, lyophilized, at least 1 month [3]; –15°C, lyophilized, at least 6 months [4]; –15°C, N_2–atmosphere [12]; 0.1 mM NAD^+, 25 mM 2-mercaptoethanol, 0.25% albumin, 20% glycerol, pH 7.0 [10]

6 CROSSREFERENCES TO STRUCTURE DATABANKS

PIR/MIPS code
PIR3:S40549 (Escherichia coli)

Brookhaven code

7 LITERATURE REFERENCES

[1] Sawada, H., Akira, H., Nakayama, T.: J. Biochem.,98,1349–1357 (1985)
[2] Martin, R., Burgos, J.: Biochim. Biophys. Acta,212,356–358 (1970)
[3] Burgos, J., Martin, R.: Biochim. Biophys. Acta,268,261–270 (1972)
[4] Diez, V., Burgos, J., Martin, R.: Biochim. Biophys. Acta,350,253–262 (1974)
[5] Sarmiento, R., Burgos, J.: Methods Enzymol.,89,516–523 (1982) (Review)
[6] Shimizu, H., Hanaichi, Y., Okada, A., Tomoyeda, M.: Agric. Biol. Chem.,41,527–532 (1977)
[7] Martin, R., Diez, V., Burgos, J.: Biochim. Biophys. Acta,429,293–300 (1976)
[8] Silber, P., Chung, H., Gargiulo, P., Schulz, H.: J. Bacteriol.,118,919–927 (1974)
[9] Burgos, R., Diez, V.: Biochim. Biophys. Acta,364,9–16 (1974)
[10] Larsen, S.H., Stormer, F.C.: Eur. J. Biochem.,34,100–106 (1973)
[11] Johansen, L., Larsen, S.H., Stormer, F.C.: Eur. J. Biochem.,34,97–99 (1973)
[12] Martin, R., Burgos, J.: Biochim. Biophys. Acta,289,13–18 (1972)
[13] Bryn, K., Hetland, O. Stormer, F.C.: Eur. J. Biochem.,18,116–119 (1971)

1 NOMENCLATURE

EC number
1.1.1.6

Systematic name
Glycerol:NAD$^+$ 2-oxidoreductase

Recommended name
Glycerol dehydrogenase

Synonymes
Glycerin dehydrogenase
NAD-linked glycerol dehydrogenase

CAS Reg. No.
9028-14-2

2 REACTION AND SPECIFICITY

Catalysed reaction
Glycerol + NAD$^+$ →
→ dihydroxyacetone + NADH

Reaction type
Redox reaction

Natural substrates

Substrate spectrum
1 Glycerol + NAD$^+$ (r) [1–5, 7, 11, 12, 14]
2 Ethanediol + NAD$^+$ (r) [1]
3 1,2-Propanediol + NAD$^+$ (r) [1, 3, 5, 7, 11, 14]
4 1,2-Butanediol + NAD$^+$ [1]
5 2,3-Butanediol + NAD$^+$ (r) [1, 4]
6 1,2,3-Butanetriol + NAD$^+$ [1]
7 DL-Glyceraldehyde + NAD$^+$ [1, 7]
8 Pyruvaldehyde + NAD$^+$ [1]

Product spectrum

1 Dihydroxyacetone + NADH [1, 4, 7]
2 Glycoaldehyde + NADH [1, 7]
3 Hydroxyacetone + NADH [1]
4 1-Hydroxybutane-2-one + NADH [1]
5 3-Hydroxybutane-2-one + NADH [1]
6 1,3-Dihydroxybutane-2-one + NADH [1]
7 3-Hydroxypyruvaldehyde + NADH [1, 7]
8 Lactaldehyde + NADH [1]

Inhibitor(s)

Thiol compounds [1]; Li^+ [1]; Na^+ [1]; 2-Mercaptoethanol [3]; Cu^{2+} [7, 8, 13]; Chelating agents (e.g. 8-hydroxyquinoline, ethyleneglycol) [1, 3, 8, 11]; Zn^{2+} [7, 13]; Fe^{3+} [7, 13]; Cd^{2+} [7]; p-Chloromercuribenzoate [7]; Tris [3]; Glycerol 3-phosphate [3]; Dihydroxyacetone phosphate [3]; D-Xylose [14]; Sucrose [14]; meso-Erythritol [14]; Cetylammonium bromide [14]; Cetylpyridinium chloride [14]; SDS [14]; Anhitol 24B [14]

Cofactor(s)/prosthetic group(s)/activating agents

NAD^+ [1–5, 7, 11, 12, 14]; NADH [1–5, 7, 11, 12, 14]

Metal compounds/salts

Zinc [5]; K^+ (oxidation enhancement) [7]; Rb^+ (oxidation enhancement) [7]; NH_4^+ (oxidation enhancement) [7]; Na^+ (reduction enhancement) [7]

Turnover number (min^{-1})

Specific activity (U/mg)

448 [3]; 138 [6]; 82 [12]; 58.1 [4]

K_m-value (mM)

0.5 (glycerol) [1]; 0.06 (dihydroxyacetone) [1]; 0.13 (NAD^+) [1]; 0.12 (NADH) [1]; More [1, 3–5, 12]

pH-optimum

9 (glycerol) [4]; 8–8.5 (propanediol) [4]; 9 (butanediol) [4]

pH-range

10–12 (glycerol oxidation) [1, 3]; 5–6 (dihydroxyacetone reduction) [1]; 7.5–8.5 (reduction) [3]

Temperature optimum (°C)

36 [3]; 50 (oxidation of glycerol [4, 7] and butanediol [4]) [4, 7]; 65 (oxidation of propanediol) [4]

Temperature range (°C)

25–65 [4, 7]

3 ENZYME STRUCTURE

Molecular weight

310000–400000 (Schizosaccharomyces pombe [1], Klebsiella aerogenes [9], gel electrophoresis) [1, 9]
160000–180000 (Bacillus megaterium [5], Klebsiella aerogenes [9], density gradient centrifugation) [5, 9]
76000–79000 (Candida valida, gel filtration) [3]

Subunits

Dimer (2 × 38000, Candida valida, SDS-PAGE [3], 2 or 4 × 40000, Klebsiella aerogenes, SDS-PAGE, dimer or tetramer, sometimes oligomers (in vitro) [9]) [3, 9]
Tetramer (4 × 36000, Bacillus megaterium, SDS-PAGE [5], 2 or 4 × 40000, Klebsiella aerogenes, SDS-PAGE, dimer or tetramer, sometimes oligomers (in vitro) [9]) [5, 9]
Octamer (8 × 47000, Schizosaccharomyces pombe, SDS-PAGE) [1]

Glycoprotein/Lipoprotein

–

4 ISOLATION/PREPARATION

Source organism

Schizosaccharomyces pombe [1, 2]; Candida valida [3]; Cellulomonas sp. [4, 7, 8, 14]; Bacillus megaterium [5]; Geotrichum candidum [6]; E. coli [13]; Klebsiella aerogenes [9]; Aerobacter aerogenes [10–12]

Source tissue

Cell [1–14]

Localisation in source

Cytoplasm (soluble) [1–14]

Purification

Schizosaccharomyces pombe [1, 2]; Candida valida [3]; Cellulomonas sp. [4, 7, 8, 14]; Bacillus megaterium [5]; Geotrichum candidum [6]; E. coli [13]; Klebsiella aerogenes [9]; Aerobacter aerogenes [10–12]

Crystallization

[8]

Cloned

–

Renaturated

–

5 STABILITY

pH

7 (stable) [12]; 5.5–10.0 [7]

Temperature (°C)

50 (10 min stable) [2]; 70 (30 min stable) [5]; 75 (10 min stable, E.coli) [13]

Oxidation

Organic solvent

Methanol (10% v/v: 25–75% loss of activity, 20% v/v, 50–95% loss of activity, depending on substrate, enzyme regains activity after removal of alcohol) [14]; Ethanol (10% v/v: 20–70% loss of activity, depending on substrate, enzyme regains activity after removal of alcohol) [14]; n-Propanol (5% v/v: 40–70% loss of activity, depending on substrate, enzyme regains activity after removal of alcohol) [14]; n-Butanol (5% v/v: 50–100% loss of activity, depending on substrate, enzyme regains activity after removal of alcohol) [14]; Isobutanol (5% v/v: 50–100% loss of activity, depending on substrate, enzyme regains activity after removal of alcohol) [14]; D-Amylalcohol (2% v/v: 50–100% loss of activity, depending on substrate, enzyme regains activity after removal of alcohol) [14]; Acetone (inactivation, enzyme regains activity after removal of ketone) [14]; Catechol (i.e. 1,2-butanediol, inactivation, enzyme regains activity after removal of alcohol) [14]; Diacetyl (inactivation, enzyme regains activity after removal of ketone) [14]; Dioxane (inactivation, enzyme regains activity after removal of ketone) [14]; Dimethylformamide (inactivation) [14]; Dimethylsulfoxide (inactivation) [14]

General stability information

KCl stabilizes [5]; $ZnCl_2$ stabilizes [5]; Glycerol stabilizes [5]

Storage

–15°C, with 5% crude cell extract [3]; –70°C, long term storage [1]

6 CROSSREFERENCES TO STRUCTURE DATABANKS

PIR/MIPS code

PIR2:JQ1474 (Bacillus stearothermophilus)

Brookhaven code

7 LITERATURE REFERENCES

[1] Marshall, J.H., May, J.W., Sloan, J.: J. Gen. Microbiol.,131,1581–1588 (1985)
[2] Kong, Y.-C., May, J.W., Marshall, J.H.: J. Gen. Microbiol.,131,1571–1579 (1985)
[3] Gärtner, G., Kopperschläger, G.: J.Gen. Microbiol.,130,3225–3233 (1984)
[4] Nishise, H., Nagao, A., Tani, Y., Yamada, H.: Agric. Biol. Chem.,48,1603–1609 (1984)
[5] Scharschmidt, M., Pfleiderer, G., Metz, H., Brümmer, W.: Hoppe-Seyler's Z. Physiol. Chem.,364,911–921 (1983)
[6] Sasaki, I., Itoh, N., Goto, H., Yamamoto, R., Tanaka, H., Yamshita, K., Yamashita J., Horio, T.: J. Biochem.,91,211–217 (1982)
[7] Yamada H., Nagao, A., Nishise, H., Tani, Y.: Agric. Biol. Chem.,46,2333–2339 (1982)
[8] Yamada, H., Nagao, A., Tani, Y.: Agric. Biol. Chem.,44,471–472 (1980)
[9] Ruch, F.E., Lin, E.C.C., Kowit, J.D., Tang, C.-T., Goldberg, A.L.: J. Bacteriol.,141, 1077–1085 (1980)
[10] McGregor, W.G., Philips, J., Suelter, C.H.: J. Biol. Chem.,249,3132–3139 (1974)
[11] Lin, E.C.C., Magasanik, B.: J. Biol. Chem.,235,1820–1823 (1960)
[12] Burton, R.M.: Methods Enzymol.,1,397–400 (1955)
[13] Asnis, R.A., Brodie, A.F.: J. Biol. Chem.,203,153–159 (1953)
[14] Nishise, H., Maehashi, S., Yamada, H., Tani, Y.: Agric. Biol. Chem.,51,3347–3353 (1987)

1 NOMENCLATURE

EC number
1.1.1.7

Systematic name
Propane-1,2-diol-1-phosphate:NAD^+ oxidoreductase

Recommended name
Propanediol-phosphate dehydrogenase

Synonymes
PDP dehydrogenase [1, 2]
1,2-Propanediol-1-phosphate:NAD^+ oxidoreductase
Propanediol phosphate dehydrogenase
Dehydrogenase, propanediol phosphate

CAS Reg. No.
9028-15-3

2 REACTION AND SPECIFICITY

Catalysed reaction
Propane-1,2-diol 1-phosphate + NAD^+ →
→ hydroxyacetone phosphate + NADH

Reaction type
Redox reaction

Natural substrates
Hydroxyacetone phosphate + NADH [1]

Substrate spectrum
1 Propane-1,2-diol 1-phosphate + NAD^+ (r) [1, 2]

Product spectrum
1 Hydroxyacetone phosphate + NADH [1]

Inhibitor(s)
Hydroxyacetone phosphate [2]; SO_4^{2-} [1, 2]; SO_3^{2-} [1, 2]; Formate [1, 2]; Br^- [1, 2]; PO_4^{3-} [2]

Cofactor(s)/prosthetic group(s)/activating agents

Metal compounds/salts

Turnover number (min^{-1})

Specific activity (U/mg)
0.1–0.43 [1]

K_m-value (mM)
0.05 (NADH) [2]; 0.1 (hydroxyacetone phosphate) [2]; 4.4 (DL-propane-1,2-diol 1-phosphate) [2]; 5.5 (hydroxyacetone phosphate) [1]

pH-optimum
7.7 (hydroxyacetone phosphate reduction) [1]; More (dehydrogenation of propane-1,2-diol 1-phosphate has alkaline pH-optimum) [1]

pH-range
5.4–9.6 (hydroxyacetone phosphate reduction, values strongly dependent on buffer system) [1]

Temperature optimum (°C)

Temperature range (°C)

---

3 ENZYME STRUCTURE

Molecular weight

Subunits

Glycoprotein/Lipoprotein
–

---

4 ISOLATION/PREPARATION

Source organism
Rabbit [1, 2]

Source tissue
Muscle [1]

Localisation in source

Purification
Rabbit (partial) [1]

Crystallization
–

Cloned
–

Renaturated
–

5 STABILITY

pH

Temperature (°C)

Oxidation

Organic solvent

General stability information

Storage

4°C, dialyzed against Tris-buffer, pH 7.7, 80% loss of activity in 1 month [1];
4°C, dialyzed against distilled water [2]

6 CROSSREFERENCES TO STRUCTURE DATABANKS

PIR/MIPS code

Brookhaven code

7 LITERATURE REFERENCES

[1] Sellinger, O.Z., Miller, O.N.: J. Biol. Chem.,234,1641–1646 (1959)
[2] Miller, O.N.: Methods Enzymol.,9,336–338 (1966)

1 NOMENCLATURE

EC number
1.1.1.8

Systematic name
sn-Glycerol-3-phosphate:NAD+ 2-oxidoreductase

Recommended name
Glycerol-3-phosphate dehydrogenase (NAD+)

Synonymes
Dehydrogenase, glycerol phosphate
alpha-Glycerol phosphate dehydrogenase (NAD)
alpha-Glycerophosphate dehydrogenase (NAD)
Glycerol 1-phosphate dehydrogenase
Glycerol phosphate dehydrogenase (NAD)
Glycerophosphate dehydrogenase (NAD)
Hydroglycerophosphate dehydrogenase
L-alpha-Glycerol phosphate dehydrogenase
L-alpha-Glycerophosphate dehydrogenase
L-Glycerol phosphate dehydrogenase
L-Glycerophosphate dehydrogenase
NAD-alpha-glycerophosphate dehydrogenase
NAD-dependent glycerol phosphate dehydrogenase
NAD-dependent glycerol-3-phosphate dehydrogenase
NAD-L-glycerol-3-phosphate dehydrogenase
NAD-linked glycerol 3-phosphate dehydrogenase
NADH-dihydroxyacetone phosphate reductase [22]

CAS Reg. No.
9075-65-4

2 REACTION AND SPECIFICITY

Catalysed reaction
sn-Glycerol-3-phosphate + NAD+ →
→ dihydroxyacetone phosphate + NADH (one active site per enzyme molecule [2], mechanism [8], random bi-bi mechanism [23], ordered ternary-complex mechanism, NADH binds first [24])

Reaction type
Redox reaction

Natural substrates

Dihydroxyacetone phosphate + NADH (cytoplasmic enzyme functions primarily in glycerol biosynthesis, not catabolism [1], key enzyme in respiration and biosynthesis of glycerophosphatides [2], reaction linking glycolysis to phospholipid and triglyceride pathways [6, 18], central reaction in insect flight metabolism, accounts for rapid reoxidation of cytoplasmic NADH during glycolysis [9], function in cellular osmoregulation by providing glycerol-3-phosphate for glycerol biosynthesis [24]) [1, 2, 4, 6, 9, 18, 24]

Substrate spectrum

1 sn-Glycerol-3-phosphate + NAD^+ (r [1, 3–7, 9–24], reverse reaction favoured direction [20], oxidation at 3% the reaction rate of dihydroxyacetone phosphate reduction at pH 7.0 [1], oxidation at 25% the rate of dihydroxyacetone phosphate reduction at optimal pH [15, 23], favoured reaction of heart isozyme $II_{6.1}$ [12], specific for L-glycerol-3-phosphate [23], equilibrium far to the side of alpha-glycerophosphate at neutral pH [2]) [1–7, 9–24]

2 More (no substrates: dihydroxyacetone [1, 15, 23, 24], glycerol [1, 15], glyceraldehyde-3-phosphate [5, 15, 23, 24], glyceraldehyde, phosphohydroxypyruvate [15], fructose-6-phosphate, fructose-1,6-bisphosphate [15, 23], glucose-6-phosphate, acetaldehyde, oxaloacetate [23]) [1, 2, 5, 15, 23, 24]

Product spectrum

1 Dihydroxyacetone phosphate + NADH (i.e. glycerone phosphate) [1, 2, 4–7, 9–21, 23, 24]

2 ?

Inhibitor(s)

NAD^+ (competitive inhibitor to NADH at physiological concentration [1, 23], non-competitive to dihydroxyacetone phosphate [23], not [18]) [1, 23]; NADH (at high concentration, not [18]) [15, 23]; Fructose-1,6-bisphosphate (at physiological concentration) [1]; ADP (at physiological concentration) [1]; ATP (at physiological concentration) [1]; NAD^+-analogs (potent inhibitors, except 4-(1-imidazolyl)-analog) [2]; L-Glycerol-3-phosphate (inhibits dihydroxyacetone phosphate reduction [3], inhibitor of bumble bee enzyme [9], strong inhibitor at high concentration [22], competitive inhibitor to dihydroxyacetone phosphate, non-competitive to NADH [23], not [1, 18]) [3, 9, 22, 23]; Dihydroxyacetone phosphate (in excess) [9, 18, 23]; Adenosine diphosphate ribose [3]; Tris [22]; Linolenic acid [22]; Phosphatidylcholine [22]; Triton X-100 [22]; Octyl glucose [22]; Thylakoid fraction (strong inhibition) [22]; Malate (at high concentration) [24]; Sulfate (at high concentration) [24]; [24]; 2,3-Dimercaptopropanol (i.e. BAL, competitive inhibitor to glycerophosphate, formation of ternary complex) [2]; PCMB (10 nM, strong inhibition [7, 12], reversible by DTT [7]) [7, 9, 11, 12]; NEM (reversible by DTT [7], pseudo first-order kinetics, reversible by NADH, inhibition increases in the presence of dihydroxyacetone phosphate and/or glycerol-3-phosphate [23]) [7, 9, 23]; Iodoacetate (reversible by DTT [7], no inhibitor of honey bee enzyme [9]) [7]; H_2PO_4 (bumble bee enzyme [9], cytoplasmic enzyme [22], competitive to dihydroxyacetone phosphate, non-competitive to NADH [23]) [9, 22, 23]; NaCl (inactivation [24]) [12, 24]; KCl (inactivation [24]) [12, 24]; High ionic strength (above 0.03 M [13]) [13, 20]; Cl^- (at high concentration [24], irreversible inactivation) [20, 24]; TEA-buffer [20]; Phosphogluconate (cytoplasm) [22]; Sedoheptulose-1,7-bisphosphate [22]; Large peptide factor (chloroplast enzyme) [22]; $(NH_4)_2SO_4$ [21, 22]; Small peptide factor (cytoplasmic enzyme) [22]; More (no inhibition by NADPH, acetaldehyde, glycerol, ethanol [1], cyclic-AMP [22]) [1, 22]

Cofactor(s)/prosthetic group(s)/activating agents

NADH (1 mol per mol enzyme [2], 2 mol per mol enzyme [3, 5, 9], spectrophoto-/fluorimetric monitoring [5]) [1–3, 5–24]; NAD^+ (not $NADP^+$ [12, 15]) [1, 2, 4–21, 23, 24]; NADPH (5% as effective as NADH [23], not NADPH [1, 10, 12, 15, 20, 24]) [23]; Deamino-NAD^+ (60% [2] or 68% [9] as effective as NAD^+) [2, 9]; 3-Acetylpyridine-NAD^+ (1.5% as effective as NAD^+) [9]; NAD^+-analogs (i.e. acetylpyridine-NAD^+, deamino-acetylpyridine-NAD^+, pyridinealdehyde-NAD^+ and deamino-pyridinealdehyde-NAD^+: 0.19%, 0.09%, 0.2% and 0.075% as effective as NAD^+, respectively) [2]; Fructose-2,6-bisphosphate (activation, cytoplasm) [22]; Reduced thioredoxin (activation, chloroplast) [22]; DTT (activation, chloroplast) [22]; Phosphogluconate (activation, chloroplast) [22]; ATP (slight activation) [22]; EDTA (activation) [2]; More (no stimulation by diphosphate [2], GSH, cysteine [2, 22], 2-mercaptoethanol, fructose-1,6-bisphosphate [22], adenosine diphosphoribose is not associated with chicken enzyme [10]) [2, 10, 22]

Metal compounds/salts

Sulfate (at low concentration, slight activation) [24]; Cl^- (at low concentration, activation) [24]; Ions (up to 0.1 M, activation) [24]; Phosphate (activation, chloroplast) [22]; Dihydrolipoic acid (activation, chloroplast) [22]; Malate (at low concentration, slight activation) [24]; Glutamate (activation) [24]; Glycine buffer (0.9 M, slight activation) [2]

Turnover number (min^{-1})

More [2]; 1920 (glycerol-3-phosphate) [13]; 11280 (dihydroxyacetone phosphate) [13]; 20670 (dihydroxyacetone phosphate, 20°C, pH 7, rabbit muscle) [2]

Specific activity (U/mg)

More [22]; 33.0 (isozyme P-V, liver) [4]; 34.0 [23]; 37.5 [15]; 41.1 [1]; 71.8 [18]; 88.2 [21]; 96.0 (isozyme P-I, liver) [4]; 103.5 [11]; 115 (heart) [12]; 117 (adipose tissue) [6]; 119 [5]; 123 (isozyme P-IV, liver) [4]; 152.4 [7]; 156 (larvae [17], isozyme 3 [19]) [17, 19]; 158 [20]; 162 (isozyme $I_{5.9}$) [16]; 175 (isozyme $I_{6.5}$) [16]; 180 (mammary gland [12], flight muscle [17], isozyme 3 [19]) [12, 17, 19]; 200 [24]; 253 (isozyme P-II, liver) [4]; 295 [14]; 310 [10]; 340 [9]; 806 (isozyme P-III, liver) [4]

K_m-value (mM)

More (various K_m-values of 6 isozymes of heart, liver and mammary gland of rabbit) [12]; 0.0043–0.0083 (NADH [6], NAD^+, adipose tissue [6], NADH [10, 18, 21, 24]) [6, 10, 18, 21, 24]; 0.01–0.012 (NADH [15, 19], isozyme 3 [19], NAD^+, NADH, isozyme $I_{6.5}$ [16], muscle [6], below: NADH [9], dihydroxyacetone phosphate [23]) [6, 9, 15, 16, 19, 23]; 0.016–0.027 (NADH [1, 13, 14, 23], dihydroxyacetone phosphate [19, 21]) [1, 13, 14, 19, 21, 23]; 0.036 (NAD^+) [21]; 0.05 (NAD^+, dihydroxyacetone phosphate, $I_{5.9}$) [16]; 0.075–0.083 (dihydroxyacetone phosphate [14], NADH, isozyme 1 [19], NAD^+ [4, 14]) [4, 14, 19]; 0.1–0.17 (NAD^+ [18], glycerol-3-phosphate [2, 21], dihydroxyacetone phosphate [7, 24], NADH, isozyme $I_{5.9}$ [16], dihydroxyacetone phosphate, muscle, Tris-buffer [6]) [2, 6, 7, 16, 18, 21, 24]; 0.19–0.3 (NAD^+ [13, 15], dihydroxyacetone phosphate, TEA-acetate-buffer, glycerol-3-phosphate, muscle [6], dihydroxyacetone phosphate [10, 16, 18], dihydroxyacetone phosphate, MOPS-buffer [9], dihydroxyacetone phosphate, Tris-buffer, adipose tissue [6], glycerol-3-phosphate [19]) [6, 9, 10, 13, 15, 16, 18, 19]; 0.30–0.35 (NAD^+ [24], isozyme 3 [19], glycerol-3-phosphate [4, 7, 16], testes, mean value [4], dihydroxyacetone phosphate, Tris/histidine-buffer [9], dihydroxyacetone phosphate [13, 15]) [4, 7, 9, 13, 15, 16, 19, 24]; 0.38–0.46 (NAD^+ [2], NAD^+, isozyme 1 [19], glycerol-3-phosphate [6], dihydroxyacetone phosphate [2]) [2, 6, 19]; 0.50–0.59 (NAD^+ [10], dihydroxyacetone phosphate, isozyme 1 [19], glycerol-3-phosphate [19]) [10, 19]; 0.74 (glycerol-3-phosphate) [13]; 0.909 (glycerol-3-phosphate) [14]; 1.20 (glycerol-3-phosphate) [24]; 1.59–1.90 (glycerol-3-phosphate [4, 15, 18], isozyme P-II [4]) [4, 15, 18]; 2.0–2.76 (glycerol-3-phosphate [10], isozymes P-III/V [4]) [4, 10]; 3.39 (glycerol-3-phosphate, liver, mean value) [4]; 5.94 (glycerol-3-phosphate, isozyme P-I) [4]; 7.26 (glycerol-3-phosphate, isozyme P-IV) [4]

pH-optimum

More (pI: 7.4 [1], various isozymes in heart, muscle, mammary gland and liver with pI: 6.1, 6.3, 6.5, major protein band at 6.5 [12], pI: 5.75 [13], major isozyme pI: 6.5 [14], 2 isozymes in brain, pI: 5.9 and 6.5, the former closely related to major heart, the latter identical to major muscle isozyme [16], pI: 5.2 (isozyme 1) and 5.6 (isozyme 3) [19], pI: 5.29 [21]) [1, 12–14, 16, 19, 21]; 6.0–6.5 (reduction) [18]; 6.6 (broad [9], reduction) [9, 13]; 6.8 (broad, reduction [15], isozyme 1, reduction [19], reduction [23]) [15, 19, 23];

6.9–7.0 (reduction) [22]; 7.2 (reduction) [7]; 7.4 (reduction, isozyme 3) [19]; 7.4–8.1 (reduction) [21]; 7.5 (reduction) [2]; 7.5–8.0 (reduction) [10]; 7.6 (reduction) [1]; 7.9–8.2 [24]; 8.2–8.6 (oxidation, isozymes P-I, II, IV, V, liver) [4]; 8.7–9.2 (oxidation) [21]; 9.2–9.3 (oxidation, isozyme P-III, liver and all isozymes of testes) [4]; 9.5 (oxidation) [15, 23]; 10.0 (oxidation) [13]; 10.0–10.5 (oxidation) [18]; 10.2 (oxidation, rate of oxidation at pH 10.2 is 8.3% of reduction rate at pH 7.5) [2]

pH-range

5.0–8.1 (about half-maximal activity at pH 5.0 and 8.1, reduction) [15]; 5.2–7.8 (about half-maximal activity at pH 5.2 and 7.8, reduction) [13]; 6.0–8.8 (about half-maximal activity at pH 6.0 and 60% of maximal activity at pH 8.0, reduction) [21]; 6.4–8.2 (about half-maximal activity at pH 6.4 and 8.2, reduction) [7]; 7.0–8.0 (rapid decrease of activity below 7.0 and above 8.0, more rapidly at acidic values) [1]; 7.5–9.5 (about half-maximal activity at pH 7.5 and 9.5, oxidation [21], isozyme P-III of liver, all isozymes of testes [4]) [4, 21]; 8.0–10.2 (about half-maximal activity at pH 8.0 and 10.2, oxidation, isozymes P-I, II, IV, V, liver) [4]; 8.1–11.0 (about half-maximal activity at pH 8.1 and 11.0, oxidation) [13]; 8.6–10.0 (about half-maximal activity at pH 8.6 and 10.0, oxidation) [15]

Temperature optimum (°C)

25 (isozymes P-I, V, liver) [4]; 30 (isozymes P-II, P-III, P-IV, liver, all isozymes of testes [4], f-isozyme [18]) [4, 18]; 31 [13]; 40 (m- and s-isozyme) [18]

Temperature range (°C)

5–40 (about half-maximal activity at 5°C and 40°C, f-isozyme) [18]; 15–40 (about half-maximal activity at 15°C, maximal activity at 40°C, m- and s-isozyme) [18]

3 ENZYME STRUCTURE

Molecular weight

62000 (rabbit, adipose tissue, gel filtration [6], human, gel filtration [21]) [6, 21]
63500 (Spinacia oleracea, gel filtration) [15]
65000 (Drosophila virilis, gel filtration) [18]
66000 (rabbit, gel filtration [6], Drosophila melanogaster, gel filtration [17]) [6, 17]
67000 (Drosophila melanogaster, gel filtration) [11]
68000 (Saccharomyces cerevisiae, gel filtration) [1]
69000–76000 (Ceratitis capitata, gel filtration) [13]
72000 (rat, gel filtration) [7]
73000 (Apis mellifera, amino acid analysis) [9]
74000–79000 (Apis mellifera, SDS-PAGE) [9]

75000 (rat, sucrose gradient ultracentrifugation) [7]
76000 (rabbit muscle, meniscus depletion method [5], Apis mellifera, gel filtration [9]) [5, 9]
78000 (rabbit muscle, gel filtration) [7]
79500 (Apis mellifera, meniscus depletion method) [9]
More (Apis mellifera, amino acid analysis [3]) [2, 3]

Subunits

Dimer (2 × 31700, Drosophila melanogaster, SDS-PAGE [11], 2 × 32000, Drosophila melanogaster, SDS-PAGE [17], 2 × 33500, Ceratitis capitata, SDS-PAGE [13], 2 × 35000–37000, Drosophila virilis, SDS-PAGE [18], 2 × 37000, rabbit brain, SDS-PAGE [16], 2 × 37500, rabbit muscle, gel filtration in 6 M guanidinium chloride [6], 2 × 38000, human, SDS-PAGE [21], 2 × 39000, Apis mellifera, gel filtration in guanidinium chloride [9], 2 × 42000, Saccharomyces cerevisiae, SDS-PAGE [1]) [1, 6, 9, 11, 13, 16–18, 21]
Oligomer (x × 33000, Chlamydomonas reinhardtii, calculated from Stokes radius [23], x × 42000, Debaryomyces hansenii, SDS-PAGE [24]) [23, 24]
More (37800–39300, rabbit muscle, minimal MW by titration with NADH at pH 6.0) [5]

Glycoprotein/Lipoprotein

–

4 ISOLATION/PREPARATION

Source organism

Saccharomyces cerevisiae (strain H44–3D [1]) [1, 20]; Debaryomyces hansenii (van Rij strain 26, salt-tolerant) [24]; Human [2, 21]; Rat (Sprague-Dawley [7]) [2, 7]; Rabbit (New Zealand white [6]) [2, 3, 5–8, 12, 14, 16]; Bovine [2]; Mouse [2]; Bumble bee [9]; Apis mellifera (honey bee) [3, 9]; Coturnix coturnix japonica (japanese quail) [4]; Chicken [10]; Drosophila melanogaster (fruit fly, Samarkand wild-type [19]) [11, 17, 19]; Drosophila virilis [18]; Ceratitis capitata (strain Wiedemann, mediterranean fruit fly) [13]; Spinacia oleracea (spinach, cv Long Standing Bloomsdale [22]) [15, 22]; Chlamydomonas reinhardtii (strain 11/32–90, green alga) [23]; More (insects, such as bumble bee species [2, 9] and yellow jackets [9]) [2, 9]

Source tissue

Cell [1, 20, 23, 24]; Blood [2]; Leukocytes [2]; Brain (low activity [2]) [2, 7, 16]; Muscle (heart, skeletal or flight muscle: high activity, smooth muscle: low activity [2], breast [10], flight muscle [17, 19]) [2, 3, 5, 6, 8–10, 12, 13, 17, 19]; Kidney (high activity) [2]; Liver (high activity) [2, 4, 12, 14]; Heart [12, 16]; Testis [4]; Placenta (term) [21]; Renal adipose tissue [6]; Mammary gland [12]; Thorax [2, 3, 9]; Fat body [19]; Tumor cells (mammary carcinoma, Ehrlich ascites carcinoma, Crocker sarcoma (mouse) [2], Brown Pearce carcinoma [12]) [2, 12]; Larva [17]; Leaf [15, 22]

Localisation in source

Cytoplasm (extramitochondrial [9]) [1–22, 24]; Chloroplast (stroma, soluble [22, 23]) [15, 22, 23]

Purification

Saccharomyces cerevisiae (affinity chromatography) [1, 20]; Debaryomyces hansenii (2 isozymes) [24]; Apis mellifera [9]; Coturnix coturnix japonica (4 isozymes in testes and 5 isozymes in liver, distinct in isoelectric focusing) [4]; Rabbit (preparative isoelectric focusing [6], several isozymes, distinct in isoelectric points pI: 6.1, 6.3, 6.5 [12, 14], 2 isozymes in brain, pI: 5.9, 6.5 [16], sequential affinity chromatography [14]) [5, 6, 12, 14, 16]; Rat (ion-exchange chromatography combined with affinity elution, 2 isozymes that differ in charge by analytical PAGE) [7]; Human (affinity chromatography) [21]; Chicken [10]; Drosophila melanogaster (affinity chromatography [11, 17, 19], 2 allelic forms of the enzyme [11], 3 isozymes in adults and 1 in larvae, distinct in isoelectric focusing [17], 3 isozymes, product of the same gene mapped to left arm of chromosome II, isozyme 1: flight muscle, isozyme 3: larval and adult fat body, isozyme 2: heterodimer of 1 and 3 [19]) [11, 17, 19]; Drosophila virilis (3 allelic forms distinguishable by electrophoresis) [18]; Ceratitis capitata [13]; Spinacia oleracea (affinity chromatography [15], 2 isozymes [22]) [15, 22]; Chlamydomonas reinhardtii (dye-affinity chromatography) [23]; More (6 to 8 isozymes of bumble bee flight muscle enzyme) [9]

Crystallization

(Rabbit (muscle) [2, 5], Apis mellifera [9], chicken [10]) [2, 5, 9, 10]

Cloned

–

Renaturated

–

---

5 STABILITY

pH

More (more stable at alkaline pH, more stable in phosphate- than in Tris-buffer [7], increasing pH from 7.0 to 8.0 causes 80% loss of activity [23]) [7, 23]; 4.8–9.9 (stable for 15 min at 21°C) [9]; 5.0 (inactivation) [13]; 5.7 (around, most stable) [2]; 6.0–9.0 (stable for 15 h at 20°C) [13]; 7.0 (less stable [2], below: rate of inactivation increases [18], stable [23]) [2, 18, 23]; 8.0 (most stable, above: inactivation) [18]; 8.5 (around, most stable) [2]; 10.0 (inactivation) [13]

Temperature (°C)

20 (stable for at least 30 min) [2]; 21 (15 min stable) [9]; 30 (30 min stable, pH 6.6) [13]; 35 (most stable at pH 8) [18]; 45 (pH 6.6 [13], at least 5 min stable) [7, 13]; 48 ($t_{1/2}$: 2 min (heart isozyme $II_{6.1}$ [12], brain isozyme $I_{5.9}$) [12, 16], 7.5 min (isozyme $I_{6.5}$) [16] and 15 min (muscle isozyme $I_{6.5}$) [12]) [12, 16]; 50 ($t_{1/2}$: 2 min (heart) and 10 min (muscle, mammary gland, liver) [12], rapid denaturation, prevented by bovine serum albumin, ammonium sulfate or phosphate buffer [7], crude 20 min stable [11]) [7, 11]; 55 ($t_{1/2}$: 1 min [2], 5 min stable in Tris-buffer plus EDTA and bovine serum albumin [9], inactivation after 30 min [13]) [2, 9, 13]; 60 (complete inactivation after 1 min [2], 5 min [7, 13]) [2, 7, 13]; 61 (complete inactivation after 5 min) [9]

Oxidation

Organic solvent

Glycerol, slight inhibition [24]

General stability information

Redistilled water prevents denaturation [2]; Charcoal inactivates rabbit enzyme, restorable by thiamic acid [2]; Freezing inactivates completely, ammonium sulfate prevents [15]; Sephadex inactivates rabbit enzyme, restorable by thiamic acid [2]; 2-Mercaptoethanol stabilizes during purification [7, 10, 13]; 2-Mercaptoethanol does not stabilize [23]; Repeated freezing and thawing results in rapid loss of activity, to some extent restorable at room temperature [7]; Bovine serum albumin stabilizes [7]; Ammonium sulfate stabilizes [7, 9]; Ammonium sulfate inactivates [21]; Lyophilization inactivates [7]; Phosphate buffer stabilizes [7]; DTT stabilizes [9, 11, 17]; DTT does not stabilize [23]; EDTA stabilizes [9, 11, 23]; Polyethylene glycol stabilizes during purification (15–17.5% w/v [23]) [23, 24]; NADH stabilizes [19]; NADH stabilizes preparations of low ionic strength [10]; NAD^+ stabilizes [11]; Phenylmethane-sulfonyl fluoride stabilizes during purification [17]; Substrates stabilize dilute preparation [24]

Storage

–80°C, unstable [1]; –20°C, unstable [1]; –20°C, stable as $(NH_4)_2SO_4$-precipitate [15]; –20 – 0°C, stable in 75% ammonium sulfate [20]; –20°C, stable for at least 2 months [23]; 0°C, $t_{1/2}$: 50 min [23]; 0°C, crude extract stable for 6 days in presence of polyethylene glycol [24]; 2°C, crystals stable at least a month in 0.1 M Tris-buffer, pH 7.6 with added EDTA, DTT and ammonium sulfate [9]; 4°C, stable [14, 24]; 4°C, dilute solution, pH 5.8, stable for weeks [2]; 4°C, purified and dilute preparation stable for several days in presence of NADH [15]; 4°C, stable for several weeks in 1 M Tris/HCl buffer with 2-mercaptoethanol, pH 7.0 [1]; 4°C, $t_{1/2}$: 6 h in absence of EDTA and polyethyleneglycol [23]; 5°C, isozyme 1: stable for over a month in 10 mM sodium phosphate buffer, pH 6.5, with added EDTA, DTT and NADH, isozyme 3: 50% loss of activity under the same conditions [19]; Crystallized, in ammonium sulfate stable for months [10]; Glycerol, 20% v/v, purified stable at 4°C [12, 23];

6 CROSSREFERENCES TO STRUCTURE DATABANKS

PIR/MIPS code

PIR2:A32937 ((clone 1A) fruit fly (Drosophila melanogaster)); PIR2:B32937 ((clone 37) fruit fly (Drosophila melanogaster)); PIR2:C32937 ((clone 411) fruit fly (Drosophila melanogaster)); PIR3:S21963 (fruit fly (Drosophila melanogaster)); PIR3:S23137 (fruit fly (Drosophila virilis)); PIR3:S31790 (fruit fly (Drosophila virilis)); PIR2:A25189 (mouse); PIR2:A26687 (mouse); PIR2:A32512 (rabbit (fragment)); PIR2:S06760 (1 fruit fly (Drosophila melanogaster)); PIR2:A28995 (1 fruit fly (Drosophila melanogaster) (fragment)); PIR2:S06759 (2 fruit fly (Drosophila melanogaster)); PIR2:B28995 (2 fruit fly (Drosophila melanogaster) (fragment)); PIR2:S06758 (3 fruit fly (Drosophila melanogaster)); PIR2:C28995 (3 fruit fly (Drosophila melanogaster) (fragment)); PIR2:JS0023 (m form fruit fly (Drosophila virilis)); PIR2:A25952 (precursor mouse); PIR3:S40059 (precursor yeast (Saccharomyces cerevisiae)); PIR2:A60985 (f allele fruit fly (Drosophila virilis)); PIR2:B60985 (s allele fruit fly (Drosophila virilis))

Brookhaven code

7 LITERATURE REFERENCES

[1] Albertyn, J., Van Tonder, A., Prior, B.A.: FEBS Lett.,308,130–132 (1992)
[2] Baranowski, T. in "The Enzymes",2nd Ed. (Boyer, P.D., Lardy, H., Myrbäck, K., eds.) 7,85–96 (1963) (Review)
[3] Brosemer, R.W., Kuhn, R.W.: Biochemistry,8,2095–2105 (1969)
[4] Yamada, M.: J. Biochem.,72,1081–1086 (1972)
[5] Bentley, P., Dickinson, F.M., Jones, I.G.: Biochem. J.,135,853–859 (1973)
[6] Warkentin, D.L., Fondy, T.P.: Eur. J. Biochem.,36,97–109 (1973)
[7] McGinnis, J., DeVellis, J.: Biochim. Biophys. Acta,364,17–27 (1974)
[8] Bentley, P., Dickinson, F.M.: Biochem. J.,143,11–17 (1974)
[9] Fink, S.C., Brosemer, R.W.: Methods Enzymol.,41B,240–245 (1975) (Review)
[10] White, H.B.: Methods Enzymol.,41B,245–249 (1975) (Review)
[11] Collier, G.E., Sullivan, D.T., MacIntyre, R.J.: Biochim. Biophys. Acta,429,316–323 (1976)
[12] Ostro, M.J., Fondy, T.P.: J. Biol. Chem.,252,5575–5583 (1977)
[13] Fernández-Sousa, J.M., Gavilanes, J.G., Municio, A.M., Pérez-Aranda, A.: Biochim. Biophys. Acta,481,6–24 (1977)
[14] McLoughlin, D.J., MacQuarrie, R.: Biochim. Biophys. Acta,527,204–211 (1978)
[15] Santora, G.T., Gee, R.W., Tolbert, N.E.: Arch. Biochem. Biophys.,196,403–411 (1979)
[16] Kornbluth, R., Tracy, P.S., Fondy, T.P.: Biochim. Biophys. Acta,568,273–286 (1979)
[17] Niesel, D.W., Bewley, G.C., Miller, S.G., Armstrong, F.B., Lee, C.-Y.: J. Biol. Chem.,255,4073–4080 (1980)
[18] Narise, S.: Biochim. Biophys. Acta,615,289–298 (1980)
[19] Niesel, D.W., Bewley, G.C., Lee, C.-Y., Armstrong, F.B.: Methods Enzymol.,89, 296–301 (1982)

[20] Merkel, J.R., Straume, M., Sajer, S.A., Hopfer, R.L.: Anal. Biochem.,122,180–185 (1982)
[21] Zolnierowicz, S., Swierczynski, J., Zelewski, L.: Eur. J. Biochem.,154,161–166 (1986)
[22] Gee, R.W., Byerrum, R.U., Gerber, D.W., Tolbert, N.E.: Plant Physiol.,87,379–383 (1988)
[23] Klöck, G., Kreuzberg, K.: Biochim. Biophys. Acta,991,347–352 (1989)
[24] Nilsson, A., Adler, L.: Biochim. Biophys. Acta,1034,180–185 (1990)

1 NOMENCLATURE

EC number
1.1.1.9

Systematic name
Xylitol:NAD^+ 2-oxidoreductase (D-xylulose-forming)

Recommended name
D-Xylulose reductase

Synonymes
Reductase, D-xylulose
NAD-dependent xylitol dehydrogenase
Xylitol dehydrogenase
Erythritol dehydrogenase [2]
2,3-cis-Polyol(DPN) dehydrogenase (C3–5) [2]
Pentitol-DPN dehydrogenase [3]
Xylitol-2-dehydrogenase [7]

CAS Reg. No.
9028-16-4

2 REACTION AND SPECIFICITY

Catalysed reaction
Xylitol + NAD^+ →
→ D-xylulose + NADH

Reaction type
Redox reaction

Natural substrates
Xylitol + NAD^+ (r [4], inducible pathway of xylose catabolism) [4, 5, 7]

Substrate spectrum
1 Xylitol + NAD^+ (r [1–4, 7]) [1–4, 6–8]
2 L-Erythrulose + NADH (reduction at the same rate as D-xylulose [1], r [2]) [1, 2]
3 D-Ribitol + NAD^+ (r [4], oxidation at 11% (Serratia marcescens) [7] and 85% [8] the rate of xylitol oxidation) [1, 4, 6–8]
4 D-Sorbitol + NAD^+ (oxidation at about 45% [4], 56% (Providencia stuartii), 67% (Morganella morganii), 95% [8] and 118% (Serratia marcescens) the rate of xylitol oxidation [4, 7]) [1, 4, 6–8]
5 L-Xylulose + NADH [2]

6 Dihydroxyacetone + NADH [2]
7 L-Arabitol + NAD^+ [1, 2]
8 D-Arabitol + NAD^+ [2]
9 L-Iditol + NAD^+ [1]
10 Glycerol + NAD^+ (poor substrate [8]) [2, 8]
11 More (specific for polyols of 5 or less carbon bearing cis-hydroxy groups in C_2 and C_3 [2], no oxidation of threitol, xylitol, sorbitol, D-iditol [1, 2], ribitol (Morganella morganii, Providencia stuartii [7]) [4, 7], mannitol [4, 8], inositol, meso-erythritol, D-(+)-arabitol [8], no reduction of L-xylulose [7], D-ribulose [1], D-ribose, D-galacturonic acid [1], D/L-xylose [1, 7], D-fructose, L-sorbose, D-tagatose [2]) [1, 2, 4, 7, 8]

Product spectrum

1 D-Xylulose + NADH [1, 3, 4, 6–8]
2 Erythritol + NAD^+ [2, 6]
3 D-Ribulose + NADH [6]
4 D-Fructose + NADH [6]
5 ?
6 ?
7 ?
8 ?
9 ?
10 ?
11 ?

Inhibitor(s)

PCMB (complete inhibition) [1]; D-Xylulose (high concentration) [1]; EDTA [2]; Lead acetate [8]; $HgCl_2$ (complete inhibition) [8]; $AgNO_3$ [8]; $ZnSO_4$ (complete inhibition) [8]; $CuSO_4$ [8]; $FeSO_4$ [8]; $MnSO_4$ [8]; NaCl (at high concentrations) [8]; NH_4Cl (at high concentrations) [8]; Cysteine (at high concentrations) [8]; Iodoacetate (complete inhibition) [8]; More (no inhibition by EDTA or iodoacetate) [1]

Cofactor(s)/prosthetic group(s)/activating agents

NAD^+ (stereochemistry of hydrogen transfer from xylitol, ribitol or sorbitol to NAD^+ is of the (R)- or A-type [6]) [1–8]; NADH [1–7]; $NADP^+$ (at high concentrations of xylitol 1.5% as effective as NAD^+ [8], no activity with $NADP^+$ [1, 2, 4], no oxidation of ribitol or sorbitol with $NADP^+$ [8]) [8]; Cysteine (slight activation [8], of partially purified preparation [1]) [1, 8]; GSH (activation) [8]; Glycine (slight activation) [8]; More (very low activity with NADPH [4], no activity with NADPH [1, 2]) [1, 2, 4]

Metal compounds/salts

Mn^{2+} (requirement) [2]; Mg^{2+} (activation at high concentration [2]) [2, 8]; NH_4Cl (activation) [8]; NaCl (activation) [8]; KCl (slight activation) [8]; More (no evidence for metal requirement [1], no activation by Zn^{2+}, Cu^{2+}, Ca^{2+}, Ni^{2+}, Co^{2+}, Fe^{2+} [2]) [1, 2]

Turnover number (min^{-1})

Specific activity (U/mg)

More (23.0, 1 unit: decrease of 1.0 per min in optical density at 340 nm) [1]; 0.11 (Morganella morganii) [7]; 0.114 (Serratia marcescens) [7]; 0.219 (Providencia stuartii) [7]; 0.9 [5]; 3.1 [2]; 15.95 [4]; 29.9 [8]

K_m-value (mM)

0.037 (NADH) [4]; 0.16 (sorbitol) [4]; 0.66 (D-xylulose) [1]; 7.1 (xylitol, Serratia marcescens) [7]; 11.0 (xylitol) [8]; 12.6 (xylitol, Providencia stuartii) [7]; 13.8 (xylulose) [4]; 16.4 (xylitol, Morganella morganii) [7]; 18.5 (xylitol) [4]; 30.0 (D-sorbitol) [8]; 50.0 (ribitol) [8]; 496 (ribitol, Serratia marcescens) [7]

pH-optimum

6.7 (reduction of erythrulose) [2]; 7.0 (reduction) [1]; 7.2 (xylitol formation) [4]; 8.6 (oxidation of erythritol [2], D-xylulose formation [4]) [2, 4]; 9.1–10.0 (oxidation, plateau) [8]

pH-range

4.0–8.0 (about 75% of maximal reduction activity at pH 4.0 and 8.0) [1]; 8.7–10.5 (about 80% of maximal oxidation activity at pH 8.7 and 10.5) [8]

Temperature optimum (°C)

55 [8]

Temperature range (°C)

35–68 (about half-maximal activity at 35°C and 68°C) [8]

3 ENZYME STRUCTURE

Molecular weight

82000 (Candida shehatae, gel filtration) [4]
120000 (Pachysolen tannophilus, gel filtration) [8]
130000 (Serratia marcescens, PAGE) [7]
142000 (Morganella morganii, PAGE) [7]
155000 (Providencia stuartii, PAGE) [7]
172000 (Pachysolen tannophilus, gel permeation chromatography) [5]

Subunits

Dimer (2 × 40000, Candida shehatae, SDS-PAGE) [4]
Tetramer (4 × 40000, Pachysolen tannophilus, SDS-PAGE [8], 2 × 40400 + 2 × 41800, Pachysolen tannophilus, SDS-PAGE [5]) [5, 8]

Glycoprotein/Lipoprotein

–

4 ISOLATION/PREPARATION

Source organism

Guinea pig [1, 6]; Aerobacter aerogenes [2]; Penicillium chrysogenum [3]; Candida shehatae [4]; Pachysolen tannophilus (strain Boidin et Adzet [5]) [5, 8]; Morganella morganii [7]; Providencia stuartii [7]; Serratia marcescens [7]

Source tissue

Liver [1, 6]; Cell [2–5, 7, 8]

Localisation in source

Cytoplasm [1, 2, 4–8]

Purification

Guinea pig (partial) [1]; Aerobacter aerogenes (partial) [2]; Candida shehatae (sequential: NAD-C_8-affinity chromatography/gel filtration/Cibacron dye ligand affinity chromatography) [4]; Pachysolen tannophilus (affinity chromatography on NAD-C_8/NAD-R–columns) [5]; Morganella morganii (partial) [7]; Providencia stuartii (partial) [7]; Serratia marcescens (partial) [7]

Crystallization

–

Cloned

–

Renaturated

–

5 STABILITY

pH

6.8–7.4 (90% of maximal activity retained at pH 6.8 and 7.4 at 2°C) [2]; 7.0 (24 h stable at 2°C) [2]; 7.0–9.0 (stable) [8]

Temperature (°C)

25 (and below stable) [8]; 60 (complete inactivation within 5 min) [8]

Oxidation

Organic solvent

Acetone, stable to precipitation with 50% v/v [1]

General stability information

Crude water extract from liver acetone-powder unstable [1]; Freezing/thawing inactivates (8 cycles lead to 89% loss of activity and 11 cycles to complete inactivation [8]) [1, 8]; 2-Mercaptoethanol stabilizes during purification [2]; EDTA stabilizes during purification [2]

Storage

–20°C, stable with 50% v/v glycerol [4]; –15°C, 60% loss of activity after 1 month [8]; –10°C, acetone precipitate stable for weeks [1]; –2°C, purified stable for a month [8]; 0°C, partially purified stable for weeks [1]; 0°C, 34% loss of activity within a month [8]; 2°C, 90% of maximal activity retained at pH 6.8 and 7.4 after 24 h [2]

6 CROSSREFERENCES TO STRUCTURE DATABANKS

PIR/MIPS code

PIR2:S13529 (yeast (Pichia stipitis))

Brookhaven code

7 LITERATURE REFERENCES

[1] Hickman, J., Ashwell, G.: J. Biol. Chem.,234,758–761 (1959)
[2] Jakoby, W.B., Fredericks, J.: Biochim. Biophys. Acta,48,26–32 (1961)
[3] Chiang, C., Knight, S.G.: Biochem. Biophys. Res. Commun.,3,554–559 (1960)
[4] Yang, V.W., Jeffries, T.W.: Appl. Biochem. Biotechnol.,26,197–206 (1990)
[5] Bolen, P.L., Roth, K.A., Freer, S.N.: Appl. Environ. Microbiol.,52,660–664 (1986)
[6] Alizade, M.A., Brendel, K., Gaede, K.: FEBS Lett.,67,41–44 (1976)
[7] Doten, R.C., Mortlock, R.P.: J. Bacteriol.,162,845–848 (1985)
[8] Morimoto, S., Matsuo, M., Azuma, K., Sinskey, A.J.: J. Ferment. Technol.,64,219–225 (1986)

1 NOMENCLATURE

EC number
1.1.1.10

Systematic name
Xylitol:$NADP^+$ 4-oxidoreductase (L-xylulose-forming)

Recommended name
L-Xylulose reductase

Synonymes
Reductase, L-xylulose
Xylitol dehydrogenase

CAS Reg. No.
9028-17-5

2 REACTION AND SPECIFICITY

Catalysed reaction
Xylitol + $NADP^+$ →
→ L-xylulose + NADPH

Reaction type
Redox reaction

Natural substrates

Substrate spectrum
1 Xylitol + $NADP^+$ (r [4, 7]) [1, 2, 4, 6, 7]

Product spectrum
1 L-Xylulose + NADPH [1, 2, 4, 6, 7]

Inhibitor(s)
Iodoacetate (not inhibitory [4]) [6]; p-Chloromercuribenzoate [4]; Phosphate buffer [7]; More (not: EDTA) [4]

Cofactor(s)/prosthetic group(s)/activating agents
$NADP^+$ (B-specific of hydrogen transfer from xylitol to $NADP^+$ [5], much more effective than NAD^+ [6]) [3–6]; NAD^+ (much less effective than $NADP^+$ [6]) [1, 3, 6]

Metal compounds/salts

Turnover number (min^{-1})

Specific activity (U/mg)

K_m-value (mM)
0.29 (L-xylulose) [4]; 48 (xylitol) [1]; 72 (xylitol) [2]

pH-optimum
7.0 [4]

pH-range

Temperature optimum (°C)

Temperature range (°C)

3 ENZYME STRUCTURE

Molecular weight
102000 (Erwinia sp. strain 4D2P, nondenaturing polyacrylamide gel electrophoresis) [2]

Subunits

Glycoprotein/Lipoprotein
–

4 ISOLATION/PREPARATION

Source organism
Erwinia uredovora [1]; Erwinia sp. (strain 4D2P) [2]; Yeast [3]; Guinea pig [4–7]

Source tissue
Liver [4–7]; Kidney [7]

Localisation in source
Mitochondria [7]

Purification
Erwinia sp. (strain 4D2P, partial) [2]; Guinea pig (partial) [4]

Crystallization
–

Cloned
–

Renaturated
–

5 STABILITY

pH

4 (2°C, 2 h stable) [6]; 9 (room temperature, several h stable) [6]

Temperature (°C)

50 (20 min stable) [6]

Oxidation

Organic solvent

General stability information

Inactivation during dialysis [6]

Storage

2°C, aqueous solution, many weeks [6]; Frozen, ammonium sulfate precipitate, several days [4]

6 CROSSREFERENCES TO STRUCTURE DATABANKS

PIR/MIPS code

Brookhaven code

7 LITERATURE REFERENCES

[1] Doten, R.C., Mortlock, R.P.: J. Bacteriol.,161,529–533 (1985)
[2] Doten, R.C., Mortlock, R.P.: J. Bacteriol.,162,845–848 (1985)
[3] Lowe, C.R., Mosbach, K., Dean, P.D.G.: Biochem. Biophys. Res. Commun.,48, 1004–1010 (1972)
[4] Hickman, J., Ashwell, G.: J. Biol. Chem.,234,758–761 (1959)
[5] Alizade, M.A., Brendl, K., Gaede, K.: FEBS Lett.,67,41–44 (1976)
[6] Hollmann, S., Touster, O.: J. Biol. Chem.,225,87–102 (1957)
[7] Touster, O., Reynolds, V.H., Hutcheson, R.M.: J. Biol. Chem.,221,697–709 (1954)

1 NOMENCLATURE

EC number
1.1.1.11

Systematic name
D-Arabinitol:NAD$^+$ 4-oxidoreductase

Recommended name
D-Arabinitol 4-dehydrogenase

Synonymes
D-Arabitol dehydrogenase [1–4]
Dehydrogenase, D-arabinitol
Arabitol dehydrogenase

CAS Reg. No.
9028-18-6

2 REACTION AND SPECIFICITY

Catalysed reaction
D-Arabinitol + NAD$^+$ →
→ D-xylulose + NADH

Reaction type
Redox reaction

Natural substrates
D-Arabitol + NAD$^+$ (catabolism of D-arabitol) [2]

Substrate spectrum
1 D-Arabitol + NAD$^+$ (r [3–6], arabitol is arabinitol) [1–6]
2 D-Mannitol + NAD$^+$ [1, 5]

Product spectrum
1 D-Xylulose + NADH [2, 5, 6]
2 D-Fructose + NADH [5]

Inhibitor(s)
2,2'-Dipyridyl (slight) [5]; 1,10-Phenanthroline [5]; Diethyldithiocarbamate [5]

Cofactor(s)/prosthetic group(s)/activating agents
NAD$^+$ [1–6]; NADH [3–6]; More (no activity with NADPH) [3, 6]

Metal compounds/salts

Turnover number (min^{-1})
2400 (D-arabitol) [1]

Specific activity (U/mg)
37.15 [1]

K_m-value (mM)
20 (D-arabitol) [1]; 0.04 (NAD^+) [1]; 70 (D-mannitol) [1]; 2 (xylulose) [3]; 0.036 (NADH) [3]; 3.5 (D-arabitol) [5]; 6.1 (D-mannitol) [5]

pH-optimum
7.0 [3, 4]; 9 (D-arabitol, D-mannitol) [5]

pH-range
5.5–8.5 (5.5: about 50% of activity maximum, 8.5: about 60% of activity maximum) [4]; 7.5–10.0 (at pH 7.5 and 10.0 about 50% of activity maximum, D-arabitol) [5]

Temperature optimum (°C)
28 (enzyme assay at) [1]; 25 (enzyme assay at) [2]

Temperature range (°C)

3 ENZYME STRUCTURE

Molecular weight
43000–44000 (Klebsiella aerogenes, gel filtration, ultracentrifugation) [1]

Subunits
Monomer (1 × 46500, Klebsiella aerogenes, SDS-PAGE) [1]

Glycoprotein/Lipoprotein
–

4 ISOLATION/PREPARATION

Source organism
Klebsiella aerogenes (gene expression in E. coli K12) [1]; Aerobacter aerogenes (PRL-R3 [2, 3]) [2–5]; Schizophyllum commune [6]

Source tissue
Cell [3]

Localisation in source
Soluble [1]

Purification
Klebsiella aerogenes (gene expression in E. coli) [1]; Aerobacter aerogenes (PRL-R3 [3]) [3–5]

Crystallization

–

Cloned

(Klebsiella aerogenes gene expressed in E. coli K12) [1]

Renaturated

–

5 STABILITY

pH

Temperature (°C)

43 (half-life: 50 min) [5]; 45 (half-life: 7 min [5], 20 min [1]) [1, 5]

Oxidation

Organic solvent

General stability information

Glycerol stabilizes [1]

Storage

–20°C, protein concentration 50 mg/ml, 50% v/v glycerol, stable over 3 months [1]

6 CROSSREFERENCES TO STRUCTURE DATABANKS

PIR/MIPS code

Brookhaven code

7 LITERATURE REFERENCES

[1] Neuberger, M.S., Patterson, R.A., Hartley, B.S.: Biochem. J.,183,31–42 (1979)
[2] Wilson, B.L., Mortlock, R.P.: J. Bacteriol.,113,1404–1411 (1973)
[3] Fossitt, D.D., Wood, W.A.: Methods Enzymol.,9,184–187 (1966) (Review)
[4] Wood, W.A., McDonough, M.J., Jacobs, L.B.: J. Biol. Chem.,236,2190–2195 (1961)
[5] Lin, E.C.C.: J. Biol. Chem.,236,31–36 (1961)
[6] Speth, J.L., Niederpruem, D.J.: Arch. Microbiol.,107,81–86 (1976)

1 NOMENCLATURE

EC number
1.1.1.12

Systematic name
L-Arabinitol:NAD^+ 4-oxidoreductase (L-xylulose-forming)

Recommended name
L-Arabinitol 4-dehydrogenase

Synonymes
Pentitol-DPN dehydrogenase [1]
L-Arabitol dehydrogenase
Dehydrogenase, L-arabinitol

CAS Reg. No.
9028-19-7

2 REACTION AND SPECIFICITY

Catalysed reaction
L-Arabinitol + NAD^+ →
→ L-xylulose + NADH

Reaction type
Redox reaction

Natural substrates

Substrate spectrum
1 L-Arabinitol + NAD^+ [1]
2 D-Adonitol + NAD^+ [1]
3 Xylitol + NAD^+ [1]

Product spectrum
1 L-Xylulose + NADH
2 ?
3 ?

Inhibitor(s)

Cofactor(s)/prosthetic group(s)/activating agents
NAD^+ [1]; NADH [1]

Metal compounds/salts

Turnover number (min^{-1})

Specific activity (U/mg)

K_m-value (mM)

pH-optimum

pH-range

Temperature optimum (°C)

Temperature range (°C)

3 ENZYME STRUCTURE

Molecular weight

Subunits

Glycoprotein/Lipoprotein

–

4 ISOLATION/PREPARATION

Source organism

Penicillium chrysogenum [1]; Aspergillus niger [2]

Source tissue

Cell [1]

Localisation in source

Purification

Crystallization

–

Cloned

–

Renaturated

–

5 STABILITY

pH

Temperature (°C)

Oxidation

Organic solvent

General stability information

Storage

6 CROSSREFERENCES TO STRUCTURE DATABANKS

PIR/MIPS code

Brookhaven code

7 LITERATURE REFERENCES

[1] Chiang, C., Knight, S.G.: Biochem. Biophys. Res. Commun.,3,554–559 (1960)
[2] Witteveen, C.F.B., Busnik, R., Van de Vondervoort, P., Dijkema, C., Swart, K., Visser, J.: J. Gen. Microbiol.,135,2163–2171 (1989)

1 NOMENCLATURE

EC number
1.1.1.13

Systematic name
L-Arabinitol:NAD^+ 2-oxidoreductase (L-ribulose-forming)

Recommended name
L-Arabinitol 2-dehydrogenase

Synonymes
Dehydrogenase, L-arabinitol (ribulose-forming)
L-Arabinitol dehydrogenase (ribulose-forming), L-arabinitol (ribulose forming)

CAS Reg. No.
9028-20-0

---

2 REACTION AND SPECIFICITY

Catalysed reaction
L-Arabinitol + NAD^+ →
→ L-ribulose + NADH

Reaction type
Redox reaction

Natural substrates
More (enzyme of pentose metabolism) [1]

Substrate spectrum
1 L-Arabinitol + NAD^+ [1]

Product spectrum
1 L-Ribulose + NADH [1]

Inhibitor(s)

Cofactor(s)/prosthetic group(s)/activating agents
NAD^+ [1]

Metal compounds/salts

Turnover number (min^{-1})

Specific activity (U/mg)

K_m**-value** (mM)

pH-optimum

pH-range

Temperature optimum (°C)

Temperature range (°C)

3 ENZYME STRUCTURE

Molecular weight

Subunits

Glycoprotein/Lipoprotein

–

4 ISOLATION/PREPARATION

Source organism

Penicillium chrysogenum [1]

Source tissue

Localisation in source

Purification

Crystallization

–

Cloned

–

Renaturated

–

5 STABILITY

pH

Temperature (°C)

Oxidation

Organic solvent

General stability information

Storage

6 CROSSREFERENCES TO STRUCTURE DATABANKS

PIR/MIPS code

Brookhaven code

7 LITERATURE REFERENCES

[1] Chiang, C., Knight, S.G.: Biochim. Biophys. Acta,46,271–278 (1961)

1 NOMENCLATURE

EC number
1.1.1.14

Systematic name
L-Iditol:NAD^+ 2-oxidoreductase

Recommended name
L-Iditol 2-dehydrogenase

Synonymes
Polyol dehydrogenase
Sorbitol dehydrogenase
L-Iditol:NAD^+ 5-oxidoreductase [1, 5]
L-Iditol (sorbitol) dehydrogenase [5]
Glucitol dehydrogenase
L-Iditol:NAD oxidoreductase
NAD^+-dependent sorbitol dehydrogenase
NAD-dependent sorbitol dehydrogenase
NAD-sorbitol dehydrogenase
Dehydrogenase, L-iditol

CAS Reg. No.
9028-21-1

2 REACTION AND SPECIFICITY

Catalysed reaction
L-Iditol + NAD^+ →
→ L-sorbose + NADH

Reaction type
Redox reaction

Natural substrates

Substrate spectrum

1 L-Iditol + NAD^+ (r [5, 6]) [5, 6, 9–11]
2 Xylitol + NAD^+ (r [1]) [1, 3, 5, 8–11]
3 Sorbitol + NAD^+ (r [1, 9, 10]) [1, 3, 8–11]
4 Ribitol + NAD^+ (r [1]) [1, 3, 8–11]
5 D-Mannitol + NAD^+ (r [1], low activity [3, 5]) [1, 3, 5, 8, 11]
6 L-Arabitol + NAD^+ [3, 8, 9, 11]
7 L-Threitol + NAD^+ [3]
8 Galactitol + NAD^+ (not [3]) [8]
9 More (no reaction with: ethanol [2], trans-2-hexen-1-ol [2], propan-2-ol [2], myo-inositol [3], D-arabitol [3], erythritol [3], glycerol [3]) [2, 3]

Product spectrum

1 L-Sorbose + NADH
2 Xylulose + NADH
3 D-Fructose + NADH [1, 9, 10]
4 D-Fructose + NADH
5 D-Ribulose + NADH
6 D-Fructose + NADH
7 ?
8 ?
9 ?

Inhibitor(s)

Sorbitol (above 49 mM, substrate inhibition) [1]; NAD^+ (above 7 mM, substrate inhibition) [1]; Fructose (above 250 mM, substrate inhibition) [1]; NADH (above 0.4 mM, substrate inhibition) [1]; Dithiothreitol (protection at low concentration, inhibition at high concentration, 100 mM) [2]; Dithioerythritol (protection at low concentration, inhibition at high concentration, 100 mM) [2]; EDTA [2, 8]; 1,10-Phenanthroline [2, 9]; Iodoacetamide [2]; Diethyldicarbonate [2]; Iodoacetate [3, 8]; p-Chloromercuribenzoate [3, 9]; Cysteine (strong effect on sorbitol oxidation, slight effect on fructose reduction, $ZnSO_4$ reverses inhibition [3]) [3, 8, 9]; Glycerol [7]; Hydroxymercuribenzoate [8]; 2,3-Dimercaptopropanol [8]; $AgNO_3$ [3]; $HgCl_2$ [3]; Glutathione [9]; Cyanide [9]

Cofactor(s)/prosthetic group(s)/activating agents

NAD^+ (specific for [5, 12]) [1–12]; NADH (specific for [5, 12]) [1–12]; $NADP^+$ (cannot replace NAD^+ [3, 5, 8], less effective than NAD^+ [2]) [2]; NADPH (less effective than NADH) [2, 8]

Metal compounds/salts

$ZnSO_4$ (reverses cysteine inhibition) [3]

Turnover number (min^{-1})

Specific activity (U/mg)
7.47 [1]; 14.4 [9]

K_m-value (mM)
0.38 (sorbitol) [1]; 0.067 (NADH) [1]; 0.082 (NAD^+) [1]; 136 (fructose) [1]; 1.1 (sorbitol) [9]; 1.8 (ribitol) [9]; 0.18 (xylitol) [9]; 0.60 (NAD^+) [9]; 9.8 (sorbitol) [11]; 9.0 (iditol) [11]; 86 (sorbitol, xylitol) [2]; 1.5 (fructose) [2]; 0.35 (sorbitol) [5]; 110 (fructose) [5]; 300 (fructose) [7]

pH-optimum
5.7–7.0 (D-fructose reduction) [2]; 5.9 (D-fructose reduction) [1, 5]; 7.0 (fructose reduction) [8]; 7.4–8.0 (fructose reduction) [8]; 8.1–8.5 (sorbitol oxidation) [1, 5]; 8.8 (fructose reduction) [12]; 9.0 (sorbitol oxidation) [8]; 9–9.5 (sorbitol oxidation) [2]; 9.6 (sorbitol oxidation) [12]; 10.0 (sorbitol oxidation) [8]

pH-range
8–11 (at pH 8 and 11 about 50% of activity maximum, sorbitol oxidation) [2]; 4.7–8 (at pH 4.7 and 8 about 50% of activity maximum, D-fructose reduction) [2]

Temperature optimum (°C)
60 [2]

Temperature range (°C)
35–65 (35°C: about 50% of activity maximum, 65°C: about 95% of activity maximum) [2]

3 ENZYME STRUCTURE

Molecular weight
95000 (rat, gel permeation chromatography, disc gel electrophoresis) [1, 5]
106000 (horse, gel chromatography) [12]
110000 (sheep, gel chromatography) [12]
115000 (sheep, ultracentrifugal analysis) [9]
140000 (sheep, gel filtration) [2]

Subunits
Tetramer (4 × 35000–40000, sheep, SDS-PAGE) [1]
? (x × 26000, horse, sheep, SDS-PAGE) [12]

Glycoprotein/Lipoprotein
–

4 ISOLATION/PREPARATION

Source organism

Malus domestica (apple) [3]; Rat [1, 4, 5, 8, 10]; Sheep (ram [11]) [1, 2, 7, 9, 11, 12]; Human [4, 6]; Dog [4]; Mouse [4]; Guinea pig [11]; Horse [12]

Source tissue

Liver (commercial preparation [7]) [1, 2, 5, 7, 9, 12]; Serum [4]; Mesocarp tissue culture [3]; Brain [8]; Seminal vesicles [10]; Coagulating gland [10]; Ventral prostate gland [10]; Dorsal prostate gland [10]; Spermatozoa [11]

Localisation in source

Purification

Rat (partial [8]) [1, 5, 8]; Sheep [2, 9]; Horse [12]

Crystallization

[9]

Cloned

–

Renaturated

–

5 STABILITY

pH

Temperature (°C)

20 (30% loss of activity after 3 days, 70% loss of activity after 6 days, 100% loss of activity after 10 days) [12]; 21 (50% loss of activity after 2 days, non-purified enzyme in serum) [6]; 22 (5 h, stable) [1, 5]; 60 (10 min, 65% loss of activity) [1, 5]

Oxidation

Organic solvent

General stability information

Freezing to –20°C, loss of activity [2]; Bovine serum albumin stabilizes at 4°C [8]; Dithiothreitol protects at low concentration, inhibits at high concentration, 100 mM [2]; Dithioerythritol protects at low concentration, inhibits at high concentration, 100 mM [2]

Storage

–20°C, 2 years [1]; –20°C, stable for 4 months [12]; –18°C, 2 days, non-purified enzyme in serum [6]; –85°C, good preservation [8]; Immobilized enzyme loses 20% of activity after 10 days at room temperature [2]

6 CROSSREFERENCES TO STRUCTURE DATABANKS

PIR/MIPS code

PIR2:A45052 (Bacillus subtilis); PIR1:S16132 (rat); PIR3:S38363 (rat); PIR1:S10065 (sheep)

Brookhaven code

1SDG (Sheep (Ovis aries) liver)

7 LITERATURE REFERENCES

[1] Leissing, N., McGuinness, E.T.: Methods Enzymol.,89,135–140 (1982) (Review)
[2] Jeffery, J., Cummins, L., Carlquist, M., Jörnvall, H.: Eur. J. Biochem.,120,229–234 (1981)
[3] Negm, F.B., Loescher, W.H.: Plant Physiol.,64,69–73 (1979)
[4] Dooley, J.F., Turnquist, L.J., Racich, L.: Clin. Chem.,25,2026–2029 (1979)
[5] Leissing, N., Mc Guinness, E.T.: Biochim. Biophys. Acta,524,254–261 (1978)
[6] Gerlach, U., Hiby, W. in "Methoden Enzym. Anal.", (Bergmeyer, H.U., Ed.) 1, 601–606 (1974)
[7] Myers, J.S., Jakoby, W.B.: Biochem. Biophys. Res. Commun.,51,631–636 (1973)
[8] Rehg, J.E., Torack, R.M.: J. Neurochem.,28,655–660 (1977)
[9] Smith, M.G.: Biochem. J.,83,135–144 (1962)
[10] Williams-Ashman, H.G., Banks, J., Wolfson, S.K.: Arch. Biochem. Biophys.,72, 485–494 (1957)
[11] King, T.E., Mann, T.: Proc. R. Soc. Lond. B Biol. Sci.,151,226–243 (1959)
[12] Bailey, J.P., Renz, C., McGuinness, E.T.: Comp. Biochem. Physiol.,69B,909–914 (1981)

1 NOMENCLATURE

EC number
1.1.1.15

Systematic name
D-Iditol:NAD^+ 2-oxidoreductase

Recommended name
D-Iditol 2-dehydrogenase

Synonymes
Dehydrogenase, D-iditol
D-Sorbitol dehydrogenase

CAS Reg. No.
9028-22-2

2 REACTION AND SPECIFICITY

Catalysed reaction
D-Iditol + NAD^+ →
→ D-sorbose + NADH

Reaction type
Redox reaction

Natural substrates

Substrate spectrum
1 D-Iditol + NAD^+ (crude enzyme extract) [1]
2 D-Gulitol + NAD^+ (crude enzyme extract) [1]
3 Sorbitol + NAD^+ (i.e. L-gulitol, D-glucitol, crude enzyme extract) [1]
4 D-Talitol + NAD^+ (crude enzyme extract) [1]
5 Xylitol + NAD^+ (crude enzyme extract) [1]

Product spectrum
1 D-Sorbose + NADH [1]
2 L-Fructose + NADH [1]
3 D-Fructose + NADH [1]
4 D-Allulose + NADH [1]
5 L-Xylulose + NADH [1]

Inhibitor(s)
KCN [1]

Cofactor(s)/prosthetic group(s)/activating agents
NAD+ [1]; NADH [1]; Methylene blue (activation, anaerobically in the presence of NAD+ with methylene blue as hydrogen acceptor) [1]

Metal compounds/salts

Turnover number (min^{-1})

Specific activity (U/mg)

K_m-value (mM)
0.073 (NAD+) [1]

pH-optimum

pH-range

Temperature optimum (°C)
18 (assay at) [1]

Temperature range (°C)

---

3 ENZYME STRUCTURE

Molecular weight

Subunits

Glycoprotein/Lipoprotein
–

---

4 ISOLATION/PREPARATION

Source organism
Pseudomonas sp. [1]

Source tissue
Cell [1]

Localisation in source

Purification

Crystallization
–

Cloned
–

Renaturated
–

5 STABILITY

pH

Temperature (°C)

Oxidation

Organic solvent

General stability information

Storage

–20°C, stable [1]; 2–3°C, 90% loss of activity after 4 days [1]

6 CROSSREFERENCES TO STRUCTURE DATABANKS

PIR/MIPS code

Brookhaven code

7 LITERATURE REFERENCES

[1] Shaw, D.R.D.: Biochem. J.,64,394–405 (1956)

1 NOMENCLATURE

EC number
1.1.1.16

Systematic name
Galactitol:NAD$^+$ 2-oxidoreductase

Recommended name
Galactitol 2-dehydrogenase

Synonymes
Dulcitol dehydrogenase [3]
Dehydrogenase, galactitol

CAS Reg. No.
9028-23-3

2 REACTION AND SPECIFICITY

Catalysed reaction
Galactitol + NAD$^+$ →
→ D-tagatose + NADH

Reaction type
Redox reaction

Natural substrates
Galactitol + NAD$^+$ (utilization of galactitols as sole carbon source) [2]

Substrate spectrum
1 Galactitol + NAD$^+$ (i.e. dulcitol) [1–3]
2 Sorbitol + NAD$^+$ [2, 3]
3 L-Iditol + NAD$^+$ [2]
4 L-Arabitol + NAD$^+$ [2]
5 Ribitol + NAD$^+$ (not [2]) [3]

Product spectrum
1 D-Tagatose + NADH
2 D-Fructose + NADH
3 L-Sorbose + NADH
4 L-Ribulose + NADH
5 ?

Inhibitor(s)

Cofactor(s)/prosthetic group(s)/activating agents
NAD+ [1–3]; More (no reaction with NADP+) [2]

Metal compounds/salts

Turnover number (min^{-1})

Specific activity (U/mg)

K_m-value (mM)
20 (galactitol) [1]

pH-optimum
8.5 (assay at) [1]; 10 (oxidation of galactitol) [2]

pH-range

Temperature optimum (°C)

Temperature range (°C)

---

3 ENZYME STRUCTURE

Molecular weight
102000 (Pseudomonas cepacia, sucrose gradient centrifugation) [1]

Subunits

Glycoprotein/Lipoprotein
–

---

4 ISOLATION/PREPARATION

Source organism
Pseudomonas cepacia [1]; Pseudomonas sp. [2]; Rhizobium trifolii [3]

Source tissue
Cell [2]

Localisation in source

Purification
Pseudomonas sp. [2]

Crystallization
–

Cloned
–

Renaturated

–

5 STABILITY

pH

Temperature (°C)

Oxidation

Organic solvent

General stability information

Storage

–20°C, 6 months, 5% loss of activity [2]; 0°C – 2°C, 16 days, 28% loss of activity [2]

6 CROSSREFERENCES TO STRUCTURE DATABANKS

PIR/MIPS code

Brookhaven code

7 LITERATURE REFERENCES

[1] Allenza, P., Lee, Y.N., Lessie, T.G.: J. Bacteriol.,150,1348–1356 (1982)
[2] Shaw, D.R.D.: Methods Enzymol.,5,323–325 (1962) (Review)
[3] Primrose, S.B., Ronson, C.W.: J. Bacteriol.,141,1109–1114 (1980)

1 NOMENCLATURE

EC number
1.1.1.17

Systematic name
D-Mannitol-1-phosphate:NAD^+ 5-oxidoreductase

Recommended name
Mannitol-1-phosphate 5-dehydrogenase

Synonymes
Hexose reductase [9]
Dehydrogenase, mannitol 1-phosphate
D-Mannitol-1-phosphate dehydrogenase
Fructose 6-phosphate reductase

CAS Reg. No.
9028-24-4

2 REACTION AND SPECIFICITY

Catalysed reaction
D-Mannitol 1-phosphate + NAD^+ →
→ D-fructose 6-phosphate + NADH (reaction is random for mannitol 1-phosphate oxidation and ordered for fructose 6-phosphate reduction [1], random Bi-Bi mechanism with 2 dead-end complexes [4])

Reaction type
Redox reaction

Natural substrates

Substrate spectrum
1 D-Mannitol 1-phosphate + NAD^+ (r [6], absolutely specific for) [3, 4, 6]
2 D-Sorbitol 6-phosphate + NAD^+ (low activity) [6]

Product spectrum
1 D-Fructose 6-phosphate + NADH
2 ?

Inhibitor(s)
Adenosine [6]; Fructose 6-phosphate [6]; Adenosine 5'-phosphate [6]; 5,5'-Dithiobis(2-nitrobenzoate) [1]; Zn^{2+} [2]; Cd^{2+} [2]; $NaHSO_3$ [9]; KCN [9]; p-Chloromercuribenzoate [9]

Cofactor(s)/prosthetic group(s)/activating agents
NAD^+ [3, 4, 6]; NADH [6]

Metal compounds/salts
No stimulating effect of metal ions [6]; No metal ion requirement [9]

Turnover number (min^{-1})

Specific activity (U/mg)
715 [3]; 0.427 [6]

K_m**-value** (mM)
0.15 (mannitol 1-phosphate) [6]; 0.2 (NAD^+) [3]; 0.8 (D-mannitol 1-phosphate) [3]; 0.066 (NAD^+) [6]; 0.016 (NADH) [6]; 1.66 (fructose 6-phosphate) [6]; 2.0 (D-mannitol 1-phosphate) [8]; 0.38 (NAD^+) [8]; 2.1 (D-fructose 6-phosphate) [8]; 0.14 (NADH) [8]

pH-optimum
9.0–9.5 [5]; 9.5 [8]

pH-range

Temperature optimum (°C)
30 (assay at) [1]; 25 (assay at) [9]

Temperature range (°C)

3 ENZYME STRUCTURE

Molecular weight
45000 (Aerobacter aerogenes, determination of sedimentation and diffusion constants [8], Streptococcus mutans, agarose molecular exclusion chromatography [6], E. coli, gel filtration [1]) [1, 6, 8]
40000 (E. coli, gel filtration [3], Aspergillus niger, gel filtration [4]) [3, 4]

Subunits
Dimer (2 × 22000, E. coli, SDS-PAGE, amino acid analysis [1], Aspergillus niger, SDS-PAGE [4]) [1, 4]
Monomer (1 × 40000, E. coli, SDS-PAGE) [3]

Glycoprotein/Lipoprotein
–

4 ISOLATION/PREPARATION

Source organism
Streptomyces lactamdurans [5]; Micromonospora sp. [5]; E. coli [1, 3, 5, 9]; Aspergillus parasiticus [2]; Aspergillus niger [2, 4]; Streptococcus mutans [6]; Aerobacter aerogenes [8]; Bacillus subtilis [5]; Aspergillus nidulans [7]

Source tissue

Cell [1, 5]

Localisation in source

Purification

E. coli [1, 3]; Streptococcus mutans [6]; Aspergillus parasiticus [2]; Aspergillus niger [2, 4]; Aerobacter aerogenes [8]

Crystallization

–

Cloned

–

Renaturated

–

5 STABILITY

pH

Temperature (°C)

30 (half-life: above 280 min, Aspergillus niger) [2]; 40 (20 min, stable) [3]; 60 (20 min, 90% loss of activity [3], 10 min, complete loss of activity [9]) [3, 9]; 55 (10 min, 45% loss of activity) [9]

Oxidation

Organic solvent

General stability information

NAD^+ protects during storage [8]; D-Mannitol 1-phosphate protects during storage [8]

Storage

4°C, stable for years as $(NH_4)_2SO_4$ precipitate [1]; –20°C, more than 4 months [3]; –80°C, presence of 2-mercaptoethanol, stable for long periods of time [6]; Stored as a 0.5–1.0% solution in 0.05 M Tris buffer of pH 7.6 containing 0.05 M NaCl and 5 mM 2-mercaptoethanol, stable for weeks [8]

6 CROSSREFERENCES TO STRUCTURE DATABANKS

PIR/MIPS code

PIR2:A26224 (Escherichia coli); PIR2:C39435 (mtlD Enterococcus faecalis); PIR3:S10394 (Escherichia coli)

Brookhaven code

7 LITERATURE REFERENCES

[1] Chase, T.: Biochem. J.,239,435–443 (1986)
[2] Foremen, J.E., Niehaus, W.G.: J. Biol. Chem.,260,10019–10022 (1985)
[3] Novotny, M.J., Reizer, J., Esch, F., Saier, M.H.: J. Bacteriol.,159,986–990 (1984)
[4] Kiser, R.C., Niehaus, W.G.: Arch. Biochem. Biophys.,211,613–621 (1981)
[5] Mehta, R.J., Fare, L.R., Shearer, M.E., Nash, C.H.: Appl. Environ. Microbiol.,33, 1013–1015 (1977)
[6] Brown, A.T., Bowles, R.D.: Infect. Immun.,16,163–167 (1977)
[7] Hankinson, O., Cove, D.J.: Can. J. Microbiol.,21,99–101 (1975)
[8] Liss, M., Horwitz, S.B., Kaplan, N.O.: J. Biol. Chem.,237,1342–1350 (1962)
[9] Wolff, J.B., Kaplan, N.O.: Methods Enzymol.,1,346–348 (1955) (Review)

1 NOMENCLATURE

EC number
1.1.1.18

Systematic name
myo-Inositol:NAD+ 2-oxidoreductase

Recommended name
myo-Inositol 2-dehydrogenase

Synonymes
Inositol 2-dehydrogenase
myo-Inositol:NAD+ oxidoreductase [4]
Inositol dehydrogenase
myo-Inositol dehydrogenase
Dehydrogenase, inositol 2-

CAS Reg. No.
9028-25-5

---

2 REACTION AND SPECIFICITY

Catalysed reaction
myo-Inositol + NAD+ →
→ 2,4,6/3,5-pentahydroxycyclohexanone + NADH (bi bi ordered mechanism with NAD+ and NADH as leading substrates [4])

Reaction type
Redox reaction

Natural substrates
myo-Inositol + NAD+ (initial step of myo-inositol catabolism) [4]

Substrate spectrum
1 myo-Inositol + NAD+ (equilibrium constant: 0.0000000000053 [4]) [1–7]
2 D-Glucose + NAD+ (reaction only with alpha-anomer, not with beta-anomer) [1]
3 D-Xylose + NAD+
4 (+)-Inositol + NAD+ [4]
5 Scyllitol + NAD+ [5]
6 D-Inositol + NAD+ [5]
7 myo-Inosose + NADPH [5]
8 DL-epi-Inosose + NADPH [5]
9 More (no reaction with scyllo-inositol, 2-deoxy-D-glucose) [1]

Product spectrum

1 2,4,6/3,5-Pentahydroxycyclohexanone + NADH (scyllo-inosose [4])
2 D-Glucono-1,5-lactone + NADH
3 D-Xylono-1,4-lactone + NADH
4 A pentahydroxycyclohexanone + NADH
5 ?
6 A pentahydroxycyclohexanone + NADH
7 ?
8 ?
9 ?

Inhibitor(s)

p-Chloromercuribenzoate [5]; Iodosobenzoate [5]; Galactinol [5]; Colchicine [5]

Cofactor(s)/prosthetic group(s)/activating agents

NAD^+ (A-specific with respect to NAD^+, i.e. transfer of pro-R hydrogen to NAD^+ [7]) [1–7]; NADH [4]; NADPH (no reaction [4]) [5]; More (no reaction with $NADP^+$) [2, 4, 5]

Metal compounds/salts

Turnover number (min^{-1})

Specific activity (U/mg)

1.68 [4]; 14.3 [5]

K_m-value (mM)

0.33 (NAD^+, pH 9) [5]; 1.25 (myo-inositol, pH 9) [5]; 0.06 (NADH, pH 7.5) [5]; 0.54 (myo-inosose, pH 7.5) [5]; 0.23 (NAD^+) [1]; 18 (myo-inositol) [1]; 56 (alpha-D-glucose) [1]; 190 (D-xylose) [1]; 1.6 (2-inosose) [1]; 0.036 (NADH) [1]; More [4]

pH-optimum

9.2–9.5 (myo-inositol + NAD^+) [1]; 9.0 (sharp decrease below and above [2], myo-inositol + NAD^+ [5]) [2, 5]; 7.3–7.7 (NADH + myo-inosose) [5]; 6.2 [6]

pH-range

6.1–6.9 (at pH 6.1 and 6.9 about 50% of activity maximum) [6]

Temperature optimum (°C)

Temperature range (°C)

---

3 ENZYME STRUCTURE

Molecular weight

155000–160000 (Bacillus subtilis, sucrose density gradient centrifugation) [1]

Subunits

Tetramer (4 × 39000, Bacillus subtilis, SDS-PAGE) [1]

Glycoprotein/Lipoprotein

–

4 ISOLATION/PREPARATION

Source organism

Cryptococcus melibiosum [4]; Aerobacter aerogenes [5]; Klebsiella pneumoniae (formerly classified as Aerobacter aerogenes) [7]; Acetomonas oxydans [6]; Bacillus subtilis (strain 60015) [1]; Serratia marcescens [2]; Streptomyces hygroscopicus forma glebosus (ATCC 14607) [3]; Streptomyces bikiniensis (ATCC 11062) [3]

Source tissue

Culture medium [1]; Mycelium [3]; Cell [5]

Localisation in source

Membrane bound [6]

Purification

Cryptococcus melibiosum (partial) [4]; Aerobacter aerogenes [5]; Bacillus subtilis (strain 60015) [1]

Crystallization

–

Cloned

–

Renaturated

–

5 STABILITY

pH

6–7 (41°C, highest stability) [1]; 6.5–7.0 (highest stability) [6]

Temperature (°C)

Oxidation

Organic solvent

General stability information

Storage

–20°C, 0.05 M potassium phosphate, pH 6.5, 0.1 mM 2-mercaptoethanol, 50% w/v glycerol, 1 year [1]; 4°C, saturated ammonium sulfate, pH 6.5, 1 year [1]

6 CROSSREFERENCES TO STRUCTURE DATABANKS

PIR/MIPS code

PIR2:JH0511 (Bacillus subtilis)

Brookhaven code

7 LITERATURE REFERENCES

[1] Ramaley, R., Fujita, Y., Freese, E.: J. Biol. Chem.,254,7684–7690 (1979)
[2] Fawole, M.O.: Z. Allg. Mikrobiol.,16,327–328 (1976)
[3] Walker, J.B.: Methods Enzymol.,43,433–439 (1975) (Review)
[4] Vidal-Leiria, M., Van Uden, N.: Biochim. Biophys. Acta,293,295–303 (1973)
[5] Larner, J.: Methods Enzymol.,5,326–328 (1962) (Review)
[6] Criddle, W.J., Fry, J.C., Keaney, M.M.: Biochem. J.,137,449–452 (1974)
[7] Alizade, M.A., Gaede, K., Brendel, K.: Z. Naturforsch.,31c,624–625 (1976)

1 NOMENCLATURE

EC number
1.1.1.19

Systematic name
L-Gulonate:NADP+ 6-oxidoreductase

Recommended name
Glucuronate reductase

Synonymes
Aldehyde reductase [1]
L-Hexonate:NADP dehydrogenase [1]
TPN-L-Gulonate dehydrogenase [2, 3]
Aldehyde reductase II [5]
NADP-L-gulonate dehydrogenase [5]
Reductase, glucuronate
D-Glucuronate dehydrogenase
D-Glucuronate reductase
More (may be identical with EC 1.1.1.2)

CAS Reg. No.
9028-29-9

2 REACTION AND SPECIFICITY

Catalysed reaction
L-Gulonate + NADP+ →
→ D-glucuronate + NADPH (ordered Bi Bi mechanism [1])

Reaction type
Redox reaction

Natural substrates
D-Glucuronate + NADPH (formation of ascorbic acid [2], catabolism of inositol [5]) [2, 5]

Substrate spectrum
1 L-Gulonate + NADP+ (r [3, 6]) [1–3, 6]
2 D-Galacturonate + NADPH [2, 3]
3 4-Nitrobenzaldehyde + NADPH (2-, and 3-isomer also reduced [5]) [1, 5]
4 Pyridine-3-aldehyde + NADPH [1, 5]
5 D-Glyceraldehyde + NADPH [1, 5]

6 More (broad specificity [1], reduction of aldehyde greatly favored compared to oxidation of alcohol [5], reduction of D-glyceraldehyde: pro-4R hydrogen is transferred from NADPH, i.e. A-specific, attacks re face of carbonyl group of the D-glyceraldehyde [1], no reaction with gamma-lactones of the substrates [3], coupled gamma-hydroxybutyrate oxidation and D-glucuronate reduction [4]) [1, 3–5]

Product spectrum

1 D-Glucuronate + NADPH [1]
2 ?
3 ?
4 ?
5 ?
6 ?

Inhibitor(s)

Statil [7]; Sorbinil [7]; Zn^{2+} [5]; NaF [5]; Valproate (poor) [7]; Diethyldicarbonate [1]; Pyridoxal 5'-phosphate [1]; Phenobarbital [1, 5]; p-Chloromercuribenzoate [1–3]; 2,3-Butanedione [1]; Phenylglyoxal [1]; 2-Mercaptoethanol [5]; Iodoacetic acid [1, 5]; Hydroxylamine [1, 5]; Sodium fluoride [1]

Cofactor(s)/prosthetic group(s)/activating agents

NADPH (reduction of D-glyceraldehyde: pro-4R hydrogen is transferred from NADPH, i.e. A-specific, attacks re face of carbonyl group of the D-glyceraldehyde [1]) [1–3, 6]; $NADP^+$ [3, 6]

Metal compounds/salts

Turnover number (min^{-1})

Specific activity (U/mg)

5.45 [1]; More [2, 5]

K_m-value (mM)

0.009 (NADPH) [1]; 7.3 (L-gulonate) [1, 3]; 0.02 ($NADP^+$) [1, 3]; 1.05 (benzaldehyde) [5]; 0.769 (2-nitrobenzaldehyde) [5]; 0.526 (3-nitrobenzaldehyde) [5]; 0.133 (4-nitrobenzaldehyde) [5]; 1.671 (DL-glyceraldehyde) [5]; 9.1 (D-glucuronate) [5]; 333.0 (D-ribose, D-galactose, D-glucose, D-xylose) [5]; 500 (2-deoxyglucose) [5]; 139.9 (D-arabinose) [5]; 0.036 (NADPH) [5]; 4.8 (D-glyceraldehyde) [1]; 8.4 (D-glucuronate) [1]; 2.6 (pyridine-3-aldehyde) [1]; 0.35 (4-nitrobenzaldehyde) [1]; 0.0014 ($NADP^+$, coupled reaction of gamma-hydroxybutyrate oxidation and D-glucuronate reduction) [4]; 0.02 ($NADP^+$, uncoupled reaction) [4]

pH-optimum

6.0–7.0 (DL-glyceraldehyde) [5]; 6.2–6.6 (D-glucuronate) [5]; 8.6–9.3 [2, 3]

pH-range

Temperature optimum (°C)
25 (assay at) [1, 5]

Temperature range (°C)

3 ENZYME STRUCTURE

Molecular weight
33000 (pig, gel filtration [1])
35000 (mouse, gel filtration) [5]

Subunits
Monomer (1 × 36700, pig, SDS-PAGE [1], 1 × 33000, mouse, SDS-PAGE [5]) [1, 5]

Glycoprotein/Lipoprotein
–

4 ISOLATION/PREPARATION

Source organism
Pig [1–3]; Hamster [4]; Mouse [5]; Schwanniomyces occidentalis [6]

Source tissue
Kidney [1–3]; Liver [4, 5]

Localisation in source
Soluble cytoplasm [5]

Purification
Pig [1–3]; Hamster [4]; Mouse [5]

Crystallization
–

Cloned
–

Renaturated
–

5 STABILITY

pH
7 (most stable) [3]

Temperature (°C)

40 (15 min, 50% loss of activity, ammonium sulfate fraction) [3]; 50 (15 min, complete inactivation, ammonium sulfate fraction) [3]

Oxidation

Organic solvent

Ethanol inactivates [3]; Acetone inactivates [3]

General stability information

Lyophilization destroys activity [3]; EDTA, 1 mM protects against heat denaturation at 40°C [3]; Ethanol inactivates [3]; Acetone inactivates [3]

Storage

–5°C, 2 weeks [2, 3]

6 CROSSREFERENCES TO STRUCTURE DATABANKS

PIR/MIPS code

Brookhaven code

7 LITERATURE REFERENCES

[1] Flynn, T.G., Cromlish, J.A., Davidson, W.S.: Methods Enzymol.,89,501–506 (1982) (Review)
[2] Bublitz, C., Lehninger, A.L.: Methods Enzymol.,6,334–337 (1963)
[3] York, J.L., Grollman, A.P., Bublitz, C.: Biochim. Biophys. Acta,47,298–306 (1961)
[4] Kaufman, E.E., Nelson, T.: J. Biol. Chem.,256,6890–6894 (1981)
[5] Tulsiani, D.R.P., Touster, O.: J. Biol. Chem.,252,2545–2550 (1977)
[6] Sivak, A., Hoffmann-Ostenhof, O.: Biochim. Biophys. Acta,53,426–428 (1961)
[7] Poulsom, R.: Biochem. Pharmacol.,35,2955–2959 (1986)

1 NOMENCLATURE

EC number
1.1.1.20

Systematic name
L-Gulono-1,4-lactone:$NADP^+$ 1-oxidoreductase

Recommended name
Glucuronolactone reductase

Synonymes
GRase [1]
Reductase, glucuronolactone
Gulonolactone dehydrogenase

CAS Reg. No.
9028-30-2

2 REACTION AND SPECIFICITY

Catalysed reaction
L-Gulono-1,4-lactone + $NADP^+$ →
→ D-glucurono-3,6-lactone + NADPH

Reaction type
Redox reaction

Natural substrates

Substrate spectrum
1 D-Glucurono-3,6-lactone + NADPH (r, oxidation activity is very low) [1]
2 D-Galacturonic acid + NADPH [1]
3 D-Glucuronic acid + NADPH [1]
4 D-Iduronic acid + NADPH [1]
5 Glyceraldehyde + NADPH [1]
6 Erythrose + NADPH [1]

Product spectrum
1 D-Gulono-1,4-lactone + $NADP^+$ [1]
2 ?
3 L-Gulonic acid + $NADP^+$
4 ?
5 ?
6 Glycerol + $NADP^+$

Inhibitor(s)
Barbital [1]; Phenobarbital [1]; Barbiturates [1]; Hg^{2+} [1]; Zn^{2+} [1]; NaF [1]; Dithiothreitol [1]; Iodoacetic acid [1]

Cofactor(s)/prosthetic group(s)/activating agents
NADPH (NADPH is 145-times as effective as NADH) [1]; NADH (NADPH is 145-times as effective as NADH) [1]; $NADP^+$ [1]

Metal compounds/salts

Turnover number (min^{-1})

Specific activity (U/mg)
More (13450 U/min × mg, 1 unit is defined as change of absorbance of 0.01/min at 340 nm) [1]

K_m-value (mM)
6 (D-glucuronic acid, L-iduronic acid) [1]; 9 (D-glucuronolactone) [1]; 4 (D-galacturonic acid) [1]; 0.045 (NADPH (+ D-glucuronic acid)) [1]

pH-optimum
6.0–7.0 (D-glucuronic acid) [1]; 5.5–7.5 (DL-glyceraldehyde) [1]; 9.5 (broad, oxidation of glycerol and L-gulonolactone) [1]

pH-range

Temperature optimum (°C)
25 (assay at) [1]

Temperature range (°C)

---

3 ENZYME STRUCTURE

Molecular weight
41000 (rat, gel filtration) [1]

Subunits
Monomer (1 × 40000, rat, SDS-PAGE) [1]

Glycoprotein/Lipoprotein
–

---

4 ISOLATION/PREPARATION

Source organism
Rat [1]

Source tissue
Kidney [1]

Localisation in source
Cytoplasm [1]

Purification
Rat [1]

Crystallization
–

Cloned
–

Renaturated
–

5 STABILITY

pH

Temperature (°C)
40 (30 min, 20% loss of activity [1], 60 min, 30% loss of activity [1], 180 min, 50% loss of activity [1]) [1]

Oxidation

Organic solvent

General stability information
Freezing and thawing: 3 times causes 25% decrease of activity, 5 times causes 40% decrease of activity [1]

Storage
0°C, 2 months, less than 10% loss of activity [1]; –80°C, 10 days, 30% loss of activity [1]

6 CROSSREFERENCES TO STRUCTURE DATABANKS

PIR/MIPS code

Brookhaven code

7 LITERATURE REFERENCES

[1] Hayashi, S., Watanabe, M., Kimura, A.: J. Biochem.,95,223–232 (1984)

1 NOMENCLATURE

EC number
1.1.1.21

Systematic name
Alditol:NAD(P)$^+$ 1-oxidoreductase

Recommended name
Aldehyde reductase

Synonymes
Aldose reductase
Polyol dehydrogenase (NADP$^+$)
ALR2
Alditol:NADP oxidoreductase [2]
Alditol:NADP$^+$ 1-oxidoreductase [3]
NADPH-aldopentose reductase
NADPH-aldose reductase

CAS Reg. No.
9028-31-3

2 REACTION AND SPECIFICITY

Catalysed reaction
Alditol + NAD(P)$^+$ →
→ aldose + NAD(P)H

Reaction type
Redox reaction

Natural substrates
D-Galactose + NADPH [2]
More (enzyme of sorbitol (polyol) pathway [4], enzyme involved in metabolism of corticosteroids [5]) [4, 5]

Substrate spectrum
1 Aldose + NAD(P)H (wide specificity [2]) [1–12]
2 2-Deoxygalactose + NADPH [2]
3 2-Deoxyribose + NADPH [2]
4 Glycoaldehyde + NADPH [2]
5 Glyoxal + NADPH [2]
6 Methylglyoxal + NADPH (i.e. pyruvaldehyde) [2]
7 Phenylglyoxal + NAD(P)H [2, 5]

8 Benzaldehyde + NAD(P)H [2, 12]
9 4-Nitrobenzaldehyde + NAD(P)H [2, 4, 5, 9, 12]
10 Acetaldehyde + NAD(P)H [2, 12]
11 Butyraldehyde + NAD(P)H [2, 4, 12]
12 n-Pentanal + NAD(P)H [2, 12]
13 Hexanal + NADPH [2]
14 D-Xylose + NAD(P)H (low activity [12]) [4, 5, 8, 9, 11, 12]
15 D-Glucuronate + NAD(P)H [1, 4, 5, 9, 11]
16 11-Deoxyisocorticosterone + NAD(P)H [4]
17 20-Isocortisol + NAD(P)H [4, 5]
18 Indole-3-acetaldehyde + NAD(P)H [4, 5]
19 4-Hydroxyphenylacetaldehyde + NAD(P)H [4, 5]
20 4-Hydroxyphenylglycolaldehyde + NAD(P)H [4, 5]
21 Decanal + NAD(P)H [4]
22 20alpha-Isocorticosterone + NAD(P)H [4]
23 2,3-Bornanedione + NAD(P)H [4]
24 D-Erythrose + NADPH [8]
25 L-Arabinose + NADPH [2, 8]
26 D-Glucuronolactone + NADPH [8, 11]
27 Dihydroxyacetone + NAD(P)H [12]
28 Glucose + NAD(P)H (not [2], low activity [12]) [1, 4–6, 8, 9, 12]
29 Galactose + NAD(P)H [1, 2, 4, 5, 8]
30 Ribose + NAD(P)H (not [2]) [1, 8, 11]
31 Propionaldehyde + NAD(P)H [1, 2, 12]
32 Pyridinecarboxaldehyde + NAD(P)H (2-, 3- and 4-pyridinecarboxaldehyde) [1, 2]
33 L-Xylose + NAD(P)H [5]
34 DL-Glyceraldehyde + NAD(P)H [2, 5, 11]
35 Isobutyraldehyde + NADPH [11]
36 More (oxidation of alcohols by $NADP^+$ is negligible at 25°C, measurable at 37°C [5], oxidation of polyols to polyolaldehydes by $NADP^+$ proceeds very slowly [4], requirement for a free aldehyde group [2]) [2, 4, 5]

Product spectrum

1 Alditol + $NAD(P)^+$
2 ?
3 ?
4 Ethyleneglycol + $NADP^+$
5 ?
6 ?
7 ?
8 Benzyl alcohol + $NAD(P)^+$
9 4-Nitrobenzyl alcohol + $NAD(P)^+$
10 Ethanol + $NAD(P)^+$
11 Butanol + $NAD(P)^+$
12 n-Pentanol + $NAD(P)^+$

13 Hexanol + $NADP^+$
14 ?
15 ?
16 ?
17 ?
18 Indol-3-ethanol + $NAD(P)^+$
19 4-Hydroxyphenylethyl alcohol + $NAD(P)^+$
20 ?
21 Decanol + $NAD(P)^+$
22 ?
23 ?
24 Erythritol + $NADP^+$
25 ?
26 ?
27 Glycerol + $NAD(P)^+$
28 Sorbitol + $NAD(P)^+$ [6]
29 ?
30 ?
31 Propanol + $NAD(P)^+$
32 Pyridylmethanol + $NAD(P)^+$ (2-, 3-, and 4-isomer)
33 ?
34 Glycerol + $NAD(P)^+$
35 Isobutanol + $NADP^+$
36 ?

Inhibitor(s)

7-Hydroxy-4-oxo-4H-chromen-2-carboxylic acid [4]; 3,3-Tetramethylene glutaric acid [1, 4, 8]; Phenobarbital [1, 2]; $NADP^+$ (competitve to NADPH [2]) [2, 6, 7]; Glycerol (uncompetitive to DL-glyceraldehyde) [2]; Pyrazol (slight) [2]; Hydroxylamine [2]; 2,2'-Dipyridyl [2]; 1,10-Phenanthroline [2]; 5,5-Diphenylhydantoin [2, 6]; Iodoacetamide (slight) [2]; $LiSO_4$ (0.4 M: essential for full activity [1], 200 mM: inhibition) [2]; Cl^- [4]; Quercetin [4, 5, 9]; Quercitrin [4]; Rutin [4, 5]; Alrestatin (1,3-dioxo-1H-benz[de]isoquinoline-2-(3H)acetic acid) [1, 4, 9]; Iodoacetate [5, 12]; Ethosuccinimide [6]; 2'-AMP [7]; NaCl [8, 11]; 3-Hydroxybutyric acid [8]; Phenobarbitone [5, 6]; Barbital [6]; p-Hydroxymercuribenzoate [1, 2, 5, 12]; N-Ethylmaleimide [1, 2]; Sorbinil [1, 9]; $HgCl_2$ [9, 12]; p-Chloromercuribenzoate [9]; Zn^{2+} [9]; Cu^{2+} [9]; $(NH_4)_2SO_4$ (0.05 mM: slight inhibition, 400 mM: activation) [9]; Ammonium acetate [11]; Oxaloacetate [11]; 3-Hydroxybutyrate [8, 11]; Acetoacetate [11]; 2-Oxopentanoate [11]; 2-Oxoglutarate [11]; 2-Oxohexanoate [11]; 2-Oxoisohexanoate [11]; 2-Oxooctanoate [11]; 2-Oxoisopentanoate [11]; 2-Oxoheptanoate [11]; 2-Oxonanoate [11]; 2-Oxodecanoate [11]; Octanoate [11]; 5'-AMP [12]; Sulfhydryl inhibitors [12]; 2'-AMP [12]; 3'-AMP [12]; c-3',5'-AMP [12]; Dithiothreitol [12]; 2-Mercaptoethanol [12]; Glyceraldehyde (substrate inhibition) [7]

Cofactor(s)/prosthetic group(s)/activating agents

NADH (10% of the NADPH-dependent activity [4], 5% [5], 50% [12]) [1, 4, 5, 12]; NADPH (NADPH-specific, pro-4R hydrogen is transferred from the nicotinamide ring of the coenzyme to the substrate [4]) [2, 4–6, 8]; $NADP^+$ [4]

Metal compounds/salts

$(NH_4)_2SO_4$ (activation [4, 8], 0.05 mM: slight inhibition [9], 400 mM: activation [9]) [4, 8, 9]; SO_4^{2-} (stimulation [5, 7, 10], ammonium and lithium salt is somewhat more effective than sodium and potassium salt [10]) [5, 7, 10]; $LiSO_4$ (stimulation [8], 0.4 M: essential for full expression of enzyme activity [1], 200 mM: inhibition [2]) [1, 2, 8]

Turnover number (min^{-1})

33 (D-glyceraldehyde) [5]; 45.6 (phenylglyoxal) [5]; 38.4 (20alpha-isocorticosterone, 20alpha-isocortisol) [5]

Specific activity (U/mg)

10.2 [6]; 5.24 [2]; 0.475 [4]; 1.68 [8]; 0.488 [10]; 11.2 [12]; More [7, 11]

K_m-value (mM)

10.6 (2-deoxygalactose) [2]; 287.2 (L-arabinose) [2]; 17.5 (2-deoxyribose) [2]; 28.1 (D-erythrose) [2]; 3.4 (DL-glyceraldehyde) [2]; 1.8 (glycoaldehyde) [2]; 18.9 (glyoxal) [2]; 0.51 (glyceraldehyde) [1]; 1.18 (glucuronate) [1]; 0.039 (pyridine-3-aldehyde) [1]; 0.011 (propionaldehyde) [1]; 155 (glucose) [1]; 142 (galactose) [1]; 102 (ribose) [1]; 0.010–0.015 (NADPH) [1, 2, 12]; 0.62 (NADH) [1]; 227.1 (D-galactose) [2]; 0.034 (hexanal) [2]; 0.087 (n-pentanal) [2]; 0.444 (butyraldehyde) [2]; 0.68 (phenylglyoxal) [2]; 0.295 (benzaldehyde) [2]; 0.003 (4-nitrobenzaldehyde) [2]; 0.099 (2-pyridinecarboxaldehyde) [2]; 0.046 (3-pyridinecarboxaldehyde) [2]; 0.004 (4-pyridinecarboxaldehyde) [2]; 0.03 (DL-glyceraldehyde) [11]; 0.4 (D-erythrose) [11]; 5 (D-xylose) [11]; 7 (D-ribose) [11]; 0.7 (D-glucuronolactone) [11]; 4 (D-glucuronate) [11]; 0.02 (isobutyraldehyde) [11]; 0.22 (NADH) [12]; 0.035 (D-glyceraldehyde) [5]; 0.018 (L-glyceraldehyde) [5]; 0.025 (DL-glyceraldehyde) [5]; 5.1 (D-xylose) [5]; 100 (L-xylose) [5]; 95 (D-glucose) [5]; 40 (D-galactose) [5]; 5 (D-glucuronate) [5]; 0.001 (below, 20alpha-isocorticosterone, 20alpha-isocortisol) [5]; 0.003 (4-hydroxyphenylglycolaldehyde) [5]; 0.010 (4-hydroxyphenylacetaldehyde) [5]; 0.016 (indole-3-acetaldehyde) [5]; 0.002 (4-nitrobenzaldehyde + NADPH) [5]

pH-optimum

5.8–6.7 [8]; 6–7 (aldehyde reduction) [4]; 6.2 [1]; 6.5–7.0 [9]; 6.6 (DL-glyceraldehyde + NADPH) [6, 10]; 6.7 (glyceraldehyde + NAD(P)H) [12]; 9.7–10.4 (glycerol + NAD(P)H) [12]; 8 (phosphate buffer, Tris-HCl buffer, substrate: DL-glyceraldehyde 3-pyridine carboxaldehyde) [2]; 8.6–9.0 (glycine-NaOH buffer, substrate: DL-glyceraldehyde 3-pyridine carboxaldehyde) [2]; 10 (glycerol + $NADP^+$) [10]

pH-range
5.5–7.5 (at pH 5.5 and 7.5 about 50% of activity maximum) [6]

Temperature optimum (°C)

Temperature range (°C)

3 ENZYME STRUCTURE

Molecular weight
67000 (Euonymus japonica, gel filtration) [2]
61000 (calf, gel filtration) [12]
39000 (human, sucrose density gradient zone centrifugation, thin layer gel filtration) [8]
38000 (human, gel filtration) [5]
37000 (bovine, gel filtration) [7]
34300 (human, gel filtration) [1]
34000 (rabbit, gel filtration) [9]
29000 (calf, isozyme AR I, gel filtration) [10]
30000 (calf, isozyme AR II, gel filtration) [10]

Subunits
Monomer (1 × 32500, human, SDS-PAGE in presence of urea and 2-mercaptoethanol [1], 1 × 38000, human, SDS-PAGE [5], 1 × 37000, bovine, SDS-PAGE in presence of 2-mercaptoethanol [7], 1 × 40200, aldose reductase 1, rabbit, SDS-PAGE [9], 1 × 41500, aldose reductase 2, rabbit, SDS-PAGE [9], 1 × 30000, isozyme AR I, calf brain, SDS-PAGE, monomer is in equilibrium with a less effective dimer [10]) [1, 5, 7, 9, 10]
Dimer (2 × 30500, calf liver, SDS-PAGE [12], 2 × 36000, Euonymus japonica, SDS-PAGE [2]) [2, 12]

Glycoprotein/Lipoprotein
No glycoprotein [9]

4 ISOLATION/PREPARATION

Source organism
Human [1, 3–6, 8]; Euonymus japonica [2]; Celastrus orbiculatus [2]; Rabbit [9]; Rat [9]; Pig [9]; Bovine (calf [10–12], 2 enzymes AR I and AR II [10]) [7, 9–12]; Euonymus sp. (distribution in various species) [2]

Source tissue
Placenta (not [3]) [4, 8]; Kidney (not [3]) [4]; Testis [4]; Muscle [3]; Aorta [3]; Erythrocytes [1, 3]; Leaf [2]; Brain [3–6, 10]; Liver [12]; Lens [4, 7, 9, 11]; Lung [4]; Heart [4]; Pancreas [4]; Skeletal muscle [9]; More (no activity in human liver) [3, 4]

Localisation in source
Cytoplasm [4]

Purification
Human (partial [3, 6]) [1, 3–6, 8]; Rabbit [9]; Euonymus japonica [2]; Bovine (calf [10–12], 2 enzymes: AR I and AR II [10]) [7, 10–12]

Crystallization
–

Cloned
–

Renaturated
–

5 STABILITY

pH
5 (4°C, unstable below) [4]; 5.1 (90 min, 50% loss of activity) [5]; 6 (unstable below) [5]; 6–9 (4°C, several days, stable) [4]; 4.8–8.6 (2 h, stable) [10]

Temperature (°C)
50 (10 min, inactivation) [12]

Oxidation

Organic solvent

General stability information
Freezing of purified enzyme results in complete loss of activity [1]; Freezing and thawing at temperatures above 50°C, loss of activity [5]; Extremely labile to freezing and thawing unless Carbowax 100 is included, 10 mg/ml [7]; Dithiothreitol, 2 mM stabilizes, other sulfhydryl compounds such as beta-mercaptoethanol, glutathione and cysteine are less effective [12]; Freezing and thawing: more than twice, unstable [10]

Storage
4°C, several weeks [5]; –20°C, 4 weeks, little loss of activity [6]; 4°C, 3 weeks, 50% loss of activity [6]; –60°C, indefinitely [7]; 4°C, 2 weeks in 1% ampholines [8]; –45°C, stable at any stage of purification [12]; –55°C, unstable to prolonged storage [10]

6 CROSSREFERENCES TO STRUCTURE DATABANKS

PIR/MIPS code
PIR2:A35452 (bovine); PIR2:A48316 (bovine (fragment)); PIR3:S23931 (fungus (Sporidiobolus salmonicolor)); PIR2:A39763 (human); PIR2:A34406 (rabbit (fragment)); PIR2:A60603 (rat)

Brookhaven code

1ADS (Human (Homo sapiens) placenta recombinant form expressed);
1DLA (Pig (Sus scrofa) lens)

7 LITERATURE REFERENCES

[1] Das, B., Srivastava, S.K.: Arch. Biochem. Biophys.,238,670–679 (1985)
[2] Negm, F.B.: Plant Physiol.,80,972–977 (1986)
[3] Srivastava, S.K., Ansari, N.H., Hair, G.A., Das, B.: Biochim. Biophys. Acta,800,220–227 (1984)
[4] Wermuth, B., von Wartburg, J.-P.: Methods Enzymol.,89,181–186 (1982)
[5] Wermuth, B., Bürgisser, H., Bohren, K., von Wartburg, J.-P.: Eur. J. Biochem.,127, 279–284 (1982)
[6] O'Brien, M.M., Schofield, P.J.: Biochem. J.,187,21–30 (1980)
[7] Sheaff, C.M., Doughty, C.C.: J. Biol. Chem.,251,2696–2702 (1976)
[8] Clements, R.S., Winegrad, A.I.: Biochem. Biophys. Res. Commun.,47,1473–1479 (1972)
[9] Cromlish, J.A., Flynn, T.G.: J. Biol. Chem.,258,3416–3424 (1983)
[10] Dons, R.F., Doughty, C.C.: Biochim. Biophys. Acta,452,1–12 (1976)
[11] Hayman, S., Kinoshita, J.H.: J. Biol. Chem.,240,877–882 (1965)
[12] Attwood, M.A., Doughty, C.C.: Biochim. Biophys. Acta,370,358–368 (1974)

1 NOMENCLATURE

EC number
1.1.1.22

Systematic name
UDPglucose:NAD$^+$ 6-oxidoreductase

Recommended name
UDPglucose 6-dehydrogenase

Synonymes
UDPglucose dehydrogenase [1]
Uridine diphosphoglucose dehydrogenase [1]
UDPG dehydrogenase [4, 11]
UDPG:NAD oxidoreductase [4]
UDP-alpha-D-glucose:NAD oxidoreductase [5]
UDPglucose:NAD$^+$ oxidoreductase [6]
Uridine diphosphate glucose dehydrogenase [9]
Dehydrogenase, uridine diphosphoglucose
UDP-D-glucose dehydrogenase
Uridine diphosphate D-glucose dehydrogenase

CAS Reg. No.
9028-26-6

2 REACTION AND SPECIFICITY

Catalysed reaction
UDPglucose + 2 NAD$^+$ + H_2O →
→ UDPglucuronate + 2 NADH

Reaction type
Redox reaction

Natural substrates
UDPglucose + NAD$^+$ + H_2O (first step of a branched pathway leading to plant cell-wall polysaccharides which contain glucuronic and galacturonic acids and the pentoses xylose, arabinose and apiose) [5]

Substrate spectrum
1 UDPglucose + NAD$^+$ + H_2O (reversal of reaction cannot be demonstrated [11]) [1–11]
2 UDP-2-deoxyglucose + NAD$^+$ + H_2O [2]

3 CDPglucose + NAD^+ + H_2O (reaction rate is 5.5% of that with UDPglucose [4], 17% [5]) [4, 5]
4 dTDPglucose + NAD^+ + H_2O (reaction rate is 16.7% of that with UDPglucose [4]) [4]
5 TDPglucose + NAD^+ + H_2O (reaction rate is 17% of that with UDPglucose) [4]
6 5-Fluorouracil + NAD^+ [9]
7 6-Azauracil + NAD^+ [9]
8 More (no activity with GTPglucose [4], ADPglucose [4], guanosine diphosphomannose [11], uridine diphosphoacetylglucosamine [11], uridine diphosphoacetylgalactosamine [11], alpha-glucose-1-phosphate [11], glucose [11], ethyl alcohol [11], cosubstrates which can replace NAD^+: 3-acetylpyridine adenine dinucleotide [4, 8, 11], 3-pyridinealdehyde adenine dinucleotide [4], thionicotinamide adenine dinucleotide [4, 8], deamino adenine dinucleotide [4, 11], nicotinamide hypoxanthine [7], no reaction with: 3-pyridinealdehyde deamino adenosine dinucleotide [4], alpha-NAD^+ [7, 8], $NADP^+$ [5, 11], deamino-NAD^+ [8], 3-formylpyridine adenine dinucleotide [8], 3-propionylpyridine adenine dinucleotide [8], ethylnicotinate adenine dinucleotide [8]) [4, 5, 7, 8, 11]

Product spectrum

1 UDPglucuronate + NADH
2 UDP-2-deoxyglucuronate + NADH
3 CDPglucuronate + NADH
4 dTDPglucuronate + NADH
5 TDPglucuronate + NADH
6 ?
7 ?
8 ?

Inhibitor(s)

NH_2OH (deactivation) [7]; UDPxylose [4–8]; NADH [4, 7]; UDPglucuronate [4, 5, 7]; UDPgalacturonic acid [5]; UDP-D-galactose (slight) [5]; UDP-D-xylose [10]; AMP [8]; ADP [8]; UDParabinose [8]

Cofactor(s)/prosthetic group(s)/activating agents

NAD^+ (specific for) [5]; beta-NAD^+ [7]; 3-Acetylpyridine adenine dinucleotide (can replace NAD^+) [4, 8, 11]; 3-Pyridinealdehyde adenine dinucleotide (can replace NAD^+) [4]; Thionicotinamide adenine dinucleotide (can replace NAD^+) [4, 8]; Deamino adenine dinucleotide (can replace NAD^+) [4, 11]; Nicotinamide hypoxanthine (can replace NAD^+) [7]

Metal compounds/salts

No metal ion requirement [6]

Turnover number (min^{-1})

Specific activity (U/mg)

More (14.0, 1 unit is defined as the amount of enzyme required to produce 0.0002 mmol of $NADH_2$ per min at 30°C [8]) [6–8, 11]

K_m-value (mM)

0.3 (UDPglucose) [5]; 0.4 (NAD^+) [5]; 1.0 (UDPglucose [4, 6], dissociated enzyme [6]) [4, 6]; 0.05 (NAD^+) [4]; 0.0756 (UDPglucose, native enzyme) [6]; More (pH-dependence of K_m [9]) [7–9]

pH-optimum

7.3–7.8 [8]; 8.4–8.8 [5]; 8.7 [11]; 8.9 [7]; 9.0 [4]; 9.4 [6]

pH-range

7.5–9.4 (at pH 7.5 and 9.4 about 50% of activity maximum) [4]; 6.3–8.6 (at pH 6.3 and 8.6 about 50% of activity maximum) [8]; 7.8–9.4 (at pH 7.8 and 9.4 about 50% of activity maximum) [11]

Temperature optimum (°C)

30 (assay at) [4]

Temperature range (°C)

3 ENZYME STRUCTURE

Molecular weight

86000 (E. coli, gel filtration) [4]
300000 (rat, gel filtration) [6]
305000 (bovine, equilibrium measurement under native conditions) [1]

Subunits

Hexamer (6 × 52000, bovine, at pH 5.5–7.8, equilibrium measurement under native and denaturing conditions) [1]
Tetramer (4 × 70000, rat, gel filtration after treatment with SDS) [6]
Dimer (2 × 47000, E. coli, SDS-PAGE) [3]

Glycoprotein/Lipoprotein

–

4 ISOLATION/PREPARATION

Source organism

Bovine (calf [2, 9–11]) [1, 2, 9–11]; Pea [10]; E. coli [3, 4]; Rat [6, 7]; Lilium longiflorum [5]; Cryptococcus laurentii [8]; Chicken [10]

Source tissue

Pollen [5]; Cell [3, 4]; Liver [2, 5–7, 10, 11]; Cartilage [10]; Skin [7]

Localisation in source

Purification

E. coli [3, 4]; Lilium longiflorum (partial) [5]; Bovine (calf) [11]; Rat [6, 7]; Cryptococcus laurentii (partial) [8]

Crystallization

–

Cloned

–

Renaturated

–

5 STABILITY

pH

Temperature (°C)

50 (without substrate, complete inactivation) [5]

Oxidation

Organic solvent

General stability information

UDPglucose stabilizes [4]; UDPglucose protects against heat inactivation [5]; 2-Mercaptoethanol stabilizes [4]; NAD^+ protects against heat inactivation [5]

Storage

–70°C, 5 mM UDPglucose, 1 mM dithiothreitol [2]; –4°C, frozen in presence of UDPglucose, stable for 2 months [4]; 0°C, 24 h [5]; –12°C, 2 weeks [5]; –10°C, little loss of activity after several weeks [11]

6 CROSSREFERENCES TO STRUCTURE DATABANKS

PIR/MIPS code

Brookhaven code

7 LITERATURE REFERENCES

[1] Jaenicke, R., Rudolph, R.: Biochemistry,25,7283–7287 (1986)
[2] Druzhinina, T.N., Kusov, Y.Y., Shibaev, V.N., Kochetkov, N.K., Biely, P.: Biochim. Biophys. Acta,381,301–307 (1975)
[3] Schiller, J.G., Lamy, F., Frazier, R., Feingold, D.S.: Biochim. Biophys. Acta,453,418–425 (1976)
[4] Schiller, J.G., Bowser, A.M., Feingold, D.S.: Biochim. Biophys. Acta,293,1–10 (1973)
[5] Davies, M.D., Dickinson, D.B.: Arch. Biochem. Biophys.,152,53–61 (1972)
[6] Sivaswami, A., Kelkar, S.M., Nadkarni, G.B.: Biochim. Biophys. Acta,276,43–52 (1972)
[7] Molz, R.J., Danishefsky, I.: Biochim. Biophys. Acta,250,6–13 (1971)
[8] Ankel, H., Ankel, E., Feingold, D.S.: Biochemistry,5,1864–1869 (1966)
[9] Goldberg, N.D., Dahl, J.L., Parks, R.E.: J. Biol. Chem.,238,3109–3114 (1963)
[10] Neufeld, E.F., Hall, C.W.: Biochem. Biophys. Res. Commun.,19,456–461 (1965)
[11] Strominger, J.L., Maxwell, E.S., Axelrod, J., Kalckar, H.M.: J. Biol. Chem.,224,79–90 (1957)

1 NOMENCLATURE

EC number
1.1.1.23

Systematic name
L-Histidinol:NAD^+ oxidoreductase

Recommended name
Histidinol dehydrogenase

Synonymes
L-Histidinol dehydrogenase [1, 2]
Dehydrogenase, histidinol

CAS Reg. No.
9028-27-7

---

2 REACTION AND SPECIFICITY

Catalysed reaction
L-Histidinol + 2 NAD^+ →
→ L-histidine + 2 NADH (bi-uni uni-bi ping-pong mechanism [3], the Neurospora enzyme also catalyzes the reactions of EC 3.5.4.19 and 3.6.1.31)

Reaction type
Redox reaction

Natural substrates
L-Histidinol + NAD^+ (last step in pathway of histidine biosynthesis) [1, 3, 6]

Substrate spectrum
1 L-Histidinol + NAD^+ (histidinal appears to be an intermediate of the reaction, however it must be tightly bound to the enzyme [3]) [1–6]

Product spectrum
1 L-Histidine + NADH

Inhibitor(s)
NADH [3]; Histidine [3]; Ba^{2+} [1]; Ca^{2+} [1]; Cd^{2+} [1]; Cu^{2+} [1]; Zn^{2+} [1]

Cofactor(s)/prosthetic group(s)/activating agents
NAD^+ [1–6]

Metal compounds/salts
Mg^{2+} (slight activation) [1]; Mn^{2+} (activation) [1]

Turnover number (min^{-1})

Specific activity (U/mg)
0.0048 [1]; 11.12 [5]

K_m-value (mM)
0.009 (L-histidinol) [1]; 0.14 (NAD^+) [1]; 0.016 (histidinol) [3]; 0.3 (NAD^+) [3]

pH-optimum
10 (assay at) [1]

pH-range

Temperature optimum (°C)
25 (assay at) [5]

Temperature range (°C)

3 ENZYME STRUCTURE

Molecular weight
84000 (Salmonella typhimurium, equilibrium centrifugation) [6]

Subunits
? (x × 45823, peptide sequencing, Salmonella typhimurium) [2]
Dimer (2 × 39600, calculation from amino acid composition, equilibrium centrifugation in presence of guanidine-HCl and 2-mercaptoethanol, Salmonella typhimurium [4], 2 × 38000–43000, equilibrium centrifugation in guanidine-HCl of native or carboxymethylated enzyme, viscometry in 6 M guanidine-HCl with 0.143 M beta-mercaptoethanol, Salmonella typhimurium [6]) [4, 6]

Glycoprotein/Lipoprotein
–

4 ISOLATION/PREPARATION

Source organism
Triticum aestivum (wheat) [1]; Raphanus sativus (radish) [1]; Cucurbita pepo (squash) [1]; Salmonella typhimurium (LT-7 [4], LT-2 [5]) [2–6]

Source tissue
Germ (wheat) [1]; Root (radish) [1]; Fruit (squash) [1]

Localisation in source

Purification
Triticum aestivum (partial) [1]; Salmonella typhimurium LT-2 [5]

Crystallization
[5]

Cloned
–

Renaturated
–

5 STABILITY

pH

Temperature (°C)

Oxidation

Organic solvent

General stability information

Storage
4°C, crystalline preparations as suspension in 67% saturated ammonium sulfate solution [5]

6 CROSSREFERENCES TO STRUCTURE DATABANKS

PIR/MIPS code
PIR2:PE0006 (Azospirillum brasilense (fragment)); PIR1:DEECHT (Escherichia coli); PIR2:S26209 (Mycobacterium smegmatis); PIR1:DEEBHT (Salmonella typhimurium); PIR3:S24815 (Streptomyces coelicolor); PIR2:PQ0112 (Streptomyces coelicolor (fragment)); PIR3:S23939 (wild cabbage); PIR2:S13894 (wild cabbage (fragment)); PIR2:A39358 (precursor chloroplast cabbage)

Brookhaven code

7 LITERATURE REFERENCES

[1] Wong, Y.-S., Mazelis, M.: Phytochemistry,20,1831–1834 (1981)
[2] Kohno, T., Gray, W.R.: J. Mol. Biol.,147,451–464 (1981)
[3] Bürger, E., Görisch, H.: Eur. J. Biochem.,116,137–142 (1981)
[4] Loper, J.C.: J. Biol. Chem.,243,3264–3272 (1968)
[5] Yourno, J., Ino, I.: J. Biol. Chem.,243,3273–3276 (1968)
[6] Martin, R.G., Berberich, M.A., Ames, B.N., Davis, W.W., Goldberger, R.F., Yourno, J.D.: Methods Enzymol.,17B,3–44 (1971) (Review)

1 NOMENCLATURE

EC number
1.1.1.24

Systematic name
Quinate:NAD+ 5-oxidoreductase

Recommended name
Quinate 5-dehydrogenase

Synonymes
Quinic dehydrogenase [1]
Dehydrogenase, quinate
Quinate:NAD oxidoreductase
Quinate dehydrogenase

CAS Reg. No.
9028-28-8

2 REACTION AND SPECIFICITY

Catalysed reaction
Quinate + NAD^+ →
→ 5-dehydroquinate + NADH

Reaction type
Redox reaction

Natural substrates
Quinate + NAD^+ (first reaction in inducible quinic acid catabolic pathway) [1]

Substrate spectrum
1 Quinate + NAD^+ (r [3]) [1–5]
2 Shikimate + NAD^+ [1, 2]

Product spectrum
1 5-Dehydroquinate + NADH
2 ?

Inhibitor(s)
p-Chloromercuribenzoate [1, 3]; 3-Hydroxybenzoic acid [3]; Sulfhydryl inhibitors [3]; Borate [3]; Molybdate [3]; 5-Dehydroquinic acid [3]; $CuSO_4$ [3]

Cofactor(s)/prosthetic group(s)/activating agents
NAD+ (specific for) [1, 3]; NADH [3]; Dithiothreitol (enhances activity) [1]

Metal compounds/salts
No metal ion requirement [3]

Turnover number (min^{-1})

Specific activity (U/mg)
86.87 [1]; 2.15 [5]

K_m-value (mM)
0.42 (NAD+ (+ shikimate)) [2]; 0.26 (NAD+ (+ quinate)) [2]; 0.32–0.37 (quinate) [1, 2]; 1.18 (shikimate) [1]; 11.5 (shikimate) [2]; 6.25 (quinate) [2]; 0.031 (NAD+) [2]

pH-optimum
9.6 [3]; 9.8 [5]

pH-range
8.5–9.8 (at pH 8.5 and 9.8 about 50% of activity maximum) [5]

Temperature optimum (°C)
25 (assay at) [5]; 30 (assay at) [3]

Temperature range (°C)

3 ENZYME STRUCTURE

Molecular weight
23000 (Aspergillus niger, gel filtration) [2]
42000 (Neurospora crassa, sucrose density gradient centrifugation) [1]
More (transfer of carrot enzyme from dark to light conditions shifts MW from 42000 to 110000, probably due to association of a regulatory subunit which may be a calciprotein) [4]

Subunits
Monomer (1 × 41000, Neurospora crassa, SDS-PAGE in presence of urea or electrophoresis in presence of cetyltrimethyl ammonium bromide) [1]

Glycoprotein/Lipoprotein
–

4 ISOLATION/PREPARATION

Source organism

Neurospora crassa (bifunctional enzyme: quinate (shikimate) dehydrogenase) [1]; Aspergillus niger (bifunctional enzyme: quinate (shikimate) dehydrogenase, different from 1.1.1.25) [2]; Mung bean (Phaseolus aureus) [3]; Carrot [4]; Aerobacter aerogenes [5]

Source tissue

Mycelium [1]; Cell suspension culture [4]

Localisation in source

Purification

Neurospora crassa (bifunctional enzyme: quinate (shikimate) dehydrogenase) [1]; Mung bean (Phaseolus aureus) [3]; Carrot [4]; Aerobacter aerogenes [5]

Crystallization

–

Cloned

–

Renaturated

–

---

5 STABILITY

pH

7.5 (–20°C, highest stability) [1]

Temperature (°C)

40 (half-life of unprotected enzyme: 20 min [1], 180 min in presence of NaCl [1], completely stable in presence of quinate, shikimate or NADH) [1]

Oxidation

Organic solvent

General stability information

High ionic strength, such as 0.1 M $(NH_4)_2SO_4$, NaCl or phosphate buffer or the presence of 0.1 M quinate or shikimate protects against thermal inactivation [1]

Storage

–20°C, 0.1 M NaCl, half-life 40 days [1]; –20°C, pH 7.5, several weeks [3]; –15°C, stable for several weeks, 30% loss of activity in 6 months [5]

6 CROSSREFERENCES TO STRUCTURE DATABANKS

PIR/MIPS code

PIR2:S04253 (Neurospora crassa); PIR3:S08499 (Emericella nidulans)

Brookhaven code

7 LITERATURE REFERENCES

[1] Barea, J.L., Giles, N.H.: Biochim. Biophys. Acta,524,1–14 (1978)
[2] Cain, R.B.: Biochem. J.,127,15P (1972)
[3] Gamborg, O.L.: Biochim. Biophys. Acta,128,483–491 (1966)
[4] Graziana, A., Ranjeva, R., Salimath, B.P., Boudet, A. M.: FEBS Lett.,163,306–311 (1983)
[5] Davies, B.D., Gilvarg, C., Mitsuhyshi, S.: Methods Enzymol.,2,307–311 (1955)

1 NOMENCLATURE

EC number
1.1.1.25

Systematic name
Shikimate:$NADP^+$ 5-oxidoreductase

Recommended name
Shikimate 5-dehydrogenase

Synonymes
Dehydroshikimic reductase [8]
Shikimate oxidoreductase [6]
Shikimate:$NADP^+$ oxidoreductase [7]
5-Dehydroshikimate reductase [4]

CAS Reg. No.
9026-87-3

2 REACTION AND SPECIFICITY

Catalysed reaction
Shikimate + $NADP^+$ →
→ 5-dehydroshikimate + NADPH

Reaction type
Redox reaction

Natural substrates
5-Dehydroshikimate + NADPH (pathway of biosynthesis of aromatic amino acids [8], enzyme of shikimic acid biosynthesis) [8]

Substrate spectrum
1 Shikimate + $NADP^+$ (r [8]) [1–10]

Product spectrum
1 5-Dehydroshikimate + NADPH

Inhibitor(s)
$ZnCl_2$ [9]; $CdSO_4$ [9]; $CuSO_4$ [9]; $HgCl_2$ [9]; Protocatechuate (moderate [3]) [2, 3, 9]; 3,5-Dihydrobenzoate (moderate) [3]; p-Hydroxymercuribenzoate (moderate) [3]; Iodoacetate (slight [8]) [4, 8]; Arsenite [4]; p-Chloromercuribenzoate [2, 3, 4, 8, 9]; N-Ethylmaleimide [2]; NH_4^+ [2]; Hg^{2+} [2]; Zn^{2+} [2]; Cu^{2+} [2]; Borate [2]; Metal ions [9]

Cofactor(s)/prosthetic group(s)/activating agents
NADP$^+$ (specific for [2, 4, 10]) [1, 2, 4, 10]; NADPH [8]; More (no activity with NAD$^+$) [2, 8]

Metal compounds/salts
No requirement for divalent ions [2]; No metal ion requirement [8]

Turnover number (min^{-1})

Specific activity (U/mg)
1100 [1]; 167 [3]; More (4.71 nkat/mg) [2]

K_m-value (mM)
0.19–0.28 (shikimate) [8]; 0.038 (shikimate) [2]; 0.01 (NADP$^+$) [2]; 0.007 (NADP$^+$) [8]; 0.025 (NADP$^+$) [3]; 0.22 (shikimate, 2 K_m-values) [3]; 0.91 (shikimate, 2 K_m-values) [3]; 0.032 (NADP$^+$) [4]; 0.43 (shikimate) [4]; More [9, 10]

pH-optimum
7.7 (Tris-hydrochloric acid buffer) [4]; 8.5 [10]; 9.1 [2]; 10.0 [3, 8]; 10.1 (glycine-sodium hydroxide buffer) [4]

pH-range
6–11 (no activity below and above) [2]; 8.5–10.5 (8.5: about 50% of activity maximum, 10.5: about 90% of activity maximum) [8]

Temperature optimum (°C)
22 (assay at) [4]; 25 (assay at) [3]; 35 (assay at) [6]

Temperature range (°C)

---

3 ENZYME STRUCTURE

Molecular weight
32000 (E. coli, gel filtration, SDS-PAGE) [1]
50000 (Euterpe oleracea, gel filtration) [9]
57000 (Phaseolus mungo, disc gel electrophoresis) [3]
73000 (tomato, gel filtration) [2]

Subunits
Monomer (1 × 32000, E. coli, SDS-PAGE) [1]

Glycoprotein/Lipoprotein
–

4 ISOLATION/PREPARATION

Source organism

Phaseolus mungo [3]; Euterpe oleracea [9]; Tea plant [4]; Pea [3, 8]; E. coli [1, 10]; Tomato [2]; Bamboo [5]; Physcomitrella patens (enzyme complex of EC 4.2.1.10 and EC 1.1.1.25) [6]; Neurospora crassa (multienzyme complex) [7]

Source tissue

Fruit [2]; Shoots [4, 5]; Etiolated epicotyls [8]; Seedlings [3]

Localisation in source

Purification

Phaseolus mungo [3]; E. coli (partial) [1, 10]; Physcomitrella patens (enzyme complex of EC 4.2.1.10 and EC 1.1.1.25) [6]; Neurospora crassa (multienzyme complex) [7]; Pea [8]; Euterpe oleracea [9]

Crystallization

–

Cloned

–

Renaturated

–

5 STABILITY

pH

Temperature (°C)

50 (5 min, 85% loss of activity) [2]

Oxidation

Organic solvent

General stability information

Storage

–20°C, 50% glycerol, 1 mM benzamidine, 0.4 mM dithiothreitol [1]

6 CROSSREFERENCES TO STRUCTURE DATABANKS

PIR/MIPS code

PIR2:S00252 (Escherichia coli)

Brookhaven code

7 LITERATURE REFERENCES

[1] Chaudhuri, S., Coggins, J.R.: Biochem. J.,226,217–223 (1985)
[2] Lourenco, E.J., Neves, V.A.: Phytochemistry,23,497–499 (1984)
[3] Koshiba, T.: Biochim. Biophys. Acta,522,10–18 (1978)
[4] Sanderson, G.W.: Biochem. J.,98,248–252 (1966)
[5] Higuchi, T., Shimada, M.: Plant Cell Physiol.,8,61ff. (1967)
[6] Polley, L.D.: Biochim. Biophys. Acta,526,259–266 (1978)
[7] Jacobson, J.W., Hart, B.A., Doy, C.H., Giles, N.H.: Biochim. Biophys. Acta,289,1–12 (1972)
[8] Balinsky, D., Davies, D.D.: Biochem. J.,80,292–296 (1961)
[9] Lemos Silva, G.M., Lourenco, E.J., Neves, V.A.: J. Food Biochem.,9,105–116 (1985)
[10] Yaniv, H., Gilvarg, C.: J. Biol. Chem.,213,787–795 (1955)

1 NOMENCLATURE

EC number
1.1.1.26

Recommended name
Glycolate:NAD^+ oxidoreductase

Systematic name
Glyoxylate reductase

Synonymes
NADH-glyoxylate reductase [1]
Glyoxylic acid reductase [4]
Reductase, glyoxylate
NADH-dependent glyoxylate reductase

CAS Reg. No.
9028-32-4

2 REACTION AND SPECIFICITY

Catalysed reaction
Glycolate + NAD^+ →
→ glyoxylate + NADH

Reaction type
Redox reaction

Natural substrates

Substrate spectrum
1 Glyoxylate + NADH (equilibrium is very far in the direction of formation of glycolate) [1–8]
2 Hydroxypyruvate + NADH [1, 7]
3 More (no activity with: pyruvate, 2-oxobutyrate, acetaldehyde, oxaloacetate, 2-oxoglutarate) [7]

Product spectrum
1 Glycolate + NAD^+ [1]
2 D-Glycerate + NAD^+ [1]
3 ?

Inhibitor(s)
Oxo acids [1]; p-Chloromercuribenzoate [1, 4]; Semicarbazide [4]; Phenylhydrazine [4]

Cofactor(s)/prosthetic group(s)/activating agents
NADH (specific for) [1, 4, 5]; More (NADPH cannot replace NADH) [7]

Metal compounds/salts

Turnover number (min^{-1})

Specific activity (U/mg)
More [1, 4]

K_m-value (mM)
0.033 (NADH) [7]; 7 (glyoxylate) [1]; 5 (hydroxypyruvate) [1]; 5.6 (glyoxylate) [4]; 9.1 (glyoxylate) [5]

pH-optimum
6.0–6.8 [1]; 6.4 [4]; 6.3–6.6 [5]

pH-range
5.1–7.8 (5.1: about 60% of activity maximum, 7.8: about 50% of activity maximum) [5]

Temperature optimum (°C)
25 (assay at) [8]

Temperature range (°C)

---

3 ENZYME STRUCTURE

Molecular weight
180000 (Pseudomonas fluorescens, thin-layer gel filtration) [1]

Subunits

Glycoprotein/Lipoprotein
–

---

4 ISOLATION/PREPARATION

Source organism
Pseudomonas fluorescens [1]; Tobacco [4, 5]; Sinapis alba [6]; Euglena gracilis (non-photosynthetic mutant) [2]; Rat [3]; Spinach [7, 8]

Source tissue
Cell [1]; Liver [3]; Leaf [4, 5, 8]; Cotyledons [6]

Localisation in source

Purification
Pseudomonas fluorescens [1]; Tobacco [4, 5]; Spinach (partial) [8]

Crystallization
[5]

Cloned
–

Renaturated
–

5 STABILITY

pH

Temperature (°C)

Oxidation

Organic solvent

General stability information

Storage
4°C, crystalline suspension in ammonium sulfate solution, 3.2 mol/l, stable for several months [7]

6 CROSSREFERENCES TO STRUCTURE DATABANKS

PIR/MIPS code

Brookhaven code

7 LITERATURE REFERENCES

[1] Hullin, R.P.: Methods Enzymol.,41, Pt. B,343–348 (1975) (Review)
[2] Yokota, A., Kitaoka, S.: Agric. Biol. Chem.,45,15–22 (1981)
[3] Suzuki, S., Suga, T., Ninobe, S.: J. Biochem.,73,1033–1038 (1973)
[4] Zelitch, I.: Methods Enzymol.,1,528–535 (1955) (Review)
[5] Zelitch, I.: J. Biol. Chem.,216,553–575 (1955)
[6] Cerff, R.: Plant Physiol.,51,76–81 (1973)
[7] Bergmeyer, H.U., Graßl, M., Walter, H.E. in "Methods Enzym. Anal.",3rd Ed. (Bergmeyer, H.U., ed.) 2,126–328 (1974)
[8] Zelitch, I.: J. Biol. Chem.,201,719–726 (1953)

1 NOMENCLATURE

EC number

1.1.1.27

Systematic name

(S)-Lactate:NAD$^+$ oxidoreductase

Recommended name

L-Lactate dehydrogenase

Synonymes

Lactic acid dehydrogenase
L(+)-nLDH [1]
L-(+)-Lactate dehydrogenase [6]
Dehydrogenase, lactate
L-Lactic dehydrogenase
L-Lactic acid dehydrogenase
Lactate dehydrogenase
Lactate dehydrogenase NAD-dependent
Lactic dehydrogenase
NAD-lactate dehydrogenase
Proteins, specific or class, anoxic stress response, p34

CAS Reg. No.

9001-60-9

2 REACTION AND SPECIFICITY

Catalysed reaction

(S)-Lactate + NAD$^+$ →
→ pyruvate + NADH

Reaction type

Redox reaction

Natural substrates

Pyruvate + NADH [14]
More (enzyme is proposed as a component of the system regulating the cellular pH and/or controlling the concentration of reducing equivalents in the cytoplasm of leaf cells) [18]

Substrate spectrum

1 (S)-Lactate + NAD^+ (r [2], virtually unidirectional, catalyzing efficiently only reduction of pyruvate [19, 23]) [1–25]
2 2-Oxobutyrate + NADH [11]
3 2-Oxopentanoate + NADH [11]
4 2-Oxoglutarate + NADH [11]
5 4-Methyl-2-oxopentanoate + NADH [11]
6 Hydroxypyruvate + NADH [18]
7 Glyoxylate + NADH [18]
8 More (not D-(-)-lactate) [6]

Product spectrum

1 Pyruvate + NADH
2 2-Hydroxybutyrate + NAD^+
3 2-Hydroxypentanoate + NAD^+
4 2-Hydroxyglutarate + NAD^+
5 2-Hydroxy-4-methylpentanoate + NAD^+
6 2,3-Dihydroxypropanoate + NAD^+
7 Glycolate + NAD^+
8 ?

Inhibitor(s)

Oxamate (activation [14]) [1, 4, 6, 11]; Oxalate [1, 11, 12, 18]; 2-Oxobutyrate (inhibition at high pyruvate concentration) [1]; Oxaloacetate [1, 11]; Lactate analogs (slight) [1]; ATP (overview [1]) [1, 4, 6, 7, 12, 16, 18]; ADP (overview [1]) [1, 5, 6, 12, 16, 18]; NAD^+ (slight [1], product inhibition [12]) [1, 12]; Cd^{2+} (Lactobacillus casei, at high concentration [1]) [1, 11]; Hg^{2+} ($HgCl_2$ [5, 21]) [1, 5, 11, 21]; Cu^{2+} ($CuSO_4$ [5]) [1, 5, 11]; Iodoacetamide (slight: Lactobacillus casei, Lactobacillus acidophilus, not: Streptococcus cremoris, Bacillus subtilis [1], not: [4]) [1, 24]; p-Chloromercuribenzoate [1, 4, 11]; p-Hydroxymercuribenzoate (Pediococcus pentosaceus, not: Lactobacillus casei, Lactobacillus acidophilus [1]) [1, 5, 8, 24]; Pyruvate (above 5 mM [2]) [2, 16]; Dihydroxyacetone phosphate [5]; 6-Phosphogluconate [5]; AMP [6, 16, 18]; Dithionitrobenzoate [8]; NAD^+ (competitive with respect to NADH) [10, 11]; Phosphoenolpyruvate [10–12]; GTP [10, 12]; 3-Fluoropyruvate [11]; Ag^+ [11]; Zn^{2+} [11]; Ni^{2+} [11]; Cd^{2+} [11]; CTP (slight [16]) [10, 16]; GTP (slight) [16]; Fructose 1,6-diphosphate (activation [1, 2, 4–6, 11, 14, 20, 22, 23, 25]) [18]; Glucose 6-phosphate [18]; Citrate [18]; 3-Phosphoglycerate [18]; Isocitrate [18]; Glycine (slight) [18]; Serine (slight) [18]; Alanine (slight) [18]

Cofactor(s)/prosthetic group(s)/activating agents

NAD^+ [1–13]; NADH [1–13]; $NADP^+$ (slow reaction [1], no reaction [6]) [1]; Fructose 1,6-diphosphate (no catalytic activity in absence of [2], stimulation [4–6, 14, 20, 22, 23, 25], 0.15 mM required for 50% of maximal activity [6], fructose 1,6-diphosphate-activated lactate dehydrogenases is found in most streptococcal species, in a few lactobacilli and in Butyrivibrio fibrisolvens, Staphylococcus epidermidis, Actinomyces viscosus [1], these enzymes are virtually nonreversible and in most cases are absolutely dependent on fructose 1,6-diphosphate [11], kinetics of activation [22], necessary to stabilize tetrameric form of MW 130000 [23], inhibition [18]) [1, 2, 4–6, 11, 14, 20, 22, 23, 25]; Glucose 1,6-diphosphate (activation) [4, 14]; alpha-Glycerophosphate (activation) [12]; Fructose 2,6-diphosphate (activation) [14]; 5-Phosphoribosyl 1-phosphate (activation) [14]; Oxamate (activation [14], inhibition [1, 4, 6, 11]) [14]; FMN (non-covalently bound prosthetic group) [21]; More (no reaction with NADPH) [2, 6]

Metal compounds/salts

$MgCl_2$ (reduces inhibitory effect of nucleotides) [10, 16]; Phosphate (no absolute requirement for phosphate ions, activity higher than in buffers of the same pH and ionic strength [11], slight activation, inhibition of fructose 1,6-diphosphate-dependent reaction [14]) [11, 14]; Mn^{2+} (stimulation) [19, 20]; Co^{2+} (stimulation) [19, 20]; Cd^{2+} (stimulation, inhibition at high concentration) [19]; Ca^{2+} (stimulation) [19]; More (no cation requirement) [2]

Turnover number (min^{-1})

140000 (pyruvate + NADH) [8]

Specific activity (U/mg)

More (overview: values up to 2460 [1]) [1, 7, 25]; 1422 [2]; 419 [3]; 2291.7 [5]; 2220 [6]; 825 [11]; 480 [8]; 1.87 [8]; 1.87 [10]; 18.5 [12]; 0.714–1.87 [16]; 27.6 [21]; 2350 (Lactobacillus plantarum) [24]; 2460 (Lactobacillus acidophilus) [24]

K_m-value (mM)

More [1, 4, 5, 8, 10, 11, 14, 16–19, 21, 25]; 360 (lactate, Lactobacillus acidophilus) [1]; 1660 (lactate, Lactobacillus acidophilus) [1]; 30 (lactate, Bacillus subtilis) [1]; 0.074 (NADH, Lactobacillus acidophilus) [1]; 0.02 (NADH, Lactobacillus plantarum) [1]; 5.0 (NAD^+, Pediococcus pentosaceus) [1]; 0.9 (NAD^+, Bacillus subtilis) [1]; 1. 75 (pyruvate) [2]; 0.024 (NADH) [2]; 109 (lactate) [2]; 230 (NAD^+) [2]

pH-optimum

4.5 (Lactobacillus casei, Lactobacillus curvatus, reduction of pyruvate) [19]; 4.5–5.5 (Lactobacillus casei, Lactobacillus curvatus, Lactobacillus acidophilus, reduction of pyruvate) [1]; 4.5–6.0 (Pediococcus pentosaceus, reduction of pyruvate) [1]; 5.0–6.2 (Actinomyces viscosus) [1]; 5.0–8.0 (Streptococcus cremoris, reduction of pyruvate in presence of fructose 1,6-diphosphate) [1]; 5.4 (sodium acetate buffer, reduction of pyruvate, fructose 1,6-diphosphate activated) [5]; 5.5–7.0 (Streptococcus mutans) [1]; 5.6 (Staphylococcus epidermidis) [5]; 6.0 (Lactobacillus plantarum, Bacillus subtilis, reduction of pyruvate) [1]; 6.5 (Acholeplasma laidlawii, absence of fructose 1,6-diphosphate) [1]; 6.9 (presence of fructose 1,6-bisphosphate) [25]; 6.9–7.0 (Streptococcus lactis, reduction of pyruvate in presence of fructose 1,6-diphosphate) [1]; 7.0 (Acholeplasma laidlawii, reduction of pyruvate in presence of fructose 1,6-diphosphate [1], reduction of pyruvate [11, 17]) [1, 11, 17]; 7.3 (reduction of pyruvate [10]) [10, 21]; 7.5 (both directions) [8]; 7.8–8.8 (oxidation of L-lactate) [10]; 8.0 (Streptococcus cremonis, Streptococcus lactis, reduction of pyruvate, absence of fructose 1,6-diphosphate, weak activity [1], presence of fructose 1,6-diphosphate [7]) [1, 7]; 8.2 (absence of fructose 1,6-diphosphate) [25]; 9.0 (oxidation of lactate) [17]; 9.8 (oxidation of lactate) [11]

pH-range

4.3–8.0 (inactive below and above) [2]; 4.5–8.5 (no activity below and above) [6]

Temperature optimum (°C)

39 [6]; 42–45 [11]

Temperature range (°C)

---

3 ENZYME STRUCTURE

Molecular weight

65000–67000 (Alcaligenes eutrophus, gel filtration) [11]

126000–130000 (Staphylococcus epidermidis, gel filtration, sedimentation velocity experiments) [6]

132000–135000 (Lactobacillus casei, gel filtration, electrophoresis in neutral polyacrylamide gels) [5]

139000 (Fundulus heteroclitus, gel filtration) [3]

140000 (about, MW of most L(+)-lactate dehydrogenases [1], Streptococcus cremoris, ultracentrifugal studies [4]) [1, 4]

144000 (Streptococcus uberis, calculation from sedimentation data) [2]

145000 (Homarus americanus, analytical ultracentrifugation) [8]

150000 (Ipomoea batatas, polyacrylamide gel electrophoresis) [10, 16]

670000 (Acinetobacter calcoaceticus, gel filtration) [21]

Subunits

Dimer (2 × 37000, Alcaligenes eutrophus, SDS-PAGE) [11]
Tetramer (4 × 35900, Streptococcus uberis, sedimentation data of guanidinium chloride denatured enzyme [2], 4 × 33700, Fundulus heteroclitus, SDS-PAGE [3], 4 × 35000, Streptococcus cremoris [4], 4 × 34000, Lactobacillus casei, SDS-PAGE [5], 4 × 36000, Staphylococcus epidermidis, SDS-PAGE [6]) [2–6]
Oligomer (x × 40000, Acinetobacter calcoaceticus, SDS-PAGE) [21]

Glycoprotein/Lipoprotein

–

4 ISOLATION/PREPARATION

Source organism

Acholeplasma laidlawii [1]; Lactobacillus acidophilus [1, 24]; Lactobacillus jugurt-helveticus [1]; Lactobacillus salivarius [1]; Lactobacillus casei [1, 5, 19, 24]; Lactobacillus curvatus [19, 24]; Lactobacillus plantarum [1, 24]; Lactobacillus brevis [1]; Lactobacillus buchnerii [1]; Lactobacillus cellobiosus [1]; Lactobacillus confusus [1]; Lactobacillus fermentum [1]; Lactobacillus vermeforme [1]; Lactobacillus viridescens [1]; Leuconostoc mesenteroides [1]; Streptococcus agalactiae [1]; Streptococcus bovis [1]; Streptococcus mutans [1]; Streptococcus thermophilus [1]; Streptococcus uberis [2]; Streptococcus cremoris [4]; Streptococcus lactis [7, 22, 25]; Pediococcus acidilactici [1]; Pediococcus damnosus [1]; Pediococcus halophilus [1]; Pediococcus inopinatus [1]; Pediococcus pentosaceus [1]; Bifidobacterium bifidum [1]; Staphylococcus sp. [1]; Staphylococcus aureus [1]; Staphylococcus epidermidis [5, 6, 23]; Mycoplasma sp. [1]; Bacillus caldolyticus [1]; Bacillus subtilis [1]; Selenomonas ruminantium [1]; Fundulus heteroclitus (minnow) [3]; Homarus americanus [8]; Mouse [9]; Ipomoea batatas (sweet potato) [10, 16]; Alcaligenes eutrophus [11]; Potato [12]; Human [13]; Thermus caldophilus GK 24 [14]; Capsella bursa-pastoris [17]; Latuca sativa [18]; Streptococcus faecalis [20]; Acinetobacter calcoaceticus [21]; Pig [15]

Source tissue

Cell [1, 3, 5, 6]; Muscle (commercial preparation [15]) [8, 15]; Testis [9]; Root [10, 16]; Tuber [12]; Liver [13]; Leaves [17, 18]; Heart (commercial preparation) [15]

Localisation in source

Soluble [11]; Cytosol [17]; Membrane-bound [20]

Purification

Streptococcus uberis [1]; Fundulus heteroclitus (3 allozymes: LDH-Ba/4, LDH-Ba/Bb, LDH-Bb/4) [3]; Lactobacillus casei [5, 19]; Staphylococcus epidermidis [6]; Homarus americanus [8]; Mouse (4 isoenzymes) [9]; Ipomoea batatas (2 isoenzymes) [10]; Alcaligenes eutrophus [11]; Potato [12]; Human (isoenzyme-5) [13]; Streptococcus lactis [7, 25]; Lactobacillus plantarum [24]; Lactobacillus acidophilus [24]; Latuca sativa [18]; Lactobacillus curvatus [19]; Acinetobacter calcoaceticus [21]

Crystallization

–

Cloned

–

Renaturated

[15]

---

5 STABILITY

pH

5.5 (more stable than at pH 7.0) [2]; 6.5 (highest thermal stability at) [6]; 5.0–7.5 (20°C, unstable below and above) [7]

Temperature (°C)

50 (5 min stable, Lactobacillus acidophilus [1], 3 min stable, Lactobacillus plantarum [1], 4 h stable, Bacillus subtilis [1]) [1]; 55 (3 min, inactivation, Pediococcus pentosaceus) [1]; 57.5 (20 min, 50% loss of activity, LDH-Ba/4) [3]; 60 (3 min, inactivation, Lactobacillus plantarum [1], 5 min, pH 5.5, no loss of activity, Lactobacillus curvatus [19]) [1, 19]; 61 (20 min, 50% loss of activity, LDH-Ba/Bb) [3]; 62.5 (20 min, 50% loss of activity, LDH-Bb/4) [3]; 70 (complete inactivation) [5]; 80 (5 min, 50% loss of activity (strain 65K), 20% loss of activity (strain A18)) [1]; More (effect of divalent metal ions on the heat stability) [19]

Oxidation

Organic solvent

General stability information

Mn^{2+} stabilizes against heat inactivation [1]; Fructose 1,6-diphosphate stabilizes against heat inactivation [1]; Fructose 1,6-diphosphate is necessary to stabilize tetrameric form of MW 130000 [23]; Phosphate stabilizes against heat inactivation [1, 4, 7]; ATP stabilizes against heat inactivation [1]; NAD^+ stabilizes against heat inactivation [3]; Glucose 6-phosphate stabilizes against heat inactivation [4]; Urea, 0.5 M, 10 min, inactivation [3]; Proteolytic susceptibility of alloenzymes [3]; Sensitive to dilution [6]; Dithiothreitol is necessary to retain enzymatic activity during purification [10]; Sucrose is necessary to retain enzymatic activity during purification [10]; Dialysis: enzyme of Lactobacillus casei is very stable during 70 h dialysis in acetate buffer, pH 5.5, in imidazole buffer pH 6.5, in phosphate buffer pH 6.5 if dithioerythritol, and in Tris-HCl buffer pH 7.5 if Mn^{2+} and fructose 1,6-diphosphate are added, the enzyme of Lactobacillus curvatus requires the addition of Mn^{2+} and fructose 1,6-diphosphate even at pH 5.5 and is very unstable at pH 7.5 altough Mn^{2+} and fructose 1,6-diphosphate are added [19]; Dilution: enzyme of Lactobacillus casei is stable in the range of 0.01–0.2 mg protein per ml, enzyme from Lactobacillus curvatus loses activity irreversibly after several h if the concentration is below 0.1 mg protein per ml [19]

Storage

–5°C [2]; –20°C, 50 mM sodium phosphate buffer, pH 7.0, several months [4]; –20°C, several months [10]; 2-Mercaptoethanol stabilizes during storage at 4°C [11]; KCl, 0.2–1 M, stabilizes during storage at 4°C [11]; Ammonium sulfate, 3.2 M, stabilizes during storage at 4°C [11]

6 CROSSREFERENCES TO STRUCTURE DATABANKS

PIR/MIPS code

PIR2:JQ2222 (chain M bovine); PIR2:A37334 (Acinetobacter calcoaceticus (fragment)); PIR2:B29704 (Bacillus caldolyticus); PIR2:S00019 (Bacillus caldotenax); PIR2:C29704 (Bacillus caldotenax); PIR1:DEBSLM (Bacillus megaterium); PIR1:DEBSLF (Bacillus stearothermophilus); PIR2:A26053 (Bacillus stearothermophilus); PIR2:A29704 (Bacillus stearothermophilus); PIR2:A25805 (Bacillus subtilis); PIR2:JQ0183 (Bifidobacterium longum); PIR1:DELBLA (Lactobacillus casei); PIR2:B40885 (Lactobacillus plantarum); PIR2:JN0449 (Lactococcus lactis); PIR2:A20629 (Lactococcus lactis subsp. cremoris (fragment)); PIR3:S33362 (Mycoplasma hyopneumoniae (SGC3)); PIR2:A38231 (sea lamprey); PIR3:S36863 (Thermotoga maritima); PIR3:S36864 (Thermotoga maritima); PIR2:JX0090 (Thermus aquaticus); PIR2:A24999 (Thermus aquaticus); PIR2:A22394 (Thermus aquaticus (fragment)); PIR2:A36070 (A barley); PIR2:JQ0471 (A12 bovine); PIR2:JQ0469 (A5 bovine (fragment)); PIR2:JQ0470 (A9 bovine); PIR2:B36070 (B barley); PIR2:S02795 (B human); PIR2:S09954 (B mouse); PIR3:A47180 (C mummichog); PIR3:S12151 (chain A chicken); PIR1:DEHULC (chain C human); PIR2:A27879 (chain C mouse); PIR2:A40488 (chain H (epsilon crystallin) duck); PIR1:DECHLH (chain H chicken); PIR2:A32430 (chain H mummichog); PIR1:DEPGLH (chain H pig); PIR2:B32957 (chain H rabbit (fragment)); PIR2:S22492 (chain Ldh1 maize); PIR1:DECHLM (chain M chicken); PIR1:DEHULM (chain M human); PIR1:DEMSLM (chain M mouse); PIR2:S06290 (chain M mouse); PIR1:DEPGLM (chain M pig); PIR2:A32957 (chain M rabbit); PIR2:A23083 (chain M rat); PIR2:A25142 (chain M rat (fragments)); PIR1:DEDFLM (chain M spiny dogfish); PIR1:DEMSLC (chain X mouse); PIR2:A26824 (chain X mouse); PIR2:A27246 (chain X mouse); PIR2:A24347 (chain X mouse (fragment)); PIR2:B27246 (chain X rat); PIR2:S08182 (P Bacillus psychrosaccharolyticus); PIR2:S08183 (X Bacillus psychrosaccharolyticus); PIR3:A45246 (fructose-1,6 bisphosphate-activated Lactococcus lactis subsp. lactis); PIR2:A43598 (fructose-1,6-diphosphate dependent Streptococcus mutans)

Brookhaven code

1LDB (Bacillus stearothermophilus); 6LDH (Dogfish (Squalus acanthias) muscle); 2LDX (Mouse (Mus musculus) testicles, swiss-*Webster strain); 5LDH (Pig (Sus scrofa) heart); 9LDB (Porcine (Sus scrofa) muscle)

7 LITERATURE REFERENCES

[1] Garvie, E.I.: Microbiol. Rev.,44,106–139 (1980) (Review)
[2] Williams, R.A., Andrews, P.: Biochem. J.,236,721–727 (1986)
[3] Place, A.R., Powers, D.A.: J. Biol. Chem.,259,1299–1308 (1984)
[4] Hillier, A.J., Jago, G.R.: Methods Enzymol.,89,362–367 (1982)
[5] Gordon, G.L., Doelle, H.W.: Eur. J. Biochem.,67,543–555 (1976)
[6] Götz, F., Schleifer, K.H.: Arch. Microbiol.,105,303–312 (1975)
[7] Mou, L., Mulvena, D.P., Jonas, H.A., Jago, G.R.: J. Bacteriol.,111,392–396 (1972)
[8] Eichner, R.D.: Methods Enzymol.,89,359–362 (1982)
[9] Lee, C.-Y., Yuan, J.H., Goldberg, E.: Methods Enzymol.,89,351–358 (1982)
[10] Oba, K., Uritani, I.: Methods Enzymol.,89,345–349 (1982)
[11] Steinbüchel, A., Schlegel, H.G.: Eur. J. Biochem.,130,321–328 (1983)
[12] Davies, D.D., Davies, S.: Biochem. J.,129,831–839 (1972)
[13] Pettit, S.M., Nealon, D.A., Henderson, A.R.: Clin. Chem.,27,88–93 (1981)
[14] Taguchi, H., Machida, M., Matsuzawa, H., Ohta, T.: Agric. Biol. Chem.,49,359–365 (1985)
[15] Jaenicke, R., Rudolph, R.: FEBS Symp.,49 (Pyridine Nucleotide-Dependent Dehydrogenases) ,351–367 (1977)
[16] Oba, K., Murakami, S., Uritani, I.: J. Biochem.,81,1193–1201 (1977)
[17] Betsche, T., Bosach, K., Gerhardt, B.: Planta,146,567–574 (1979)
[18] Betsche, T.: Biochem. J.,195,615–622 (1981)
[19] Hensel, R., Mayr, U., Stetter, K.O., Kandler, O.: Arch. Microbiol.,112,81–93 (1977)
[20] Hardman, M.J., Pritchard, G.G.: Biochim. Biophys. Acta,912,185–190 (1987)
[21] Allison, N., Fewson, C.A.: FEMS Microbiol. Lett.,36,183–186 (1986)
[22] Hardman, M.J., Crow, V.L., Criuckshank, D.S., Pritchard, G.G.: Eur. J. Biochem.,146,179–183 (1985)
[23] Götz, F., Schleifer, K.H.: Eur. J. Biochem.,90,555–561 (1978)
[24] Hensel, R., Mayr, U., Fujiki, H., Kandler, O.: Eur. J. Biochem.,80,83–92 (1977)
[25] Crow, V.L., Pritchard, G.G.: J. Bacteriol.,131,82–91 (1977)

1 NOMENCLATURE

EC number
1.1.1.28

Systematic name
(R)-Lactate:NAD$^+$ oxidoreductase

Recommended name
D-Lactate dehydrogenase

Synonymes
Lactic acid dehydrogenase [1]
D-Specific lactic dehydrogenase [1]
Dehydrogenase, D-lactate
D-(-)-Lactate dehydrogenase (NAD)
D-Lactic acid dehydrogenase
D-Lactic dehydrogenase

CAS Reg. No.
9028-36-8

2 REACTION AND SPECIFICITY

Catalysed reaction
(R)-Lactate + NAD$^+$ →
→ pyruvate + NADH (binary-binary mechanism with either NAD$^+$ or NADH as the first substrate [2])

Reaction type
Redox reaction

Natural substrates

Substrate spectrum
1 (R)-Lactate + NAD$^+$ (r [2], stereospecific for D-isomer [9], stereospecific transfer of 4alpha-hydrogen of NADH to pyruvate [9]) [1–12]

Product spectrum
1 Pyruvate + NADH (r [2])

Inhibitor(s)

Pyruvate (above 2 mM, substrate inhibition [9], some salt solutions influence the susceptibility to inhibition by pyruvate [3], Haliotis cracherodii: no substrate inhibition [7], Helix aspera: substrate inhibition [7]) [3, 6, 7, 9]; $HgCl_2$ [8]; p-Chloromercuriphenylsulfonic acid [8, 12]; GTP [2]; ATP [2, 8, 11, 12]; ITP [2]; p-Hydroxymercuribenzoate [9, 10, 12]; Iodoacetamide (in some experiments inhibitory in others not [12]) [9, 12]; Iodoacetate (in some experiments inhibitory in others not [12]) [9, 12]; NAD^+ (product inhibition) [11, 12]; Oxamate (in some experiments inhibitory in others not [12]) [8, 12]; Oxalate [12]; ADP (in some experiments inhibitory in others not) [12]; Hg^{2+} [12]; Cu^{2+} [12]; Sodium arsenite [12]; $MgCl_2$ [12]; p-Chloromercuribenzoate [12]

Cofactor(s)/prosthetic group(s)/activating agents

NAD^+ [1–12]; NADH [1–12]; Fructose 1,6-diphosphate (streptococcal enzyme practically inactive in absence of [2], Streptococcus bovis: fructose 1,6-bisphosphate required for activity [4]) [2, 4]; alpha-Ketoglutarate (stimulation) [8]

Metal compounds/salts

Turnover number (min^{-1})

38000 (NADH) [9]

Specific activity (U/mg)

36.54 [8]; More [2, 7, 9, 10]

K_m-value (mM)

0.045 (NADH) [8]; 0.070 (pyruvate) [9]; 0.021 (NADH) [9]; 3.2–3.8 (D-lactate) [6, 9]; 0.00044 (NAD^+) [9]; 1250 (lactate, Lactobacillus acidophilus) [12]; 70 (lactate, Lactobacillus leichmannii, Lactobacillus jensenii) [12]; 20 (lactate, Leuconostoc lactis, Lactobacillus fermentum) [12]; 0.16 (NADH, Lactobacillus acidophilus) [12]; 0.071 (NADH, Lactobacillus plantarum) [12]; 0.001 (NADH, Lactobacillus casei) [12]; 2.9 (NAD^+, Leuconostoc lactis) [12]; 2.2 (NAD^+, Leuconostoc mesenteroides) [12]; 1.05 (NAD^+, Pediococcus pentosaceus) [12]; 0.43 (pyruvate) [6]; 0.02 (NAD^+) [6]; 0.015 (NADH) [6]; More [7, 12]

pH-optimum

3.6 (or 8.0, Pediococcus pentosaceus, changes with pyruvate concentration) [12]; 5.8–6.5 [8]; 6.0 (pyruvate reduction) [7]; 6.4 (pyruvate reduction) [6]; 6.4–7.5 (E. coli) [12]; 6.6–8.0 (Leuconostoc lactis) [12]; 7.0 (mammalian muscle) [12]; 7.3 (Lactobacillus acidophilus) [12]; 7.6 (Lactobacillus leichmannii) [12]; 7.8 (Lactobacillus jensenii) [12]; 8.0 (3.6 or 8.0, Pediococcus pentosaceus, changes with pyruvate concentration) [12]; 8.2–9.7 (broad, D-lactate oxidation) [7]; 8.6 (Lactobacillus fermentum) [12]; 9.0 (lactate oxidation) [6]

pH-range

Temperature optimum (°C)
25 (assay at) [3, 7]

Temperature range (°C)

3 ENZYME STRUCTURE

Molecular weight
68000–91000 (Helix aspera, meniscus depletion sedimentation equilibrium) [7]
68000–70000 (Cardium edule, gel filtration, analytical ultracentrifugation) [1]
70000–80000 (Limulus polyphemus, gel filtration, equilibrium sedimentation) [9, 10]
79000–88000 (Haliotis cracherodii, meniscus depletion sedimentation equilibrium) [7]
98000–115000 (Phytium debaryanum, sucrose density gradient sedimentation, gel filtration) [2]
140000 (Helix aspera, gel filtration) [6]
145000 (Butyribacterium rettgeri, gel filtration) [8]

Subunits
Dimer (2 × 35000, Cardium edule [1], 2 × 35000, Limulus polyphemus [10], equilibrium centrifugation of acid dissociated enzyme [1, 10], 2 × 35000–40000, Helix aspera [7], Haliotis cracherodii [7]) [1, 7, 10]
Tetramer (4 × 34000, Helix aspera, SDS-PAGE) [6]

Glycoprotein/Lipoprotein
–

4 ISOLATION/PREPARATION

Source organism
Helix aspera (gastropod mollusc) [6, 7]; Cardium edule (bivalve mollusc) [1]; Aerobacter aerogenes [5]; Butyribacterium rettgeri [8, 12]; Polyspondylium pallidum [3]; Streptococcus sp. [4]; Phytium debaryanum [2]; Phytium undulatum [2]; Sapromyces elongatus [2]; Limulus polyphemus (horseshoe crab) [9, 10]; Nereis virens (seaworm) [10]; Allomyces sp. [11]; Lactobacillus acidophilus [12]; Lactobacillus bulgaricus [12]; Lactobacillus delbrueckii [12]; Lactobacillus jensenii [12]; Lactobacillus jugurt-helveticus [12]; Lactobacillus lactis [12]; Lactobacillus leichmannii [12]; Lactobacillus casei [12]; Lactobacillus plantarum [12]; Lactobacillus brevis [12]; Lactobacillus buchneri [12]; Lactobacillus cellobiosus [12]; Lactobacillus confusus [12]; Lactobacillus fermentum [12]; Lactobacillus vermeforme [12]; Lactobacillus viri-

descens [12]; Leuconostoc cremoris [12]; Leuconostoc dextranicum [12]; Leuconostoc lactis [12]; Leuconostoc mesenteroides [12]; Leuconostoc paramesenteroides [12]; Leuconostoc venosus [12]; Streptococcus lactis [12]; Pediococcus acidilactici [12]; Pediococcus damnosus [12]; Pediococcus inopinatus [12]; Pediococcus pentosaceus [12]; Pediococcus aerogenes [12]; Pediococcus agalactiae [4]; Pediococcus thermophilus [4]; Pediococcus faecalis [4]; E. coli [5, 12]; Staphylococcus sp. [12]; Staphylococcus aureus [12]; Mycoplasma sp. [12]; Staphylococcus rettgeri [12]; Selenomonas ruminantium [12]; Haliotis cracherodii [7]; Staphylococcus haemolyticus [12]; Staphylococcus hominis [12]; Staphylococcus warneri [12]

Source tissue

Gills [1]; Foot plus adductor muscle [1]; Mycelium [2]; Cells [8]; Skeletal muscle [9]; Vegetative cells [3]; Foot [6]; Heart [6]; Kidney [6]; Brain [6]; Hepatopancreas [6]

Localisation in source

Purification

Phytium debaryanum [2]; Phytium undulatum [2]; Sapromyces elongatus [2]; Butyribacterium rettgeri [8]; Limulus polyphemus [9, 10]; Helix aspera [7]; Nereis virens [10]; Allomyces sp. (partial) [11]; Haliotis cracherodii [7]

Crystallization

–

Cloned

–

Renaturated

–

5 STABILITY

pH

6.5 (unstable above) [8]; 5.8–6.7 (highest stability) [9]

Temperature (°C)

35 (stable below, Tris-chloride, pH 8.5, Leuconostoc mesenteroides) [12]; 40 (stable below, phosphate buffer, pH 7.5, Leuconostoc mesenteroides) [12]; 45 (5 min, 98% inactivation, Leuconostoc lactis, Lactobacillus acidophilus, 3 min, inactivation, Lactobacillus plantarum) [12]; 50 (3 min, 5% loss of activity [9], 5 min, 25% loss of activity, Lactobacillus jensenii [12], 5 min, 50% loss of activity, Lactobacillus leichmannii [12], 10 min, inactivation, Staphylococcus haemolyticus, Staphylococcus hominis, Staphylococcus warneri [12], 3 min, inactivation, Lactobacillus plantarum [12], 5 min, 98% inactivation, Lactobacillus fermentum [12]) [9, 12]; 55 (3 min, inactivation, Pediococcus pentosaceus) [12]; 60 (3 min, 90% loss of activity) [9]; 80 (5 min, 100% loss of activity, Lactobacillus jensenii, Lactobacillus leichmannii) [12]

Oxidation

Reducing agents stabilize [11]

Organic solvent

General stability information

Loss of activity after repeated freezing and thawing [8]; Low ionic strength causes loss of activity [8]; EDTA stabilizes [9]; 2-Mercaptoethanol stabilizes [9]; Glycerol stabilizes [11]; Reducing agents stabilize [11]

Storage

4°C, several weeks [2]; –20°C, for at least 2 years [2]; –20°C, 50% glycerol, 2 months [11]

6 CROSSREFERENCES TO STRUCTURE DATABANKS

PIR/MIPS code

PIR2:C37334 (Acinetobacter calcoaceticus (fragment)); PIR2:A60843 (Atlantic horseshoe crab (fragments)); PIR1:DEECDL (Escherichia coli); PIR2:A38094 (Lactobacillus delbrueckii subsp. bulgaricus); PIR2:JN0245 (Lactobacillus delbrueckii subsp. bulgaricus); PIR3:S17556 (Lactobacillus delbrueckii subsp. bulgaricus); PIR3:S24969 (Lactobacillus helveticus); PIR3:S29296 (Lactobacillus helveticus); PIR2:A40885 (Lactobacillus plantarum)

Brookhaven code

7 LITERATURE REFERENCES

[1] Jaenicke, R., Müller, K., Gäde, G.: Naturwissenschaften,68,205–206 (1981)
[2] LeJohn, H.B., Stevenson, R.M.: Methods Enzymol.,41, Pt. B,293–298 (1975) (Review)
[3] Garland, R.C., Kaplan, N.O.: Biochem. Biophys. Res. Commun.,26,679–685 (1967)
[4] Wolin, M.J.: Science,146,775–776 (1964)
[5] Pascal, M.-C., Pichinoty, F.: Ann. Inst. Pasteur,107,55–62 (1964)
[6] Storey, K.B.: Comp. Biochem. Physiol.,56B,181–187 (1977)
[7] Long, G.L., Ellington, W.R., Duda, T.F.: J. Exp. Zool.,207,237–248 (1979)
[8] Wittenberger, C.L.: Methods Enzymol.,41B,299–303 (1975)
[9] Long, G.L.: Methods Enzymol.,41, Pt. B,313–323 (1975) (Review)
[10] Long, G.L., Kaplan, N.O.: Arch. Biochem. Biophys.,154,696–710 (1973)
[11] Purohit, K., Turian, G.: Arch. Mikrobiol.,84,287–300 (1972)
[12] Garvie, E.L.: Microbiol. Rev.,44,106–139 (1980) (Review)

1 NOMENCLATURE

EC number
1.1.1.29

Systematic name
(R)-Glycerate:NAD^+ oxidoreductase

Recommended name
Glycerate dehydrogenase

Synonymes
D-Glycerate dehydrogenase
Hydroxypyruvate reductase
Dehydrogenase, glycerate
More (see also EC 1.1.1.81 for enzymes catalyzing the same reaction. Only those that are specific for NAD^+ are included in EC 1.1.1.29, those that can use either NAD^+ or $NADP^+$ and those without information of coenzyme specificity are summarized under EC 1.1.1.81)

CAS Reg. No.
9028-37-9

2 REACTION AND SPECIFICITY

Catalysed reaction
(R)-Glycerate + NAD^+ →
→ hydroxypyruvate + NADH (bi bi-mechanism [2])

Reaction type
Redox reaction

Natural substrates

Substrate spectrum
1 Hydroxypyruvate + NADH (ir [2], r [3]) [1–3]
2 Glyoxylate + NADH (reduced at 3.5% the rate of hydroxypyruvate) [2]

Product spectrum
1 D-Glycerate + NAD^+ [3]
2 Glycolate + NAD^+

Inhibitor(s)
Dihydroxyfumarate [2]; $HgCl_2$ [2]; $AgNO_3$ [2]; p-Chloromercuribenzoate [2]; Phenylhydrazine [2]; $CuSO_4$ (slight) [2]; $ZnCl_2$ (slight) [2]; More (not inhibitory: citrate, $NADP^+$, EDTA, 1,10-phenanthroline, iodoacetate, iodoacetamide, N-ethylmaleimide, hydroxylamine, semicarbazide, NaN_3, NaF, KCN, L-ascorbate, dithiothreitol, pyridoxal 5-phosphate) [2]

Cofactor(s)/prosthetic group(s)/activating agents
NADH (specific for, not replaceable by NADPH [2, 3]) [1–3]; NAD^+ [3]

Metal compounds/salts

Turnover number (min^{-1})

Specific activity (U/mg)
1086 [2]

K_m-value (mM)
0.055 (NADH) [2]; 0.09 (hydroxypyruvate) [1]; 0.175 (hydroxypyruvate) [2]; 10.8 (glyoxylate) [2]

pH-optimum
6.8 [2]

pH-range

Temperature optimum (°C)
45 [2]

Temperature range (°C)

---

3 ENZYME STRUCTURE

Molecular weight
69000–70000 (Hyphomicrobium methylovorum, gel permeation HPLC, gel filtration) [2]

Subunits
Dimer (2 × 38000, Hyphomicrobium methylovorum, SDS-PAGE) [2]

Glycoprotein/Lipoprotein
–

---

4 ISOLATION/PREPARATION

Source organism
Zea mays [1]; Hyphomicrobium methylovorum [2]; Pisum sativum [3]; Beta vulgaris [3]; Lycopersicon esculentum [3]; Raphanus sativus [3]; Spinacia oleracea [3]; Parsley [3]; Lettuce [3]; Kohlrabi [3]; Carrot [3]

Source tissue
Leaves [1, 3]

Localisation in source

Purification
Hyphomicrobium methylovorum [2]

Crystallization
[2]

Cloned
–

Renaturated
–

5 STABILITY

pH
5.0–9.0 [2]

Temperature (°C)
25 (up to) [2]; 30 (30 min, 13.3% loss of activity) [2]; 35 (30 min, 30.8% loss of activity) [2]; 40 (30 min, 100% loss of activity) [2]

Oxidation

Organic solvent

General stability information

Storage
–20°C, 10 mM potassium phosphate buffer, pH 7.0, 45% v/v glycerol, 1.0 M DTT, 18 months [2]

6 CROSSREFERENCES TO STRUCTURE DATABANKS

PIR/MIPS code
PIR1:DEKVG (cucumber)

Brookhaven code
1GDH (Hyphomicrobium methylovorum)

7 LITERATURE REFERENCES

[1] Kleczkowski, L.A., Edwards, G.E.: Plant Physiol.,91,278–286 (1989)
[2] Izumi, Y., Yoshida, T., Kanzaki, H., Toki, S.-i., Miyazaki, S.S., Yamada, H.: Eur. J. Biochem.,190,279–284 (1990)
[3] Stafford, H.A., Magaldi, A., Vennesland, B.: J. Biol. Chem.,207,621–629 (1954)

1 NOMENCLATURE

EC number
1.1.1.30

Systematic name
(R)-3-Hydroxybutanoate:NAD$^+$ oxidoreductase

Recommended name
3-Hydroxybutyrate dehydrogenase

Synonymes
NAD-beta-hydroxybutyrate dehydrogenase
Hydroxybutyrate oxidoreductase
Dehydrogenase, 3-hydroxybutyrate
beta-Hydroxybutyrate dehydrogenase
D-beta-Hydroxybutyrate dehydrogenase
D-3-Hydroxybutyrate dehydrogenase
D-(-)-3-Hydroxybutyrate dehydrogenase
beta-Hydroxybutyric acid dehydrogenase
3-D-Hydroxybutyrate dehydrogenase
beta-Hydroxybutyric dehydrogenase

CAS Reg. No.
9028-38-0

2 REACTION AND SPECIFICITY

Catalysed reaction
(R)-3-Hydroxybutanoate + NAD$^+$ →
→ acetoacetate + NADH

Reaction type
Redox reaction

Natural substrates
(R)-3-Hydroxybutanoate + NAD$^+$ (first step of energy production in degradation of poly-beta-hydroxybutanoate) [15]

Substrate spectrum
1 (R)-3-Hydroxybutanoate + NAD$^+$ (r [18], specific for D-(-)-3-hydroxybutanoate, no oxidation of L-(+)-3-hydroxybutanoate [15, 20]) [1, 3–9, 11–24]
2 Acetonylsulfonate + NAD$^+$ [18]
3 2-Methylacetoacetate + NADH [18]

Product spectrum

1 Acetoacetate + NADH [18]
2 ?
3 3-Hydroxy-2-methylbutanoate + NAD^+

Inhibitor(s)

NAD^+ [1, 13, 18]; Polyclonal antibodies to purified enzyme [1]; N-Ethylmaleimide [1, 6, 9, 16]; L-3-Hydroxybutanoate [5, 6, 18]; Propionate [5]; L-Lactate [5]; Butanoate [5]; ADP-ribose [5]; D-Lactate [5, 15]; DL-2-Hydroxybutanoate [5, 15]; $(NAD)_2$ [5]; Methylmalonate [6, 17, 21]; AMP [13]; ADP [13]; ATP [13]; Acetoacetate [13, 17]; NADH [13, 17, 18]; Dimethoxyphosphinylacetate [13]; Methyl-2-methoxyphosphinylacetate [13]; Methylacetonylphosphonate [13]; Acetyl-CoA [15]; Mn^{2+} [15]; Cu^{2+} [15]; p-Chloromercuribenzoic acid [15, 20]; 5,5'-Dithiobis(2-nitrobenzoic acid) [15]; $HgCl_2$ [15, 20]; Butanedione [9]; Phenylglyoxal [9]; Diethyldicarbonate [9]; p-Chloromercuriphenylsulfonate [9, 16]; Phenylarsine oxide [9, 16]; Sulfite [9, 16]; Sulfide [9, 16]; Cyanide [16]; Cd^{2+} [25]; EDTA [18, 22]; Alkyl phosphonates [18]; Alkyl phosphates [18]; 2-Epoxypropionate [18]; Pyruvate [18]; Malonate [21]; Tartronate [21]; Mesoxalate [21]

Cofactor(s)/prosthetic group(s)/activating agents

NAD^+ (specific for, no oxidation with $NADP^+$ [15, 20]) [5, 7, 11, 15, 20]; Phosphatidylcholine [1–3, 6, 8, 11, 23–25]; NADH [18]

Metal compounds/salts

Turnover number (min^{-1})

Specific activity (U/mg)

19.2 [12]; 149 (bovine heart) [1]; 91 (rat heart) [1]

K_m-value (mM)

0.091 (NADH) [6]; 0.59 (acetoacetate) [6]; 0.24–0.36 (NAD^+) [1]; 0.84–1.32 (3-hydroxybutanoate) [1]; More (anomalous kinetics, hysteresis) [5, 7]

pH-optimum

7.5–8.0 (rat liver) [1]; 6.8–7.2 (bovine heart) [1]; 9.0 [4]; 8.0 (oxidation) [15, 18, 20]; 6–7 (reduction) [15, 18, 20]; 7.0 (acetoacetate) [5]; 8.5 (3-hydroxybutanoate) [5]

pH-range

5.5–10.0 [1, 5, 15]

Temperature optimum (°C)

Temperature range (°C)

3 ENZYME STRUCTURE

Molecular weight

132000 (Paracoccus denitrificans, gel filtration) [7]
110000 (bovine heart mitochondria, radiation inactivation) [11]
89000 (Rhodopseudomonas sphaeroides, gel filtration) [12]
112000 (Zoogloea ramigera, gel filtration) [15]

Subunits

Tetramer (4 × 32000, Paracoccus denitrificans [7], bovine [11], 4 × 23000, Rhodopseudomonas sphaeroides [12], 4 × 28000, Zoogloea ramigera [15], SDS-PAGE [7, 11, 12, 15]) [7, 11, 12, 15]

Glycoprotein/Lipoprotein

–

4 ISOLATION/PREPARATION

Source organism

Bovine [1–3, 6, 8, 9, 11, 16, 19, 23–25]; Rat [1, 3, 10, 14, 17]; Bacillus cereus [22]; Rhodopseudomonas sphaeroides [4, 12]; Paracoccus denitrificans [5, 7]; Pseudomonas lemoignei [13, 18]; Zoogloea ramigera [15]; Mycobacterium phlei [20]; Guinea pig [21]

Source tissue

Heart [1, 2, 6, 8, 9, 11, 16, 17, 19, 21, 23]; Liver [1, 14, 17, 21]; Cells [7, 12, 15, 20]; Stomach mucosa [10]

Localisation in source

Inner face of mitochondrial inner membrane [1, 3, 11, 16, 19]; Membrane-bound [4, 7, 8, 21]; Soluble [22]

Purification

Bovine [1, 2, 6, 11, 23]; Rat [1, 14]; Pseudomonas denitrificans [7]; Rhodopseudomonas sphaeroides [12]; Zoogloea ramigera [15]; Mycobacterium phlei [20]

Crystallization

–

Cloned

–

Renaturated

–

5 STABILITY

pH

Temperature (°C)

37 (15 min, 70% loss of activity) [20]

Oxidation

Organic solvent

General stability information

1–3 mg protein/ml in 400 mM LiBr, 5 mM HEPES, 5 mM dithiothreitol pH 7–8 [6, 11, 23]; 1–10 mM EDTA stabilizes [7]; Freezing/thawing inactivates [1]; Mn^{2+} protects [4]

Storage

4°C, lyophilized, several months [7]; –20°C, several months [7, 15], -80°C, 50 mM Tris/HCl buffer, pH 7.8, 0.1 M KCl, 5 mM DTT, 50% glycerol, long term storage [8]

6 CROSSREFERENCES TO STRUCTURE DATABANKS

PIR/MIPS code

PIR3:B42845 (bovine (fragments)); PIR3:A42845 (human (fragment)); PIR2:A42345 (precursor rat)

Brookhaven code

7 LITERATURE REFERENCES

[1] McIntyre, J.O., Latruffe, N., Brenner, S.C., Fleischewr, S.: Arch. Biochem. Biophys., 262,85–98 (1988)
[2] Wang, S., Martin, E., Cimino, J., Omann, G., Glaser, M.: Biochemistry,27,2033–2039 (1988)
[3] Cortese, J.D., Fleischer, S.: Biochemistry,26,5283–5293 (1987)
[4] Worrall, E.B., Gassain, S., Cox, D.J., Sugden, M.C., Palmer, T.N.: Biochem. J.,241,297–300 (1987)
[5] Kovár, J., Matyskovà, I., Matyska, L.: Biochim. Biophys. Acta,871,302–309 (1986)
[6] Dubois, H., Fritzsche, T.M., Trommer, W.E., McIntyre, J.O., Fleischer, S.: Biol. Chem. Hoppe-Seyler,367,343–353 (1986)
[7] Matysková, I., Kovár, J., Racek, P.: Biochim. Biophys. Acta,839,300–307 (1985)
[8] Burnett, B.K., Khorana, H.G.: Biochim. Biophys. Acta,815,51–56 (1985)
[9] Phelps, D.C., Hatefi, Y.: Biochemistry,20,459–463 (1981)
[10] MacGill, A.K., Anderson, N. G., Trotman, C.N.A., Carrington, J.M., Hanson, P.J.: Biochem. Soc. Trans.,12,788 (1984)
[11] McIntyre, J.O., Churchill, P., Maurer, A., Berenski, C.J., Jung, C.Y., Fleischer, S.: J. Biol. Chem.,258,953–959 (1982)
[12] Scawen, M.D., Darbyshire, J., Harvey, M.J., Atkinson, T.: Biochem. J.,203,699–705 (1982)

[13] Kluger, R., Tsui, W.-C.: Can. J. Biochem.,59,810–815 (1981)
[14] Miyahara, M., Utsumi, K., Deamer, D.W.: Biochim. Biophys. Acta,641,222–231 (1981)
[15] Nakada, T., Fukui, T., Saito, T., Miki, K., Oji, C., Matsuda, S., Ushijima, A., Tomita, K.: J. Biochem.,89,625–635 (1981)
[16] Phelps, D.C., Hatefi, Y.: Biochemistry,20,453–458 (1981)
[17] Tucker, G.A., Dawson, A.P.: Biochem. J.,179,579–581 (1979)
[18] Kluger, R., Nakaoka, K., Tsui, W.-C.: J. Am. Chem. Soc.,100,7388–7392 (1978)
[19] McIntyre, J.O., Bock, H.-G., Fleisher, S.: Biochim. Biophys. Acta,513,255–267 (1978)
[20] Dhariwal, K.R., Venkitasubramanian, T.A.: J. Gen. Microbiol.,104,123–126 (1978)
[21] Tan, A.W.H., Smith, C.M., Aogaichi, T., Plaut, G.W.E. : Arch. Biochem. Biophys., 166,164–173 (1975)
[22] Thompson, E.D., Nakata, H.M.: Can. J. Microbiol.,19,673–677 (1973)
[23] Nielsen, N.C., Fleischer, S.: J. Biol. Chem.,248,2549–2555 (1973)
[24] Nielsen, N.C., Zahler, W.L., Fleischer, S.: J. Biol. Chem.,248,2556–2562 (1973)
[25] Sekuzu, I., Jurtshuk, P.Jr., Green, D.E.: J. Biol. Chem.,238,975–982 (1963)

1 NOMENCLATURE

EC number
1.1.1.31

Systematic name
3-Hydroxy-2-methylpropanoate:NAD$^+$ oxidoreductase

Recommended name
3-Hydroxyisobutyrate dehydrogenase

Synonymes
Dehydrogenase, 3-hydroxyisobutyrate
beta-Hydroxyisobutyrate dehydrogenase

CAS Reg. No.
9028-39-1

2 REACTION AND SPECIFICITY

Catalysed reaction
3-Hydroxy-2-methylpropanoate + NAD$^+$ →
→ 2-methyl-3-oxopropanoate + NADH (mechanism [1], stereochemistry [8])

Reaction type
Redox reaction

Natural substrates

Substrate spectrum
1 (S)-3-Hydroxyisobutyrate + NAD$^+$ (oxidation of S-isomer 350 times faster than of R-isomer [1]) [1, 3, 4]
2 (R)-3-Hydroxyisobutyrate + NAD$^+$ (oxidation of S-isomer 350 times faster than of R-isomer [1]) [1, 3, 4]
3 More (no substrates: ethanol, 1-propanol, L-lactate, L-malonate, acetoacetate, DL-3-hydroxybutyrate, 2-hydroxybutyrate, DL-2-hydroxyisovalerate, 4-hydroxybutyrate, DL-2-hydroxycaproate, DL-2-hydroxy-n-valerate, 3-aminobutyrate, 3-chloropivalic acid, 3-chloropropionate [1], glycolate, 3-hydroxypropionate, 3-hydroxyisobutyryl-CoA, DL-homoserine, 2-methyl-4-hydroxybutyrate, 2-hydroxy-3,3-dimethylbutyrate, pantoate, 4-hydroxyvalerate [6]) [1, 6]

Product spectrum
1 Methylmalonyl semialdehyde + NADH [4]
2 Methylmalonyl semialdehyde + NADH [4]
3 ?

Inhibitor(s)

(R)-3-Hydroxyisobutyrate [1]; NADH [1]; p-Chloromercuribenzoate [1, 4, 6]; $AgNO_3$ [1, 4]; Iodoacetate [1, 6]; $HgCl_2$ [4, 6]

Cofactor(s)/prosthetic group(s)/activating agents

NAD^+ (specific for, not replaceable by $NADP^+$ [1, 4, 6]) [1, 3, 4, 6]

Metal compounds/salts

No cations required [1, 4, 6]

Turnover number (min^{-1})

58.8 ((R)-3-hydroxyisobutyrate) [1]; 846 ((S)-3-hydroxyisobutyrate) [1]

Specific activity (U/mg)

10.7 [1]; 36.1 ((S)-3-hydroxyisobutyrate) [4]; 8.94 ((R)-3-hydroxyisobutyrate) [4]; More [6]

K_m-value (mM)

0.022 (NAD^+) [2]; 0.054 (NAD^+) [6]; 0.059 ((S)-3-hydroxyisobutyrate) [1]; 0.1 (DL-3-hydroxyisobutyrate) [2, 6]; 0.37 ((S)-3-hydroxyisobutyrate) [4]; 1.25 ((R)-3-hydroxyisobutyrate) [4]; 1.44 ((R)-3-hydroxyisobutyrate) [1]; 15 (NAD^+ (+ (S)-3-hydroxyisobutyrate)) [4]; 20 (NAD^+ (+ (R)-3-hydroxyisobutyrate)) [4]

pH-optimum

8.6 [2]; 9.0 [6]; 9.0–11.0 [1]; 9.3–10.0 [4]

pH-range

7–11.5 [1, 6]

Temperature optimum (°C)

40 [4]

Temperature range (°C)

55 (up to) [4]

---

3 ENZYME STRUCTURE

Molecular weight

75000–80000 (Candida rugosa, calculation from diffusion coeffient and partial specific volume, gel filtration) [4]

Subunits

Dimer (identical, 2 × 40000, Candida rugosa, SDS-PAGE, N-terminal amino acid analysis [4], 2 × 34000, rabbit, SDS-PAGE [1]) [1, 4]

Glycoprotein/Lipoprotein

–

4 ISOLATION/PREPARATION

Source organism

Rabbit [1, 3, 6]; Pig [3, 6]; Pseudomonas aeruginosa [3, 7]; Pseudomonas putida [5, 8]; Rhodopseudomonas sphaeroides [2]; Candida rugosa [3, 4]; Candida parapsilosis [3]; Endomyces reessii [3]; Rhodotorula rubra [3]; Trichosporon aculeatum [3]; Corynebacterium hydrocarbons [3]; Corynebacterium paurometabolum [3]; Micrococcus flavus [3]; Micrococcus luteus [3]; Pseudomonas dacunhae [3]; Nocardia lyena [3]; Bovine [3, 6]; Pigeon [6]; Tetrahymena pyriformis [6]; Neurospora crassa [6]

Source tissue

Liver [1, 3]; Kidney [6]; Commercial preparation (of 3-hydroxybutyrate dehydrogenase, EC 1.1.1.30) [2]

Localisation in source

Purification

Rabbit [1]; Pseudomonas aeruginosa [7]; Candida rugosa [4]; Pig [6]

Crystallization

[4]

Cloned

–

Renaturated

–

5 STABILITY

pH

4.5–7.3 [4]

Temperature (°C)

0 (stable at) [2]; 35 (10 min stable) [4]; 37 (rapid inactivation) [2]

Oxidation

Organic solvent

General stability information

Storage

–20°C, 50 mM HEPES buffer, pH 7.5, 40% v/v glycerol, 5 mM DTT, 0.5 mM EDTA, 12 months, 25% loss of activity [1]

6 CROSSREFERENCES TO STRUCTURE DATABANKS

PIR/MIPS code

PIR2:C42902 (Pseudomonas aeruginosa); PIR2:JQ0613 (homolog Escherichia coli); PIR2:A32867 (precursor rat (fragment))

Brookhaven code

7 LITERATURE REFERENCES

[1] Rougraff, P.M., Paxton, R., Kuntz, M.J., Crabb, D.W., Harris, R.A.: J. Biol. Chem.,263,327–331 (1988)
[2] Worrall, E.B., Gassain, S., Cox, D.J., Sugden, M.C., Palmer, T.N.: Biochem. J.,241,297–300 (1987)
[3] Hasegawa, J.: Agric. Biol. Chem.,45,2899–2901 (1981)
[4] Hasegawa, J.: Agric. Biol. Chem.,45,2805–2814 (1981)
[5] DePaul Marshall, V., Sokatch,J.R.: J. Bacteriol.,110,1073–1081 (1972)
[6] Robinson, W.G., Coon, M.J.: J. Biol. Chem.,225,511–521 (1957)
[7] Bannerjee, D., Sanders, L.E., Sokatch, J.R.: J. Biol. Chem.,245,1828–1835 (1970)
[8] Aberhart, D.J., Hsu, C.-T.: J. Chem. Soc. Perkin Trans.,1,1404–1406 (1979)

1 NOMENCLATURE

EC number
1.1.1.32

Systematic name
(R)-Mevalonate:NAD$^+$ oxidoreductase

Recommended name
Mevaldate reductase

Synonymes
Mevalonic dehydrogenase

CAS Reg. No.
9028-33-5

2 REACTION AND SPECIFICITY

Catalysed reaction
(R)-Mevalonate + NAD$^+$ →
→ mevaldate + NADH

Reaction type
Redox reaction

Natural substrates

Substrate spectrum
1 Mevaldic acid + NADH (ir [2]) [1, 2]
2 More (no substrates: glyoxylate, 3-hydroxypropionate, 3-hydroxyisobutyrate, 4-hydroxybutyrate, gluconate) [2]

Product spectrum
1 Mevalonic acid + NAD$^+$ [2]
2 ?

Inhibitor(s)

Cofactor(s)/prosthetic group(s)/activating agents
NADH (A-specific i.e. transfer of pro-R hydrogen from cofactor [3], preferred [1], 2–3 times as effective as NADPH [2]) [1–3]; NADPH (2–3 times less effective than NADH) [2]

Metal compounds/salts

Turnover number (min^{-1})

Specific activity (U/mg)
0.62 [2]; 0.2 [3]

K_m-value (mM)
0.003 (NADH) [2]; 0.09 (NADPH) [2]; 4 ((+)-mevalonic acid) [2]

pH-optimum
5.4 [2]

pH-range
4.6–6.9 [2]

Temperature optimum (°C)

Temperature range (°C)

---

3 ENZYME STRUCTURE

Molecular weight

Subunits

Glycoprotein/Lipoprotein
–

---

4 ISOLATION/PREPARATION

Source organism
Pig [1–3]; Chicken [2]; Rat [2]; More (not in carrot, Tetrahymena pyriformis, Eremothecium ashbyii, E. coli) [2]

Source tissue
Liver [1–3]; Kidney [1, 2]; Heart [1, 2]; More (not in mammalian brain) [2]

Localisation in source
Soluble part of cell [1]

Purification
Pig (partial) [2]

Crystallization
–

Cloned
–

Renaturated
–

5 STABILITY

pH

Temperature (°C)

Oxidation

Organic solvent

General stability information

Storage

6 CROSSREFERENCES TO STRUCTURE DATABANKS

PIR/MIPS code

Brookhaven code

7 LITERATURE REFERENCES

[1] Coon, M.J., Kupiecki, F.P., Dekker, E.E., Schlesinger, M.J., del Campillo, A. in "CIBA Symposium on the Biosynthesis of Terpenes and Sterols" (Wolstenholme, G.E.W., O'Connor, M., eds) pp.62–74, Churchill London (1959)
[2] Schlesinger, M.J., Coon, M.J.: J. Biol. Chem.,236,2421–2424 (1961)
[3] Nigan, H.-L., Popjak, G.: Bioorg. Chem.,4,166–180 (1975)

1 NOMENCLATURE

EC number
1.1.1.33

Systematic name
(R)-Mevalonate:$NADP^+$ oxidoreductase

Recommended name
Mevaldate reductase (NADPH)

Synonymes
Reductase, mevaldate (reduced nicotinamide adenine dinucleotide phosphate)
More (may be identical with EC 1.1.1.2)

CAS Reg. No.
9028-34-6

2 REACTION AND SPECIFICITY

Catalysed reaction
(R)-Mevalonate + $NADP^+$ →
→ mevaldate + NADPH

Reaction type
Redox reaction

Natural substrates
Aldehydes + $NADP^+$ (probably not mevaldate, may be identical with EC 1.1.1.2) [1]

Substrate spectrum
1 (3R)-Mevaldate + NADPH (ir, (3S)-isomer is reduced at a lower rate) [1]
2 Pyridine-3-carboxaldehyde + NADPH (ir) [1]
3 4-Cyanobenzaldehyde + NADPH (ir) [1]

Product spectrum
1 Mevalonate + $NADP^+$
2 3-Pyridinemethanol + $NADP^+$
3 4-Cyanobenzyl alcohol + $NADP^+$

Inhibitor(s)
Sodium aminobarbitone [1]; Thiobarbituric acid [1]

Cofactor(s)/prosthetic group(s)/activating agents
NADPH [1]; NADH (20% of NADPH activity) [1]

Metal compounds/salts

Turnover number (min^{-1})

Specific activity (U/mg)

K_m-value (mM)

pH-optimum

pH-range

Temperature optimum (°C)

Temperature range (°C)

3 ENZYME STRUCTURE

Molecular weight
27000–30000 (rat, gel filtration) [1]

Subunits

Glycoprotein/Lipoprotein
–

4 ISOLATION/PREPARATION

Source organism
Rat [1]

Source tissue
Liver [1]

Localisation in source

Purification
Rat [1]

Crystallization
–

Cloned
–

Renaturated
–

5 STABILITY

pH

Temperature (°C)

Oxidation

Organic solvent

General stability information

Storage
-20°C, at least 3 months [1]

6 CROSSREFERENCES TO STRUCTURE DATABANKS

PIR/MIPS code

Brookhaven code

7 LITERATURE REFERENCES

[1] Beedle, A.S., Rees, H.H., Goodwin, T.W.: Biochem. J.,139,205-209 (1974)

1 NOMENCLATURE

EC number
1.1.1.34

Systematic name
(R)-Mevalonate:$NADP^+$ oxidoreductase (CoA-acylating)

Recommended name
Hydroxymethylglutaryl-CoA reductase (NADPH)

Synonymes
Hydroxymethylglutaryl coenzyme A reductase (reduced nicotinamide adenine dinucleotide phosphate)
3-Hydroxy-3-methylglutaryl-CoA reductase
beta-Hydroxy-beta-methylglutaryl coenzyme A reductase [2]
Hydroxymethylglutaryl CoA reductase (NADPH)
S-3-Hydroxy-3-methylglutaryl-CoA reductase
NADPH-hydroxymethylglutaryl-CoA reductase
HMGCoA reductase-mevalonate:NADP-oxidoreductase (acetylating-CoA) [3]
3-Hydroxy-3-methylglutaryl CoA reductase (NADPH)

CAS Reg. No.
9028-35-7

---

2 REACTION AND SPECIFICITY

Catalysed reaction
(S)-3-Hydroxy-3-methylglutaryl-CoA + 2 NADPH →
→ (R)-mevalonate + CoA + 2 $NADP^+$

Reaction type
Redox reaction

Natural substrates
3-Hydroxy-3-methylglutaryl-CoA + NADPH [1–18]

Substrate spectrum
1 3-Hydroxy-3-methylglutaryl-CoA + NADPH (ir) [1–18]
2 More (not: 3-hydroxy-3-methylglutaric acid, 3-hydroxy-3-methylglutaryl-glutathione) [2]

Product spectrum
1 Mevalonic acid + CoA + $NADP^+$ [1–18]
2 ?

Inhibitor(s)

p-Hydroxymercuribenzoate [2]; Arsenite [2]; Coenzyme A (large amounts) [2, 3]; Acetyl-CoA [3]; Acetoacetyl-CoA [3]; 1,10-Phenanthroline [3]; Sodium deoxycholate [3]; Digitonin [3]; Fe^{2+} [3]; Fe^{3+} [3]; CoASH [3, 10]; 5,5'-Dithiobis(2-nitrobenzoate) [10]; Glutathione disulfide [10]; 3-Hydroxy-3-methylglutaryl-CoA (substrate inhibition) [10, 17]; Hydroxymethylglutarate [12]; p-Chloromercuribenzoate [12]; Mg-ATP [12, 15]; Mg-ADP [15]; Adenosine-2'-monophospho-5'-diphosphoribose [17]; NADPH [17]; Compactin [18]; Mevinolin [18]; More (preparation from mitochondria of rat liver [4], not: arsenate [2]) [2, 4]

Cofactor(s)/prosthetic group(s)/activating agents

NADPH [2, 5, 7, 12]; NADH (slight activity) [7]

Metal compounds/salts

Turnover number (min^{-1})

Specific activity (U/mg)

More [2, 3, 5, 10, 12]

K_m-value (mM)

0.012 (3-hydroxy-3-methylglutaryl-CoA) [3]; 0.087 (NADPH) [3]; 0.0211 (3-hydroxy-3-methylglutaryl-CoA) [7]; 0.0196 (NADH) [7]; 0.01 (NADPH) [7]; 0.025 (3-hydroxy-3-methylglutaryl-CoA) [7]; More [11, 12]

pH-optimum

7 (purified reductase) [3]; 7.3–7.7 [3]; 7.4 [7]; 6.6–6.9 [12]; 6.5–7.5 [12]

pH-range

6–9 [7]

Temperature optimum (°C)

25 [2]; 30 [12]; 37 [3, 4, 7–9, 11, 13–17]

Temperature range (°C)

15–50 [7]

---

3 ENZYME STRUCTURE

Molecular weight

217000–226000 (rat, gel filtration) [3]
11000 (Fusarium oxysporum, gel electrophoresis, gel filtration) [6, 7]
100000 (rat, electrophoresis, enzyme exists in 2 MW-forms) [16]
200000 (rat, electrophoresis, enzyme exists in 2 MW-forms) [16]

Subunits

Manomer (1 × 11000, Fusarium oxysporum, SDS-PAGE) [7]

Glycoprotein/Lipoprotein

–

4 ISOLATION/PREPARATION

Source organism

Rat [1, 3, 5, 9, 11, 13, 16–18]; Brewer's yeast [4]; Saccharomyces cerevisiae [2, 4]; Fusarium oxysporum [6, 7]; Pigeon [7]; Pseudomonas sp. [7]; Neurospora crassa [7]; Hamster [8]; Chicken [8]; Hevea brasiliensis [12]; Pisum sativum [12]; Nepeta cataria [14]; Tetrahymena pyriformis [14]; Pig [14]; Spinach [14]; Human [15]; Baker's yeast [10]

Source tissue

Liver [1, 3, 9, 16, 17]; Adrenal [8]; Small intestine [11]; Latex [12]; Leaf [14]; Leukocytes [15]

Localisation in source

Microsomes [1, 3, 5, 7, 9, 11, 13, 15, 16, 18]; Mitochondria [7, 8]; Endoplasmic reticulum [12]; Chloroplast [14]

Purification

Rat [1, 3, 5, 9, 11, 13]; Brewer's yeast [4]; Saccharomyces cerevisiae [2, 4]; Fusarium oxysporum [6, 7]; Hamster [8]; Hevea brasiliensis [12]; Human [15]

Crystallization

–

Cloned

–

Renaturated

–

5 STABILITY

pH

Temperature (°C)

Oxidation

Organic solvent

General stability information

Unstable in absence of a thiol [12]

Storage

–10°C, 10–20% loss of activity per week [3]; –20°C, 10% loss of activity per week [5]; –15°C, 30% activity loss in 24 h, 50% loss in 1 week [12]

6 CROSSREFERENCES TO STRUCTURE DATABANKS

PIR/MIPS code

PIR2:PQ0761 ((HMGR 10) wheat (fragment)); PIR2:PQ0762 ((HMGR 18) wheat (fragment)); PIR2:PQ0763 ((HMGR 23) wheat (fragment)); PIR2:A35728 (African clawed frog); PIR2:S07551 (Arabidopsis thaliana); PIR1:RDHYE (Chinese hamster); PIR2:A34416 (fluke (Schistosoma mansoni)); PIR3:S17343 (Gibberella fujikuroi); PIR2:A42149 (Haloferax volcanii); PIR1:RDHUE (human); PIR2:A43533 (mouse (fragment)); PIR3:S17345 (Phycomyces blakesleeanus); PIR3:S29622 (radish); PIR3:S29623 (radish); PIR3:S33175 (rat (fragment)); PIR2:S12554 (rat (fragments)); PIR2:A31898 (sea urchin (Strongylocentrotus purpuratus)); PIR2:A28367 (sea urchin (Strongylocentrotus purpuratus) (fragment)); PIR2:S25316 (tomato); PIR2:PQ0187 (tomato (fragment)); PIR2:S24760 (wood tobacco); PIR2:S14955 (1 Para rubber tree); PIR2:S14953 (2 Para rubber tree (fragment)); PIR2:B24317 (2 yeast (Saccharomyces cerevisiae)); PIR2:S22521 (hmg3 Para rubber tree)

Brookhaven code

7 LITERATURE REFERENCES

[1] Bucher, N.L.R., Overath, P., Lynen, F.: Biochim. Biophys. Acta,40,491–501 (1960)
[2] Durr, I.F., Rudney, H.: J. Biol. Chem.,235,2572–2578 (1960)
[3] Kawachi, T., Rudney, H.: Biochemistry,9,1700–1705 (1970)
[4] Madhosingh, C., Migicovsky, B.B., Wood, I.M.: FEBS Lett.,46,20–22 (1974)
[5] Heller, R.A., Gould, R.G.: Biochim. Biophys. Acta,388,254–259 (1975)
[6] Madhosingh, C., Orr, W.: Agric. Biol. Chem.,41,1519–1521 (1977)
[7] Madhosingh, C., Orr, W.: Biochim. Biophys. Acta,523,283–296 (1978)
[8] Preiss, B., Lehoux, J.-G.: Biochem. Biophys. Res. Commun.,88,1140–1146 (1979)
[9] Ingebritsen, T.S., Parker, R.A., Gibson, D.M.: J. Biol. Chem.,256,1138–1144 (1981)
[10] Gilbert, H.F., Steward, M.D.: J. Biol. Chem.,256,1782–1785 (1981)
[11] Young, N.L., Saudek, C.D., Crawford, S.A., Zuckerbrod, S.L.: J. Lipid Res.,23,257–265 (1982)
[12] Sipat, A.B.: Phytochemistry,21,2613–2618 (1982)
[13] van Heusden, G.P.H., Wirtz, K.W.A.: J. Lipid Res.,25,27–32 (1984)
[14] Arebalo, R.E., Mitchell, E.D.: Phytochemistry,23,13–18 (1984)
[15] Harwood, H.J., Schneider, M., Stacpole, P.W.: Biochim. Biophys. Acta,805,245–251 (1984)
[16] Ness, G.C., Eales, S.J., Pedleton, L.C., Smith, M.: J. Biol. Chem.,260,12391–12393 (1985)
[17] Feingold, K.R., Moser, A.M.: Arch. Biochem. Biophys.,249,46–52 (1986)
[18] Gibson, D.M., Parker, R.A. in "The Enzymes", 3rd Ed. (Boyer, P.D.; Krebs, E.G., eds.) vol. 18,179–215 (1987)

1 NOMENCLATURE

EC number
1.1.1.35

Systematic name
(S)-3-Hydroxyacyl-CoA:NAD$^+$ oxidoreductase

Recommended name
3-Hydroxyacyl-CoA dehydrogenase

Synonymes
3-Keto reductase
3-Hydroxyacyl coenzyme A dehydrogenase
beta-Hydroxyacyl-coenzyme A synthetase
beta-Hydroxyacylcoenzyme A dehydrogenase
beta-Hydroxybutyrylcoenzyme A dehydrogenase
3-Hydroxyacetyl-coenzyme A dehydrogenase
L-3-Hydroxyacyl coenzyme A dehydrogenase
L-3-Hydroxyacyl CoA dehydrogenase
beta-Hydroxyacyl CoA dehydrogenase
3beta-Hydroxyacyl coenzyme A dehydrogenase
3-Hydroxybutyryl-CoA dehydrogenase
beta-Ketoacyl-CoA reductase
beta-Hydroxy acid dehydrogenase
3-L-Hydroxyacyl-CoA dehydrogenase
3-Hydroxyisobutyryl-CoA dehydrogenase
beta-Hydroxyacyl dehydrogenase
beta-Keto-reductase
1-Specific DPN-linked beta-hydroxybutyric dehydrogenase [1]
More (cf. EC 1.1.1.211)

CAS Reg. No.
9028-40-4

2 REACTION AND SPECIFICITY

Catalysed reaction
(S)-3-Hydroxyacyl-CoA + NAD$^+$ →
→ 3-oxoacyl-CoA + NADH

Reaction type
Redox reaction

Natural substrates

L-3-Hydroxybutyryl-CoA + NAD^+ [1, 2, 12, 13, 15, 17]

Substrate spectrum

1 3-Hydroxybutyryl-CoA + NAD^+ (r) [1, 2, 12, 13, 15, 17]
2 3-Hydroxyhexanoyl-CoA + NAD^+ (r) [2]
3 3-Hydroxyoctanoyl-CoA + NAD^+ (r) [2]
4 3-Hydroxylauryl-CoA + NAD^+ (r) [2]
5 3-Hydroxy-4-hexenoyl-CoA + NAD^+ [3]
6 S-Acetoacetyl-N-acetylcysteamine + NAD^+ [4, 12]
7 S-Acetoacetylpantetheine + NAD^+ [4, 9, 12]
8 S-Acetoacetyl-CoA + NAD^+ [4, 9, 12]
9 3-Hydroxyhexadecanoyl-CoA + NAD^+ (r) [13, 17]
10 More (3-hydroxyacyl derivatives with chain length C_4-C_{12}, only D-isomers [2], only L-isomers [1, 12, 13]) [1, 2, 12, 13]

Product spectrum

1 3-Oxobutyryl-CoA + NADH (acetoacetyl-CoA) [1, 2, 12, 13, 15, 17]
2 3-Oxohexanoyl-CoA + NADH [2]
3 3-Oxooctanoyl-CoA + NADH [2]
4 3-Oxolauryl-CoA + NADH [2]
5 3-Oxo-4-hexenoyl-CoA + NADH [3]
6 ?
7 ?
8 ?
9 3-Oxohexadecanoyl-CoA + NADH [17]
10 ?

Inhibitor(s)

Fluoride [1]; 3-Hydroxylauryl-CoA [2]; Acetoacetyl-CoA [4, 12, 13]; S-Acetoacetylpantetheine [4]; p-Chloromercuribenzoate [4, 9]; Urea [9]; Guanidine-HCl [9]; Iodoacetamide [9]; Iodoacetic acid [9]; N-Ethylmaleimide [9]; 5,5'-Dithiobis(2-nitrobenzoic acid) [9]; N-Bromosuccinimide [9]; p-Chloromercuriphenylsulfonic acid [12]

Cofactor(s)/prosthetic group(s)/activating agents

NAD^+ [1–4, 6–9, 12, 13, 15–17]; $NADP^+$ [8]

Metal compounds/salts

Turnover number (min^{-1})

Specific activity (U/mg)

More [2, 4, 8, 9, 12–15]

K_m-value (mM)

0.31 (3-hydroxybutyryl-CoA) [2]; 0.34 (3-hydroxyhexanoyl-CoA) [2]; 0.35 (3-hydroxylauryl-CoA) [2]; 10 (S-acetoacetyl-N-acetylcysteamine) [4, 9]; 0.4 (S-acetoacetylpantetheine) [4, 9]; 0.06 (S-acetoacetyl-CoA) [4, 9]; 0.036 (acetoacetyl-CoA) [12]; 0.083 (3-hydroxybutyryl-CoA) [12]; 1.19 (S-acetoacetylpantetheine) [12]; 44.4 (S-acetoacetyl-N-acetylcysteamine) [12]; 0.066 (acetoacetyl-CoA) [15]; 0.35 (3-hydroxyoctanoyl-CoA) [2]; More [13]

pH-optimum

10 [2]; 6–7 (reduction) [3]; 9.6 (oxidation) [3]; 7.3 [4]; 7.6 [8]; 4.5–6.2 (NADH oxidation) [12, 13]; 9–10 (NAD reduction) [12]

pH-range

Temperature optimum (°C)

25 (assay at) [3, 4, 9, 12, 15]; 30 (assay at) [13, 17]

Temperature range (°C)

3 ENZYME STRUCTURE

Molecular weight

65000 (pig, sedimentation equilibrium) [5, 9]
67000 (pig, amino acid analysis) [5]
75000 (pig, gel filtration) [4, 5]
71000 (rat, enzyme with preference for short-chain acyl-CoA substrates, gel filtration) [17]
186000 (rat, enzyme with preference for long-chain acyl-CoA substrates, gel filtration) [17]
50300 (Mycobacterium smegmatis, gel electrophoresis) [12]
270000 (E. coli, fatty acid oxidation multienzyme complex, gel electrophoresis) [15]

Subunits

Dimer (2 × 31000, pig, SDS-PAGE, sedimentation equilibrium of guanidine-HCl treated enzyme) [5]
More [13]

Glycoprotein/Lipoprotein

–

4 ISOLATION/PREPARATION

Source organism

Rat [1, 11, 13, 17]; Sheep [2]; Bovine [2]; Pig [3–7, 9, 10, 16, 18–20]; Clostridium kluyveri [8]; Mycobacterium smegmatis [12]; Mycobacterium tuberculosis [12]; E. coli [14, 15]

Source tissue

Liver [1, 2, 13, 17]; Kidney [1, 17]; Heart [3–7, 9, 10, 17, 19, 20]; Brain [17]

Localisation in source

Mitochondria [1, 2, 4, 7, 10, 11]; Peroxisomes [13]

Purification

Rat [1, 17]; Bovine [2]; Pig [3–5, 9, 16]; Clostridium kluyveri [8]; Mycobacterium smegmatis [12]; E. coli [14, 15]

Crystallization

[3, 4, 6, 9, 16]

Cloned

–

Renaturated

---

5 STABILITY

pH

Temperature (°C)

0–45 [8]

Oxidation

Organic solvent

General stability information

Storage

–15°C, 4 weeks [1]; 4°C, 6 months [9]; –20°C, several months [12, 13]; –76°C, several months, multienzyme complex [15]

---

6 CROSSREFERENCES TO STRUCTURE DATABANKS

PIR/MIPS code

PIR1:DEPGC (pig); PIR2:A39867 (/enoyl-CoA hydratase (EC 4.2.1.17) /3-hydroxyacyl-CoA epimerase trifunctional protein Neurospora crassa (fragment))

Brookhaven code

7 LITERATURE REFERENCES

[1] Lehninger, A.L., Greville, G.D.: Biochim. Biophys. Acta,12,188–202 (1953)
[2] Wakil, S.J., Green, D.E., Mii, S., Mahler, H.R.: J. Biol. Chem.,207,631–638 (1954)
[3] Stern, J.R.: Biochim. Biophys. Acta,26,448–449 (1957)
[4] Noyes, B.E., Bradshaw, R.A.: J. Biol. Chem.,248,3052–3059 (1973)
[5] Noyes, B.E., Bradshaw, R.A.: J. Biol. Chem.,248,3060–3066 (1973)
[6] Weininger, M.: J. Mol. Biol.,90,409–413 (1974)
[7] Noyes, B.E., Glatthaar, B.E., Garavelli, J.S., Bradshaw, R.A.: Proc. Natl. Acad. Sci. USA,71,1334–1338 (1974)
[8] Hillmer, P., Gottschalk, G.: Biochim. Biophys. Acta,334,12–23 (1974)
[9] Bradshaw, R.A., Noyes, B.E.: Methods Enzymol.,35, Pt. B,122–128 (1975)
[10] Middleton, B.: Biochem. Soc. Trans.,6,80–83 (1978)
[11] El-Hakhri, M., Middleton, B.: Biochem. Soc. Trans.,7,392–393 (1979)
[12] Shimakata, T., Fujita, Y., Kusaka, T.: J. Biochem.,86,1191–1198 (1979)
[13] Osumi, T., Hashimoto, T.: Arch. Biochem. Biophys.,203,372–383 (1980)
[14] Pawar, S., Schulz, H.: J. Biol. Chem.,256,3894–3899 (1981)
[15] Binstock, J.F., Schulz, H.: Methods Enzymol.,71,403–411 (1981)
[16] Holden, H.M., Banaszak, L.J., Frieden, C., McLoughlin,, D.J.: FEBS Lett.,132,15–18 (1981)
[17] El-Fakhri, M., Middleton, B.: Biochim. Biophys. Acta,713,270–279 (1982)
[18] Holden, H.M., Banaszak, L.J.: J. Biol. Chem.,258,2383–2389 (1983)
[19] Sumegi, B., Srere, P.A.: J. Biol. Chem.,259,8748–8752 (1984)
[20] Birktoft, J.J., Holden, H.M., Hamlin, R., Xuong, N.H., Banaszak, L.J.: Proc. Natl. Acad. Sci. USA,84,8262–8266 (1987)

1 NOMENCLATURE

EC number
1.1.1.36

Systematic name
(R)-3-Hydroxyacyl-CoA:NADP$^+$ oxidoreductase

Recommended name
Acetoacetyl-CoA reductase

Synonymes
Acetoacetyl coenzyme A reductase
Hydroxyacyl coenzyme-A dehydrogenase
NADP-linked acetoacetyl CoA reductase [2]
NADPH:acetoacetyl-CoA reductase [3]
D(-)-beta-Hydroxybutyryl CoA-NADP oxidoreductase [2]
Short chain beta-ketoacetyl(acetoacetyl)-CoA reductase [5, 6]
beta-Ketoacyl-CoA reductase [5]
D-3-Hydroxyacyl-CoA reductase [8]
(R)-3-Hydroxyacyl-CoA dehydrogenase [8]

CAS Reg. No.
9028-41-5

2 REACTION AND SPECIFICITY

Catalysed reaction
(R)-3-Hydroxyacyl-CoA + NADP$^+$ →
→ 3-oxoacyl-CoA + NADPH

Reaction type
Redox reaction

Natural substrates
Acetoacetyl-CoA + NADPH (biosynthesis of poly(D-3-hydroxybutyrate)) [8]

Substrate spectrum

1 Acetoacetyl-CoA + NADPH (r [1, 2, 4–8], ir [3]) [1–9]
2 Acetoacetyl-S-(D-pantetheine)11-pivalate + NADPH [8]
3 Propionylacetyl-CoA + NADPH [8]
4 Butyrylacetyl-CoA + NADPH [8]
5 2-Methylacetoacetyl-CoA + NADPH [8]
6 S-Acetoacetyl-N-acetylcysteamine + NADPH [3, 4]
7 More (poor substrate: D(-)-3-hydroxyvaleryl-CoA [7], not: crotonyl-CoA [1, 4], octanoyl-CoA [1], beta-hydroxybutyryl-CoA [1], trans-2-tetradecenoyl-CoA [6], D(-)-3-hydroxyacyl-CoA's with chain length exceeding 6 [7], L(+)-3-hydroxyacyl-CoA's [7]) [1, 4, 6, 7]

Product spectrum

1 (3R)-D-3-Hydroxybutyryl-CoA + $NADP^+$ [8]
2 3-Hydroxybutyryl-S-(D-pantetheine)11-pivalate + $NADP^+$
3 3-Hydroxypentanoyl-CoA + $NADP^+$
4 3-Hydroxyhexanoyl-CoA + $NADP^+$
5 2-Methyl-3-hydroxybutyryl-CoA + $NADP^+$
6 3-Hydroxybutyryl-N-acetylcysteamine + $NADP^+$
7 ?

Inhibitor(s)

p-Hydroxymercuribenzoate [1]; N-Ethylmaleimide [1, 3]; N-Phenylmaleimide [3]; Acetoacetyl-CoA (not inhibitory [4]) [2, 7]; Acetyl-CoA [3]; Butyryl-CoA [3]; Methylmalonyl-CoA [3]; 2-Methylcrotonyl-CoA [3]; Iodoacetamide [3]

Cofactor(s)/prosthetic group(s)/activating agents

$NADP^+$ [1–9]; NADPH (specific for [3, 7, 8], B-specific, i.e. transfer of pro-S-hydrogen from NADPH to acetoacetyl-CoA [8]) [1–9]; NADH [4–6, 9]

Metal compounds/salts

Turnover number (min^{-1})

5400 (D-3-hydroxybutyryl-CoA) [8]; 7440 (propionylacetyl-CoA) [8]; 18000 (acetoacetyl-CoA) [8]

Specific activity (U/mg)

608 [8]; 412 [7]; 0.25 [1]; More [2, 6]

K_m-value (mM)

0.002 (acetoacetyl-CoA, propionylacetyl-CoA) [8]; 0.005 (acetoacetyl-CoA) [3]; 0.0083 (acetoacetyl-CoA) [2]; 0.01 (NADPH [4], butyrylacetyl-CoA [8]) [4, 8]; 0.018 (acetoacetyl CoA [4], NADH [4, 5]) [4, 5]; 0.02–0.021 (NADPH [2, 8], acetoacetyl-CoA [5]) [2, 5, 8]; 0.026 (D-3-hydroxybutyryl-CoA) [8]; 0.04 (S-acetoacetyl-N-acetylcysteamine) [3]; 0.045 (NADH) [4]; 0.74 (acetoacetyl-CoA [1], 2-methylacetoacetyl-CoA [8]) [1, 8]; 0.82 (acetoacetyl-pantetheine) [3]; 0.99 (acetoacetyl-S-(D-pantetheine)11-pivalate) [8]

pH-optimum
6.5–8.0 [5]; 8.1 (reduction of acetoacetyl-CoA) [2]

pH-range
6.5–8.5 [1]

Temperature optimum (°C)
37 [1, 5]; 30 [7]

Temperature range (°C)

3 ENZYME STRUCTURE

Molecular weight
250000 (Streptomyces coelicolor, gel filtration) [4]
92000 (Zoogloea ramigera, gel electrophoresis) [7]

Subunits
Tetramer (4 × 25500, Zoogloea ramigera, SDS-PAGE [7], calculation from gene sequence [8], 2 × 55000 + 2 × 65000, Streptomyces coelicolor, SDS-PAGE [4]) [4, 7, 8]

Glycoprotein/Lipoprotein
–

4 ISOLATION/PREPARATION

Source organism
Pigeon [1, 3]; Zoogloea ramigera (gene cloned in E. coli [8]) [2, 7, 8]; Azotobacter beijerinckii [2]; Rat [3, 5]; Bovine (female) [3]; Goose [3]; Chicken [3]; Yeast [3]; Pig [6]; Streptomyces coelicolor [4]; Syntrophomonas wolfei (ssp. wolfei) [9]

Source tissue
Liver [1, 3, 5]; Mammary gland [3]; Adipose tissue [3]; Uropygial gland [3]; Heart [6]; Cell [8, 9]

Localisation in source
Microsomes [5]; Cytosol [5]; Soluble [8, 9]

Purification
Pigeon [1]; Zoogloea ramigera (gene cloned in E. coli [8]) [2, 7, 8]; Bovine [3]; Rat [5]; Syntrophomonas wolfei (partial) [9]

Crystallization
–

Cloned

–

Renaturated

–

5 STABILITY

pH

6.0–8.5 [5]

Temperature (°C)

Oxidation

Organic solvent

General stability information

Glycerol stabilizes [7]

Storage

–20°C, 10 mM Tris/HCl buffer, pH 7.5, 10 mM 2-mercaptoethanol, 50% glycerol, several months [7]

6 CROSSREFERENCES TO STRUCTURE DATABANKS

PIR/MIPS code

PIR1:RDALAE (Alcaligenes eutrophus); PIR2:S06998 (Zoogloea ramigera); PIR2:S02631 (Zoogloea ramigera (fragment))

Brookhaven code

7 LITERATURE REFERENCES

[1] Wakil, S.J., Bressler, R.: J. Biol. Chem.,237,687–693 (1962)

[2] Saito, T., Fukui, T., Ikeda, F., Tanaka, Y., Tomita, K.: Arch. Microbiol.,114,211–217 (1977)

[3] Dodds, P.F., Guzman, M.G.F., Chalberg, S.C., Anderson, G.J., Kumar, S.: J. Biol. Chem.,256,6282–6290 (1981)

[4] Packter, N.M., Flatman, S.: Biochem. Soc. Trans.,11,598–599 (1973)

[5] Prasad, M.R., Cook, L., Vieth, R., Cinti, D.L.: J. Biol. Chem.,259,7460–7467 (1984)

[6] Cook, L., Prasad, M.R., Cook, W.R., Cinti, D.L.: Arch. Biochem. Biophys.,246, 206–216 (1986)

[7] Fukui, T., Ito, M., Saito, T., Tomita, K.: Biochim. Biophys. Acta,917,365–371 (1987)

[8] Ploux, O., Masamune, S., Walsh, C.T.: Eur. J. Biochem.,174,177–182 (1988)

[9] Amos, D.A., McInerney, M.J.: Arch. Microbiol.,159,16–20 (1993)

1 NOMENCLATURE

EC number
1.1.1.37

Systematic name
(S)-Malate:NAD^+ oxidoreductase

Recommended name
Malate dehydrogenase

Synonymes
L-Malate dehydrogenase
Malic dehydrogenase
NAD-L-malate dehydrogenase
Malic acid dehydrogenase
NAD-dependent malic dehydrogenase
NAD-malate dehydrogenase
NAD-malic dehydrogenase
Malate (NAD) dehydrogenase
NAD-dependent malate dehydrogenase
NAD-specific malate dehydrogenase
NAD-linked malate dehydrogenase
MDH
L-Malate-NAD^+ oxidoreductase [20]

CAS Reg. No.
9001-64-3

2 REACTION AND SPECIFICITY

Catalysed reaction
(S)-Malate + NAD^+ →
→ oxaloactetate + NADH

Reaction type
Redox reaction (mechanism [13])

Natural substrates
Oxaloacetate + NADH (citric acid cycle [12], in vivo reaction rates [15]) [12, 15]

Substrate spectrum

1 Oxaloacetate + NADH (r [3, 5, 6, 8, 9, 14, 24, 30, 32, 35], preferred direction [23], reverse reaction too slow to be measured [20], reverse reaction at 0.54% the rate of forward reaction [45]) [3–11, 13–17, 19, 20, 23–25, 27, 29, 30, 32, 35, 36, 45]
2 Ketomalonate + NADH (r [32]) [6, 32]
3 2-Oxobutyrate + NADH [6]
4 meso-Tartrate(S,R) + NAD^+ (0.82% of L-malate activity [6]) [6, 32]
5 (-)-Tartrate(S,S) + NAD^+ [32]
6 L-(-)-2-Oxoglutarate + NADH (extremly low rate) [32]
7 More (no reaction with D(+)-malate (R), (+)-tartrate (R,R), succinate, L-(+)-aspartate, D,L-alpha-hydroxybutyrate, 2-oxobutyrate) [32]

Product spectrum

1 L-Malate + NAD^+ [3, 5, 6, 8, 9, 14, 24, 30, 32, 35]
2 2-Hydroxymalonate + NAD^+ [32]
3 2-Hydroxybutyrate + NAD^+
4 ? + NADH
5 ? + NADH
6 2-Hydroxyglutarate + NAD^+
7 ?

Inhibitor(s)

NADH (product inhibition) [6, 13, 24]; Oxaloacetate (product inhibition [6]) [6, 13, 25, 32]; Iodoacetamide (mitochondrial MDH) [8]; Phenylmethanesulfonylfluoride (mitochondrial MDH) [8]; Dicarbonate [8]; Pyridoxal 5'-phosphate [8]; Citrate [9]; 2-Oxoglutarate (slight [9], mitochondrial MDH [17]) [9, 17, 23]; ATP [9, 17, 23, 33]; ADP [9, 23, 33]; AMP [9, 33]; NAD^+ [13, 17]; $NADP^+$ (slight) [17]; cis-Aconitate (mitochondrial MDH) [17]; Isocitrate (cytoplasmic MDH) [17]; Malate (noncompetitive with respect to oxalacetate [13]) [13, 20, 23, 24]; S-Acetyl-CoA [23]; 2',3'-cyclic AMP [23]; Enol-oxaloacetate [25]; Hg^{2+} [27, 30]; p-Chloromercuribenzoate [27]; Zn^{2+} [27]; Mg^{2+} [27]; Cu^{2+} [27]; Ni^{2+} [27]; Co^{2+} [27]; ADP-ribose [33]; Pt^{2+} [33]; Diethyldicarbonate [33]; Iodoacetate [33]; $AgNO_3$ [30]; More (not inhibitory: 5,5'-dithiobis(2-nitrobenzoate), 4,4'-bisdimethylaminodiphenylcarbinol, soluble MDH is not affected by thiol modifying agents, mitochondrial MDH is affected by thiol modifying agents and thus possesses a thiol group near coenzyme binding site [33], substrate inhibition [35]) [33, 35]

Cofactor(s)/prosthetic group(s)/activating agents

NAD+ (not replaceable by NADP+ [6, 45]) [3, 6, 7, 9, 45]; NADH (A-side specific, i.e. pro-R stereospecificity [32, 49], not replaceable by NADPH [6–8]) [3–9, 32, 45, 49]; NADPH (A-side specific, i.e. pro-R stereospecificity [5, 45], 10% of NADH activity [5], 20–30% of NADH activity [45]) [5, 45]; Urea (activation) [21]; Acetone (5–20% (v/v), activation, depending on organism) [21]; Guanidine-HCl (activation of enzyme from thermophilic organisms) [21]; Cysteine (activation) [27]; 2-Mercaptoethanol (activation) [27]; 3-Acetylpyridine adenine dinucleotide (activation) [35]; 3-Acetylpyridine hypoxanthine dinucleotide (activation) [35]

Metal compounds/salts

KCl (activation) [21]; K^+ (3.3 mM: activation) [27]; Na^+ (3.3 mM: activation) [27]; NH_4^+ (3.3 mM: activation) [27]; Phosphate (content: soluble MDHa 1.3–1.8 mol per mol enzyme, soluble MDHb 0.3–0.6 mol per mol enzyme) [33]

Turnover number (min^{-1})

6480 (NADH) [5]; 4600 (L-malate) [24]

Specific activity (U/mg)

348 [6]; 1280 [9]; 550 [14]; 1.24 [23]; More [3, 5, 7, 13, 16, 27, 29, 32, 35, 36]

K_m-value (mM)

0.0065 (oxaloacetate) [5]; 0.011–0.019 (NADH [5], oxaloacetate [9]) [5, 9]; 0.023 (oxaloacetate, mitochondrial MDH) [8]; 0.025–0.03 (NADH, (soluble MDH [27]) [6, 8, 9, 27], oxaloacetate [23]) [6, 8, 9, 23, 27]; 0.034–0.036 (oxaloacetate, soluble MDH [8], mitochondrial MDH [10]) [8, 10]; 0.048–0.049 (NAD+ [6], oxaloacetate [27]) [6, 27]; 0.052 (NADH, soluble MDH) [8]; 0.063 (NADH) [10]; 0.08–0.09 (oxaloacetate [6, 17], NADH [23]) [6, 17, 23]; 0.14–0.2 (NADH) [17]; 0.3 (NAD+, soluble MDH) [8]; 0.459 (NAD+) [27]; 0.6 (NAD+, mitochondrial MDH, L-malate, soluble MDH) [8]; 0.8–0.9 (L-malate, (mitochondrial MDH [8])) [8, 27]; 2.1 (malate) [6]; 7.0 (malate) [9]; More [24, 30, 35, 36, 45]

pH-optimum

4.3–7.0 (reduction of ketomalonate) [6]; 6.0–7.6 (Glycine max, reduction of oxaloacetate) [7]; 7.2 [10]; 7.3–8.5 (reduction of oxaloacetate) [6]; 7.5–9.5 (soluble MDH) [8]; 7.6 (reduction of oxaloacetate) [35]; 8.0 (mitochondrial MDH) [8]; [8]; 8.0–8.5 (Rhizobium japonicum, reduction of oxaloacetate) [7]; 8.1 (reduction of oxaloacetate) [5]; 8.2–9.4 (reduction of oxaloacetate) [9]; 8.3–8.6 (Glycine max, oxidation of L-malate) [7]; 8.5 (reduction of oxaloacetate) [36]; 8.8 (reduction of oxaloacetate) [32]; 8.6–9.0 (Rhizobium japonicum, oxidation of L-malate) [7]; 9.0 (oxidation of L-malate) [27]; 9.0–9.5 (oxidation of malate) [36]; 9.2–9.6 (reduction of oxaloacetate) [14, 27]; 9.5 (oxidation of L-malate) [6]; 9.7–10.6 (oxidation of malate) [9, 32, 35]; More (effects of buffer composition) [9]

pH-range

6.5–9.0 (85% of maximal activity at pH 6.5, 83% of maximal activity at pH 9.0) [23]

Temperature optimum (°C)

40 [10]; 65 [5]

Temperature range (°C)

3 ENZYME STRUCTURE

Molecular weight

56000–60000 (Rhodocyclus purpureus, E. coli, pig, gel filtration, sucrose gradient centrifugation) [3]

61000–61700 (Methanospirillum hungatii, gel filtration [23], Moraxella lwoffi, gel filtration, ultracentrifugation [30]) [23, 30]

64000–65000 (Anthocidaris crassispina, gel filtration [9], Methanospirillum hungatii, sedimentation equilibrium centrifugation [20]) [9, 20]

64000–186000 (Spinacia oleracea, gel filtration, values depend on presence of Mg^{2+}, K^{+}, dithioerythritol) [34]

68000 (Aspergillus niger, gel filtration [17], Drosophila virilis, gel filtration [36]) [17, 36]

69000–70000 (Euglena gracilis, mitochondrial isozyme, gel filtration [4], Glycine max, native PAGE [7], Dictyostelium discoideum, gel filtration [14], Physarum polycephalum, gel filtration, sedimentation equilibrium centrifugation [31]) [4, 7, 14, 31]

70000–80000 (Thermomonospora fuscha, gel filtration, sedimentation equilibrium centrifugation [21], Pseudomonas testosteroni, gel filtration, sedimentation equilibrium centrifugation [32]) [21, 32]
71000–72400 (Euglena gracilis, cytosolic isozyme, gel filtration [4], human liver, cytosolic enzyme, gel filtration [13], rat liver, cytosolic enzyme, gel filtration [15], Thermus aquaticus, gel filtration, sedimentation equilibrium centrifugation [21]) [4, 13, 15, 21]
76000 (Lycopersicon esculentum, gel filtration) [2]
80000 (Rhodobacter capsulatus, gel filtration) [6]
84000 (Halobacterium sp., sedimentation equilibrium centrifugation) [29]
86860 (Mycobacterium phlei, gel filtration) [27]
105000–144000 (Thermoplasma acidophilum, gel filtration, sedimentation equilibrium centrifugation, cross-linkage with dimethylsuberimidate) [5]
122000–135000 (Rhodomicrobium vannielii, gel filtration, sucrose gradient centrifugation [3], Bacillus caldotenax, Thermoactinomyces sacchari, gel filtration, sedimentation equilibrium centrifugation [21], Sulfolobus acidocaldarius, sedimentation equilibrium centrifugation, electrophoresis after cross-linkage with dimethylsuberimidate [45]) [3, 21, 45]
130000–142000 (Rhodobacter capsulatus, Rhodospirillum rubrum, gel filtration, sucrose gradient centrifugation, native PAGE [3], Rhizobium japonicum, native PAGE [7], Bacillus subtilis, Bacillus stearothermophilus, gel filtration sedimentation equilibrium centrifugation [21]) [3, 7, 21]
More (comparison of amino acid composition of MDH from various species [20], overview [33]) [20, 33]

Subunits

Tetramer (4 × 35000–37000, Rhodobacter capsulatus, Rhodospirillum rubrum, Rhodomicrobium vannielii, SDS-PAGE [3], Thermoplasma acidophilum, SDS-PAGE [5], Rhizobium japonicum, SDS-PAGE [7], Bacillus caldotenax, Bacillus stearothermophilus, Bacillus subtilis, SDS-PAGE [21]) [3, 5, 7, 21]
Dimer (2 × 31350, Methanospirillum hungatii, SDS-PAGE [20], 2 × 34000, Euglena gracilis, mitochondrial isozyme, SDS-PAGE [4], 2 × 36000, Euglena gracilis, cytosolic isozyme, SDS-PAGE [4], Thermus aquaticus, SDS-PAGE [21], 2 × 35000, Rhodobacter capsulatus, SDS-PAGE [6], Anthocidaris crassispina, SDS-PAGE [9], Aspergillus niger, SDS-PAGE [17], Thermomonospora fuscha, SDS-PAGE [21], 2 × 37000, Glycine max, SDS-PAGE [7]) [4, 6, 7, 9, 17, 20, 21]

Glycoprotein/Lipoprotein

–

4 ISOLATION/PREPARATION

Source organism

Rhodobacter capsulatus (formerly Rhodopseudomonas) [1, 3, 6]; Rhodospirillum rubrum [1, 3]; Rhodomicrobium vannielii [1, 3]; Rhodocyclus purpureus [1, 3]; Bacillus megaterium [3]; Lycopersicon esculentum (tomato) [2]; E. coli [3, 16, 41, 42]; Euglena gracilis [4, 8, 10]; Horse [47]; Chicken [48]; Thermoplasma acidophilum [5, 49]; Sulfolobus acidocaldarius [45, 49]; Halobacterium marismortui [18, 19]; Glycine max [7]; Rhizobium japonicum [7]; Anthocidaris crassispina (sea urchin) [9]; Phycomyces blakesleeanus [11]; Arabidopsis thaliana [12]; Human [13]; Dictyostelium discoideum [14]; Rat [15]; Bacillus subtilis [16, 21, 40–42]; Bacillus caldotenax [16, 21]; Thermus aquaticus [16, 21]; Aspergillus niger [17]; Bacillus stearothermophilus [21, 41, 42]; Chromobacterium violaceum [42]; Methanospirillum hungatii [20, 23]; Neurospora crassa [21]; Thermoactinomyces sacchari [21]; Pseudomonas indigofera [21]; Thermus flavus [22]; Pig [3, 21, 24–26, 38, 43, 44]; Mycobacterium phlei [27]; Moraxella lwoffi [30]; Physarum polycephalum [31, 35]; Spinacia oleracea (spinach) [34]; Thermomonospora fuscha [21]; Halobacterium sp. [28, 29]; Drosophila virilis [36]; Drosophila melanogaster [36]; Bovine (male [37]) [37, 39]; Saccharomyes cerevisiae [46]; Pseudomonas testosteroni [32]; More (overview) [33]

Source tissue

Liver [13, 15]; Heart [3, 24–26, 37–39, 43, 44, 47, 48]; Root nodules (Glycine max) [7]; Leaf [12, 33]; Fruit [2]; Eggs (unfertilized) [9]; Mycelium [11, 17]

Localisation in source

Cytoplasm [4, 7, 8, 11, 13, 15, 17, 24, 25, 27, 30, 33, 35, 36]; Mitochondria (matrix [10]) [3, 4, 7–10, 14, 17, 24–26, 35, 37]

Purification

Lycopersicon esculentum [2]; Rhodobacter capsulatus [3, 6]; Rhodospirillum rubrum [3]; Rhodomicrobium vannielii [3]; Rhodocyclus purpureus [3]; Euglena gracilis (cytosolic and mitochondrial isozyme) [4]; Thermoplasma acidophilum [5]; Rhizobium japonicum [7]; Glycine max [7]; Anthocidaris crassispina [9]; Human [13]; Dictyostelium discoideum [14]; Rat [15]; Bacillus subtilis [16, 40]; Bacillus caldotenax [16]; Thermus aquaticus [16]; Aspergillus niger [17]; Halobacterium marismortui [19]; Phycomyces blakesleeanus [11]; E. coli [16, 41]; Methanospirillum hungatii [20, 23]; Mycobacterium phlei [27]; Halobacterium sp. [29]; Moraxella lwoffi [30]; Physarum polycephalum [35]; Pseudomonas testosteroni [32]; Drosophila virilis [36]; Pig [38]; Bovine [39]

Crystallization

(crystallographic properties [26], crystal structure [33, 43, 44]) [5, 26, 32, 33, 43–45]

Cloned

–

Renaturated

–

5 STABILITY

pH

7–9 [45]; 7–10 [5]

Temperature (°C)

40 (Rhodobacter capsulatus, 20 min, 60% loss of activity, Rhodomicrobium vannielii, 10 min, 90% loss of activity [1], 10 min, inactivation, stabilization by NAD^+, NADH, oxaloacetate, ATP, ADP [6]) [1, 6]; 45 (50 min, 90% loss of activity, but only 20% loss in presence of NADH) [9]; 50 (Rhodocyclus purpureus, inactivation [1], 15 min, stable [27], 60 min, mitochondrial MDH: 80% loss of activity, soluble MDH: 35% loss of activity [35]) [1, 27, 35]; 51 (mitochondrial MDH, half-life 15 min) [35]; 55 (20 min, up to 25% loss of activity, stabilization by NADH, destabilization by oxaloacetate [7], 15 min, 40% loss of activity [27], soluble MDH: half-life 15 min [35]) [7, 27, 35]; 60 (10 min, stable [5], 15 min, 75% loss of activity [27]) [5, 27]; 65 (Rhodospirillum rubrum, 2 h, stable) [1]; 70 (Rhodospirillum rubrum, 1 h, 80–90% loss of activity) [1]; 77 (10 min, 50% loss of activity) [5]; More (thermostability of enzyme from cold- and warm-adapted plants [12], stability increases with increasing salt concentration and decreasing temperature [18]) [12, 18]

Oxidation

Organic solvent

General stability information

Sulfate stabilizes [19]; Chloride stabilizes [19]; Citrate: stabilizes [19]; Na^+ stabilizes [19]; K^+: stabilizes [19]; Cs^+: stabilizes [19]; 2-Mercaptoethanol stabilizes [35]; NAD^+ stabilizes against heat inactivation [6]; NADH stabilizes against heat inactivation [6]; Oxaloacetate stabilizes against heat inactivation [6]; ATP stabilizes against heat inactivation [6]; ADP stabilizes against heat inactivation [6]

Storage

4°C, 25 mM phosphate buffer, pH 7.4, 1 mM EDTA, 1 mM DTT, 5 mM $MgCl_2$, 1 month, Rhizobium japonicum less than 20% loss of activity, Glycine max 80% loss of activity [7]; –20°C, several months [9]; 4°C [13]; 4°C, up to 3 months, but inactivation at –20°C [15]; 4°C, 0.01 M Tris-HCl buffer, pH 8.0, 1 M NaCl, at least 4 weeks [23]; 0°C, 0.05 M potassium phosphate buffer, pH 7.5, 0.001 M mercaptoethanol, up to 2 months [35]

6 CROSSREFERENCES TO STRUCTURE DATABANKS

PIR/MIPS code

PIR2:S04959 (Actinoplanes missouriensis (fragment)); PIR2:S03352 (diatom (Nitzschia alba) (fragment)); PIR1:DEECM (Escherichia coli); PIR3:A49496 (Haloarcula marismortui); PIR2:S04961 (Kibdelosporangium aridum (fragment)); PIR3:S08981 (Methanothermus fervidus); PIR2:S08689 (Methanothermus fervidus (fragment)); PIR2:S04958 (Microtetraspora glauca (fragment)); PIR2:S07574 (Phenylobacterium immobile (fragment)); PIR2:S04957 (Planomonospora venezuelensis (fragment)); PIR1:DEEBM (Salmonella typhimurium); PIR2:JN0504 (Salmonella typhimurium); PIR2:S04960 (Streptomyces atratus (fragment)); PIR2:S04956 (Streptosporangium roseum (fragment)); PIR2:S03958 (Sulfolobus acidocaldarius (fragment)); PIR2:A60689 (Thermoleophilum album (fragment)); PIR1:DETWMA (Thermus aquaticus (strain AT-62)); PIR1:DETWMB (Thermus aquaticus (strain B)); PIR1:DEPUGW (precursor glyoxysomal watermelon); PIR1:DEMSMM (precursor mitochondrial mouse); PIR1:DERTMM (precursor mitochondrial rat); PIR1:DEPUMW (precursor mitochondrial watermelon); PIR1:DEBYMM (precursor mitochondrial yeast (Saccharomyces cerevisiae)); PIR1:DEMSMC (cytosolic mouse); PIR2:S00192 (cytosolic pig (fragment)); PIR1:DEBYMC (cytosolic yeast (Saccharomyces cerevisiae)); PIR2:S05770 (cytosolic yeast (Saccharomyces cerevisiae) (fragment)); PIR1:DEPGMM (mitochondrial pig); PIR1:DEBYMP (peroxisomal yeast (Saccharomyces cerevisiae))

Brookhaven code

1EMD (Escherichia coli); 2CMD (Escherichia coli); 4MDH (Porcine (Sus scrofa) heart)

7 LITERATURE REFERENCES

[1] Tayeh, M.A., Madigan, M.T.: Biochem. J.,252,595–600 (1988)

[2] Jeffery, D., Goodenough, P.W., Weitzmann, P.D.J.: Phytochemistry,27,41–44 (1988)

[3] Tayeh, M.A., Madigan, M.T.: J. Bacteriol.,169,4196–4202 (1987)

[4] Miyatake, K., Washio, K., Yokota, A., Nakano, Y., Kitaoka, S.: Agric. Biol. Chem.,50,2651–2653 (1986)

[5] Grossebüter, W., Hartl, T., Görisch, H., Stezowski, J. J.: Biol. Chem. Hoppe-Seyler, 367,457–463 (1986)

[6] Oshima, T., Sakuraba, H.: Biochim. Biophys. Acta,869,171–177 (1986)

[7] Waters, J.K., Karr, D.B., Emerich, D.W.: Biochemistry,24,6479–6486 (1985)

[8] Miyatake, K., Washio, K., Yokota, A., Nakano, Y., Kitaoka, S.: Agric. Biol. Chem.,49,859–860 (1985)

[9] Okabayashi, K., Nakano, E.: J. Biochem.,95,1625–1632 (1984)

[10] Isegawa, Y., Nakano, Y., Kitaoka, S.: Agric. Biol. Chem.,48,549–552 (1984)

[11] De Arriaga, D., Teixido, F., Busto, F., Soler, J.: Biochim. Biophys. Acta,784,158–163 (1984)

[12] Potvin, C., Simon, J.-P., Blanchard, M.-H.: Plant Sci. Lett.,31,35–47 (1983)
[13] Crow, K.E., Braggins, T.J., Hardman, M.J.: Arch. Biochem. Biophys.,225,621–629 (1983)
[14] Emyanitoff, R.G., Kelly, P.J.: J. Gen. Microbiol.,128,1767–1771 (1982)
[15] Crow, K.E., Braggins, T.J., Batt, R.D., Hardmann, M.J.: J. Biol. Chem.,257,14217–14225 (1982)
[16] Smith, K., Sundaram, T.K., Kernick, M., Wilkinson, A. E.: Biochim. Biophys. Acta,708,17–25 (1982)
[17] Ma, H., Kubicek, C., Röhr, M.: FEMS Microbiol. Lett.,12,147–151 (1981)
[18] Pundak, S., Aloni, H., Eisenberg, H.: Eur. J. Biochem.,118,471–477 (1981)
[19] Pundak, S., Eisenberg, H.: Eur. J. Biochem.,118,463–470 (1981)
[20] Storer, A.C., Sprott, G.D., Martin, W.G.: Biochem. J.,193,235–244 (1981)
[21] Sundaram, T.K., Wright, I.P., Wilkinson, A.E.: Biochemistry,19,2017–2022 (1980)
[22] Iijima, S., Saiki, T., Beppu, T.: Biochim. Biophys. Acta,613,1–9 (1980)
[23] Sprott, G.D., McKellar, R.C., Shaw, K.M., Giroux, J., Martin, W.G.: Can. J. Microbiol.,25,192–200 (1978)
[24] Mueggler, P.A., Wolfe, R.G.: Biochemistry,17,4615–4620 (1978)
[25] Bernstein, L.H., Grisham, M.B., Cole, K.D., Everse, J.: J. Biol. Chem.,253,8697–8701 (1978)
[26] Weininger, M.S., Banaszak, L.J.: J. Mol. Biol.,119,443–449 (1978)
[27] Tyagi, A.K., Siddiqui, F.A., Venkitasubramanian, T.A. : Biochim. Biophys. Acta,485,255–267 (1977)
[28] Mevarech, M., Neumann, E.: Biochemistry,16,3786–3792 (1977)
[29] Mevarech, M., Eisenberg, H., Neumann, E.: Biochemistry,16,3781–3785 (1977)
[30] Fernandes, R., Jones, M., King, H.K.: Biochem. Soc. Trans.,4,1080 (1976)
[31] Teague, W.M., Henney, H.R.: Biochim. Biophys. Acta,434,118–125 (1976)
[32] You, K.-S., Kaplan, N.O.: J. Bacteriol.,123,704–716 (1975)
[33] Banaszak, L.J., Bradshaw, R.A. in "The Enzymes",3rd Ed. (Boyer, P.D., ed.) 11,369–396 (1975) (Review)
[34] Ziegler, I.: Phytochemistry,13,2403–2410 (1974)
[35] Teague W.M., Henney, H.R.: J. Bacteriol.,116,673–684 (1973)
[36] McReynolds, M.S., Kitto, G.B.: Biochim. Biophys. Acta,198,165–175 (1970)
[37] Davies, D.D., Kun, E.: Biochem. J.,66,307–316 (1957)
[38] Thorne, C.J.R., Cooper, P.M.: Biochim. Biophys. Acta,81,397–399 (1963)
[39] Guha, A., Engelhard, S., Listowsky, I.: J. Biol. Chem.,243,609–615 (1968)
[40] Yoshida, A.: J. Biol. Chem.,240,1113–1117 (1965)
[41] Murphey, W.H., Barnaby, C., Lin, F.J., Kaplan, N.O.: J. Biol. Chem.,242,1548–1559 (1967)
[42] Murphey, W.H., Kitto, G.B., Everse, J., Kaplan, N.O.: Biochemistry,6,603–609 (1967)
[43] Hill, E., Tsernoglou, D., Webb, C., Banaszak, L.J.: J. Mol. Biol.,72,577–591 (1972)
[44] Webb, L.E., Hill, E., Banaszak, L.J.: Biochemistry,12,5101–5109 (1973)
[45] Hartl, T., Grossebüter, W., Görisch, H., Stezowski, J.J.: Biol. Chem. Hoppe-Seyler, 368,259–260 (1987)
[46] Hägele, E., Neeff, J., Mecke, D.: Eur. J. Biochem.,83,67–76 (1978)
[47] Thorne, C.J.R.: Biochim. Biophys. Acta,59,624–633 (1962)
[48] Kitto, G.B., Kaplan, N.O.: Biochemistry,5,3966–3980 (1966)
[49] Görisch, H., Hartl, T., Grossebüter, W., Stezowski, J.J.: Biochem. J.,226,885–888 (1985)

1 NOMENCLATURE

EC number
1.1.1.38

Systematic name
(S)-Malate:NAD^+ oxidoreductase (oxaloacetate-decarboxylating)

Recommended name
Malate dehydrogenase (oxaloacetate-decarboxylating)

Synonymes
Malic enzyme
Pyruvic-malic carboxylase
NAD-specific malic enzyme
NAD-malic enzyme
NAD-linked malic enzyme

CAS Reg. No.
9080-52-8

2 REACTION AND SPECIFICITY

Catalysed reaction
(S)-Malate + NAD^+ →
→ pyruvate + CO_2 + NADH
Oxaloacetate →
→ pyruvate + CO_2

Reaction type
Redox reaction
Oxidative decarboxylation

Natural substrates
L-Malate + NAD^+ (production of reducing equivalents for the synthesis of fatty acids and other functions, production of pyruvate for gluconeogenesis) [2]

Substrate spectrum
1 L-Malate + NAD^+ (r) [1–12]
2 Oxaloactetate [7, 12, 13]
3 Oxaloacetate + NAD(P)H [7]
4 Pyruvate + NAD(P)H [7]

Product spectrum

1 Pyruvate + CO_2 + NADH [1–12]
2 Pyruvate + CO_2 [7, 12, 13]
3 L-Malate + $NAD(P)^+$ [7]
4 L-Lactate + $NAD(P)^+$ [7]

Inhibitor(s)

Coenzyme-A [6, 8, 10]; Acetyl-CoA [6, 8]; ATP [6, 8]; 5,5'-Dithiobis(2-nitrobenzoate) [1, 5, 7]; 4,4'-Dithiodipyridine [1, 5]; 2,3-Butanedione [1]; p-Hydroxymercuribenzoate [7, 13]; N-Ethylmaleimide [7]; Diethyldicarbonate [3]; Urea [11]; Guanidine [11]

Cofactor(s)/prosthetic group(s)/activating agents

NAD^+ [1–12]; $NADP^+$ [12, 13, 16]; NADH [1–12]

Metal compounds/salts

Mn^{2+} (required for activity [1–12], K_m: 0.14 mM [12]) [1–12]; Mg^{2+} (required for activity) [1–12]; Divalent cations (required for activity) [2, 10, 13, 16]; NH_4^+ (enhances activity) [16]; K^+ (enhances activity) [16]

Turnover number (min^{-1})

Specific activity (U/mg)

170 [7, 11]; 17.5 [11]; 0.2 [12]; More [7, 11, 14]

K_m-value (mM)

50 (L-malate) [12]; 0.5 (NAD^+) [12]; More [9–11]

pH-optimum

8.0 (oxidative decarboxylation of malate) [16]; 7.4 [12]; 7.5 (oxidative decarboxylation of L-malate, decarboxylation of oxaloacetate [7]) [7, 13]; 6.0 (reduction of pyruvate) [16]; 5.0 (reduction of pyruvate) [7]; 4.0 (reduction of oxaloacetate) [7]

pH-range

5.5–7.5 (5.5: about 40% of activity maximum, 7.5: about 50% of activity maximum, pyruvate reduction) [16]; 6–9 (6: about 45% of activity maximum, 9: about 60% of activity maximum, malate oxidation) [16]

Temperature optimum (°C)

55 [16]

Temperature range (°C)

3 ENZYME STRUCTURE

Molecular weight

150000 (Lactobacillus plantarum, gel filtration) [11]
130000–140000 (Leuconostoc mesenteroides, gel filtration) [11]
203000 (E. coli, density gradient centrifugation) [13]
200000 (10 bacterial strains, overview [9], Bacillus stearothermophilus (gene cloned in E. coli), HPLC gel filtration [16]) [9, 16]

Subunits

Tetramer (4 × 57500, E. coli, SDS gel-electrophoresis [13], 4 × 48000, Bacillus stearothermophilus (gene cloned in E. coli), SDS-PAGE [16]) [13, 16]

Glycoprotein/Lipoprotein

–

4 ISOLATION/PREPARATION

Source organism

Ascaris suum [1–5, 15]; E. coli [6–10, 13, 16]; Lactobacillus plantarum [11]; Lactobacillus arabinosus [14]; Leuconostoc mesenteroides [11]; Saccharomyces cerevisiae [12]; Escherichia freundii [9]; Aerobacter aerogenes [9]; Erwinia aroideae [9]; Pseudomonas vulgaris [9]; Agrobacterium tumefaciens [9]; Clostridium tetanomorphum [9]; Bacillus stearothermophilus [16]

Source tissue

Localisation in source

Mitochondria [1]

Purification

E. coli [13]; Ascaris suum [15]; Lactobacillus arabinosus [14]; Lactobacillus plantarum (partial) [11]; Leuconostoc mesenteroides (partial) [11]; Saccharomyces cerevisiae (partial) [12]; Bacillus stearothermophilus (gene cloned in E. coli) [16]

Crystallization

–

Cloned

(Bacillus stearothermophilus gene cloned in E. coli) [16]

Renaturated

–

5 STABILITY

pH

6–10 (4°C, 24 h, stable) [16]; 7.5 [12]

Temperature (°C)

60 (20 min, 20% loss of activity) [16]

Oxidation

Organic solvent

General stability information

Stable to freezing/thawing [2]

Storage

–20°C, 15 mM triethanolamine-maleate buffer, pH 7.5, 1 mM EDTA, 10 mM DTT, 5% v/v glycerol, several months [2, 3]; –20°C, N_2–atmosphere, trisethanolamine, EDTA, dithiothreitol, glycerol, several months [5]; 4°C, 100 mM Tris-buffer, pH 7.4, 10 mM $MgCl_2$, 5 mM mercaptoethanol, 5 mM L-aspartate, 3–4 months [6, 13]

---

6 CROSSREFERENCES TO STRUCTURE DATABANKS

PIR/MIPS code

Brookhaven code

OMMD (Porcine (Sus scrofa) heart)

---

7 LITERATURE REFERENCES

[1] Rao, G.S.J., Kong, C.-T., Benjamin, R.C., Harris, B.G., Cook, P.F.: Arch. Biochem. Biophys.,255,8–13,(1987)

[2] Park, S.-H., Harris, B.G., Cook, P.F.: Biochemistry,25,3752–3759,(1986)

[3] Rao, J.G.S., Harris, B.G., Cook, P.F.: Arch. Biochem. Biophys.,241,67–74,(1985)

[4] Park, S.-H., Kiick, D.M., Harris, B.G., Cook, P.F.: Biochemistry,23,5446–5453,(1984)

[5] Kiick, D.M., Allen, B.L., Rao, J.G.S., Cook, P.F.: Biochemistry,23,5454–5459,(1984)

[6] Cook, R.A.: Biochim. Biophys. Acta,749,198–203,(1983)

[7] Yamaguchi, M.: J. Biochem.,86,325–333,(1979)

[8] Milne, J.A., Cook, R.A.: Biochemistry,18,3604–3610,(1979)

[9] Iwakura, M., Tokushige, M., Katsuki, H., Muramatsu, S. : J. Biochem.,83,1387–1394,(1978)

[10] Yamaguchi, M., Tokushige, M., Takeo, K., Katsuki, H.: J. Biochem.,76,1259–1268,(1974)

[11] Schütz, M., Radler, F.: Arch. Mikrobiol.,91,183–202,(1973)

[12] Fuck, E., Stärk, G., Radler, F.: Arch. Mikrobiol.,89,223–231,(1973)

[13] Yamaguchi, M., Tokushige, M., Katsuki, H.: J. Biochem.,73,169–180,(1973)

[14] Kaufmann, S., Korkes, S., Del Campillo, A.: J. Biol. Chem.,192,301–312,(1951)

[15] Allen, B.L., Harris, B.G.: Mol. Biochem. Parasitol.,2,367ff.(1981)

[16] Kobayashi, K., Doi, S., Negoro, S., Urabe, I., Okada, H.: J. Biol. Chem.,264,3200–3205 (1989)

1 NOMENCLATURE

EC number
1.1.1.39

Systematic name
(S)-Malate:NAD$^+$ oxidoreductase (decarboxylating)

Recommended name
Malate dehydrogenase (decarboxylating)

Synonymes
Malic enzyme
Pyruvic-malic carboxylase
NAD-specific malic enzyme
NAD-malic enzyme

CAS Reg. No.
9028-46-0

2 REACTION AND SPECIFICITY

Catalysed reaction
(S)-Malate + NAD$^+$ →
→ pyruvate + CO_2 + NADH (does not decarboxylate added oxaloacetate)

Reaction type
Redox reaction
Oxidative decarboxylation

Natural substrates

Substrate spectrum
1 L-Malate + NAD$^+$ (r) [1–6, 8–17]

Product spectrum
1 Pyruvate + CO_2 + NADH [1–6, 8–17]

Inhibitor(s)
Citrate [8]; N-Ethylmaleimide [6]; KCNO [6]; Bicarbonate [14]; ATP [16]

Cofactor(s)/prosthetic group(s)/activating agents
NAD$^+$ [1–6, 8–17]; NADP$^+$ (5–30% of NAD$^+$-activity) [7, 14]; CoA (activation) [5, 7, 8]; NADH [1–6, 8–17]

Metal compounds/salts

Mn^{2+} (required) [1–4, 6–8, 11–16]; Mg^{2+} (required) [1–4, 6–8, 11–16]; Divalent cations (required) [1, 6]

Turnover number (min^{-1})

1400 (malate, with Mg^{2+}) [8]; 1000 (malate, with Mn^{2+}) [8]

Specific activity (U/mg)

637 [18]; 81.5 [3]; More [11, 13]

K_m-value (mM)

12.54 (malate) [12]; 0.3 (NAD^+) [14]; More [1, 6, 8, 11, 14, 16]

pH-optimum

6.7 [2]; 6.8 [11]; 7.2 [10, 14]; 7–7.5 [5]; 7.6–7.9 [13]

pH-range

5.5–7.5 [2]; 6–7 [8, 11]

Temperature optimum (°C)

Temperature range (°C)

3 ENZYME STRUCTURE

Molecular weight

200000–400000 (Brassica oleracea, gel filtration) [12, 19]
115000–490000 (Solanum tuberosum, gel filtration) [11, 20, 21]
279000 (Atriplex spongiosa, gel filtration) [14]
120000–490000 (Crassula argentea, gel electrophoresis) [8]

Subunits

Dimer (1 × 58000 and 1 × 61000, C_4-plants, association of subunits depending on isolation method, conceivable that the 58000 MW polypeptide is a proteolytic product of the 62000 MW subunit, enzyme is active as dimer, tetramer and octamer, plants, overview) [1]

Tetramer (2 × 58000 and 2 × 61000, association of subunits depending on isolation method, conceivable that the 58000 MW polypeptide is a proteolytic product of the 62000 MW subunit, enzyme is active as dimer, tetramer and octamer, plants, overview) [1]

Octamer (4 × 58000 and 4 × 61000, association of subunits depending on isolation method, conceivable that the 58000 MW polypeptide is a proteolytic product of the 62000 MW subunit, enzyme is active as dimer, tetramer and octamer, plants, overview [1]) [1, 8, 11]

Glycoprotein/Lipoprotein

–

4 ISOLATION/PREPARATION

Source organism

C_4-plants [1, 7, 14]; Brassica oleracea (cauliflower) [1, 5, 6, 12, 19]; Solanum tuberosum (potato) [1–3, 5, 11, 20, 21]; Crassula argentea [1, 3, 4, 8, 10]; Amaranthus retroflexus [9]; Glossina morsitans (tse-tse fly) [13]; Human [15]; Bovine [15]; Salmon [15]; Cod [15]; Rabbit [16]; Ascaris lumbricoides [18]; Pseudomonas aeruginosa [17]; Atriplex spongiosa [8, 14]; Panicum miliaceum [14]; Amaranthus edulis [14]; More (distribution in insects) [13]

Source tissue

Tuber [1–3, 5, 11]; Leaf [4, 8, 9, 14]; Muscle [13]; Florets [6]; Spermatozoa [15]; Heart [16]

Localisation in source

Mitochondria [1–16, 18–21]

Purification

C_4-plants [14]; Crassula argentea [8, 10]; Solanum tuberosum [11]; Amaranthus retroflexus (partial) [9]; Glossina morsitans (partial) [13]; Rabbit [16]

Crystallization

–

Cloned

–

Renaturated

–

5 STABILITY

pH

Temperature (°C)

40 (unstable above) [17]

Oxidation

Organic solvent

General stability information

Freezing destroys plant enzyme [10, 14]

Storage

22°C, N_2–atmosphere, Mn^{2+}, dithiothreitol [14]

6 CROSSREFERENCES TO STRUCTURE DATABANKS

PIR/MIPS code

PIR2:A60683 (finger millet (fragment)); PIR2:B60683 (millet (fragment)); PIR3:S40269 (potato); PIR2:C60683 (tampala (fragment)); PIR3:S40242 (59K chain precursor mitochondrial potato)

Brookhaven code

7 LITERATURE REFERENCES

[1] Artus, N.N., Edwards, G.E.: FEBS Lett.,182,225–233 (1985) (Review)
[2] Willeford, K.O., Wedding, R.T.: Plant Physiol.,84,1084–1087 (1987)
[3] Willeford, K.O., Wedding, R.T.: J. Biol. Chem.,262,8423–8429 (1987)
[4] Willeford, K.O., Wedding, R.T.: Plant Physiol.,80,792–795 (1986)
[5] Day, D.A., Neuburger, M., Douce, R.: Arch. Biochem. Biophys.,231,233–242 (1984)
[6] Canellas, P.F., Wedding, R.T.: Arch. Biochem. Biophys.,229,414–425 (1984)
[7] Outlaw, W.H., Springer, S.A. in "Methods Enzym. Anal." (3rd Ed.) , Vol.3,176–183 (1983)
[8] Wedding, R.T., Black, M.K.: Plant Physiol.,72,1021–1028 (1983)
[9] Hatch, M.D., Tsuzuki, M., Edwards, G.E.: Plant Physiol.,69,483–491 (1982)
[10] Wedding, R.T., Canellas, P.F., Black, M.K.: Plant Physiol.,68,1416–1423 (1981)
[11] Grover, S.D., Canellas, P.F., Wedding, R.T.: Arch. Biochem. Biophys.,209,396–407 (1981)
[12] Canellas, P.F., Wedding, R.T.: Arch. Biochem. Biophys.,199,259–264 (1980)
[13] Hoek, H.B., Pearson, D.J., Olembo, N.K.: Biochem. J.,160,253–262 (1976)
[14] Hatch, M.D., Mau, S.-L., Kagawa, T.: Arch. Biochem. Biophys.,165,188–200 (1974)
[15] Mounib, M.S.: FEBS Lett.,48,79–84 (1974)
[16] Lin, R.C., Davis, E.J.: J. Biol. Chem.,249,3867–3875 (1974)
[17] Eyzaguirre, J., Cornwell, E.., Borie, G., Ramirez, B. : J. Bacteriol.,116,215–221 (1973)
[18] Saz, H.J., Hubbard, J.A.: J. Biol. Chem.,225,921–933 (1957)
[19] Davies, D.D., Patil, K.D.: Planta,126,197–211 (1975)
[20] Day, D.A., Neuburger, M., Douce, R.: Arch. Biochem. Biophys.,229,253–258 (1984)
[21] Grover, S.D., Wedding, R.T.: Plant Physiol.,70,1169–1172 (1982)

1 NOMENCLATURE

EC number
1.1.1.40

Systematic name
(S)-Malate:NADP+ oxidoreductase (oxaloacetate-decarboxylating)

Recommended name
Malate dehydrogenase (oxaloacetate-decarboxylating, NADP+)

Synonymes
'Malic'enzyme
Pyruvic-malic carboxylase
Malate dehydrogenase (decarboxylating, NADP)
NADP-linked decarboxylating malic enzyme
NADP-malic enzyme
NADP-specific malic enzyme
NADP-specific malate dehydrogenase [3]
Malate dehydrogenase (NADP, decarboxylating) [18]
L-Malate:NADP oxidoreductase [3]

CAS Reg. No.
9028-47-1

2 REACTION AND SPECIFICITY

Catalysed reaction
(S)-Malate + NADP+ →
→ pyruvate + CO_2 + NADPH [2, 9, 11]
Oxaloacetate →
→ pyruvate + CO_2 [2]

Reaction type
Redox reaction
Decarboxylation

Natural substrates
Malate + NADP+ [2, 3, 9, 11, 16]
Oxaloacetate [3]
Pyruvate + CO_2 + NADPH [8]

Substrate spectrum

1 (S)-Malate + $NADP^+$ (r) [2, 8, 9, 11]
2 Oxaloacetate [2]
3 More (not: malate (doubly-ionized) [2], NAD^+ [2, 8], 3'$NADP^+$ [2], NADH [8], oxaloacetate [15]) [2, 8, 15]

Product spectrum

1 Pyruvate + CO_2 + NADPH [2, 8, 9, 11]
2 Pyruvate + CO_2 [2]
3 ?

Inhibitor(s)

NAD^+ [2]; Arsenite (weak) [2]; Iodoacetate (weak) [2]; Diphenyliodonium chloride (weak) [2]; o-Iodosobenzoate (strong) [2]; p-Substituted mercuribenzoate (strong) [2]; Hg^{2+} [2]; Cu^{2+} (weak) [2]; p-Mercuriphenylsulfonate [2]; Acetyl-CoA [5, 18]; Pyruvate [7, 17]; Sulfite [8]; EDTA [8]; 2-Mercaptoethanol (absence of Mg^{2+}) [10]; Glutathione (absence of Mg^{2+}) [10]; Malate [11]; Oxalate [12]; Fumarate (weak) [13]; Succinate (weak) [13]; L(+)-Tartrate [13]; Phenylglyoxal [16]; 2,3-Butanedione [16]; Oxaloacetate [17]; 2-Oxoglutarate [17]; Malonyl-CoA (weak) [18]

Cofactor(s)/prosthetic group(s)/activating agents

$NADP^+$ [2, 8, 9, 11]; NADPH [2, 3, 7, 9, 11]

Metal compounds/salts

Mn^{2+} (required [2, 5, 7–9], K_m: 0.008 mM [7], 0.0018 mM [8]) [2, 5, 7–9]; Mg^{2+} (required) [2, 5, 9]; Cu^{2+} (weak activation) [2]; Zn^{2+} (weak activation) [2]; Ni^{2+} (weak activation) [2]

Turnover number (min^{-1})

0.09 [16]; 0.08 [16]

Specific activity (U/mg)

0.983 [5]; 2.6 [17]; More [1, 2, 4, 10, 13, 15, 16, 18]

K_m-value (mM)

3.3 (malate, pH 8.5) [2]; 0.39 (malate, pH 7.5) [2]; 0.041 (malate, pH 6.5) [2]; 0.016 ($NADP^+$, pH 7.5) [2]; 0.02 ($NADP^+$, pH 4.5) [2]; 0.03 (oxaloacetate) [3]; 0.07 (NADPH) [3]; 0.04 ($NADP^+$) [3]; 22 (HCO_3^-) [8]; 0.015 ($NADP^+$) [8]; 0.045 (NADPH) [8]; 3 (pyruvate) [8]; 1.25 (malate, mitochondrial enzyme) [18]; 25 (pyruvate, mitochondrial enzyme) [18]; 0.00188 ($NADP^+$, mitochondrial enzyme) [18]; 0.14 (malate, cytosolic enzyme) [18]; 10 (pyruvate, cytosolic enzyme) [18]; 0.00118 ($NADP^+$, cytosolic enzyme) [18]; More [3, 7, 8]

pH-optimum

5.2 (decarboxylation of oxaloacetate) [1]; 7.3 (malic enzyme, wheat germ) [1]; 4.5 (malic enzyme, pigeon liver) [1]; 7.2 (0.1 mM malate) [2]; 8.5–9.0 (0.1 M malate [2], oxidation of malate [3]) [2, 3]; 7.0 (reduction of oxaloacetate [3], carboxylation/decarboxylation, extramitochondrial enzyme [5]) [3, 5]; 6.6 (carboxylation, mitochondrial enzyme) [5]; 7.6 (decarboxylation, mitochondrial enzyme) [5]; 8.0 [7]; 7.4 (carboxylation of pyruvate) [8]; 8.0 (decarboxylation of malate) [8]; 6.5–6.6 [9]; 7.2 [9]; 6.8–7.5 (decarboxylation) [13]; 8.75 (decarboxylation of malate) [9]

pH-range

6.5–8.0 [3]; 7.5–9.5 [3]; 5.5–8.0 [3]

Temperature optimum (°C)

25 [1, 3, 5, 13, 17]; 22–23 [2]; 27 [7]; 29 [11]; 30 [12]

Temperature range (°C)

---

3 ENZYME STRUCTURE

Molecular weight

110000 (spinach, gel filtration) [3]
227000 (pig, intramitochondrial enzyme, density gradient centrifugation) [5, 18]
230000–240000 (Zea mays, gel filtration, monomeric form) [8]
460000–480000 (Zea mays, gel filtration, dimeric form) [8]
260000 (pigeon) [11, 12, 16]
258000 (Mangifera indica, gel electrophoresis) [13]
224000 (Pyrus communis) [13]
260000–265000 (tomato, gel electrophoresis) [17]
216000 (pig, cytosolic enzyme, density gradient centrifugation) [18]

Subunits

Tetramer (4 × 64900, Magnifera indica, SDS-PAGE [13], 4 × 65000, pigeon liver, SDS-PAGE [16], 4 × 64000–65000, tomato, denaturing gel electrophoresis [17]) [13, 16, 17]

Glycoprotein/Lipoprotein

–

4 ISOLATION/PREPARATION

Source organism

Pigeon [1, 2, 11, 12, 16]; Wheat [1]; Lactobacillus arabinosus [1]; Chicken [4]; Spinach [3]; Pig [5, 18]; Bovine [5]; Rat [6, 10]; Guinea pig [6]; Rabbit [6]; Seagull [6]; Locusta migratoria [6]; Pennisetum purpureum [7]; Zea mays [3, 8]; Streptomyces aureofaciens [9]; Apple [9]; Mangifera indica [13]; Pyrus communis [13]; E. coli [13]; Trypanosoma cruzi [13]; Digitaria sanguinalis [14]; Echinochloa crussgali [14]; Gomphrena globosa [14]; Sorghum sudanense [14]; Saccharomyces bailii [15]; Saccharomyces cerevisiae [15]; Tomato [17]

Source tissue

Liver [1, 2, 4, 10–12, 16]; Leaf (stroma) [3, 7]; Heart [5, 18]; Mycelium [9]; Fruit [17]

Localisation in source

Chloroplast [3, 7]; Mitochondria [5, 6, 18]; Cytosol [5, 6, 17, 18]

Purification

Wheat [1]; Pigeon [2]; Spinach [3]; Chicken [4]; Pig [5, 18]; Pennisetum purpureum [7]; Streptomyces aureofaciens [9]; Mangifera indica [13]; Tomato [17]

Crystallization

[4, 10]

Cloned

–

Renaturated

–

5 STABILITY

pH

6.0–9.2 [12]

Temperature (°C)

Oxidation

Organic solvent

General stability information

Mg^{2+} increases heat stability of pigeon liver enzyme [2]; Mn^{2+} increases heat stability of Lactobacillus arabinosus enzyme [2]; Partially purified enzyme very labile [15]

Storage

4°C, 3 weeks [8]; 0°C [10]; –20°C, several months [12]; –70°C, several months [16]

6 CROSSREFERENCES TO STRUCTURE DATABANKS

PIR/MIPS code

Brookhaven code

7 LITERATURE REFERENCES

[1] Harary, I., Korey, S.R., Ochoa, S.: J. Biol. Chem.,203,595–604 (1953)
[2] Rutter, W.J., Lardy, H.A.: J. Biol. Chem.,233,374–382 (1958)
[3] Ting, I.P., Rocha, V.: Arch. Biochem. Biophys.,147,156–164 (1971)
[4] Clifford, K.H., Cornforth, J.W., Donninger, C., Mallaby, R.: Eur. J. Biochem.,26, 401–406 (1972)
[5] Bartholome, K., Brdiczka, D.G., Pette, D.: Hoppe-Seyler's Z. Physiol. Chem.,353,1487–1495 (1972)
[6] Nolte, J., Brdiczka, D., Pette, D.: Biochim. Biophys. Acta,284,497–507 (1972)
[7] Coombs, J., Baldry, C.W., Bucke, C.: Planta,110,109–120 (1973)
[8] Ziegler, I.: Biochim. Biophys. Acta,364,28–37 (1974)
[9] Jechova, V., Hostalek, Z., Vanek, Z.: Folia Microbiol.,20,137–141 (1975)
[10] Wada, F., Numata, M., Eguchi, Y., Sakamoto, Y.: Biochim. Biophys. Acta,410,237–242 (1975)
[11] Reynolds, C.H., Hsu, R.Y., Matthews, B., Pry, T.A., Dalziel, K.: Arch. Biochem. Biophys.,189,309–316 (1978)
[12] Pry, T.A., Hsu, R.Y.: Biochemistry,19,951–962 (1980)
[13] Dubery, I.A., Schabort, J.C.: Biochim. Biophys. Acta,662,102–110 (1981)
[14] Hatch, M.D., Tsuzuki, M., Edwards, G.E.: Plant Physiol.,69,483–491 (1982)
[15] Kuczynski, J.T., Radler, F.: Arch. Microbiol.,131,266–270 (1982)
[16] Vernon, C.M., Hsu, R.Y.: Arch. Biochem. Biophys.,225,296–305 (1983)
[17] Goodenough, P.W., Prosser, I.M., Young, K.: Phytochemistry,24,1157–1162 (1985)
[18] Bartholome, K., Brdiczka, D.G., Pette, D.: Hoppe-Seyler's Z. Physiol. Chem.,353,1487–1495 (1972)

1 NOMENCLATURE

EC number
1.1.1.41

Systematic name
Isocitrate:NAD$^+$ oxidoreductase (decarboxylating)

Recommended name
Isocitrate dehydrogenase (NAD$^+$)

Synonymes
Isocitric dehydrogenase
beta-Ketoglutaricisocitric carboxylase
Isocitric acid dehydrogenase
NAD dependent isocitrate dehydrogenase
NAD isocitrate dehydrogenase
NAD-linked isocitrate dehydrogenase
NAD-specific isocitrate dehydrogenase
NAD isocitric dehydrogenase

CAS Reg. No.
9001-58-5

2 REACTION AND SPECIFICITY

Catalysed reaction
Isocitrate + NAD$^+$ →
→ 2-oxoglutarate + CO_2 + NADH

Reaction type
Redox reaction
Oxidative decarboxylation [3]

Natural substrates

Substrate spectrum
1 Isocitrate + NAD$^+$ [1–20]

Product spectrum
1 2-Oxoglutarate + CO_2 + NADH [1–20]

Inhibitor(s)

Cyanide [1]; Thiocyanate [1]; p-Hydroxymercuribenzoate [1]; Azide [1]; Molybdate [1]; ATP [1, 2, 8, 10, 13, 16, 18]; Glyoxylate (+ oxaloacetate, concomitantly added) [2, 6]; Pyruvate (+ oxaloacetate, concomitantly added, less effective than glyoxylate + oxaloaceetate) [2]; NADH [2, 8, 13, 14, 16, 18]; NADPH [14, 16, 17]; 2-Oxoglutarate [2, 18]; Ca^{2+} [5, 15]; Bilirubin [12]; Congo Red [12]; Chicago Sky Blue [12]; Sulfobromophthalein [12]; Fluorescin (halogenated derivatives) [12]; Mg^{2+} (free form) [15]; EDTA [17]; EGTA [17]; Nitrilotriacetate [17]; 2-Hydroxyethylamine-N,N-diacetate [17]; 2-Oxopropane-1,1,3-tricarboxylate (only in presence of ADP) [17]; L-Glutamate [18]; Citrate [18]; Oxaloacetate [18]

Cofactor(s)/prosthetic group(s)/activating agents

NAD^+ [1, 8, 16]; Adenosine-5'-phosphate (K_m: 0.009 mM) [1]

Metal compounds/salts

Mn^{2+} (required) [1, 3]; Mg^{2+} (requirement, active substrate is Mg^{2+}-DL-isocitrate chelate [3, 15], free form inhibits [15]) [1, 3, 10, 15]; Co^{2+} (less effective than Mg^{2+} or Mn^{2+}) [3, 10]; Zn^{2+} (less effective than Mn^{2+} or Mg^{2+}) [3]; Ca^{2+} (less effective than Mn^{2+} or Mg^{2+} [10], K_m: 0.001 mM [14]) [10, 14]

Turnover number (min^{-1})

Specific activity (U/mg)

More [2, 6, 7, 18]

K_m-value (mM)

0.11 (isocitrate, yeast) [1]; 0.45 (isocitrate, bovine) [1]; 0.03 (isocitrate) [2]; 0.3 (NAD^+) [2]; 0.25 (NAD^+) [8]; 0.06–0.09 (NAD^+) [12]; More [8, 9, 10, 13, 18]

pH-optimum

7.5 [1]; 6.5 [1]; 7.5 (invertebrates) [5]; 8.1 (vertebrate muscle) [5]; 7.0–7.2 [5]; 8.1–8.3 [6]; 8.3 [8]; 8.4 [9]; 7.2–8.2 [18]

pH-range

6.5–8.5 [1]; 6.0–8.0 [3]; 5.5–8.6 [4]; 6.0–8.5 [9]

Temperature optimum (°C)

22 [4]; 25 [5, 9, 10, 13]; 30 [6, 8, 19]

Temperature range (°C)

12–37 [4]

---

3 ENZYME STRUCTURE

Molecular weight

300000–340000 (pig, gel filtration) [7]
224000 (pig, ultracentrifugation) [11]
245000 (pig, sedimentation equilibrium analysis) [11]
191800 (Rhodosporidium toruloides, gel filtration) [18]

Subunits

Dimer (1 × 39000 + 1 × 41000, pig, SDS-PAGE) [7]
Tetramer (4 × 39000–41000, pig, 3 types of subunits, ratio 2:1:1) [11]

Glycoprotein/Lipoprotein

Glycoprotein [7]

4 ISOLATION/PREPARATION

Source organism

Guinea pig [1]; Bovine (calf [4]) [1, 4, 9, 10, 12, 14, 16, 17, 20]; Pigeon [1]; Human [1]; Aspergillus niger [1]; Yeast [1]; Pea [1]; Rat [1, 19]; Hydrogenomonas eutropha [2]; Pig [3, 7, 11]; Lepas antifera [5]; Waterbug [5]; Rhinoceros beetle [5]; Locust [5]; Cockroach [5]; Sarcophaga barbata [5]; Sheep [5]; Candida tropicalis [6]; Pseudomonas sp. W 6 [8]; Solanum tuberosum [13]; Phormia regina [15]; Rhodosporidium toruloides [18]

Source tissue

Heart [1, 3, 4, 7, 11, 12, 14, 17, 19]; Placenta [1]; Muscle (insect, flight muscle [5, 15], breast [1]) [1, 5, 15]; Brain [9, 10]

Localisation in source

Mitochondria [1, 6, 13, 14, 18, 19]

Purification

Hydrogenomonas eutropha [2]; Pig [3, 7, 11]; Candida tropicalis [6]; Pseudomonas sp. W 6 [8]; Solanum tuberosum [13]; Phormia regina [15]; Bovine (calf [4]) [4, 9, 10, 17]; Rhodosporidium toruloides [18]

Crystallization

–

Cloned

–

Renaturated

–

5 STABILITY

pH

5.5–8.5 [9]

Temperature (°C)

48 [2]; 4 (16% activity loss after 10 min) [5]; 0 (activity loss in a few h) [18]

Oxidation

Organic solvent

General stability information

20 h in ADP containing medium [5]; Stable in 5 M glycerol [5]

Storage

-90°C, no loss of activity after 6 weeks [3, 4]

6 CROSSREFERENCES TO STRUCTURE DATABANKS

PIR/MIPS code

PIR2:A33418 (pig (fragment)); PIR2:A35834 (alpha chain pig (fragments)); PIR2:B35834 (beta chain pig (fragments)); PIR2:S31264 (chain IDH1 precursor yeast (Saccharomyces cerevisiae)); PIR2:A39309 (chain IDH2 precursor yeast (Saccharomyces cerevisiae)); PIR2:A34317 (gamma chain pig (fragment)); PIR2:C35834 (gamma chain pig (fragments)); PIR3:S40020 (gamma chain rat)

Brookhaven code

7 LITERATURE REFERENCES

[1] Plaut, G.W.E. in "The Enzymes",2nd Ed. (Boyer, P.D., Lardy, H., Myrbäck, K., eds.) 7,105–126 (1963)
[2] Glaeser, H., Schlegel, H.G.: Arch. Microbiol.,86,327–337 (1972)
[3] Cohen, P.F., Colman, R.F.: Eur. J. Biochem.,47,35–45 (1974)
[4] Silinski, J.M., Colman, R.F.: Biochim. Biophys. Acta,370,1–25 (1974)
[5] Alp, P.R., Newsholme, E.A., Zammit, V.A.: Biochem. J.,154,689–700 (1976)
[6] Nabeshima, S., Ishiyama, S., Tanaka, A., Fukui, S.: Agric. Biol. Chem.,41,509–516 (1977)
[7] Ramachandran, N., Colman, R.F.: Proc. Natl. Acad. Sci. USA,75,252–255 (1978)
[8] Hoffmann, K.H., Babel, W.: Z. Allg. Mikrobiol.,20,399–404 (1980)
[9] Willson, V.J.C., Tripton, K.F.: Eur. J. Biochem.,109,411–416 (1980)
[10] Willson, V.J.C., Tripton, K.F.: Eur. J. Biochem.,113,477–483 (1981)
[11] Ehrlich, R.S., Hayman, S., Ramachandran, N., Colman, R.F.: J. Biol. Chem.,256,10560–10564 (1981)
[12] Gabriel, J.L., Plaut, G.W.E.: J. Biol. Chem.,257,8021–8029 (1981)
[13] Tezuka, T., Laties, G.G.: Plant Physiol.,72,959–963 (1982)
[14] Gabriel, J.L., Plaut, G.W.E.: Biochemistry,23,2773–2778 (1984)
[15] Bulos, B.A., Thomas, B.J., Sacktor, B.: J. Biol. Chem.,259,10232–10237 (1984)
[16] Gabriel, J.L., Milner, R., Plaut, G.W.E.: Arch. Biochem. Biophys.,240,128–134 (1985)
[17] Gabriel, J.L., Plaut, G.W.E.: Biochem. J.,229,817–822 (1985)
[18] Evans, C.T., Ratledge, C.: Can. J. Microbiol.,31,845–850 (1985)
[19] Rutter, G.A., Denton, R.M.: Biochem. J.,252,181–189 (1988)
[20] Gabriel, J.L., Plaut, G.W.E.: J. Biol. Chem.,259,1622–1628 (1984)

1 NOMENCLATURE

EC number
1.1.1.42

Systematic name
Isocitrate:$NADP^+$ oxidoreductase (decarboxylating)

Recommended name
Isocitrate dehydrogenase ($NADP^+$)

Synonymes
Oxalosuccinate decarboxylase
Isocitrate dehydrogenase (NADP)
Oxalsuccinic decarboxylase
Isocitrate (NADP) dehydrogenase
Isocitrate (nicotinamide adenine dinucleotide phosphate) dehydrogenase
NADP-specific isocitrate dehydrogenase
NADP-linked isocitrate dehydrogenase
NADP-dependent isocitrate dehydrogenase
NADP isocitric dehydrogenase
Isocitrate dehydrogenase (NADP-dependent)
NADP-dependent isocitric dehydrogenase
TPN-dependent isocitrate dehydrogenase
TPN-specific isocitrate dehydrogenase
TPN-linked isocitrate dehydrogenase
Isocitrate dehydrogenase
TPN-isocitrate dehydrogenase
$NADP^+$-linked isocitrate dehydrogenase

CAS Reg. No.
9028-48-2

2 REACTION AND SPECIFICITY

Catalysed reaction
Isocitrate + $NADP^+$ →
→ 2-oxoglutarate + CO_2 + NADPH

Reaction type
Redox reaction
Reductive carboxylation
Oxidative decarboxylation [1]

Natural substrates

Isocitrate + $NADP^+$ (Mg^{2+}- or Mn^{2+}-isocitrate complex [8, 9, 11]) [1, 4, 5, 7–9, 11, 16–18, 20]

Substrate spectrum

1 Isocitrate + $NADP^+$ (r, Mg^{2+}- or Mn^{2+}-isocitrate complex [8, 9, 11]) [1, 4, 5, 7–9, 11, 16–18, 20]
2 3-Fluoroisocitrate + $NADP^+$ [20]
3 3-Hydroxyisocitrate + $NADP^+$ [20]

Product spectrum

1 2-Oxoglutarate + NADPH + CO_2 (oxalosuccinate is an intermediate enzyme-bound product, which decarboxylates to 2-oxoglutarate [1]) [1, 4, 5, 7, 16–18, 20]
2 3-Fluoro-2-oxoglutarate + NADPH + CO_2
3 3-Hydroxy-2-oxoglutarate + NADPH + CO_2

Inhibitor(s)

Ag^+ (decarboxylation of isocitrate and oxalosuccinate) [1, 7]; Isocitrate (decarboxylation of oxalosuccinate) [1]; Zn^{2+} [7]; Hg^{2+} [18, 19]; Cu^{2+} [1, 19]; Phenylmercuric nitrate [1]; Phenarsazines [1]; Diphenylchloroarsine [1]; Glyoxylate (plus oxaloacetate) [2, 3, 4, 10, 11, 16, 18]; Oxalomalate [2, 18]; Urea (low molecular weight form) [4]; Oxaloacetate [4, 7, 19]; 2-Oxoglutarate [4, 7, 9, 16]; 5'-ADP [9]; 2',5'-ADP [9]; 5'-ATP [9, 19]; ATP [4, 16]; Nicotinamide mononucleotide [9]; Citrate [7, 9, 16]; Monoiodoacetate [7, 18]; Adenosine-3',5'-cyclic monophosphate [7]; Maleate [7]; Propanetricarboxylate [9]; threo-L-Isocitrate [9]; NADPH [16, 19]; Dithiothreitol [18]; 2-Mercaptoethanol [18]; Glutathione [18]; Glyceraldehyde-3-phosphate [19]; 3-Phosphoglycerate [19]; Phosphoenolpyruvate [19]; cis-Aconitate [19]; N-Ethylmaleimide [19]; p-Hydroxymercuribenzoate (decarboxylation of isocitrate and oxalosuccinate) [1]; p-Chloromercuribenzoate [7, 11, 18, 19]

Cofactor(s)/prosthetic group(s)/activating agents

$NADP^+$ (natural coenzyme [1]) [1, 3, 4, 10, 11, 14, 16–19]; NADPH [1, 6, 8]

Metal compounds/salts

Mn^{2+} (absolute requirement for divalent cations, K_m: 0.01 mM, oxidative decarboxylation of isocitrate, 0.3 mM, decarboxylation of oxalosuccinate [1]) [1, 4, 7, 10–12, 14–20]; Mg^{2+} (absolute requirement for divalent cations) [1, 4, 7, 10–12, 14–20]; Co^{2+} (absolute requirement for divalent cations) [7, 11, 19]; Zn^{2+} (absolute requirement for divalent cations) [11, 18–20]; Cd^{2+} (absolute requirement for divalent cations) [11, 19, 20]; Ni^{2+} (absolute requirement for divalent cations) [20]

Turnover number (min^{-1})
3500 (oxidative decarboxylation of isocitrate) [1]

Specific activity (U/mg)
More [2, 5–8, 10–12, 15, 17, 19]

K_m-value (mM)
0.0026 (isocitrate) [1]; 25–26 (oxalosuccinate, decarboxylase) [1]; 1.2 (oxalosuccinate, reductase) [1]; 0.13 (2-oxoglutarate) [1]; 0.0092 (NADPH, detritiation) [1]; 0.0097 (isocitrate) [2]; 0.0073 ($NADP^+$) [19]; 0.01 ($NADP^+$) [17]; 0.011 ($NADP^+$) [11]; More [2, 4, 5, 9–11, 17, 18]

pH-optimum
8.0 (low molecular weight form) [4]; 7.5 (high molecular weight form) [4]; 7.4 [5]; 7.8 (isocitrate + $NADP^+$) [7]; 6.6 (2-oxoglutarate + CO_2 + NADPH) [7]; 7.6–8.0 (isoenzyme I) [10]; 8.0–8.6 (isoenzyme II) [10]; 7.2–8.2 [11]; 6.0–8.0 [16]; 7.5–9.5 [17]; 8.5 [18]; 7.9–8.4 [19]

pH-range
5.6–8.5 [1]; 6.0–9.5 [11]; 6.3–7.8 [16]; 6.5–10.0 [17, 18]

Temperature optimum (°C)
20 (isoenzyme II) [10, 14]; 23 [7]; 25 [2]; 25–30 [18]; 28 [16]; 37 [2]; 40 (isoenzyme I) [10, 14]; 70 (Thermus aquaticus) [2]; 75 [19]

Temperature range (°C)
30–80 [3]

3 ENZYME STRUCTURE

Molecular weight
61000–64000 (pig) [1]
60000–70000 (Thermus aquaticus, gel filtration) [2]
75000–80000 (Bacillus subtilis, gel filtration) [2]
90000–100000 (Chlamydomonas reinhardtii, gel filtration) [2]
87000–92600 (Bacillus stearothermophilus) [2]
80000 (Azotobacter vinelandii, E. coli) [2]
360000 (Acinetobacter calcoaceticus, gel filtration, high molecular weight form) [4]
90000 (Acinetobacter calcoaceticus, gel filtration, low molecular weight form [4], bovine, gel filtration, sedimentation equilibrium, gel electrophoresis [6, 8], Fundulus heteroclitus, 3 isoenzymes [15], Pisum sativum, gel filtration [17], Bacillus sp., gel filtration [19]) [4, 6, 8, 15, 17, 19]
81000–83000 (Zea mays, permeation chromatography) [5]
68000 (pig) [5]
105000 (Rhodopseudomonas sphaeroides) [5]
86000 (Bombyx mori, gel filtration) [7]

53000 (human, gel electrophoresis) [9]
85000 (Vibrio sp., gel filtration, isoenzyme I: dimer, isoenzyme II: monomer) [10, 14]
55000 (bovine, gel electrophoresis) [11]
45000 (Mytilis edulis, gel electrophoresis) [12]
60000 (Aspergillus niger, gel electrophoresis) [16]

Subunits

Dimer (2 × 43000–45000, bovine, SDS-PAGE [6, 8], 2 × 44000, Bombyx mori, SDS-PAGE [7], 2 × 42000, Vibrio sp., isoenzyme I, SDS-PAGE [14], 2 × 45000, Fundulus heteroclitus, SDS-PAGE [15], 2 × 46000, Pisum sativum, SDS-PAGE [17], 2 × 44000, Bacillus sp., SDS-PAGE [19]) [6–8, 14, 15, 17, 19]

Glycoprotein/Lipoprotein

–

4 ISOLATION/PREPARATION

Source organism

Aspergillus niger [1, 16]; Trypanosoma cruzi [1]; Escherichia freundii [1]; Mycobacterium tuberculosis [1]; Saccharomyces cerevisiae [1]; Pig [1, 5, 20]; Thermus aquaticus [2]; Bacillus subtilis [2]; Chlamydomonas reinhardtii [2]; Bacillus stearothermophilus [2, 5]; Thermus flavus [3]; Acinetobacter calcoaceticus [4]; Zea mays [5]; Azotobacter sp. [5]; Rhodopseudomonas sphaeroides [5]; Bombyx mori [7]; Human [9]; E. coli [9, 13]; Vibrio sp. (strain ABE-1) [10, 14]; Bovine [6, 8, 11]; Mytilis edulis [12]; Leucothrix mucor [13]; Halobacterium centirubrum [14]; Fundulus heteroclitus [15]; Phycomyces blakesleeanus [16]; Blastocladiella emersonii [16]; Pisum sativum [17]; Mycobacterium phlei [18]; Brevibacterium flavum [18]; Thiobacillus novellus [18]; Crithidia fasciculata [18]; Bacillus sp. [19]

Source tissue

Heart [1, 5, 6, 8, 9, 20]; Seed [5]; Pupa [7]; Liver [8, 15]; Mammary gland [11]; Brain [15]; Intestine [15]; Leaf [17]

Localisation in source

Mitochondria [5, 6, 8, 15, 16]; Cytosol [6, 8, 17]; Cytoplasm [12, 15, 16]

Purification

Thermus aquaticus [2]; Bacillus subtilis [2]; Chlamydomonas reinhardtii [2]; Thermus flavus [3]; Zea mays [5]; Bombyx mori [7]; Human [9]; Vibrio sp. (strain ABE-1) [10, 14]; Bovine [6, 8, 11]; Mytilus edulis [12]; E. coli [13]; Fundulus heteroclitus [15]; Aspergillus niger [16]; Pisum sativum [17]; Mycobacterium phlei [18]; Bacillus sp. [19]

Crystallization

[6, 8]

Cloned

–

Renaturated

(overnight dialysis against guanidine-HCl, GDP buffer [5], with 2-mercaptoethanol after inactivation at 40°C, isoenzyme II, 40% [14]) [5, 14]

5 STABILITY

pH

6–9 (Thermus aquaticus) [2]; 6.0 [11]; 8.0–10.0 [19]

Temperature (°C)

30 (Chlamydomonas reinhardtii, 50% loss of activity in 1 h) [2]; 54 (Bacillus subtilis, 50% loss of activity in 1 h) [2]; 88 (Thermus aquaticus, 50% loss of activity in 1 h) [2]; 70 (Thermus flavus) [3]; 45 [7, 19]; 20 (isoenzyme II) [10]; 15 (isoenzyme II) [14]; 35 (isoenzyme C, T_{50}-value) [15]; 49 (isoenzyme A, T_{50}-value) [15]; 54 (isoenzyme B, T_{50}-value) [15]; 52 (isoenzyme C, T_{50}-value, presence of isocitrate) [15]; 58 (isoenzyme A, T_{50}-value, presence of isocitrate) [15]; 69 (isoenzyme B, T_{50}-value, presence of isocitrate) [15]

Oxidation

Organic solvent

General stability information

Instable in solutions of low ionic strength, stabilization by ammonium sulfate [1, 5]; Low molecular weight form of Acinetobacter calcoaceticus enzyme more heat stable than high molecular weight form [4]; Stabilization by dithiothreitol [11]

Storage

2°C, several weeks [6]; –20°C, 6 months, 10% loss of activity [7]

6 CROSSREFERENCES TO STRUCTURE DATABANKS

PIR/MIPS code

PIR2:A10759 (Azotobacter vinelandii (fragments)); PIR3:S31390 (bovine); PIR3:S33859 (bovine); PIR3:S42892 (common tobacco); PIR1:DCECIS (Escherichia coli); PIR2:A43294 (pig); PIR2:A27371 (pig (fragment)); PIR2:A27372 (pig (fragment)); PIR3:S20292 (Synechocystis sp. (PCC 6803) (fragment)); PIR3:A43934 (Thermus aquaticus); PIR2:B49341 (I Vibrio sp. (strain ABE-1)); PIR2:A49341 (II Vibrio sp. (strain ABE-1)); PIR1:DCBYIS (precursor mitochondrial yeast (Saccharomyces cerevisiae)); PIR2:A39333 (mitochondrial pig (fragments))

Brookhaven code

3ICD (Escherichia coli); 4ICD (Escherichia coli)

7 LITERATURE REFERENCES

[1] Plaut, G.W.E. in " The Enzymes",2nd ed. (Boyer, P.D., Lardy, H., Myrbäck, K., eds.) 7,105–126 (1963) (Review)
[2] Ramaley, R.F., Hudock, M.O.: Biochim. Biophys. Acta,315,22–36 (1973)
[3] Saiki, T., Arima, K.: J. Biochem.,77,233–240 (1975)
[4] Kleber, H.-P.: Z. Allg. Mikrobiol.,15,431–435 (1975)
[5] Curry, R.A., Ting, I.P.: Arch. Biochem. Biophys.,176,501–509 (1976)
[6] Macfarlane, N., Mathews, B., Dalziel, K.: Eur. J. Biochem.,74,553–559 (1977)
[7] Miake, F., Torikata, T., Koga, K., Hayashi, K.: J. Biochem.,82,449–454 (1977)
[8] Macfarlane, N., Reynolds, H., Matthews, B., Dalziel, K.: Biochem. Soc. Trans.,5,790–793 (1977)
[9] Seelig, G.F., Colman, R.F.: Arch. Biochem. Biophys.,188,394–409 (1978)
[10] Ochiai, T., Fukunaga, N., Sasaki, S.: J. Biochem.,86,377–384 (1979)
[11] Farrell, H.M.: Arch. Biochem. Biophys.,204,551–559 (1980)
[12] Head, E.J.H.: Eur. J. Biochem.,111,575–579 (1980)
[13] Vasquez, B., Reeves, H.C.: Biochim. Biophys. Acta,660,16–22 (1981)
[14] Ochiai, T., Fukunaga, N., Sasaki, S.: J. Gen. Appl. Microbiol.,30,479–487 (1984)
[15] Gonzalez-Villasenor, L.I., Powers, D.A.: J. Biol. Chem.,260,9106–9113 (1985)
[16] Meixner-Monori, B., Kubicek, C.P., Harrer, W., Schreferl, G., Rohe, M.: Biochem. J.,236,549–557 (1986)
[17] Ni, W., Robertson, E.F., Reeves, H.C.: Plant Physiol.,83,785–788 (1987)
[18] Dhariwal, K.R., Venkitasubramanian, T.A.: J. Gen. Microbiol.,133,2457–2460 (1987)
[19] Shikata, S., Ozaki, K., Kawai, S., Ito, S., Okamoto, K.: Biochim. Biophys. Acta,952,282–289 (1988)
[20] Grissom, C.B., Cleland, W.W.: Biochemistry,27,2934–2943 (1988)

1 NOMENCLATURE

EC number
1.1.1.43

Systematic name
6-Phospho-D-gluconate:NAD(P)$^+$ 2-oxidoreductase

Recommended name
Phosphogluconate 2-dehydrogenase

Synonymes
6-Phosphogluconic dehydrogenase
Dehydrogenase, phosphogluconate
Phosphogluconate dehydrogenase
Gluconate 6-phosphate dehydrogenase
6-Phosphogluconate dehydrogenase (NAD)
2-Keto-6-phosphogluconate reductase [1]

CAS Reg. No.
9001-82-5

2 REACTION AND SPECIFICITY

Catalysed reaction
6-Phospho-D-gluconate + NAD(P)$^+$ →
→ 6-phospho-2-dehydro-D-gluconate + NAD(P)H

Reaction type
Redox reaction

Natural substrates

Substrate spectrum
1 2-Keto-6-phosphogluconate + NAD(P)H [1]
2 2-Ketogluconate + NAD(P)H [1]

Product spectrum
1 6-Phosphogluconate + NAD(P)$^+$ [1]
2 Gluconate + NAD(P)$^+$

Inhibitor(s)
Mg^{2+} [1]; Zn^{2+} [1]; p-Chloromercuribenzoate [1]; More (not inhibitory: F^-, EDTA) [1]

Cofactor(s)/prosthetic group(s)/activating agents
NADPH (preferred) [1]; NADH (72% of NADPH activity) [1]

Metal compounds/salts
More (no metal ion requirement) [1]

Turnover number (min^{-1})

Specific activity (U/mg)
More [1]

K_m-value (mM)
0.38 (2-keto-6-phosphogluconate (+ NADPH)) [1]; 0.66 (2-keto-6-phosphogluconate (+ NADH)) [1]

pH-optimum
7.4 (cofactor NADPH) [1]; 7.2–8.0 (cofactor NADH) [1]

pH-range
7.2–7.6 (cofactor NADPH) [1]

Temperature optimum (°C)

Temperature range (°C)

---

3 ENZYME STRUCTURE

Molecular weight

Subunits

Glycoprotein/Lipoprotein
–

---

4 ISOLATION/PREPARATION

Source organism
Pseudomonas fluorescens [1]

Source tissue

Localisation in source
Soluble [1]

Purification
Pseudomonas fluorescens [1]

Crystallization
–

Cloned

–

Renaturated

–

5 STABILITY

pH

Temperature (°C)

Oxidation

Organic solvent

General stability information

Diluted enzyme solutions unstable [1]

Storage

–14°C, 1 month [1]

6 CROSSREFERENCES TO STRUCTURE DATABANKS

PIR/MIPS code

Brookhaven code

7 LITERATURE REFERENCES

[1] Frampton, E.W., Wood, W.A.: J. Biol. Chem.,236,2571–2577 (1961)

1 NOMENCLATURE

EC number
1.1.1.44

Systematic name
6-Phospho-D-gluconate:$NADP^+$ 2-oxidoreductase (decarboxylating)

Recommended name
Phosphogluconate dehydrogenase (decarboxylating)

Synonymes
6-Phosphogluconate dehydrogenase (decarboxylating)
6-Phospho-D-gluconate dehydrogenase
Dehydrogenase, phosphogluconate (decarboxylating)

CAS Reg. No.
9073-95-4

2 REACTION AND SPECIFICITY

Catalysed reaction
6-Phospho-D-gluconate + $NADP^+$ →
→ D-ribulose 5-phosphate + CO_2 + NADPH

Reaction type
Redox reaction

Natural substrates
6-Phospho-D-gluconate + $NADP^+$ (second step of pentose phosphate pathway) [1, 4]

Substrate spectrum
1 6-Phospho-D-gluconate + $NADP^+$ (r [33], reverse reaction at low rate [17], isotope effects [38]) [1, 2, 6–8, 11–14, 17, 20, 30, 33, 38]
2 2-Deoxy-6-phosphogluconate + $NADP^+$ [16]

Product spectrum
1 D-Ribulose 5-phosphate + CO_2 + NADPH [17, 33]
2 3-Oxo-2-deoxy-6-phosphogluconate + NADPH [16]

Inhibitor(s)
Monovalent cations (above 0.1 M) [2]; NADPH (competitive to $NADP^+$ [13, 36]) [3, 13, 14, 21, 36, 47]; Fructose 1,6-diphosphate (not [8]) [3, 12, 15, 39]; Iodoacetamide [3, 33]; 5,5'-Dithiobis(2-nitrobenzoate) [5, 21, 33, 41]; p-Hydroxymercuribenzoate [5]; Iodoacetate [7]; Oxaloacetate [12]; Citrate [12, 20]; Nucleoside 5'-diphosphate [12]; Nucleoside 5'-triphosphate [12];

Nucleoside 5'-monophosphate [12]; Nucleoside 3'-monophosphate [12]; Fructose 1-phosphate [12]; Fructose 6-phosphate [12]; Glucose 6-phosphate [12]; Glucose 1,6-diphosphate [24, 27, 39]; Phosphate (competitive to 6-phosphogluconate) [13]; $MgCl_2$ (above 50 mM) [15]; Diphosphate [20]; p-Chloromercuribenzoate [21, 33, 41]; Arsenate [23]; Sulfate [23]; Molybdate [23]; Tungstate [23]; Permanganate [23]; Periodate [23]; Chlorogenic acid [27]; Scopolin [27]; Scopoletin [27]; Esculin [27]; Esculetin [27]; Ferulic acid [27]; Caffeic acid [23]; p-Coumaric acid [27]; Periodate oxidized $NADP^+$ [29]; 7-Chloro-4-nitrobenzo-2-oxa-1,3-diazole [33]; Mg^{2+} [33]; ATP [36, 47]; Hg^{2+} [41]; N-Ethylmaleimide [41]; Ribulose 5-phosphate [47]; Diethyldicarbonate [42]

Cofactor(s)/prosthetic group(s)/activating agents

$NADP^+$ (specific for, no activity with NAD^+ [3, 12, 14, 15, 17, 21, 30, 31, 35, 41, 44, 47, 51], coenzyme binding [30, 35]) [3, 6, 12, 14, 15, 17, 21, 30, 31, 33, 35, 41, 43, 44, 47, 51]; Deamino $NADP^+$ [3]; NAD^+ (slight activity) [6]; NADPH [33]

Metal compounds/salts

$MgCl_2$ (activation [6, 48], inhibition above 50 mM [15]) [6, 48]; Ca^{2+} (activation) [41]; Mn^{2+} (activation) [41]; More (influence of ionic strength [8, 13], no divalent cations required [12], no metal requirement [16, 17]) [8, 12, 13, 16, 17]

Turnover number (min^{-1})

780 (6-phosphogluconate) [13]

Specific activity (U/mg)

17.4 [2]; 21.2 [5]; 87 [6]; More (assay method [11], temperature dependence [33]) [7, 8, 11–14, 17, 20, 30, 31, 33]

K_m-value (mM)

0.0022 (NADPH) [12]; 0.0057–0.0086 ($NADP^+$, depending on buffer) [2]; 0.0068–0.0069 ($NADP^+$, 6-phosphogluconate) [12]; 0.01–0.03 (6-phosphogluconate [1, 6, 13–15, 33, 41, 47], $NADP^+$ [6, 13, 15, 16, 20, 33, 41, 47], NADPH [14]) [1, 6, 13–16, 20, 33, 41, 47]; 0.05–0.08 (6-phosphogluconate) [1, 16, 20, 21]; 34 (CO_2) [12]; More (kinetic studies [43, 46, 49], mutant enzyme [4], phosphorylation [32]) [4, 26, 32, 37, 43, 46, 48, 49]

pH-optimum

7.5–8.0 [17, 44]; 7.6–7.8 [2]; 7.6–8.0 [15]; 7.8–8.0 [48]; 7.8–8.2 [21]; 7.8–9.0 [3]; 8.0 [6, 14, 16, 33, 41]; 8.3 [8, 13]

pH-range

6–9 [33]; 7.8–9.0 [14]

Temperature optimum (°C)

40 [41]; 55 (mesophilic cyanobacteria) [36]; 55–60 [21]; 65 (thermophilic cyanobacteria) [36]; More (Arrhenius plots) [2, 6, 7]

Temperature range (°C)

3 ENZYME STRUCTURE

Molecular weight

94000–110000 (Spinacia oleracea, sedimentation velocity [1], Neurospora crassa, gel filtration [3, 14], sheep, gel filtration [5], sedimentation equilibrium centrifugation [12], Bacillus stearothermophilus, sucrose density centrifugation [6, 33], sedimentation equilibrium centrifugation [7], human, sedimentation equilibrium centrifugation [9, 13], Streptococcus faecalis, sedimentation equilibrium centrifugation [15, 17], rabbit, gel filtration [20]) [1, 3, 5–7, 9, 12–15, 17, 20, 33]
69000–70000 (Nicotiana tabacum, gel filtration, SDS-PAGE) [27]

Subunits

Dimer (2 × 47000–57000, Neurospora crassa, SDS-PAGE [3, 14], sheep, SDS-PAGE [5], Bacillus stearothermophilus, SDS-PAGE [7, 33], sucrose density centrifugation after treatment with SDS [6], human, SDS-PAGE [9, 13], Streptococcus faecalis, SDS-PAGE [17], rabbit, SDS-PAGE [20], Phormidium sp., SDS-PAGE [41], Candida boidinii, SDS-PAGE [47], Ricinus communis, SDS-PAGE [48], pig, SDS-PAGE [49]) [3, 5–7, 9, 13, 14, 17, 20, 33, 41, 47–49]
Monomer (1 × 69000–70000, Nicotiana tabacum, SDS-PAGE) [27]

Glycoprotein/Lipoprotein

–

4 ISOLATION/PREPARATION

Source organism

Spinacia oleracea (spinach) [1]; Sheep [2, 5, 12, 21, 22, 31, 34, 42, 43]; Neurospora crassa (mutant enzymes [4]) [3, 4, 14]; Bacillus stearothermophilus [6, 7, 18, 19, 21, 33, 37]; Streptococcus faecalis [15, 17]; Rabbit [20]; Human [8, 9, 11, 13, 40]; Pseudomonas multivorans [10]; Candida utilis [16, 23, 29]; Yeast [35, 37, 38]; E. coli [18, 19, 21, 32]; Rat [24, 25, 45]; Avena fatua (wild oat) [26]; Nicotiana tabacum (tobacco) [27, 28]; Acer pseudoplatanus (sycamore) [30]; Synechococcus sp. (strains 6307, 6716) [36]; Synechocystis sp. (strain 6714) [36]; Ricinus communis [39, 48]; Phormidium sp. [41]; Brevibacterium flavum [44]; Dicentrarchus labrax (bass) [46]; Candida boidinii [47]; Pig [49]; Dictyostelium discoideum [50]; Corylus avellana (hazel) [51]; Streptomyces griseus [52]

Source tissue

Leaf [1]; Liver [2, 5, 12, 21, 22, 24, 25, 31, 34, 42, 43, 46, 49]; Erythrocytes [8, 9, 11, 13, 40]; Mammary gland [20]; Testis [24]; Leg muscle [24]; Diaphragm [24]; Fat pad [24]; Seeds [26]; Endosperm [39, 48]; Cell suspension culture [30]; Tissue culture [27, 28]; Commercial preparation [37]; Hepatocytes [45]; Nut [51]

Localisation in source

Cytosol [1, 48]; Chloroplasts [1]; Cytoplasm [25, 36, 49]; Microsomes [25]; Proplastid [48]

Purification

Spinacia oleracea (2 isozymes) [1]; Sheep [2, 5, 12, 31]; Neurospora crassa [3, 14]; Bacillus stearothermophilus [6, 7, 33]; Human [8, 13, 40]; Streptococcus faecalis [15, 17]; Candida utilis [16]; Rabbit [20]; E. coli [21]; Nicotiana tabacum (2 isozymes) [28]; Acer pseudoplatanus [30]; Phormidium sp. [41]; Rat [24]; Dicentrarchus labrax [46]; Ricinus communis [48]; Pig [49]

Crystallization

(crystal structure [22, 34, 35]) [5, 7, 12, 16, 22, 34, 35]

Cloned

–

Renaturated

–

5 STABILITY

pH

4.5–11 (Bacillus stearothermophilus) [37]; 5–8.5 (yeast) [37]; 6–9 [41]

Temperature (°C)

48 (yeast, half-life 20 min) [37]; 50 (up to) [21, 41]; 60 (15 min stable [33], 2 h stable [6], stable up to [7]) [6, 7, 33]; 67 (Bacillus stearothermophilus, half-life 20 min) [37]

Oxidation

Photooxidation in presence of Rose Bengal [14, 21]

Organic solvent

Acetone: 50% v/v, Bacillus stearothermophilus 90 min stable, complete inactivation of yeast enzyme [37]; Dioxane: 60% v/v, Bacillus stearothermophilus, 90 min stable, complete inactivation of yeast enzyme [37]; Dimethylformamide: 70% v/v, Bacillus stearothermophilus, 90 min stable [37]

General stability information

Bacillus stearothermophilus enzyme stable in 6 M urea, E. coli enzyme not [19]; Phormidium sp. enzyme stable in 1 M urea [41]; 3.5 M urea: Bacillus stearothermophilus half-life 1 h, complete inactivation of yeast enzyme [37]; Stablization by KCl [44]; Stable to lyophilization [6, 33]; Slight inactivation during ultrafiltration [6]

Storage

4°C, 50 mM Tris-HCl buffer, pH 7.2, 1 mM EDTA, 1 mM 2-mercaptoethanol, several days, or several months as ammonium sulfate suspension [6]; 0–4°C, 0.1 M phosphate buffer, pH 6.0, 1 mM EDTA, 0.1% v/v 2-mercaptoethanol, 0.01 mM $NADP^+$, N_2–atmosphere, several months [8]; 5°C, up to 3 months [14]; 5°C, 50 mM Tris-HCl buffer, pH 7.5, 10 mM DTT, 20% glycerol, 2 weeks [15]; Crystalline suspension, several months [16]; –20°C [31], 4°C, saturated ammonium sulfate suspension, several months [33]

6 CROSSREFERENCES TO STRUCTURE DATABANKS

PIR/MIPS code

PIR3:S39755 (Trypanosoma brucei); PIR1:DEECGC (Escherichia coli); PIR2:JH0531 (fruit fly (Drosophila melanogaster)); PIR3:S15279 (pig); PIR2:A48325 (pig (fragment)); PIR2:S04397 (Salmonella typhimurium); PIR3:S15315 (Salmonella typhimurium); PIR1:DESHGC (sheep); PIR3:S15280 (sheep); PIR3:S14628 (Synechococcus sp.); PIR3:A48565 (Trypanosoma brucei)

Brookhaven code

1PGD (Sheep (Ovis orientalis aries))

7 LITERATURE REFERENCES

[1] Schnarrenberger, C., Oeser, A., Tolbert, N. E.: Arch. Biochem. Biophys.,154, 438–448 (1973)
[2] Dyson, J. E. D., D'Orazio, R. E., Hanson, W. H.: Arch. Biochem. Biophys.,154, 623–635 (1973)
[3] Scott, W. A., Abramsky, T.: J. Biol. Chem.,248,3535–3541 (1973)
[4] Scott, W. A., Abramsky, T.: J. Biol. Chem.,248,3542–3545 (1973)
[5] Silverberg, M., Dalziel, K.: Eur. J. Biochem.,38,229–238 (1973)
[6] Veronese, F. M., Boccu, E., Fontana, A., Benassi, C. A., Scoffone, E.: Biochim. Biophys. Acta,334,31–44 (1974)
[7] Pearse, B. M. F., Harris, J. I.: FEBS Lett.,38,49–52 (1973)
[8] Pearse, B. M. F., Rosemeyer, M. A.: Eur. J. Biochem.,42,213–223 (1974)
[9] Pearse, B. M. F., Rosemeyer, M. A.: Eur. J. Biochem.,42,225–235 (1974)
[10] Lee, Y. N., Lessie, T. G.: J. Bacteriol.,120,1043–1057 (1974)
[11] King, J. in " Methods Enzym. Anal." (Bergmeyer, H. U., ed.) 1,668–672 (1974)
[12] Silverberg, M., Dalziel, K.: Methods Enzymol.,41, Pt.B,214–220 (1975)
[13] Pearse, B. M. F., Rosemeyer, M. A.: Methods Enzymol.,41, Pt.B,220–226 (1975)
[14] Scott, W. A., Abramsky, T.: Methods Enzymol.,41, Pt.B,227–231 (1975)
[15] Bridges, R. B., Wittenberger, C. L.: Methods Enzymol.,41, Pt.B,232–237 (1975)
[16] Rippa, M., Signorini, M.: Methods Enzymol.,41, Pt.B,237–240 (1975)
[17] Bridges, R. B., Palumbo, M. P., Wittenberger, C. L.: J. Biol. Chem.,250,6093–6100 (1975)

[18] Fontana, A., Grandi, C., Boccu, E., Veronese, F. M.: Hoppe-Seyler's Z. Physiol. Chem.,356,1191–1193 (1975)
[19] Veronese, F. M., Boccu, E., Fontana, A.: Int. J. Pept. Protein Res.,7,341–343 (1975)
[20] Betts, S. A., Mayer, R. J.: Biochem. J.,151,263–270 (1975)
[21] Veronese, F. M., Boccu, E., Fontana, A.: Biochemistry,15,4026–4033 (1976)
[22] Adams, M. J., Helliwell, J. R., Bugg, C. E.: J. Mol. Biol.,112,183–197 (1977)
[23] Rippa, M., Signorini, M., Bellini, T., Dallocchio, F.: Arch. Biochem. Biophys.,189, 516–523 (1978)
[24] Beitner, R., Nordenberg, J.: Biochim. Biophys. Acta,583,266–269 (1979)
[25] Bublitz, C.: Biochem. Biophys. Res. Commun.,98,588–594 (1981)
[26] Adkins, S. W., Gosling, P. G., Ross, J. D.: Phytochemistry,19,2523–2525 (1980)
[27] Quadan, F. A., Wender, S. H., Smith, E. C.: Phytochemistry,20,961–964 (1981)
[28] Quadan, F. A., Wender, S. H., Smith, E. C.: Phytochemistry,20,1201–1203 (1981)
[29] Dallocchio, F., Matteuzzi, M., Bellini, T.: J. Biol. Chem.,256,10778–10780 (1981)
[30] Jessup, W., Dean, P. D. G.: J. Chromatogr.,219,419–426 (1981)
[31] Carne, A.: Anal. Biochem.,121,227–229 (1982)
[32] Dallocchio, F., Matteuzzi, M., Bellini, T.: Biochem. J.,203,401–404 (1982)
[33] Veronese, F. M., Boccu, E., Fontana, A.: Methods Enzymol.,89,282–291 (1982)
[34] Adams, M. J., Archibald, I. G., Bugg, C. E., Carne, A., Gover, S., Helliwell, J. R., Pickersgill, R. W., White, S. W.: EMBO J.,2,1009–1014 (1983)
[35] Adams, M. J., Gover, S., Pickersgill, R. W., Helliwell, J. R.: Biochem. Soc. Trans.,11,429–435 (1983)
[36] Lubberding, H. J., Bot, P. V. M.: Arch. Microbiol.,137,115–120 (1984)
[37] Veronese, F. M., Boccu, E., Schiavon, O., Grandi, C., Fontana, A.: J. Appl. Biochem.,6,39–47 (1984)
[38] Rendina, A. R., Hermes, J. D., Cleland, W. W.: Biochemistry,23,6257–6262 (1984)
[39] Miernyk, J. A., MacDougall, P. S., Dennis, D. T.: Plant Physiol.,76,1093–1094 (1984)
[40] Dallocchio, F., Matteuzzi, M., Bellini, T.: Biochem. J.,227,305–310 (1985)
[41] Sawa, Y., Suzuki, K., Ochiai, H.: Agric. Biol. Chem.,49,2543–2549 (1985)
[42] Topham, C. M., Dalziel, K.: Eur. J. Biochem.,155,87–94 (1986)
[43] Topham, C. M., Matthews, B., Dalziel, K.: Eur. J. Biochem.,156,555–567 (1986)
[44] Sugimoto, S., Shio, I.: Agric. Biol. Chem.,51,1257–1263 (1987)
[45] Hansen, R. J., Jungermann, K.: Biol. Chem. Hoppe-Seyler,368,955–962 (1987)
[46] Medina-Puerta, M., Gallego-Iniesta, M., Garrido-Pertierra, A.: Arch. Biochem. Biophys.,262,130–141
[47] Kato, N., Sahm, H., Schütte, H., Wagner, F.: Biochim. Biophys. Acta,566,1–11 (1979)
[48] Simcox, P. D., Dennis, D. T.: Plant Physiol.,62,287–290 (1978)
[49] Toews, M. L., Kanji, M. I., Carper, W. R.: J. Biol. Chem.,251,7127–7131 (1976)
[50] Edmundson, T. D., Ashworth, J. M.: Biochem. J.,126,593–600 (1972)
[51] Gosling, P.G., Ross, J.D.: Phytochemistry,18,1441–1445 (1979)
[52] Gräfe, U., Bormann, E.J., Truckenbrodt, G.: Z. Allg. Mikrobiol.,20,607–611 (1980)

1 NOMENCLATURE

EC number
1.1.1.45

Systematic name
L-Gulonate:NAD$^+$ 3-oxidoreductase

Recommended name
L-Gulonate 3-dehydrogenase

Synonymes
L-3-Aldonate dehydrogenase
Dehydrogenase, L-gulonate
L-3-Aldonic dehydrogenase
L-Gulonic acid dehydrogenase
L-beta-Hydroxyacid dehydrogenase [1]
L-beta-Hydroxy-acid-NAD-oxidoreductase [1]
L-3-Hydroxyacid dehydrogenase [4]

CAS Reg. No.
9028-51-7

2 REACTION AND SPECIFICITY

Catalysed reaction
L-Gulonate + NAD$^+$ →
→ 3-dehydro-L-gulonate + NADH

Reaction type
Redox reaction

Natural substrates

Substrate spectrum
1 L-Gulonic acid + NAD$^+$ (r [7], i.e. xylosecarboxylic acid, stereochemical specificity for 3-OH position in L-configuration [1]) [1–4, 6–8]
2 L-3-Hydroxybutyric acid + NAD$^+$ (ir [8], not oxidized [6, 7]) [3, 8]
3 L-Lyxonic acid + NAD$^+$ [6, 8]
4 L-Idonic acid + NAD$^+$ [6, 8]
5 D-Xylonic acid + NAD$^+$ [8]
6 D-Gluconic acid + NAD$^+$ [6, 8]
7 L-Erythronic acid + NAD$^+$ [8]
8 L-Threonic acid + NAD$^+$ [8]
9 D-Talonic acid + NAD$^+$ [8]

10 L-Altronic acid + NAD^+ [8]
11 D-Mannonic acid + NAD^+ [8]
12 L-Ribonic acid + NAD^+ [8]
13 D-Galactonic acid + NAD^+ [8]
14 More (not oxidized: DL-2-hydroxybutyrate, DL-4-hydroxybutyrate, 2-butanal, DL-3-hydroxybutyrate, L-threonine) [7]

Product spectrum

1 3-Keto-L-gulonic acid + NADH [6]
2 Acetoacetate + NADH [8]
3 ?
4 ?
5 ?
6 ?
7 ?
8 ?
9 ?
10 ?
11 ?
12 ?
13 ?
14 ?

Inhibitor(s)

Acetoacetate [3]; NADH [3]; p-Chloromercuribenzoate [6, 8]; Phenylmercurinitrate [6]; Iodoacetate [6]; More (not inhibitory: EDTA, 2,2'-bipyridine, 8-hydroxyquinoline, 1,10-phenanthroline, diethyldithiocarbamate, thiamine diphosphate, Zn^{2+}, Mg^{2+}, Cd^{2+}, Cu^{2+}, Fe^{2+}, Mn^{2+}, Ni^{2+}, Co^{2+}, As^{3+}) [8]

Cofactor(s)/prosthetic group(s)/activating agents

NAD^+ [1–8]; $NADP^+$ (10% of NAD^+ activity) [3]; Cysteine (activation) [8]; NADPH [7]

Metal compounds/salts

Turnover number (min^{-1})

Specific activity (U/mg)

63.8 [1]; 0.1 [2]; 12–13 [3]; More [4, 7, 8]

K_m-value (mM)

0.04 (NAD^+) [8]; 0.085 (NAD^+ (+ L-gulonic acid)) [3]; 0.25 (NAD^+) [7]; 0.45 (3-keto-L-gulonic acid) [8]; 0.85 (L-gulonic acid) [3]; 1.1 (NAD^+) [6]; 5.1 (L-gulonic acid) [8]; 5.3 (L-idonic acid, L-3-hydroxybutyrate) [8]; 6.4 (acetoacetate) [8]; 7.3 (diketo-L-gulonate) [8]; 14.8 (L-3-hydroxybutyrate) [3]; 19 (D-gluconic acid) [6]; 55–67 (L-gulonic acid, pH-range 7.5–8.5) [6]

pH-optimum

6.3 (reduction of beta-keto acids) [8]; 8.0 [2]; 8.2 [7]; 8.5 (oxidation of beta-hydroxy acids) [8]

pH-range

7–10 [7]

Temperature optimum (°C)

Temperature range (°C)

3 ENZYME STRUCTURE

Molecular weight

63000 (Drosophila melanogaster, sedimentation coefficient, Stokes radius, partial specific radius) [7]

Subunits

Dimer (2 × 33300–35000, Drosophila melanogaster, SDS-PAGE) [1]
? (x × 23000, Drosophila melanogaster, SDS-PAGE, in presence of mercaptoethanol value rises to 40000) [3]

Glycoprotein/Lipoprotein

–

4 ISOLATION/PREPARATION

Source organism

Drosophila melanogaster [1, 3, 4, 7], Salmonella typhimurium [2]; E. coli [2]; Sheep [5]; Schwanniomyces occidentalis [6]; Pig [8]

Source tissue

Whole flies [1, 3, 4, 7]; Malpighian tube [4]; Intestine [4]; Carcass [4]; Brain [4]; Fat body [4]; Kidney [8]; Liver [8]

Localisation in source

Purification

Drosophila melanogaster [1, 7]; Schwanniomyces occidentalis [6]; Pig [8]

Crystallization

–

Cloned

–

Renaturated

–

5 STABILITY

pH

Temperature (°C)

Oxidation

Organic solvent

General stability information

Cyanide stabilizes [6]; Urea: 3 M, 7 min, 83% inactivation, partially protected by DL-beta-hydroxybutyrate [3]; Unstable in solutions [6]

Storage

Frozen, several weeks [8]

6 CROSSREFERENCES TO STRUCTURE DATABANKS

PIR/MIPS code

Brookhaven code

7 LITERATURE REFERENCES

[1] Menotti-Raymond, M., Sullivan, D.T.: Biochim. Biophys. Acta,841,15–21 (1985)
[2] Cooper, R.A.: FEBS Lett.,115,63–67 (1980)
[3] Cannistraro, V.J., Borack, L.I., Chase, T.: Biochim. Biophys. Acta,569,1–5 (1979)
[4] Borack, L.I.: Experientia,30,31 (1974)
[5] Koundakjian, P.P., Snoswell, A.M.: Biochem. J.,127,449–452 (1972)
[6] Dworsky, P., Hoffmann-Ostenhof, O.: Acta Biochim. Pol.,11,269–277 (1964)
[7] Borack, L.I., Sofer, W.: J. Biol. Chem.,246,5345–5350 (1971)
[8] Smiley, J.D., Ashwell, G.: J. Biol. Chem.,236,357–364 (1961)

1 NOMENCLATURE

EC number
1.1.1.46

Systematic name
L-Arabinose:NAD^+ 1-oxidoreductase

Recommended name
L-Arabinose 1-dehydrogenase

Synonymes
Dehydrogenase, L-arabinose

CAS Reg. No.
9028-52-8

2 REACTION AND SPECIFICITY

Catalysed reaction
L-Arabinose + NAD^+ →
→ L-arabinono-1,4-lactone + NADH

Reaction type
Redox reaction

Natural substrates

Substrate spectrum
1 L-Arabinose + $NAD(P)^+$ [1–7]

Product spectrum
1 L-Arabinono-1,4-lactone + NAD(P)H [1, 2, 4, 5, 7]

Inhibitor(s)
ATP [5]

Cofactor(s)/prosthetic group(s)/activating agents
$NADP^+$ (greater affinity for $NADP^+$ than for NAD^+ [2], no activity [5]) [1, 2, 6, 7]; NAD^+ (specific for [5], preferred [7]) [2, 3, 5–7]

Metal compounds/salts
More (no activation by divalent cations) [2]

Turnover number (min^{-1})

Specific activity (U/mg)

4.12 [2]; 0.068 [6]; 0.07 (assay with NAD$^+$) [7]; 0.03 (assay with NADP$^+$) [7]

K_m-value (mM)

0.075 (L-arabinose (+ NADP$^+$)) [2]; 0.14 (L-arabinose (+ NAD$^+$)) [2]

pH-optimum

9.5 [2]

pH-range

More (increase of activity up to pH 12, no reaction at pH 13) [5]

Temperature optimum (°C)

Temperature range (°C)

3 ENZYME STRUCTURE

Molecular weight

175000 (Azospirillum brasiliense, gel filtration) [2]

Subunits

Glycoprotein/Lipoprotein

–

4 ISOLATION/PREPARATION

Source organism

Rhizobium leguminosarum (strain MNF 300 [1]) [1, 3]; Rhizobium sp. (strain NGR 234 [1], strain 32111 [3]) [1, 3]; Rhizobium meliloti [3]; Rhizobium trifolii [3]; Rhizobium phaseolii [3]; Rhizobium japonicum [3, 6]; Azospirillum brasiliense [2, 7]; Pseudomonas fluorescens [4]; Pseudomonas saccharophila [5]

Source tissue

Localisation in source

Purification

Azospirillum brasiliense (partial) [2]

Crystallization

–

Cloned

–

Renaturated

–

5 STABILITY

pH

Temperature (°C)
55 (5 min, no loss of activity) [2]

Oxidation

Organic solvent

General stability information

Storage

6 CROSSREFERENCES TO STRUCTURE DATABANKS

PIR/MIPS code

Brookhaven code

7 LITERATURE REFERENCES

[1] Dilworth, M.J., Arwas, R., McKay, I.A., Saroso, S., Glenn, A.R.: J. Gen. Microbiol., 132,2733–2742 (1986)
[2] Novick, N.J., Tyler, M.E.: Can. J. Microbiol.,29,242–246 (1983)
[3] Duncan, M.J.: J. Gen. Microbiol.,113,177–179 (1979)
[4] Schimz, K.-L., Kurz, G.: Biochem. Soc. Trans.,3,1087–1089 (1975)
[5] Weimberg, R., Doudoroff, M.: J. Biol. Chem.,217,607–624 (1955)
[6] Pedrosa, F.O., Zancan, G.T.: J. Bacteriol.,119,336–338 (1974)
[7] Novick, N.J., Tyler, M.E.: J. Bacteriol.,149,364–367 (1982)

1 NOMENCLATURE

EC number
1.1.1.47

Systematic name
beta-D-Glucose:NAD(P)$^+$ 1-oxidoreductase

Recommended name
Glucose 1-dehydrogenase

Synonymes
D-Glucose dehydrogenase (NAD(P))
Hexose phosphate dehydrogenase
Dehydrogenase, glucose

CAS Reg. No.
9028-53-9

2 REACTION AND SPECIFICITY

Catalysed reaction
beta-D-Glucose + NAD(P)$^+$ →
→ D-glucono-1,5-lactone + NAD(P)H

Reaction type
Redox reaction

Natural substrates

Substrate spectrum
1 beta-D-Glucose + NAD(P)$^+$ (reverse reaction 2.2–5.9% of forward reaction [15]) [1, 2, 4, 5, 8, 11–13, 15, 16, 18, 22–24]
2 D-Glucose-6-phosphate + NAD(P)$^+$ [1, 8, 11, 13, 22, 23]
3 D-Idose + NAD(P)$^+$ [4]
4 D-Xylose + NAD(P)$^+$ [4, 12, 15, 22]
5 2-Deoxy-D-glucose + NAD(P)$^+$ [4, 18, 20]
6 D-Mannose + NAD(P)$^+$ [11, 16]
7 D-Galactose + NAD(P)$^+$ [11, 22, 24]
8 Gentiobiose + NAD(P)$^+$ [12, 15]
9 Cellobiose + NAD(P)$^+$ [12, 15]
10 2-Deoxy-glucose-6-phosphate + NAD(P)$^+$ [22]

Product spectrum

1 beta-D-Glucono-1,5-lactone + NAD(P)H
2 6-Phospho-1,5-gluconolactone + NAD(P)H
3 D-Idono-1,5-lactone + NAD(P)H [4]
4 D-Xylono-1,4-lactone + NAD(P)H [4, 12, 15, 22]
5 2-Deoxy-D-glucono-1,5-lactone + NAD(P)H [4, 18, 20]
6 D-Galactono-1,5-lactone + NAD(P)H [11]
7 D-Mannono-1,5-lactone + NAD(P)H [11, 16]
8 6-O-(beta-D-Glucopyranosyl)-D-glucono-1,5-lactone + NAD(P)H [12, 15]
9 4-O-(beta-D-Glucopyranosyl)-D-glucono-1,5-lactone + NAD(P)H [22]
10 2-Deoxy-6-phospho-glucono-1,5-lactone + NAD(P)H [22]

Inhibitor(s)

EDTA [4]; Mercaptoethanol [4]; N-Ethylmaleimide [4]; 5,5'-Dithiobis(2-nitrobenzoic acid) [4]; Sulfhydryl reagents [11, 15, 16]; Divalent heavy metal ions [11, 15, 16]

Cofactor(s)/prosthetic group(s)/activating agents

NAD^+ [1, 4, 8, 12, 13, 15, 18, 20, 22–24]; $NADP^+$ (3 times more effective than NAD^+ [5], specific for [16]) [1, 2, 4, 5, 8, 11, 13, 15, 16, 18, 20, 22–24]

Metal compounds/salts

Mg^{2+} (activation [4], required with $NADP^+$ as cofactor not with NAD^+ [22]) [4, 22]; Mn^{2+} (activation) [4]; Ca^{2+} (activation) [4]; Zn^{2+} (activation) [4]; K^+ (activation) [18]

Turnover number (min^{-1})

Specific activity (U/mg)

550 [20]; 437 [4]; 368 [9]; More [1, 5, 8, 11, 13, 15, 16, 18, 22, 23]

K_m-value (mM)

0.0022 ($NADP^+$, pH 7.5) [1]; 0.0028 (NAD^+, pH 7.5) [1]; 0.0036 (glucose 6-phosphate, pH 7.5) [1]; 0.01 ($NADP^+$) [11]; 0.67 (NAD^+) [12]; 3.2 (glucose, pH 7.5) [1]; 5.6 (gentiobiose) [12]; 9.8 (cellobiose) [12]; 51–59 (D-glucose) [12]; 148 (D-xylose) [12]; More [5, 8, 9, 11, 15–18, 22, 23]

pH-optimum

8.0 (Tris/HCl buffer [20]) [4, 20]; 8.5–9 [11, 16]; 9.0 (acetate/borate buffer [20]) [17, 20]; 9.5 [13]; 9.8 [15]; 10 [8, 12, 23]; 10.7 [14]

pH-range

6.5–10.5 [13, 22]; 7–9 [20]; 7–11 [23]

Temperature optimum (°C)
77 [4]; 37 [8]; 50 [11, 16, 17]; 30–35 [15]; 30 [23]

Temperature range (°C)
20–40 [15]; 20–50 [23]

3 ENZYME STRUCTURE

Molecular weight
54000 (Corynebacterium sp., gel filtration) [12, 15]
80000 (Bacillus sp., gel filtration) [17]
116000–126000 (Bacillus subtilis, gel filtration [9], Bacillus megaterium, gel filtration [20]) [9, 20]
130000 (Sulfolobus solfataricus, gel filtration) [4]
153000 (Gluconobacter suboxydans, gel filtration) [11]
155000 (Thermoplasma acidophilum, gel filtation) [24]
180000 (Salmo gairdneri, gel filtration) [22]
222400–235000 (pig, amino acid analysis [8], gel filtration [23], bovine, gel filtration [13]) [8, 13, 23]
More (secondary, tertiary, quarternary structure) [3]

Subunits
Monomer (1 × 51000, Corynebacterium sp. No 93–1, SDS-PAGE) [15]
Tetramer (4 × 40000, Gluconobacter suboxydans, SDS-PAGE [11, 16], 4 × 30000, Sulfolobus solfataricus [4], Bacillus subtilis [9], Bacillus megaterium [20], SDS-PAGE [4, 9, 20], 4 × 38000, identical, Thermoplasma acidophilum, SDS-PAGE, N-terminal amino acid sequence [24], 4 × 59000, bovine, SDS-PAGE [13]) [4, 9, 11, 13, 16, 20, 24]
More (structural prediction) [3]

Glycoprotein/Lipoprotein
Lipoprotein (lipid content 1.7% [8]) [1, 8, 13]

4 ISOLATION/PREPARATION

Source organism
Pig [1, 8, 10, 23]; Bovine [13]; Salmo gairdneri (rainbow trout) [22]; Bacillus megaterium [2, 3, 7, 19, 20]; Bacillus subtilis [3, 9, 18]; Sulfolobus solfataricus [4]; Nostoc sp. [5, 21]; Klebsiella aerogenes [6]; Gluconobacter suboxydans [11, 16]; Corynebacterium sp. (strain No. 93–1 [12, 15], No. 150–1 [12, 14]) [12, 14, 15]; Lyngbya lagerheimii [21]; Schizothrix calcicola [21]; Chlorogloea fritschii [21]; Bacillus sp. [17]; Thermoplasma acidophilum [24, 25]; More (overview trout and salmon) [21]

Source tissue

Liver [1, 8, 10, 13, 22, 23]; Sporulating cells [18]

Localisation in source

Endoplasmic reticulum (luminal side) [1, 13]; Cytoplasmic membrane (periplasmic side) [6]; Membrane (associated [6, 8], bound [22]) [6, 8, 22]; Soluble [22]

Purification

Pig [1, 8, 23]; Salmo gairdneri [22]; Bacillus megaterium [20]; Bacillus subtilis [9, 18]; Sulfolobus solfataricus [4]; Gluconobacter suboxydans [11, 16]; Corynebacterium sp. [12, 15]; Bovine [13]; Thermoplasma acidophilum [24, 25]

Crystallization

[11, 16, 23, 25]

Cloned

–

Renaturated

–

5 STABILITY

pH

5–9 [4]; 6–8 [15]; 6.5 [7, 18]; 6–6.5 [9]

Temperature (°C)

21 (24 h) [1]; 37 (half-life 40 days) [4]; 40 (up to) [15]; 45 [22]; 50 [17]; 70 (half-life 45 h) [4]; 75 (half-life 3 h) [24]

Oxidation

Organic solvent

Methanol: 50% v/v, stable to [4, 24]; Acetone: 50% v/v, stable to [4, 24]; Ethanol: 50% v/v, stable to [24]

General stability information

NaCl stabilizes tetrameric form [7]; KH_2PO_4 stabilizes tetrameric form [7]; Urea: 2 M, stable, 4 M, inactivation [4]; Urea: 4 M, stable [24]; Tetrahydrofuran inactivates [4]; Guanidinium chloride inactivates [4]; Polyvinylpyrrolidone stabilizes [20]; NAD^+ and $NADP^+$ protect during chromatography [17]

Storage

−20°C, 0.05 M imidazole buffer, pH 6.5, 20% v/v glycerol, 3 months to several years [9, 15]; 4°C, 20 mM $MgCl_2$, 20% v/v ethylene glycol, several months [4]; 5°C, crystalline suspension, several months [11] 5°C, crude enzyme, inactivation in 3 days [17]

6 CROSSREFERENCES TO STRUCTURE DATABANKS

PIR/MIPS code

PIR2:A33528 (Bacillus megaterium); PIR2:JS0385 (Bacillus megaterium); PIR2:S02299 (Bacillus megaterium); PIR2:S36090 (Bacillus subtilis); PIR2:A39019 (fruit fly (Drosophila melanogaster)); PIR2:B39019 (fruit fly (Drosophila pseudoobscura)); PIR2:S05410 (Thermoplasma acidophilum (fragment)); PIR2:S00812 (A Bacillus megaterium); PIR2:S01227 (B Bacillus megaterium); PIR2:A20238 (major peptide AH-2-TA3 Bacillus megaterium (fragment)); PIR2:B20238 (minor peptide AH2-TA2 Bacillus megaterium (fragment)); PIR2:C20238 (minor peptide AH2-TA4 Bacillus megaterium (fragment)); PIR2:D20238 (minor peptide AH2-TA5 Bacillus megaterium (fragment))

Brookhaven code

7 LITERATURE REFERENCES

[1] Carper, W.R., Groutas, W.C., Coffin, D.B.: Experientia,44,29–32,(1988)

[2] Maurer, E., Pfleiderer, G.: Z. Naturforsch.,42c,907–915,(1987)

[3] Hönes, J., Jany, K.-D., Pfleiderer, G., Wagner, A.F.V. : FEBS Lett.,212,193–198,(1987)

[4] Giardina, P., De Biasi, M.-G., De Rosa, M., Gambacorta, A., Buonocore, V.: Biochem. J.,239,517–522,(1986)

[5] Juhász, A., Csizmadia, V., Borbély, G., Uvardy, J., Farkas, G.L.: FEBS Lett.,194,121–125,(1986)

[6] Hommes, R.W.J., van Hell, B., Postma, P.W., Neijssel, O.M., Tempest, D.W.: Arch. Microbiol.,143,163–168,(1985)

[7] Maurer, E., Pfleiderer, G.: Biochim. Biophys. Acta,827,381–388,(1985)

[8] Carper, W.R., Campbell, D.P., Morrical, S.W., Thompson, R.E.: Experientia,39, 1295–1297,(1983)

[9] Ramaley, R.F., Vasantha, N.: J. Biol. Chem.,258,12558–12565,(1983)

[10] Thompson, R.E., Morrical, S.W., Campbell, D.P., Carper, W.R.: Biochim. Biophys. Acta,745,279–284,(1983)

[11] Adachi, O., Ameyama, M.: Methods Enzymol.,89,159–163,(1982)

[12] Kobayashi, Y., Ueyama, H., Horikoshi, K.: Agric. Biol. Chem.,46,2139–2142,(1982)

[13] Campbell, D.P., Carper, W.R., Thompson, R.E.: Arch. Biochem. Biophys.,215, 289–301,(1982)

[14] Kobayashi, Y., Ueyama, H., Horikoshi, K.: Agric. Biol. Chem.,44,2837–2841,(1980)
[15] Kobayashi, Y., Horikoshi, K.: Agric. Biol. Chem.,44,2261–2269,(1980)
[16] Adachi, O., Matsushita, K., Shinagawa, E., Ameyama, M.: Agric. Biol. Chem.,44,301–308,(1980)
[17] Yokota, A., Sasajima, K., Yoneda, M.: Agric. Biol. Chem.,43,271–278,(1979)
[18] Yasutaro, F., Ramaley, R., Freese, E.: J. Bacteriol.,132,282–293,(1977)
[19] Pauly, H.E., Pfleiderer, G.: Biochemistry,16,4599–4604,(1977)
[20] Pauly, H.E., Pfleiderer, G.: Hoppe-Seyler's Z. Physiol. Chem.,356,1613–1623,(1975)
[21] Pulich, W.M., van Baalen, C.: J. Bacteriol.,114,28–33,(1973)
[22] Shatton, J.B., Halver, J.E., Weinhouse, S.: J. Biol. Chem.,246,4878–4885,(1971)
[23] Thompson, R.E., Carper, W.R.: Biochim. Biophys. Acta,198,397–406,(1970)
[24] Smith, D., Budgen, N., Bungard, S.J., Danson, M.J., Hough, D.W.: Biochem. J.,261,973–977 (1989)
[25] Bright, J.R., Mackness, R., Danson, M.J., Hough, D.W., Taylor, G.L., Towner, P.: J. Mol. Biol.,222,143–144 (1991)

1 NOMENCLATURE

EC number
1.1.1.48

Systematic name
D-Galactose:NAD^+ 1-oxidoreductase

Recommended name
Galactose 1-dehydrogenase

Synonymes
D-Galactose dehydrogenase
beta-Galactose dehydrogenase
NAD-dependent D-galactose dehydrogenase

CAS Reg. No.
9028-54-0

2 REACTION AND SPECIFICITY

Catalysed reaction
D-Galactose + NAD^+ →
→ D-galactono-1,4-lactone + NADH (stereochemistry of hydrogen transfer [6, 7])

Reaction type

Natural substrates

Substrate spectrum
1 D-Galactose + NAD^+ [1–16]
2 L-Arabinose + NAD^+ [10, 11]
3 2-Deoxy-D-galactose + NAD^+ [11, 13]
4 6-Deoxy-D-galactose + NAD^+ [11, 13]
5 3-Deoxy-D-galactose + NAD^+ [13]
6 4-Deoxy-D-galactose + NAD^+ [11]
7 D-Talose + NAD^+ [11]
8 2-Deoxy-2-amino-D-galactose + NAD^+ [11]

Product spectrum

1 D-Galactono-1,5-lactone + NADH [1–16]
2 L-Arabinono-1,5-lactone + NADH [10, 11]
3 2-Deoxy-D-galactono-1,5-lactone + NADH [11, 13]
4 6-Deoxy-D-galactono-1,5-lactone + NADH [11, 13]
5 3-Deoxy-D-galactono-1,5-lactone + NADH [13]
6 4-Deoxy-D-galactono-1,5-lactone + NADH [11]
7 D-Talono-1,5-lactone + NADH [11]
8 2-Deoxy-2-amino-galactono-1,5-lactone + NADH [11]

Inhibitor(s)

5,5'-Dithiobis(2-nitrobenzoic acid) [12]

Cofactor(s)/prosthetic group(s)/activating agents

NAD^+ [1–16]; $NADP^+$ (Pseudomonas fluorescens) [11]

Metal compounds/salts

Turnover number (min^{-1})

Specific activity (U/mg)

849 [3, 11]; 202 [13]

K_m**-value** (mM)

0.14 (NAD^+) [13]; 1.0 (galactose) [13]

pH-optimum

7.5–10 [9]; 9.1–9.5 [11]; 8–9 [13]

pH-range

Temperature optimum (°C)

Temperature range (°C)

---

3 ENZYME STRUCTURE

Molecular weight

64000–67000 (Pseudomonas fluorescens, gel chromatography) [3, 11]
102000 (Pseudomonas saccharophila, gel filtration, electrophoresis) [8, 13]

Subunits

Dimer (2 × 32000, SDS-PAGE, Pseudomonas fluorescens) [3, 11]
Tetramer (4 × 25000, SDS-PAGE, Pseudomonas saccharophila) [8, 13]

Glycoprotein/Lipoprotein

–

4 ISOLATION/PREPARATION

Source organism

Rhizobium meliloti [1]; Pseudomonas fluorescens (gene expressed in E. coli [4]) [2–4, 6, 7, 11]; Pseudomonas saccharophila [8, 10, 12–14, 16]; Pseudomonas sp. [9, 15]; Caulobacter crescentus [5]

Source tissue

Cell [1–16]

Localisation in source

Purification

Pseudomonas fluorescens [3, 11]; Pseudomonas saccharophila [13]

Crystallization

–

Cloned

(expression of Pseudomonas fluorescens gene in E. coli) [4]

Renaturated

–

5 STABILITY

pH

5–9 [3]; 4.6–9.2 [11]

Temperature (°C)

60 [13]

Oxidation

Organic solvent

General stability information

Storage

–20°C, phosphate buffer, 20–50% glycerol, several months [3, 11]; 4°C, mercaptoethanol, ammonium sulfate, several months [2, 13]

6 CROSSREFERENCES TO STRUCTURE DATABANKS

PIR/MIPS code

PIR2:S04853 (Pseudomonas fluorescens)

Brookhaven code

7 LITERATURE REFERENCES

[1] Arias, A., Cervenansky, C.: J. Bacteriol.,167,1092–1094 (1986)
[2] Fujimura, Y. in "Methods Enzym. Anal.",3rd Ed. (Bergmeyer, H.U., ed.) 6,288–296 (1984)
[3] Maier, E., Kurz, G.: Methods Enzymol.,89,176–181 (1982)
[4] Buckel, P., Zehelein, E.: Gene,16,149–159 (1981)
[5] Kurn, N., Contreras, I., Shapiro, L.: J. Bacteriol.,135,517–520 (1978)
[6] Brendel, K., Bressler, R., Alizade, M.A.: Biochim. Biophys. Acta,397,1–4 (1975)
[7] Ueberschär, H.-H., Blachnitzky, E.-O., Lehmann, J., Kurz, G.: Biochim. Biophys. Acta,391,15–18 (1975)
[8] Wengenmayer, F., Kurz, G.: Biochim. Biophys. Acta,386,590–602 (1975)
[9] Kurz, G., Wallenfels K. in "Methods Enzym. Anal."3rd Ed. (Bergmeyer, H.U., ed.) 2,1324–1327 (1974)
[10] Ueberschär, K.-H., Blachnitzky, E.-O., Kurz, G.: Eur. J. Biochem.,48,389–405 (1974)
[11] Blachnitzky, E.-O., Wengenmayer, F., Kurz, G.: Eur. J. Biochem.,47,235–250 (1974)
[12] Wengenmayer, F., Ueberschär, K.-H., Kurz, G.: Eur. J. Biochem.,43,49–58 (1974)
[13] Wengenmayer, F., Ueberschär, K.-H., Kurz, G.: Eur. J. Biochem.,40,49–61 (1973)
[14] Wengenmayer, F., Blachnitzky, E.-O., Kurz, G.: Hoppe-Seyler's Z. Physiol. Chem.,354,131–135 (1973)
[15] Hu, A.S.L., Cline, A.L.: Biochim. Biophys. Acta,93,237–245 (1964)
[16] De Ley, J., Doudoroff, M.: J. Biol. Chem.,227,745–757 (1957)

1 NOMENCLATURE

EC number
1.1.1.49

Systematic name
D-Glucose-6-phosphate:$NADP^+$ 1-oxidoreductase

Recommended name
Glucose-6-phosphate 1-dehydrogenase

Synonymes
NADP-glucose-6-phosphate dehydrogenase
Zwischenferment [10]
D-Glucose 6-phosphate dehydrogenase
Glucose 6-phosphate dehydrogenase (NADP)
NADP-dependent glucose 6-phosphate dehydrogenase
6-Phosphoglucose dehydrogenase
Entner-Doudoroff Enzyme [10]

CAS Reg. No.
9001-40-5

2 REACTION AND SPECIFICITY

Catalysed reaction
D-Glucose 6-phosphate + $NADP^+$ →
→ D-glucono-1,5-lactone 6-phosphate + NADPH

Reaction type
Redox reaction

Natural substrates
D-Glucose 6-phosphate + $NADP^+$ [1–10, 12, 14, 16–24, 26]

Substrate spectrum
1 D-Glucose 6-phosphate + $NADP^+$ (r) [1–10, 12, 14, 16–24, 26]

Product spectrum
1 D-Gluconate 6-phosphate + NADPH (r, 6-phosphoglucono-delta-lactone) [1–10, 12, 14, 16–24, 26]

Inhibitor(s)

Pyridoxal 5'-phosphate [1, 8]; p-Chloromercuribenzoate [5]; Estrogen [5]; Progesterone [5]; ATP [6, 8, 10, 11, 14, 21, 22]; Acetyl-CoA [6, 10]; CoA [6, 10]; NADPH [8, 10–12, 23, 24]; Phosphate ions [8]; ADP [10, 21]; AMP [10, 12]; GTP [10, 12]; ITP [10]; NADH [10, 14]; Phosphoenolpyruvate [10]; Ribulose 1,5-diphosphate [11]; Glyceraldehyde 3-phosphate [12]; Zn^{2+} [13, 20]; Hg^{2+} [13]; Cd^{2+} [13, 20]; Cu^{2+} [13]; Blue Dextran [13]; Co^{2+} [20]; Isocitrate [22]; Glyoxylate [22]; Palmitoyl-CoA [24]

Cofactor(s)/prosthetic group(s)/activating agents

$NADP^+$ [1–7, 10, 12, 14, 16–20, 22, 26]; NAD^+ [5, 6, 10, 14, 17–19, 26]; NADPH [1–7, 10, 12, 14, 16–20, 22, 26]

Metal compounds/salts

Mg^{2+} (required [8, 11], activation [5, 17, 26]) [5, 8, 11, 17, 26]; Mn^{2+} (activation) [17]; Ca^{2+} (activation) [17]

Turnover number (min^{-1})

83000 (NAD^+) [10]; 67000 ($NADP^+$) [10]

Specific activity (U/mg)

More [1, 2, 10, 18, 22, 20, 26, 23]

K_m-value (mM)

0.067 ($NADP^+$) [1]; 0.23 (glucose 6-phosphate) [1]; 0.037 (glucose 6-phosphate) [3]; 0.012 ($NADP^+$) [3]; 0.0056 ($NADP^+$) [5]; 0.047 (NAD^+) [5]; 0.042 (glucose 6-phosphate, (+ $NADP^+$)) [5]; 0.12 (NAD^+) [6]; More [6, 7, 9, 11–14, 16, 17, 19, 21, 23, 26]

pH-optimum

9.1 [1, 19]; 9.2 [2, 17]; 7.4–8.2 [3]; 8–9 [5, 7, 12]; 8.9 [10]; 7.4 [11, 20, 24]; 7.0 [13]; 8.5–9.5 [14]; 8.0–8.2 [22]; 7.5–9 [25]; 8.5 [26]

pH-range

7.4–8.2 [3]

Temperature optimum (°C)

25 (assay at [1–4, 6–9, 17]; 30 (assay at) [10, 14, 18–20, 24]; 37 [5]; 40 [13]; 45 [19]; 55 [12]

Temperature range (°C)

3 ENZYME STRUCTURE

Molecular weight

103000–110000 (Candida utilis, gel filtration [1, 2, 8], Neurospora crassa, gel electrophoresis [3], Leuconostoc mesenteroides, sedimentation equilibrium method [6], Methylomonas sp. M15, sedimentation equilibrium method [19]) [1–3, 6, 8, 19]
206000 (Neurospora crassa, gel electrophoresis) [3]
238700 (cow ultracentrifugation) [5]
126000 (Penicillium duponti, gel filtration) [7]
210000 (human) [9]
220000 (Pseudomonas fluorescens, Entner-Doudoroff enzyme, gel chromatography) [10, 18]
265000 (Pseudomonas fluorescens, Zwischenferment, gel chromatography) [10, 18]
345000 (Anabaena sp., three principal forms with MW of 120000, 240000 and 345000, gel electrophoresis) [11]
240000 (Anabaena sp., three principal forms with MW of 120000, 240000 and 345000, gel electrophoresis [11], Lactobacillus buchneri, gel filtration [13], mouse [16], Pisum sativum, gel filtration [23]) [11, 13, 16, 23]
118000–120000 (Anabaena sp., three principal forms with MW of 120000, 240000 and 345000, gel electrophoresis [11], Candida boidinii, gel filtration [12], Bacillus subtilis, sucrose density gradient centrifugation [17]) [11, 12, 17]
150000 (Pseudomonas sp. C, gel filtration) [14]
180000 (Aspergillus parasiticus, gel permeation chromatography) [20]
209000 (Zea mays, presence of $NADP^+$, gel chromatography) [21]
138000 (Agaricus bisporus, gel filtration) [24]
200000 (Bacillus stearothermophilus, gel filtration) [26]

Subunits

Dimer (2 × 57000, Neurospora crassa, SDS-PAGE [3], 2 × 54800, Leuconostoc mesenteroides, SDS gel electrophoresis [6], 2 × 61000, Candida boidinii, SDS gel electrophoresis [12], 2 × 58000, Bacillus subtilis, SDS gel electrophoresis [17], 2 × 55000, Methylomonas sp. M 15, SDS-PAGE [19], 2 × 54000, Zea mays, SDS gel electrophoresis [21], 2 × 55300, Agaricus bisporus, SDS-PAGE [24]) [3, 6, 12, 17, 19, 21, 24]
Tetramer (4 × 57000, Neurospora crassa, SDS-PAGE [3], 4 × 53000, human, SDS-PAGE [9], 4 × 60000, Lactobacillus buchneri, SDS-PAGE [13], Pisum sativum, SDS-PAGE [23], 4 × 60000, mouse [16], 4 × 55000, Pseudomonas fluorescens, Entner-Doudoroff enzyme [18], 4 × 65000, Pseudomonas fluorescens, Zwischenferment [18], 4 × 48000, Aspergillus parasiticus, SDS-PAGE [20], 4 × 54000, Zea mays, SDS-PAGE [21], 4 × 52000, Bacillus stearothermophilus, SDS-PAGE [26]) [3, 9, 13, 18, 20, 21, 23, 26]

Glycoprotein/Lipoprotein

–

4 ISOLATION/PREPARATION

Source organism

Candida utilis [1, 2, 8]; Brewer's yeast [2]; Human [2, 9, 27]; Rat [2, 27]; Bovine [4, 5]; Pseudomonas fluorescens [10, 18]; Neurospora crassa [3]; Leuconostoc mesenteroides [6]; Penicillium duponti [7]; Anabaena sp. [11]; Candida boidinii [12]; Lactobacillus buchneri [13]; Pseudomonas sp. C [14]; Mouse [16, 27]; Bacillus subtilis [17]; Methylomonas sp. M 15 [19]; Aspergillus parasiticus [20]; Zea mays [21]; Candida maltosa [22]; Saccharomyces cerevisiae [15, 22]; Pisum sativum [23]; Agaricus bisporus [24]; Plasmodium falciparum [25]; Bacillus stearothermophilus [26]; Dog [27]; Bacillus licheniformis [24]

Source tissue

Liver [2, 27]; Mycelium [3, 20]; Mammary gland [4]; Adrenal cortex [5]; Erythrocytes [9, 27]; Kidney [16]; Leaf [21, 23]; Commercial product [15]

Localisation in source

Cytoplasm [21, 23]; Chloroplast [21]; Cytosol [23]

Purification

Candida utilis [1, 2, 8]; Neurospora crassa [3]; Bovine [4, 5]; Leuconostoc mesenteroides [6]; Penicillium duponti [7]; Human [9]; Pseudomonas fluorescens [10, 18]; Anabaena sp. [11]; Candida boidinii [12]; Pseudomonas sp. C [14]; Mouse [16]; Bacillus subtilis [17]; Methylomonas sp. M 15 [19]; Aspergillus parasiticus [20]; Zea mays [21]; Candida maltosa [22]; Pisum sativum [23]; Agaricus bisporus [24]; Plasmodium falciparum [25]; Bacillus stearothermophilus [26]

Crystallization

[1, 2, 4–6, 8]

Cloned

–

Renaturated

(inactivated by potassium myristate) [15]

5 STABILITY

pH

5.4–7 (highest stability) [7]; 6.5–10 (Entner-Doudoroff enzyme stable) [10, 18]; 6.0–10.5 (stable at 5°C for 48 h) [12, 18]; 3–10 (stable) [13]; 7.5–9 (stable) [19]

Temperature (°C)

25 ($t_{1/2}$: 15 min) [23]; 55 (15 min, complete inactivation) [13]

Oxidation

Organic solvent

General stability information

Thioglycolate stabilizes [4]; EDTA stabilizes [4]

Storage

–20°C, 2 months [6]; –20°C or –40°C, impure state, more than 1 week [7]; 4°C, several months [9, 10, 18]; 5°C, pH 6.0–10.5, stable for 48 h [12]; 30°C, Bacillus stearothermophilus enzyme, more than 80% of activity maintained after 1 month [26]

6 CROSSREFERENCES TO STRUCTURE DATABANKS

PIR/MIPS code

PIR3:S37053 (Erwinia chrysanthemi); PIR2:A38174 (Escherichia coli); PIR1:DEFFG6 (fruit fly (Drosophila melanogaster)); PIR2:A47740 (fruit fly (Drosophila melanogaster) (strain OK93)); PIR2:A40309 (human); PIR2:A23949 (human); PIR2:A24935 (human); PIR2:A25756 (human); PIR2:B32902 (human); PIR2:B23949 (human (fragment)); PIR2:A32902 (human (fragment)); PIR2:A22002 (human (fragment)); PIR2:A45758 (human (fragment)); PIR2:A39864 (Leuconostoc mesenteroides); PIR2:A29027 (Leuconostoc mesenteroides (fragment)); PIR3:S31406 (mouse); PIR2:S27830 (Plasmodium falciparum); PIR3:S40259 (Plasmodium falciparum); PIR2:A44520 (rabbit); PIR2:S07083 (rat); PIR1:DEYCG6 (Synechococcus sp. (PCC 7942)); PIR3:S31337 (yeast (Kluyveromyces marxianus var. lactis)); PIR3:S29381 (yeast (Pichia jadinii)); PIR2:A32485 (yeast (Pichia jadinii) (fragment)); PIR2:S11078 (yeast (Pichia jadinii) (fragment)); PIR2:A33126 (yeast (Saccharomyces cerevisiae)); PIR2:S13744 (yeast (Saccharomyces cerevisiae)); PIR2:A21823 (yeast (Saccharomyces cerevisiae) (fragment)); PIR2:S12509 (yeast (Saccharomyces cerevisiae) (fragment)); PIR2:B37855 (Zymomonas mobilis); PIR2:S01233 (precursor rat); PIR3:S35910 (Escherichia coli (fragment)); PIR3:S21809 (human); PIR3:S17293 (mouse)

Brookhaven code

7 LITERATURE REFERENCES

[1] Engel, H.J., Domschke, W., Alberti, M., Domagk, G.F.: Biochim. Biophys. Acta,191,509–516 (1969)
[2] Chilla, R., Doering, K.M., Domagk, F., Rippa, M.: Arch. Biochem. Biophys.,159, 235–239 (1973)
[3] Scott, W.A.: Methods Enzymol.,41,177–182 (1975)
[4] Julian, G.R., Reithel, F.J.: Methods Enzymol.,41,183–188 (1975)
[5] McKerns, K.W.: Methods Enzymol.,41,188–196 (1975)
[6] Olive, C., Levy, H.R.: Methods Enzymol.,41,196–201 (1975)
[7] Shepherd, M.G.: Methods Enzymol.,41,201–205 (1975)
[8] Domagk, G.F., Chilla, R.: Methods Enzymol.,41,205–208 (1975)
[9] Cohen, P., Rosemeyer, M.A.: Methods Enzymol.,41,208–214 (1975)
[10] Lessmann, D., Schimz, K.-L., Kurz, G.: Eur. J. Biochem.,59,545–559 (1975)
[11] Schaeffer, F., Stanier, R.Y.: Arch. Microbiol.,116,9–19 (1978)
[12] Kato, N., Sahm, H., Schütte, H., Wagner, F.: Biochim. Biophys. Acta,566,1–11 (1979)
[13] Kawai, K., Eguchi, Y.: J. Ferment. Technol.,57,369–371 (1979)
[14] Ben-Bassat, A., Goldberg, I.: Biochim. Biophys. Acta,611,1–10 (1980)
[15] Tortora, R., Burlini, N., Hanozet, G.M., Guerritore, A.: Experientia,38,427–429 (1982)
[16] Lee, C.-Y.: Methods Enzymol.,89,252–257 (1982)
[17] Ujita, S., Kimura, K.: Methods Enzymol.,89,258–261 (1982)
[18] Maurer, P., Lessmann, D., Kurz, G.: Methods Enzymol.,89,261–270 (1982)
[19] Steinbach, R.A., Schütte, H., Sahm, H.: Methods Enzymol.,89,271–275 (1982)
[20] Niehaus, W.G., Dilts, R.P.: Arch. Biochem. Biophys.,228,113–119 (1984)
[21] Valenti, V., Stanghellini, M.A., Pupillo, P.: Plant Physiol.,75,521–526 (1984)
[22] Röber, B., Stolle, J., Reuter, G.: Z. Allg. Mikrobiol.,24,629–636 (1984)
[23] Fickenscher, K., Scheibe, R.: Arch. Biochem. Biophys.,247,393–402 (1986)
[24] Hammond, J.B.W.: J. Gen. Microbiol.,131,321–328 (1985)
[25] Yoshida, A., Roth, E.F.: Blood,69,1528–1530 (1987)
[26] Okuno, H., Nagata, K., Nakajima, H.: J. Appl. Biochem.,7,192–201 (1985)
[27] Schraven, E., Gruber, C.: Arzneim. Forsch.,38,36–39 (1988)

1 NOMENCLATURE

EC number
1.1.1.50

Systematic name
3alpha-Hydroxysteroid:NAD(P)$^+$ oxidoreductase (B-specific)

Recommended name
3alpha-Hydroxysteroid dehydrogenase (B-specific)

Synonymes
Dehydrogenase, 3alpha-hydroxy steroid
3alpha-Hydroxysteroid oxidoreductase
Sterognost 3alpha
More (see also EC 1.1.1.213 for enzymes catalyzing the same reaction, but which are A-specific with respect to hydrogen transfer to/from NAD$^+$/NADH. Enzymes without information of stereospecificity are summerized under EC 1.1.1.50 but may in fact belong to EC 1.1.1.213)

CAS Reg. No.
9028-56-2

2 REACTION AND SPECIFICITY

Catalysed reaction
Androsterone + NAD(P)$^+$ →
→ 5alpha-androstan-3,17-dione + NAD(P)H

Reaction type
Redox reaction

Natural substrates
5alpha-Dihydrotestosterone + NAD(P)H (discussion of physiological role [7]) [4, 7]

Substrate spectrum
1 Androsterone + NAD(P)$^+$ (i.e. 3alpha-hydroxy-5alpha-androstan-17-one, r) [1, 5, 6, 9, 12, 13, 18]
2 5alpha-Dihydrotestosterone + NAD(P)H (i.e. 5alpha-androstan-17beta-ol-3-one, r) [1, 4, 7, 8, 10, 11]
3 5beta-Dihydrotestosterone + NAD(P)H [7, 11]
4 3alpha-Hydroxy-5alpha-pregnan-20-one + NADP$^+$ [9]
5 3alpha-Hydroxy-5beta-pregnan-20-one + NADP$^+$ (r) [6]
6 3alpha-Hydroxy-5beta-androstan-17-one + NADP$^+$ [6, 18]

7 3alpha,21-Dihydroxy-5beta-pregnan-20-one + $NADP^+$ [6]
8 5beta-Androstan-3-alpha,17beta-diol + $NADP^+$ [6, 18]
9 5alpha-Androstan-3alpha,17beta-diol + $NADP^+$ (r) [6, 18]
10 Glycolithocholic acid + $NADP^+$ (i.e. 3alpha-hydroxy-5beta-cholanoyl glycine) [5, 6]
11 3alpha,17alpha,21-Trihydroxy-5beta-pregnan-11,20-dione + $NADP^+$ (r) [6]
12 Cholic acid + $NADP^+$ (i.e. 3alpha,7alpha,12alpha-trihydroxy-5beta-cholan-24-oic acid) [1]
13 Chenodeoxycholic acid + $NADP^+$ (i.e. 3alpha,7alpha-dihydroxy-5beta-cholan-24-oic acid) [1]
14 Deoxycholic acid + $NADP^+$ (i.e. 3alpha,12alpha-dihydroxy-5beta-cholan-24-oic acid) [1, 5]
15 Lithocholic acid + $NADP^+$ (i.e. 3alpha-hydroxy-5beta-cholan-24-oic acid) [1, 5]
16 Taurocholic acid + $NADP^+$ (i.e. 3alpha,7alpha,12alpha-trihydroxy-5beta-cholanoyl taurine) [1]
17 Etiocholan-3alpha-ol-17-one + $NADP^+$ [1]
18 3alpha-Hydroxy-12-oxo-9,11-cholenic acid + $NADP^+$ [1]
19 5beta-Androstan-3alpha,17alpha-diol + NAD^+ [18]
20 5alpha-Androstan-3alpha,17alpha-diol + NAD^+ [18]
21 3alpha,7alpha-Dihydroxy-5beta-cholanoyl glycine + NAD^+ [18]
22 3alpha,7alpha,12alpha-Trihydroxy-5beta-cholanoyl glycine + NAD^+ [18]
23 3alpha,12alpha-Dihydroxy-5beta-cholanoyl glycine + NAD^+ [18]
24 More (absolutely specific for 3alpha-hydroxy group of C_{19}, C_{21} and C_{24} steroids [1, 6], acts faster on A/B cis hydroxysteroids than on A/B trans hydroxysteroids [1], substrate specificity of various isozymes [5], reduction of carbonyl compounds e.g. quinones, aliphatic short-chain diketones [6]) [1, 5, 6]

Product spectrum

1 5alpha-Androstan-3,17-dione + NAD(P)H [13]
2 5alpha-Androstan-3alpha,17beta-diol + $NAD(P)^+$ [1, 7, 8, 10, 11]
3 5beta-Androstan-3alpha,17beta-diol + $NAD(P)^+$ [11]
4 5alpha-Pregnan-3,20-dione + NADPH
5 5beta-Pregnan-3,20-dione + NADPH
6 5-beta-Androstan-3,17-dione + NADPH
7 21-Hydroxy-5beta-pregnan-3,20-dione + NADPH
8 17beta-Hydroxy-5beta-androstan-3-one + NADPH
9 17beta-Hydroxy-5alpha-androstan-3-one + NADPH [6]
10 3-Oxo-glycocholanoic acid + NADPH (i.e. 3-oxo-5beta-cholanoyl glycine)
11 17alpha,21-Dihydroxy-5beta-pregnan-3,11,20-trione + NADPH [6]
12 7alpha,12alpha-Dihydroxy-3-oxo-5beta-cholan-24-oic acid + NADPH
13 7alpha-Hydroxy-3-oxo-5beta-cholan-24-oic acid + NADPH
14 12alpha-Hydroxy-3-oxo-5beta-cholan-24-oic acid + NADPH

15 3-Oxo-5beta-cholan-24-oic acid + NADPH
16 2[[7alpha,12alpha-Dihydroxy-3,24-dioxo-5beta-cholan-24-yl]amino]ethane sulfonic acid + NADPH
17 Etiocholan-3,17-dione + NADPH
18 3,12-Dioxo-9,11-cholenic acid + NADPH
19 5beta-Androstan-17alpha-ol-3-one + NADH
20 5alpha-Androstan-17alpha-ol-3-one + NADH
21 7alpha-Hydroxy-3-oxo-5beta-cholanoyl glycine + NADH
22 7alpha,12alpha-Dihydroxy-3-oxo-5beta-cholanoyl glycine + NADH
23 12alpha-Hydroxy-3-oxo-5beta-cholanoyl glycine + NADH
24 ?

Inhibitor(s)

Heavy metal ions [1]; p-Chloromercuribenzoate [1, 5]; Indomethacin (not [14]) [4, 6]; Fenamates [4]; 1-Methylpyrrole acetic acid [4]; Arylpropionic acid [4]; Salicylates [4]; Acetaminophen [4]; Medroxyprogesterone acetate [4, 6]; Progesterone [4]; Norethindrone [4]; Hg^{2+} [6]; 2alpha-Bromo-17beta-hydroxy-5alpha-androstan-3-one [6, 8]; Hexestrol [6]; Dienstrol [6]; Stilbestrol [6]; Zomepirac [6]; Dexamethasone [6]; Triamcinolone [6]; Bethamethasone [6]; Prostaglandin A_1 [6]; Prostaglandin F_{2alpha} [6]; Prostaglandin E_2 [6]; Prostaglandin D_2 [6]; Deoxycorticosterone [6]; 4-Pregnen-3,20-dione [6]; 17alpha-Hydroxy-4-pregnen-3,20-dione [6]; 17beta-Hydroxy-4-androsten-3-one [6]; 6alpha-Methylprednisolone [6]; 17alpha-Hydroxy-4-androsten-3-one [6]; 4-Androsten-3,17-dione [6]; 17alpha,21-Dihydroxy-1,4-pregnadien-3,11,20-trione [6]; Prednisolone [6]; 11beta,21-Dihydroxy-4-pregnen-3,20-dione [6]; 17alpha,21-Dihydroxy-4-pregnen-3,11,20-trione [6]; 11beta,17alpha,21-Trihydroxy-4-pregnen-3,20-dione [6]; 5alpha-Androstan-3-one [8]; 5alpha-Androstan-3,16-dione [8]; 5alpha-Androstan-3,17-dione [8]; 17alpha-Methyl-5alpha-dihydrotestosterone [8]; 5alpha-Dihydrotestosterone hemisuccinate [8]; 2alpha-Cyano-17beta-methoxy-5alpha-androstan-3-one [8]; 2alpha-Hydroxymethylene-17beta-methoxy-5alpha-androstan-3-one [8]; 2alpha-Bromo-5alpha-androstan-3,17-dione [8]; 2alpha-Iodo-5alpha-androstan-3,17-dione [8]; 2alpha-Fluoro-5alpha-androstan-3,17-dione [8]; 2alpha-Cyano-5alpha-dihydrotestosterone [8]; 5alpha-Pregnan-3,20-dione [8]; 21-Hydroxy-5alpha-pregnan-3,20-dione [8]; 17,21-Dihydroxy-5alpha-pregnan-3,20-dione [8]; 5beta-Dihydrotestosterone [8]; 5beta-Pregnan-3,20-dione [8]; 2alpha-Cyano-17beta-methoxy-4-androsten-3-one [8]; More (product inhibition [6, 12], steroids [7], not inhibitory: phenobarbital, pyrazole [5]) [5–7, 12]

Cofactor(s)/prosthetic group(s)/activating agents

NAD^+ (specific for [1, 3, 15], NAD^+ and $NADP^+$ equally effective with androsterone [5], preferred [8], 3beta-hydrogen of androsterone is transferred to side II of the pyrimidine ring [13]) [1–3, 5, 8, 9, 11–13, 15, 18]; $NADP^+$ (not [1, 3, 9], preferred with lithocholic acid as substrate [5]) [4, 5, 7, 10, 11]; NADH (specific for [1, 3, 15], NAD^+ and $NADP^+$ equally effective with androsterone [5], preferred [8]) [1, 3, 5, 8, 12–13, 15]; NADPH (8-fold higher activity with NADPH than with NADH [4], not [1, 3, 9], preferred with lithocholic acid as substrate [5]) [4, 5, 7, 10, 11]; More (see also EC 1.1.1.213 for enzymes catalyzing the same reaction but with A-specificity concerning the cofactor. All enzymes without information of stereospecificity for cofactors are included in EC 1.1.1.50, but some of those may in fact belong to EC 1.1.1.213)

Metal compounds/salts

Turnover number (min^{-1})

26.64–156.0 (androsterone) [5]; More (overview, values for various substrates and 6 different isozymes) [5]

Specific activity (U/mg)

63.0 [2]; 6.1 [6]; 2.2 [4]; 0.7 [14]; 0.358 [10]

K_m-value (mM)

0.00035–0.00049 (5beta-androstan-3,17-dione, 5beta-pregnan-3,20-dione, 5alpha-androstan-3,17-dione) [6]; 0.00062–0.00065 (5alpha-dihydrotestosterone) [8]; 0.0007–0.00075 (3-oxo-5beta-cholan-24-oic acid [6], 5beta-dihydrotestosterone, rat testis [7]) [6, 7]; 0.00097 (NADPH) [6]; 0.0012–0.0017 (17beta-hydroxy-5beta-androstan-3-one, 17alpha,21-dihydroxy-5beta-pregnan-3,11,20-trione, acenaphthequinone [6], 5beta-dihydrotestosterone, rat epididymis [7]) [6, 7]; 0.002–0.0024 (17beta-hydroxy-5alpha-androstan-3-one [6], 3alpha,7alpha-dihydroxy-5beta-cholanoyl glycine [18]) [6, 18]; 0.0052–0.0059 (glycolithocholic acid [6], 3alpha-hydroxy-5alpha-pregnan-20-one [9]) [6, 9]; 0.007–0.0083 (5alpha-dihydrotestosterone [4], 3alpha-hydroxy-5beta-pregnan-17-one [6], 3alpha-hydroxy-5alpha-androstan-17-one [9], androstanedione [12], 3alpha,12alpha-dihydroxy-5beta-cholanoyl glycine [18]) [4, 6, 9, 12, 18]; 0.0095 (androsterone) [6]; 0.01–0.025 (cholic acid [1], 3alpha-hydroxy-5beta-androstan-17-one, 3alpha,21-dihydroxy-5beta-pregnan-20-one, 5beta-androstan-3alpha,17beta-diol, 5alpha-androstan-3alpha,17beta-diol, $NADP^+$ [6], NAD^+, 3alpha,7alpha,12alpha-trihydroxy-5beta-cholanoyl glycine [18]) [1, 6, 18]; 0.03–0.077 (camphorquinone, 4-nitrobenzaldehyde, 2,3-heptanedione) [6]; 0.2–0.8 (4-nitroacetophenone, 2,3-hexanedione, 1-phenyl-1,2-propanedione, 2,3-pentanedione, pyridine-4-aldehyde, 2,3-butanedione) [6]; 0.87 (NAD^+) [1]; More (values for various substrates and different isozymes) [5]

pH-optimum

6.5 (reduction with NADPH) [6]; 7.0 (reduction of 3-oxo-steroid) [6]; 7.9 (reduction of etiocholan-3,17-dione) [1]; 8.5 (oxidation with NAD^+) [5]; 9.5 (oxidation of 3alpha-hydroxysteroids) [6]; 10 (oxidation with $NADP^+$) [5]; 11.4 [18]

pH-range

Temperature optimum (°C)

45–47.5 [10]; 50 [15]

Temperature range (°C)

16–36 (increase of activity over range) [6]

---

3 ENZYME STRUCTURE

Molecular weight

205000 (Eubacterium lentum, gel filtration) [16]
100000 (Pseudomonas testosteroni, gel filtration, concentration dependent monomer-dimer association) [3]
47100 (Pseudomonas testosteroni [15], gel filtration, concentration dependent monomer-dimer association [3]) [3, 15]
45000 (Pseudomonas putida, sedimentation equilibrium centrifugation) [1]
32500–34000 (rat, gel filtration [4, 5, 10], mouse, gel filtration [6]) [4–6, 10]

Subunits

Monomer (1 × 31000, rat, SDS-PAGE [4], 1 × 34000, mouse, SDS-PAGE [6], 1 × 45000, Pseudomonas putida, sedimentation equilibrium, presence of 2-mercaptoethanol [1], 1 × 47100, Pseudomonas testosteroni, concentration dependent monomer-dimer association [3]) [1, 3, 4, 6]
Dimer (2 × 47100, Pseudomonas testosteroni, concentration dependent monomer-dimer association) [3]

Glycoprotein/Lipoprotein

–

---

4 ISOLATION/PREPARATION

Source organism

Pseudomonas putida [1, 2]; Pseudomonas testosteroni [3, 12, 13, 15]; Rat [4, 5, 7, 9–11, 14]; Mouse [6]; Dog [8]; Eubacterium lentum [16, 18]; Clostridium irregularis [17]

Source tissue

Brain [4]; Liver [5, 6, 11, 14]; Testis [7]; Epididymis [7]; Prostate [8, 10]; Ovary [9]; Cell [1–3]

Localisation in source
Cytosol [4–10]; Microsomes [8, 11, 14]

Purification
Pseudomonas putida [2]; Rat (multiple forms [5], partial [10, 11, 14]) [4, 5, 10, 11, 14]; Mouse [6]

Crystallization
(association to dimer during crystallization [2]) [1, 2]

Cloned
–

Renaturated
–

5 STABILITY

pH
5.0–10.5 (at 30°C stable for 30 min) [1]

Temperature (°C)
40 (15 min stable in presence of mercaptoethanol, 10% loss of activity in absence of mercaptoethanol) [1]; 50 (denaturation above [10], 30–60 min stable [18]) [10, 18]

Oxidation

Organic solvent

General stability information
Glycerol: 30%, partially stabilizes [6]; Mercaptoethanol stabilizes [1]

Storage
–80°C, 20 mM potassium phosphate buffer, pH 7.0, 0.1 M EDTA, 1 mM mercaptoethanol, 20% glycerol, 3 months, 10–15% loss of activity [4]; 0–3°C, lyophilized, at least 6 months [1]; –20°C, 25% v/v glycerol, at least 1 year or 4°C, 1 week, stability depends on isozyme [5]; –20°C, purified enzyme, at least 2 weeks, crude extract loses 80% activity at 4°C in 3 days [6]; 4°C, 14 days [11]; 4°C, 2 days [18]

6 CROSSREFERENCES TO STRUCTURE DATABANKS

PIR/MIPS code
PIR3:JN0829 (Pseudomonas sp. (strain B-0831)); PIR2:A39350 (rat)

Brookhaven code
1RAL (Rat (Rattus norvegicus, sprague-Dawley strain))

7 LITERATURE REFERENCES

[1] Uwajima, T., Terada, O.: Agric. Biol. Chem.,43,1521–1528 (1979)
[2] Uwajima, T., Takayama, K., Terada, O.: Agric. Biol. Chem.,42,1577–1583 (1978)
[3] Skalhegg, B.A.: Eur. J. Biochem.,46,117–125 (1974)
[4] Penning, T.M., Sharp, R.B., Krieger, N.R.: J. Biol. Chem.,260,15266–15272 (1985)
[5] Ikeda, M., Hattori, H., Ikeda, N., Hayakawa, S., Ohmori, S.: Hoppe-Seyler's Z. Physiol. Chem.,365,377–391 (1984)
[6] Hara, A., Inoue, Y., Nakagawa, M., Naganeo, F., Sawada, H.: J. Biochem.,103, 1027–1034 (1988)
[7] Hastings, C., Hansson, V.: J. Steroid Biochem.,14,705–711 (1981)
[8] Jacobi, G.H., Moore, R.J., Wilson, J.D.: J. Steroid Biochem.,8,719–723 (1977)
[9] Nimrod, A., Lamprecht, S.A., Lindner, H.R.: J. Steroid Biochem.,6,1205–1209 (1975)
[10] Inano, H., Hayashi, S., Tamaoki, B.-i.: J. Steroid Biochem.,8,41–46 (1977)
[11] Golf, S.W., Graef, V., Nowotny, E.: Hoppe-Seyler's Z. Physiol. Chem.,357,35–40 (1976)
[12] Skalhegg, B.A.: Eur. J. Biochem.,50,603–609 (1975)
[13] Jarabak, J., Talalay, P.: J. Biol. Chem.,235,2147–2151 (1960)
[14] Boutin, J.A., Shikata, M., Talalay, P.: Biochem. Biophys. Res. Commun.,135,795–801 (1986)
[15] Squire, P.G., Dehlin, S., Porath, J.: Biochim. Biophys. Acta,89,409–421 (1964)
[16] MacDonald, I.A., Jellett, J.F., Mahony, D.E., Holdeman, L.V.: Appl. Environ. Microbiol.,37,992–1000 (1979)
[17] Mahony, D.E., Meier, C.E., MacDonald, I.A., Holdeman, L.V.: Appl. Environ. Microbiol.,34,419–423 (1977)
[18] MacDonald, D.E., Mahony, D.E., Jellet, J.F., Meier, C.E.: Biochim. Biophys. Acta,489,466–476 (1977)

1 NOMENCLATURE

EC number
1.1.1.51

Systematic name
3(or 17)beta-Hydroxysteroid:NAD(P)$^+$ oxidoreductase

Recommended name
3(or 17)beta-Hydroxysteroid dehydrogenase

Synonymes
Dehydrogenase, beta-hydroxy steroid
17-Ketoreductase
17beta-Hydroxy steroid dehydrogenase
3beta-Hydroxysteroid dehydrogenase
Dehydrogenase, 17beta-hydroxy steroid
Dehydrogenase, 3beta-hydroxy steroid
More (see also EC 1.1.1.62, EC 1.1.1.63, EC 1.1.1.64 for enzymes catalyzing 17beta-hydroxysteroid dehydrogenation, enzymes without detailed investigation concerning coenzyme specificity are included in EC 1.1.1.51, and EC 1.1.1.145 for enzymes catalyzing 3beta-hydroxysteroid dehydrogenation but no 17beta-dehydrogenation. In some cases there is no clear evidence in the literature whether 3beta- and 17beta-dehydrogenation are catalyzed by a single enzyme molecule. All doubtful enzymes are included in EC 1.1.1.51)

CAS Reg. No.
9015-81-0

2 REACTION AND SPECIFICITY

Catalysed reaction
Testosterone + NAD(P)$^+$ →
→ androst-4-ene-3,17-dione + NAD(P)H

Reaction type
Redox reaction

Natural substrates
17-Oxo steroids + NADPH (maintenance of balance between oxidized and reduced steroids in blood) [9]

Substrate spectrum

1 Testosterone + NAD(P)$^+$ [1, 3, 6, 7, 9–11]
2 5beta-Pregnan-3,20-dione + NADH [4]
3 5alpha-Androst-16-en-3-one + NADH [5]
4 5alpha-Dihydrotestosterone + NAD(P)H [6, 7]
5 Estrone + NADPH (r) [7, 9, 11]
6 Dehydroepiandrosterone + NADPH (r [9], not [6]) [7, 9, 11]
7 Androsterone + NADPH [7]
8 5alpha-Androstane-3beta,17beta-diol + NADP$^+$ [7]
9 3beta-Hydroxy-5alpha-androstan-17-one + NADP$^+$ [7]
10 5alpha-Dihydrotestosterone + NADP$^+$ [7]
11 Dehydroepiandrosterone sulfate + NADPH [9]
12 Estrone sulfate + NADPH [9]

Product spectrum

1 Androst-4-en-3,17-dione + NAD(P)H [3, 9, 11]
2 3beta-Hydroxy-5beta-pregnan-20-one + NAD$^+$ [4]
3 3beta-Hydroxy-5alpha-androst-16-ene + NAD$^+$ [5]
4 5alpha-Androstane-3beta,17beta-diol + NADP$^+$ [7]
5 Estradiol + NADP$^+$ [7, 9, 11]
6 Androst-5-ene-3beta,17beta-diol + NADP$^+$ [7, 9, 11]
7 5alpha-Androstane-3alpha,17beta-diol + NADP$^+$ [7]
8 3beta-Hydroxy-5alpha-androstan-17-one + NADPH [7]
9 5alpha-Androstane-3,17-dione + NADPH [7]
10 5alpha-Androstane-3,17-dione + NADPH [7]
11 Androstenediol sulfate + NADP$^+$ [9]
12 Estradiol sulfate + NADP$^+$ [9]

Inhibitor(s)

p-Substituted mercuribenzoate [3]; Ethanol [7]; Propyleneglycol [7]; Dioxane [7]; Dimethylsulfoxide [7]; Glycerol [7]; Detergents (e.g. Tween-80, Triton X-100) [7]; p-Chloromercuribenzoate [9, 10]; Chloromercuribenzene-p-sulfonic acid [9]; Albumin [9]; $ZnCl_2$ [10]; Iodosobenzoate [10]; More (not: sodium azide, ammonium sulfate) [7]

Cofactor(s)/prosthetic group(s)/activating agents

NAD$^+$ (preferentially [5], substituted NAD$^+$ [2], not [9]) [1, 2, 5, 10]; NADH (not NADPH [4], preferentially [5], not [9]) [1, 2, 4, 5]; NADP$^+$ (specific for [9], 3 times greater activity with NADP$^+$ than with NAD$^+$ [10]) [6, 7, 9, 10]; NADPH (specific for [9], higher activity with NADPH than with NADH [11]) [7, 9, 11]

Metal compounds/salts

Phosphate (0.1 mol per mol of subunit) [3]

Turnover number (min^{-1})

Specific activity (U/mg)
201 [3]; More [6]

K_m-value (mM)
0.0018 (NADPH (+ 5alpha-dihydrotestosterone) i.e. 3-beta-hydroxysteroid dehydrogenase activity) [7]; 0.0021 (NADPH (+ estrone) i.e. 17beta-hydroxysteroid dehydrogenase activity) [7]; 0.0082 ($NADP^+$ (+ testosterone) i.e. 17beta-hydroxysteroid dehydrogenase activity) [7]; 0.009 ($NADP^+$ (+ 3beta-hydroxy-5alpha-androstan-17-one) i.e. 3beta-hydroxysteroid dehydrogenase activity) [7]; 0.063–0.068 (3-pyridinealdehyde adenine dinucleotide [2], 5alpha-androst-16-en-3-one, microsomal enzyme [5]) [2, 5]; 0.083 (NAD^+, thionicotinamide adenine dinucleotide) [2]; 0.107 (5alpha-androst-16-en-3-one, mitochondrial enzyme) [5]; 0.14 (5alpha-dihydrotestosterone) [7]; 0.17 (testosterone) [9]; 0.176 (5alpha-androst-16-en-3-one, cytosolic enzyme) [5]; 0.2–0.24 (testosterone [6, 10], 5alpha-dihydrotestosterone [6], NAD^+ [6], $NADP^+$ [6, 9], 5beta-pregnan-3,20-dione [4]) [4, 6, 9, 10]; 0.31 (testosterone) [7]; 0.685 (3-acetylpyridine adenine dinucleotide) [2]; More [11]

pH-optimum
6–7 (reductase activity) [7]; 6.0–7.5 (reductase activity) [4]; 8.2 (dehydrogenase activity) [7]; 8.7 (testosterone oxidation) [10]; 10.6 [6]

pH-range

Temperature optimum (°C)

Temperature range (°C)

3 ENZYME STRUCTURE

Molecular weight
98500 (Pseudomonas testosteroni, native PAGE) [3]
80000 (Clostridium innocuum, HPLC gel filtration) [4]
70000 (rat, gel filtration) [7]
64000 (human erythrocytes, gel filtration) [9]
32000 (guinea pig, gel filtration) [6]

Subunits
Tetramer (nonidentical subunits, similar in size, different in charge, stoichiometry A3B, 4 × 23500–26700, Pseudomonas testosteroni, sedimentation of urea-treated protein, SDS-PAGE) [3]
Monomer (1 × 34000, female guinea pig, SDS-PAGE) [6]

Glycoprotein/Lipoprotein
No glycoprotein [3]

4 ISOLATION/PREPARATION

Source organism

Pseudomonas testosteroni [1–3]; Clostridium innocuum [4]; Pig (boar) [5]; Guinea pig (male [8], female only after admission of testosterone [6]) [6, 8]; Rat [7, 10]; Human [9]; Monkey [11]

Source tissue

Cell [3, 4]; Testis [5]; Kidney [6, 8]; Erythrocytes [7, 9]; Adrenal gland [10]; Liver [11]

Localisation in source

Cytosol (80% of activity [6]) [3–6, 9–11]; Mitochondria [5]; Microsomes [5]

Purification

Pseudomonas testosteroni [3]; Rat [7]; Guinea pig [8]; Human (partial) [9]

Crystallization

–

Cloned

–

Renaturated

[3]

---

5 STABILITY

pH

More (unstable under slightly alkaline conditions) [7]

Temperature (°C)

Oxidation

Organic solvent

Ethanol, 1.7% v/v: inactivation [7]; Propyleneglycol, 3.3% v/v: inactivation [7]; Dioxane, 3.3% v/v: inactivation [7]; Dimethylsulfoxide, 3.3% v/v: inactivation [7]; Glycerol, 5% v/v: inactivation [7]

General stability information

Storage

–70°C, very labile in all extracts, inactivation in 5 days, but only 30% loss of activity if 20% glycerol is included [4]; –20°C, 5 mM potassium phosphate buffer, pH 7.5, 1 mM dithiothreitol, 50% glycerol, at least 7 months [3]; –15°C, a few percent loss of activity in several months [9]

6 CROSSREFERENCES TO STRUCTURE DATABANKS

PIR/MIPS code

Brookhaven code

7 LITERATURE REFERENCES

[1] Minard, P., Legoy, M.-D., Thomas, D.: FEBS Lett.,188,85–90 (1985)
[2] Schultz, R.M., Groman, E.V., Engel, L.L.: J. Biol. Chem.,252,3784–3790 (1977)
[3] Schultz, M., Groman, E.V., Engel, L.L.: J. Biol. Chem.,252,3775–3783 (1977)
[4] Stokes, N.A., Hylemon, P.B.: Biochim. Biophys. Acta,836,255–261 (1985)
[5] Kwan, T.K., Gower, D.B.: Biochem. Soc. Trans.,12,842–843 (1984)
[6] Shen, C.C., Kochakian, C.D.: J. Steroid Biochem.,10,187–193 (1979)
[7] Heyns, W., de Moor, P.: Biochim. Biophys. Acta,358,1–13 (1974)
[8] Shen, C.C., Kochakian, C.D.: Endocrinology,103,2040–2052 (1978)
[9] Mulder, E., Lamers-Stahlhofen, G.J.M., van der Molen, H.J.: Biochem. J.,127,649–659 (1972)
[10] Dahm, K., Breuer, H.: Hoppe-Seyler's Z. Physiol. Chem.,336,63–68 (1964)
[11] Fan, D.-F., VanCura, S., Hartsock, R.J.: Gen. Comp. Endocrinol.,35,465–474 (1978)

1 NOMENCLATURE

EC number
1.1.1.52

Systematic name
3alpha-Hydroxy-5beta-cholanate:NAD$^+$ oxidoreductase

Recommended name
3alpha-Hydroxycholanate dehydrogenase

Synonymes
Dehydrogenase, 3alpha-hydroxycholanate
alpha-Hydroxy-cholanate dehydrogenase

CAS Reg. No.
9028-57-3

2 REACTION AND SPECIFICITY

Catalysed reaction
3alpha-Hydroxy-5beta-cholanate + NAD$^+$ →
→ 3-oxo-5beta-cholanate + NADH

Reaction type
Redox reaction

Natural substrates

Substrate spectrum
1 3alpha-Hydroxy-5beta-cholanate + NAD$^+$ (i.e. lithocholic acid, r) [1]
2 Cholic acid + NAD$^+$ (oxidation at the same rate as hydroxycholanic acid) [1]
3 Deoxycholic acid + NAD$^+$ (oxidation at the same rate as hydroxycholanic acid) [1]
4 Dehydrocholic acid + NADH (i.e. 3,7,12-trioxocholan-24-oic acid) [1]
5 3alpha-Hydroxy-bis-norcholanic acid + NAD$^+$ (oxidation at the same rate as hydroxycholanic acid) [1]
6 3alpha-Hydroxy-norcholanic acid + NAD$^+$ (oxidation at the same rate as hydroxycholanic acid) [1]
7 More (no substrates: androsterone and tetrahydrocortisone, methylation of carboxyl group or reduction to alcohol inactivates) [1]

Product spectrum

1 3-Oxo-5beta-cholanate + NADH [1]
2 7alpha,12alpha-Dihydroxy-3-oxo-5beta-cholan-24-oic acid + NADH
3 12alpha-Hydroxy-3-oxo-5beta-cholan-24-oic aicd + NADH
4 3-Hydroxy-7,12-dihydroxy-5beta-cholan-24-oic acid + NAD^+
5 3-Oxo-bis-norcholanic acid + NADH
6 3-Oxo-norcholanic acid + NADH
7 ?

Inhibitor(s)

Cofactor(s)/prosthetic group(s)/activating agents

NAD^+ ($NADP^+$ cannot replace NAD^+) [1]; NADH [1]; More (SH-groups are required for activity) [1]

Metal compounds/salts

Turnover number (min^{-1})

Specific activity (U/mg)

K_m-value (mM)

pH-optimum

10.4 (oxidation, assay at) [1]

pH-range

More (oxidation proceeds at alkaline and reduction favorably at neutral pH-values) [1]

Temperature optimum (°C)

26 (assay at) [1]

Temperature range (°C)

---

3 ENZYME STRUCTURE

Molecular weight

Subunits

Glycoprotein/Lipoprotein

–

---

4 ISOLATION/PREPARATION

Source organism

Escherichia sp. (tentative identification: E. freundii) [1]

Source tissue
Cell [1]

Localisation in source
Cytoplasm [1]

Purification
Escherichia sp. (partial) [1]

Crystallization
–

Cloned
–

Renaturated
–

5 STABILITY

pH

Temperature (°C)

Oxidation

Organic solvent

General stability information

Storage

6 CROSSREFERENCES TO STRUCTURE DATABANKS

PIR/MIPS code

Brookhaven code

7 LITERATURE REFERENCES

[1] Hayaishi, O., Sato, Y., Jacoby, W.B., Stohlman, E.F: Arch. Biochem. Biophys.,56, 554–555 (1955)

1 NOMENCLATURE

EC number
1.1.1.53

Systematic name
3alpha(or 20beta)-Hydroxysteroid:NAD^+ oxidoreductase

Recommended name
3alpha(or 20beta)-Hydroxysteroid dehydrogenase

Synonymes
Cortisone reductase
(R)-20-Hydroxysteroid dehydrogenase
Dehydrogenase, 20beta-hydroxy steroid
DELTA4–3-Ketosteroid hydrogenase
20beta-Hydroxysteroid dehydrogenase
3alpha,20beta-Hydroxysteroid:NAD^+-oxidoreductase
NADH-20beta-hydroxysteroid dehydrogenase
20beta-HSD [2, 7]

CAS Reg. No.
9028-42-6

2 REACTION AND SPECIFICITY

Catalysed reaction
Androstan-3alpha,17beta-diol + NAD^+ →
→ 17beta-hydroxyandrostan-3-one + NADH

Reaction type
Redox reaction

Natural substrates
17-Hydroxyprogesterone + NADPH [2]

Substrate spectrum
1 Progesterone + NADH [1, 2, 5, 8–11]
2 17alpha-Hydroxyprogesterone + NADPH [2, 8, 15]
3 Pregnenolone + NADPH (i.e. 3-hydroxypregn-5-en-20-one) [2]
4 Deoxycorticosterone + NADPH (i.e. 21-hydroxypregn-4-ene-3,20-dione, 21-hydroxyprogesterone) [2, 10]
5 17alpha-Hydroxypregnenolone + NADPH (i.e. 3beta,17alpha-dihydroxypregn-5-ene-20-one) [2]
6 Deoxycortisol + NADPH [2]

7 Cortisone + NADH (i.e. 17beta,21-dihydroxy-pregn-4-ene-3,11,20-trione) [5, 8, 12]
8 17beta-[(1S)-1-Hydroxy-2-propynyl]-androst-4-ene-3-one + NAD^+ [7]
9 5alpha-Pregnane-3alpha,20beta-dione + NADH [8]
10 6beta-(Bromoacetoxy)-progesterone + NADH [8]
11 17-Hydroxyprogesterone + NADH [8]
12 21-Hemisuccinyloxyprogesterone + NADH [10]
13 21-Acetoxyprogesterone + NADH [10]
14 11alpha-Hemisuccinyloxyprogesterone + NADH [10]
15 17-Hydroxy-3,20-dioxo-4-pregnen-21-oate + NADH [9]
16 3,20-Dioxo-4-pregnen-21-oic acid + NADH [9]
17 17-Hydroxy-3,11,20-trioxo-4-pregnen-21-oate + NADH [9]
18 5alpha-Dihydrotestosterone + NADPH [15]
19 17alpha,20beta-Dihydroxypregn-4-ene-3-one + $NADP^+$ [16]
20 20beta-Hydroxypregn-4-ene-3-one + $NADP^+$ [16]
21 3beta,20alpha-Dihydroxypregn-5-ene + $NADP^+$ [16]
22 3beta,17alpha,20alpha-Trihydroxypregn-5-ene + $NADP^+$ [16]

Product spectrum

1 20beta-Hydroxy-pregn-4-ene-3-one + NAD^+ [1]
2 17alpha,20beta-Dihydroxy-pregn-4-ene-3-one + $NADP^+$ [2]
3 3beta,20beta-Dihydroxypregn-5-ene + $NADP^+$
4 20,21-Dihydroxypregn-4-ene-3-one + $NADP^+$
5 3beta,17alpha,20beta-Trihydroxypregn-5-ene + $NADP^+$
6 ?
7 17alpha,20beta,21-Trihydroxy-pregn-4-ene-3,11-dione + NAD^+ [5]
8 17beta-(1-Oxo-2-propynyl)-androst-4-ene-3-one + NADH [7]
9 ?
10 ?
11 ?
12 ?
13 ?
14 ?
15 17,20-Dihydroxy-4-pregnen-3-one-21-oate + NAD^+
16 20beta-Hydroxy-3-oxo-4-pregnen-21-oic acid + NAD^+ [9]
17 17,20-Dihydroxy-3,11-dioxo-4-pregnen-21-oate + NAD^+
18 5alpha-Androstan-3alpha,17beta-diol + 5alpha-androstan-3beta,17beta-diol + $NADP^+$ (ratio 4:3) [15]
19 17alpha-Hydroxypregn-4-en-3,20-dione + NADPH
20 Pregn-4-ene-3,20-dione + NADPH
21 3beta-Hydroxypregn-4-en-20-one + NADPH
22 3beta,17alpha-Dihydroxypregn-5-en-20-one + NADPH

Inhibitor(s)

Hg^{2+} [2], Cu^{2+} [2]; Cysteine (slight) [2]; Dithiothreitol (slight) [2]; 17-Acetoxyprogesterone [8]; 5alpha-Dihydrotestosterone bromoacetate [8]; 17-Bromoacetoxyprogesterone [8]; Cortisone 21-iodoacetate (in absence of NADH, slight) [11]; 16alpha-Bromoacetoxyprogesterone [11]; 17beta-Bromoacetoxyandrostan-3-one [11]; 20beta-Hydroxycortisol [12]; 20beta-Hydroxyprogesterone [12]; More (substrate and product inhibition) [12]

Cofactor(s)/prosthetic group(s)/activating agents

NADH (not NADPH [9]) [1, 7, 9, 11, 16]; NADPH (not [9]) [2, 14, 15]; $NADP^+$ (preferentially beta-$NADP^+$) [16]; NAD^+ (preferentially beta-NAD^+) [16]; beta-3'-$NADP^+$ [16]

Metal compounds/salts

Turnover number (min^{-1})

400 (NADH + progesterone) [5]; 20 (NADH + cortisone) [5]

Specific activity (U/mg)

0.004 [2]

K_m-value (mM)

0.0011–0.0025 (progesterone in reversed micelles, value depends on type of detergent) [1]; 0.0015 (progesterone) [2]; 0.0028 (progesterone) [11]; 0.0028–0.005 (progesterone in aqueous solution, value increases with increasing ionic strength) [1]; 0.0035–0.0057 (NADH in aqueous solution, value increases with increasing ionic strength) [1]; 0.0039 (progesterone) [8]; 0.00395 (progesterone) [10]; 0.004–0.016 (NADH, in reversed micelles, value depends on type of detergent) [1]; 0.004 (pregnonolone) [2]; 0.0045 (5alpha-pregnane-3alpha,20beta-dione) [8]; 0.0061 (progesterone) [9]; 0.007 (NADH, free enzyme [3]) [3, 5]; 0.0086 (deoxycorticosterone) [2]; 0.009–0.01 (NADH) [11]; 0.0105 (17-hydroxyprogesterone) [8]; 0.018 (NADH, immobilized enzyme) [3]; 0.02 (6beta-(bromoacetoxy)-progesterone) [8]; 0.024 (cortisone) [11]; 0.0251 (21-hydroxyprogesterone) [10]; 0.029 (NADH) [12]; 0.042–0.048 (cortisone, free enzyme) [3, 5]; 0.46 (17beta-hydroxyandrostan-2-one) [11]; 0.085–0.098 (cortisone, immobilized enzyme) [3, 5]; 0.14 (3,20-dioxo-4-pregnen-21-oic acid [9], 17beta-hydroxyandrostan-3-one [11]) [9, 11]; 0.318 (cortisone) [8]; 0.49 (17-hydroxy-3,11,20-trioxo-4-pregnen-21-oate) [9]; 0.64 (17-hydroxy-3,20-dioxo-4-pregnen-21-oate) [9]; 1.43 (21-hemisuccinyloxyprogesterone) [10]; 2.19 (11alpha-hemisuccinyloxyprogesterone) [10]; 2.23 (21-acetoxyprogesterone) [10]; More [12]

pH-optimum

5.1 (reduction of oxo-steroids) [9]; 5.8 (reduction of progesterone in aqueous solution) [1]; 5.5 (reduction of oxosteroid) [2]; 6.4 (reduction of progesterone in reversed micelles) [1]; 7.5 [17]; 9.2 (oxidation of hydroxysteroid) [7]

pH-range

4.5–5.5 [2]

Temperature optimum (°C)

50 [2]

Temperature range (°C)

35–60 (50% of maximal activity at 35°C, 10.4% of maximal activity at 60°C) [2]

3 ENZYME STRUCTURE

Molecular weight

106000–111000 (Streptomyces hydrogenans, agarose gel filtration, density gradient centrifugation) [17]
30500 (pig, gel filtration) [2]

Subunits

Tetramer (4 × 27000, Streptomyces hydrogenans) [17]
Monomer (1 × 30500, pig, SDS-PAGE with and without 2-mercaptoethanol) [2]

Glycoprotein/Lipoprotein

No glycoprotein [11]

4 ISOLATION/PREPARATION

Source organism

Streptomyces hydrogenans [1, 3–13, 17]; Pig [2, 15, 16]; Guinea pig [14]

Source tissue

Commercial product [1, 3–12]; Testis [2, 15–16]; Mycelium [13]

Localisation in source

Cytosol [1, 14]

Purification

Pig [2]; Streptomyces hydrogenans [11, 13]

Crystallization
[11]

Cloned
–

Renaturated
[3, 6]

5 STABILITY

pH
6.0–10.0 [2]; 9.0 (in Tris-buffer, best value for stability) [8]

Temperature (°C)
45 (up to) [2]; 55 (inactivation at) [2]

Oxidation

Organic solvent

General stability information
Urea: investigation of denaturation process [4]

Storage
–20°C, 5 mM phosphate buffer, pH 7.6 [1]; –20°C, 50 mM potassium phosphate buffer, pH 7.4 [2]

6 CROSSREFERENCES TO STRUCTURE DATABANKS

PIR/MIPS code
PIR2:PC1129 (human (fragments)); PIR2:A42912 (pig)

Brookhaven code
2HSD (Streptomyces hydrogenans)

7 LITERATURE REFERENCES

[1] Tyrakowska, B., Verhaert, R.M.D., Hilhorst, R., Veeger, C.: Eur. J. Biochem.,187, 81–88 (1990)
[2] Nakajin, S., Ohno, S., Shinoda, M.: J. Biochem.,104,565–569 (1988)
[3] Carrea, G., Pasta, P.: Methods Enzymol.,135,475–483 (1987)
[4] Vecchio, G., Pasta, P., Mazzola, G., Carrea, G.: Biochim. Biophys. Acta,914,122–126 (1987)
[5] Carrea, G., Cremonesi, P.: Methods Enzymol.,136,150–157 (1987)

[6] Carrea, G., Pasta, P., Vecchio, G.: Biochim. Biophys. Acta,784,16–23 (1984)
[7] Strickler, R.C., Covey, D.F., Tobias, B.: Biochemistry,19,4950–4954 (1980)
[8] Sweet, F., Samant, B.R.: Biochemistry,19,978–986 (1980)
[9] Monder, C.: J. Steroid Biochem.,9,1229–1231 (1978)
[10] Sweet, F., Patrick, T.B., Judd, M.: Biochem. Biophys. Res. Commun.,83,187–194 (1978)
[11] Edwards, C.A.F., Orr, J.C.: Biochemistry,17,4370–4376 (1978)
[12] Szymanski, E.S., Furfine, C.S.: J. Biol. Chem.,252,205–211 (1977)
[13] Hübner, H.J., Sahrholz, F.G.: Biochem. Z.,333,95–105 (1960)
[14] Lynn, W.S., Brown, R.H.: J. Biol. Chem.,232,1015–1030 (1958)
[15] Ohno, S., Nakajin, S., Shinoda, M.: J. Steroid Biochem.,38,787–794 (1991)
[16] Ohno, S., Nakajin, S., Shinoda, M.: Chem. Pharm. Bull.,39,972–975 (1991)
[17] Blomquist, C.H.: Arch. Biochem. Biophys.,159,590–595 (1973)

1 NOMENCLATURE

EC number
1.1.1.54

Systematic name
Allyl-alcohol:$NADP^+$ oxidoreductase

Recommended name
Allyl-alcohol dehydrogenase

Synonymes
Dehydrogenase, allyl alcohol
Allyl alcohol dehydrogenase

CAS Reg. No.
9028-58-4

2 REACTION AND SPECIFICITY

Catalysed reaction
Allyl alcohol + $NADP^+$ →
→ acrolein + NADPH

Reaction type
Redox reaction

Natural substrates

Substrate spectrum
1 Allyl alcohol + $NADP^+$ [1]
2 2,3-Butylene glycol + $NADP^+$ [1]
3 More (no activity with other primary alcohols) [1]

Product spectrum
1 Acrolein + NADPH
2 ?
3 ?

Inhibitor(s)
Sodium cyanide [1]; EDTA [1]; Dipyridyl [1]; p-Chloromercuribenzoate [1]; More (not: monoiodoacetate) [1]

Cofactor(s)/prosthetic group(s)/activating agents
$NADP^+$ (specific for) [1]

Metal compounds/salts

Turnover number (min^{-1})

Specific activity (U/mg)

K_m-value (mM)
0.04 ($NADP^+$ (+ allyl alcohol)) [1]; 5 (allyl alcohol) [1]

pH-optimum
7.7 [1]

pH-range

Temperature optimum (°C)

Temperature range (°C)

---

3 ENZYME STRUCTURE

Molecular weight

Subunits

Glycoprotein/Lipoprotein
–

---

4 ISOLATION/PREPARATION

Source organism
E. coli [1]

Source tissue
Cell [1]

Localisation in source

Purification
E. coli (partial) [1]

Crystallization
–

Cloned

–

Renaturated

–

5 STABILITY

pH

Temperature (°C)

Oxidation

Organic solvent

General stability information

Storage

6 CROSSREFERENCES TO STRUCTURE DATABANKS

PIR/MIPS code

Brookhaven code

7 LITERATURE REFERENCES

[1] Otsuka, K.: J. Gen. Appl. Microbiol.,4,211–215 (1958)

1 NOMENCLATURE

EC number
1.1.1.55

Systematic name
Propane-1,2-diol:$NADP^+$ oxidoreductase

Recommended name
Lactaldehyde reductase (NADPH)

Synonymes
Reductase, lactaldehyde (reduced nicotinamide adenine dinucleotide phosphate)
NADP-1,2-propanediol dehydrogenase
Propanediol dehydrogenase [1]
1,2-Propanediol:$NADP^+$ oxidoreductase

CAS Reg. No.
9028-43-7

2 REACTION AND SPECIFICITY

Catalysed reaction
Propane-1,2-diol + $NADP^+$ →
→ L-lactaldehyde + NADPH

Reaction type
Redox reaction

Natural substrates

Substrate spectrum
1 Propane-1,2-diol + $NADP^+$ [1]

Product spectrum
1 L-Lactaldehyde + NADPH [1]

Inhibitor(s)

Cofactor(s)/prosthetic group(s)/activating agents
$NADP^+$ [1]

Metal compounds/salts

Turnover number (min^{-1})

Specific activity (U/mg)

K_m-value (mM)

pH-optimum

pH-range

Temperature optimum (°C)

Temperature range (°C)

3 ENZYME STRUCTURE

Molecular weight

Subunits

Glycoprotein/Lipoprotein

–

4 ISOLATION/PREPARATION

Source organism

Bacillus macerans [1]

Source tissue

Localisation in source

Purification

Crystallization

–

Cloned

–

Renaturated

–

5 STABILITY

pH

Temperature (°C)

Oxidation

Organic solvent

General stability information

Storage

6 CROSSREFERENCES TO STRUCTURE DATABANKS

PIR/MIPS code

PIR2:S04702 (Escherichia coli); PIR3:S30352 (Escherichia coli)

Brookhaven code

7 LITERATURE REFERENCES

[1] Weimer, P.J.: Appl. Environ. Microbiol.,47,263–267 (1984)

1 NOMENCLATURE

EC number
1.1.1.56

Systematic name
Ribitol:NAD^+ 2-oxidoreductase

Recommended name
Ribitol 2-dehydrogenase

Synonymes
Adonitol dehydrogenase
Ribitol dehydrogenase A (wild type)
Ribitol dehydrogenase B (mutant enzyme with different properties) [1, 6]
Ribitol dehydrogenase D (mutant enzyme with different properties) [1]

CAS Reg. No.
9014-23-7

2 REACTION AND SPECIFICITY

Catalysed reaction
Ribitol + NAD^+ →
→ D-ribulose + NADH

Reaction type
Redox reaction

Natural substrates
Ribitol + NAD^+ [8]

Substrate spectrum
1 Ribitol + NAD^+ (r) [6–9]
2 Xylitol + NAD^+ (r) [6, 9]
3 L-Arabitol + NAD^+ (r) [6, 9]

Product spectrum
1 D-Ribulose + NADH [9]
2 Xylulose + NADH [9]
3 L-Xylulose + NADH [9]

Inhibitor(s)
Heavy metal ions [7]; p-Chloromercuribenzoate [7]; Dithiothreitol [7]; beta-Mercaptoethanol (competitive to ribitol) [7]; Chelating agents (Mycobacterium sp.) [3]

Cofactor(s)/prosthetic group(s)/activating agents
NAD^+ [6–10]; NADH [6–10]; Sulfhydryl compounds (requirement) [9]

Metal compounds/salts
Zn^{2+} (activation, Mycobacterium sp.) [3]; Mg^{2+} (0.02 M, requirement) [9]; More (no requirement for metal ions [7], bacterial enzyme has no requirement for Zn^{2+}: [5]) [5, 7]

Turnover number (min^{-1})
17940 (ribitol) [6]; 2160 (xylitol) [6]; 1680 (arabitol) [6]

Specific activity (U/mg)
90.5 [10]; 43.6 [6]; 83 [7]

K_m-value (mM)
5 (ribitol) [7]; 1000 (xylitol, pH 11) [7]; 0.12 (NAD^+, pH 11) [7]; 8.0 (ribitol, pH 7) [6]; 147 (arabitol, pH 7) [6]; 579 (xylitol, pH 7) [6]

pH-optimum
11 (unstable at optimum) [7]; 8.5–9 (alcohol oxidation) [9]; 6.0–6.5 (sugar reduction) [9]; 8.0 (alcohol oxidation) [4]; 7.4 (sugar reduction) [4]

pH-range
7–11 [7]

Temperature optimum (°C)
28 [4, 7, 10]

Temperature range (°C)

3 ENZYME STRUCTURE

Molecular weight
100000–110000 (Klebsiella aerogenes, gel filtration) [7]

Subunits
Tetramer (4 × 27000, Klebsiella aerogenes, SDS-PAGE) [5, 7]

Glycoprotein/Lipoprotein
–

4 ISOLATION/PREPARATION

Source organism
Klebsiella aerogenes (inducible enzyme [6, 8], mutants with constitutive enzyme [6, 8]) [1, 2, 6–8]; Aerobacter aerogenes [4, 9, 10]; Mycobacterium sp. (similar enzyme) [3]; Bacteria [5, 10]; Mammals [10]

Source tissue
Cell [1–10]

Localisation in source
Cytoplasm

Purification
Klebsiella aerogenes [1, 6, 7]; Aerobacter aerogenes [9, 10]

Crystallization
[9]

Cloned
[2]

Renaturated
–

5 STABILITY

pH
10–11 (unstable at high pH values) [7]; 5.5 (unstable below)

Temperature (°C)
37 (rapid thermal inactivation) [10]

Oxidation

Organic solvent

General stability information
NAD^+ improves stability and reverses inhibition [7, 10]; Unstable in dilute solutions [7]; Glutathione increases stability [9]

Storage
Frozen, pH 7.0, 6 months [10]; 2°C, presence of NAD^+, several days [2]

6 CROSSREFERENCES TO STRUCTURE DATABANKS

PIR/MIPS code
PIR1:DEKBR (Enterobacter aerogenes); PIR3:S07134 (Klebsiella pneumoniae); PIR3:S07135 (Klebsiella pneumoniae)

Brookhaven code

7 LITERATURE REFERENCES

[1] Dothie, J.M., Giglio, J.R., Moore, C.B., Tayler, S.S., Hartley, B.S.: Biochem. J.,230,569–578 (1985)
[2]Loving, T., Norton, P.M., Hartley, B.S.: Biochem. J.,230,579–585 (1985)
[3] Szumiko, T., Byra, A.: Acta Biochim. Pol.,31,401–408 (1984)
[4] Fromm, H.J. in " Methods Enzym. Anal.",3rd Ed. (Bergmeyer, H.U., ed.) 6,432–437 (1984)
[5] Jörnvall, H., v. Bahr-Lindström, H., Jany, K.-D., Ulmer, W., Fröschle, M.: FEBS Lett.,165,190–196 (1984)
[6] Burleigh, B.D., Rigby, P.W.J., Hartley, B.S.: Biochem. J.,143,341–352 (1974)
[7] Taylor, S.S., Rigby, P.W.J., Hartley, B.S.: Biochem. J.,141,693–700 (1974)
[8] Charnetzky, W.T., Mortlock, R.P.: J. Bacteriol.,119,162–169 (1974)
[9] Fossitt, D.D., Wood, W.A.: Methods Enzymol.,9,180–184 (1966)
[10] Nordlie, R.C., Fromm, H.J.: J. Biol. Chem.,234 ,2523–2531 (1959)

1 NOMENCLATURE

EC number
1.1.1.57

Systematic name
D-Mannonate:NAD^+ 5-oxidoreductase

Recommended name
Fructuronate reductase

Synonymes
Reductase, fructuronate
Mannonate oxidoreductase
Mannonic dehydrogenase [1]
D-Mannonate dehydrogenase [2]
D-Mannonate:NAD oxidoreductase [4]

CAS Reg. No.
9028-44-8

2 REACTION AND SPECIFICITY

Catalysed reaction
D-Fructuronate + NADH →
→ D-mannonate + NAD^+ (sequential mechanism [5], reaction is consistent with a bi-bi and dead end EBQ complex [2, 5], random mechanism [6]) [1, 2, 5, 6]

Reaction type
Redox reaction

Natural substrates

Substrate spectrum
1 D-Fructuronate + NADH (r [1, 2, 4], approximately equal activity towards both D-fructuronate and D-tagaturonate in the forward reaction, and towards D-mannonate and D-altronate in the reverse reaction [1]) [1–6]
2 D-Tagaturonate + NADH (r [1, 2], approximately equal activity towards both fructuronate and tagaturonate in the forward reaction, and towards D-mannonate and D-altronate in the reverse reaction [1], D-tagaturonate and D-altronate are 20% as active as D-fructuronate and D-mannonate [2]) [1, 2, 6]
3 D-Glucuronate + NADH ("hit-and-run" mechanism) [4]

Product spectrum
1 D-Mannonate + NAD^+ [1–6]
2 D-Altronate + NAD^+ [1, 2, 6]
3 L-Gulonate + NAD^+

Inhibitor(s)
Mannonate [2]; p-Chloromercuribenzoate [1, 2, 6]; Fructuronate [2]; NAD^+ [2]; NADH [2]; Altronate [2]; ATP [2]; Tagaturonate [2]

Cofactor(s)/prosthetic group(s)/activating agents
NAD^+ [1–6]; NADPH (inactive with [1], about 5% as active as NADH [2]) [2, 6]; NADH [1–6]

Metal compounds/salts
No metal ion requirement [2]

Turnover number (min^{-1})

Specific activity (U/mg)
320 [2]; More [1, 6]

K_m-value (mM)
0.2 (NAD^+) [6]; 0.7 (NADPH) [6]; 1 (D-fructuronic acid [1], NAD^+ [3]) [1, 3]; 1.7 (mannonate) [3]; 0.5 (fructuronate) [6]; 0.03 (NADH) [6]; 1.8 (mannonate) [6]

pH-optimum
6.0 [1]; 6.3 (fructuronate reduction [6], NADH oxidation [2]) [2, 6]; 8.4 (mannonate oxidation [6], NAD^+ reduction [2]) [2, 6]

pH-range

Temperature optimum (°C)
37 (enzyme assay) [2]

Temperature range (°C)
More (Q_{10}: 1.5, at 30°C - 40°C) [6]

3 ENZYME STRUCTURE

Molecular weight

Subunits

Glycoprotein/Lipoprotein
–

4 ISOLATION/PREPARATION

Source organism
E. coli (inducible [6]) [1–6]

Source tissue
Cell [1, 2]

Localisation in source

Purification
E. coli [1, 2, 6]

Crystallization
–

Cloned
–

Renaturated
–

5 STABILITY

pH
6.5 (unstable below) [1]

Temperature (°C)
53 (pH 7.0, half-life: 10 min [2]

Oxidation

Organic solvent

General stability information
NADH: effective stabilizer, stabilization proportional to concentration [6]; NADH protects against heat inactivation [2, 6]; NADPH protects against heat inactivation [2, 6]; Fructuronate protects against heat inactivation [2, 6]; Tagaturonate protects against heat inactivation [2, 6]; NAD^+ protects against heat inactivation [2, 6]; Mannonate protects against heat inactivation [2, 6]; Altronate protects against heat inactivation [2, 6]; Rapid loss of activity after dialysis [1]; Freezing and thawing completely destroys activity [1]; Cysteine stabilizes [1]

Storage
4°C, sodium phosphate buffer, pH 7.0, 3 weeks [2]; 3°C, 1 mM cysteine, pH 7.0, 1–2 weeks [1]

6 CROSSREFERENCES TO STRUCTURE DATABANKS

PIR/MIPS code

Brookhaven code

7 LITERATURE REFERENCES

[1] Hickman, J., Ashwell, G.: J. Biol. Chem.,235,1566–1570 (1960)
[2] Portalier, R., Stoeber, F.: Methods Enzymol.,89,210–218 (1982) (Review)
[3] Portalier, R.C., Stoeber, F.R.: Biochim. Biophys. Acta,289,19–27 (1972)
[4] Mandrand-Berthelot, M.-A., Lagarde, A.E.: Biochim. Biophys. Acta,483,6–23 (1977)
[5] Portalier, R.C.: Eur. J. Biochem.,30,220–233 (1972)
[6] Portalier, R.C., Stoeber, F.R.: Eur. J. Biochem.,26,290–300 (1972)

1 NOMENCLATURE

EC number
1.1.1.58

Systematic name
D-Altronate:NAD^+ 3-oxidoreductase

Recommended name
Tagaturonate reductase

Synonymes
Reductase, tagaturonate
Altronic oxidoreductase
Altronate oxidoreductase
D-Tagaturonate reductase
TagUAR [1]
Altronate dehydrogenase [5]
D-Tagaturonate reductase [1]

CAS Reg. No.
9028-45-9

2 REACTION AND SPECIFICITY

Catalysed reaction
D-Altronate + NAD^+ →
→ D-tagaturonate + NADH (ordered bi-bi mechanism [4, 5])

Reaction type
Redox reaction

Natural substrates
D-Tagaturonate + NADH (degradation of D-galacturonate) [2, 5]

Substrate spectrum
1 D-Altronate + NAD^+ (r [4, 6], at equilibrium approximately 10% of the keturonic acid is formed [6], other carbohydrates tested are unsuitable as substrates [2], specific for D-tagaturonate and D-altronate [5, 6], highly specific for altronic acid [7]) [1–7]

Product spectrum
1 D-Tagaturonate + NADH (i.e. 5-oxo-L-galactonate, D-arabino-5-hexulosonate) [1–7]

Inhibitor(s)

Galacturonate (slight) [2]; Glucuronate (slight) [2]; p-Chloromercuribenzoate [2, 5, 6]; Substrate analogues [2]; Tagaturonate [2, 4, 5]; Fructuronate [2, 4, 5]; Mannonate [2, 4, 5]; ATP [4, 5]; Altronate [2, 4, 5]; NAD^+ [2, 4, 5]; NADH [2, 4, 5]

Cofactor(s)/prosthetic group(s)/activating agents

NADH [2, 4–6]; NAD^+ [2, 4–6]; NADPH (can replace NADH [2], 10% of the activity with NADH [5], inactive [6]) [2, 5]; More (no activity with $NADP^+$) [5]

Metal compounds/salts

No metal ion requirement [5, 7]

Turnover number (min^{-1})

Specific activity (U/mg)

328 [5]; More [2, 6, 7]

K_m-value (mM)

0.400 (NADPH) [2]; 0.350 (tagaturonate) [2]; 0.060 (NADH) [2]; 0.090 (altronate) [2, 7]; 0.110 (NAD^+) [2]; 0.085 (NADH) [4, 5]; 0.670 (tagaturonate) [4, 5]; 0.053 (altronate) [4, 5]; 0.075 (NAD^+) [4, 5]; 0.1 (D-tagaturonate) [6]; 0.088 (NAD^+) [7]

pH-optimum

6.0 [6]; 6.3 (tagaturonate reduction) [2, 5]; 6.9–7.0 (D-tagaturonate reduction) [1]; 8.9 (altronate oxidation) [2, 5]

pH-range

5.3–8.0 (5.3: about 75% of activity maximum, 8.0: about 50% of activity maximum) [2]; 6–8 (at pH 6 and 8: about 50% of activity maximum) [2]

Temperature optimum (°C)

37 (assay at) [5]

Temperature range (°C)

More (Q_{10}: 1.5) [2]

---

3 ENZYME STRUCTURE

Molecular weight

Subunits

Glycoprotein/Lipoprotein

–

4 ISOLATION/PREPARATION

Source organism

E. coli (K12 [2, 4, 7], inducible [2]) [2–7]; Bacillus polymyxa [1]

Source tissue

Cell [2, 5, 6]

Localisation in source

Purification

E. coli (K12 [2]) [2, 5, 6]

Crystallization

–

Cloned

More (construction of a hybrid plasmid containing the E. coli uxa B gene encoding for altronate oxidoreductase) [3]

Renaturated

–

5 STABILITY

pH

Temperature (°C)

58.5 (half-life: 10.5 min) [5]; More [2]

Oxidation

Organic solvent

General stability information

NADH protects against heat denaturation [2, 5]; NADPH protects against heat denaturation [2, 5]; Altronate protects against heat denaturation [2, 5]; Mannonate protects against heat denaturation [2, 5]; NAD^+ does not protect against heat denaturation [2, 5]; Fructuronate does not protect against heat denaturation [2, 5]; Tagaturonate does not protect against heat denaturation [2, 5]

Storage

4°C or at –20°C for several months [2]; 4°C, sodium phosphate buffer, pH 7.0, 2 months [5]; 3°C or frozen at –10°C, 1 month [6]

6 CROSSREFERENCES TO STRUCTURE DATABANKS

PIR/MIPS code

Brookhaven code

7 LITERATURE REFERENCES

[1] Gierschner, K., Endreß, H.-U. in "Methods Enzym. Anal.",3rd Ed. (Bergmeyer, H.U., ed.) 6,313–320 (1984) (Review)
[2] Portalier, R.C., Stoeber, F.R.: Eur. J. Biochem.,26,50–61 (1972)
[3] Blanco, C., Mata-Gilsinger, M., Ritzenthaler, P.: J. Bacteriol.,153,747–755 (1983)
[4] Portalier, R. C.: Eur. J. Biochem.,30,211–219 (1972)
[5] Portalier, R., Stoeber, F.: Methods Enzymol.,89,210–218 (1982) (Review)
[6] Hickman, J., Ashwell, G.: J. Biol. Chem.,235,1566–1570 (1960)
[7] Portalier, R.C., Stoeber, F.R.: Biochim. Biophys. Acta,289,19–27 (1972)

1 NOMENCLATURE

EC number
1.1.1.59

Systematic name
3-Hydroxypropanoate:NAD^+ oxidoreductase

Recommended name
3-Hydroxypropionate dehydrogenase

Synonymes

CAS Reg. No.
9028-59-5

---

2 REACTION AND SPECIFICITY

Catalysed reaction
3-Hydroxypropanoate + NAD^+ →
→ 3-oxopropanoate + NADH

Reaction type
Redox reaction

Natural substrates
3-Hydroxypropanoate + NAD^+ [1, 2]

Substrate spectrum
1 3-Hydroxypropanoate + NAD^+ (r) [1, 2]

Product spectrum
1 3-Oxopropanoate + NADH [1, 2]

Inhibitor(s)

Cofactor(s)/prosthetic group(s)/activating agents
NAD^+ [1, 2]

Metal compounds/salts

Turnover number (min^{-1})

Specific activity (U/mg)
More (1.000, 1 unit defined as 0.001 increase in optical density) [2]

K_m-value (mM)
0.34 (NAD^+) [2]; 20 (3-hydroxypropanoate) [2]

pH-optimum

10 [2]

pH-range

Temperature optimum (°C)

Temperature range (°C)

---

3 ENZYME STRUCTURE

Molecular weight

Subunits

Glycoprotein/Lipoprotein

–

---

4 ISOLATION/PREPARATION

Source organism

Candida rugosa [1]; Candida catenulata [1]; Candida gropengiesseri [1]; Debaryomyces polymorphus [1]; Pichia haplophila [1]; Pig [2]; Chicken [2]; Propionibacterium shermanii [2]; Tetrahymena pyriformis [2]; Eremothecium ashbyii [2]; E. coli [2]

Source tissue

Kidney (pig) [2]; Heart (pig) [2]; Liver (pig) [2]; Breast muscle (chicken) [2]

Localisation in source

Purification

Pig (partially) [2]

Crystallization

–

Cloned

–

Renaturated

–

5 STABILITY

pH

Temperature (°C)

Oxidation

Organic solvent

General stability information

Storage

Several weeks, frozen [2]

6 CROSSREFERENCES TO STRUCTURE DATABANKS

PIR/MIPS code

Brookhaven code

7 LITERATURE REFERENCES

[1] Miyakoshi, S., Uchiyama, H., Someya, T., Satoh, T., Tabuchi, T.: Agric. Biol. Chem.,51,2381–2387 (1987)
[2] Den, H., Robinson, W.G., Coon, M.J.: J. Biol. Chem.,234,1666–1671 (1959)

1 NOMENCLATURE

EC number
1.1.1.60

Systematic name
(R)-Glycerate:$NAD(P)^+$ oxidoreductase

Recommended name
2-Hydroxy-3-oxopropionate reductase

Synonymes
Tartronate semialdehyde reductase

CAS Reg. No.
9028-68-6

2 REACTION AND SPECIFICITY

Catalysed reaction
(R)-Glycerate + $NAD(P)^+$ →
→ 2-hydroxy-3-oxopropanoate + NAD(P)H

Reaction type
Redox reaction

Natural substrates
Tartronate semialdehyde + NAD(P)H [1–5]

Substrate spectrum
1 Tartronate semialdehyde + NAD(P)H (r, NADH 3 times more effective than NADPH [1]) [1–5]
2 Hydroxypyruvate + NAD(P)H (r) [3–5]
3 Malonic semialdehyde + NADH [4, 5]
4 Mesoxalic semialdehyde + NADH [4, 5]

Product spectrum
1 (R)-Glycerate + NA(P)D+ [1–5]
2 Glycerate + $NAD(P)^+$ [3–5]
3 3-Hydroxypropanoate + NAD^+ [4, 5]
4 Hydroxypyruvate + NAD^+ [4, 5]

Inhibitor(s)
Glyoxylate [1, 4, 5]; Glycolate [1, 3–5]; beta-Mercaptopyruvate [3]; Malonic semialdehyde [3]; Succinic semialdehyde [3]; Dihydroxyfumarate [3]; Fluoroacetate [4, 5]

Cofactor(s)/prosthetic group(s)/activating agents

NADH (3 times more effective than NADPH [1]) [1–5]; NADPH [1–5]; NAD^+ [1–5]; $NADP^+$ [1–5]

Metal compounds/salts

Turnover number (min^{-1})

14600 (NADH) [4, 5]

Specific activity (U/mg)

50 [3]; 160 [4, 5]

K_m-value (mM)

0.05–0.4 (tartronate semialdehyde) [1, 3–5]; 0.02–0.027 (NADH) [3–5]; 0.036–0.05 (NADPH) [3–5]; 0.05–0.4 (D-glycerate) [3–5]; 0.04 (NAD^+) [3]; 0.2 (malonic semialdehyde) [5]

pH-optimum

7.1 [1]; 6.0 [3]; 6.2–8.7 [4, 5]

pH-range

Temperature optimum (°C)

Temperature range (°C)

---

3 ENZYME STRUCTURE

Molecular weight

104000 (Pseudomonas putida, sedimentation velocity) [3]
91000 (Pseudomonas ovalis, sedimentation equilibrium) [4, 5]

Subunits

Tetramer (4 × 26000, Pseudomonas putida, sedimentation velocity after treatment with guanidine hydrochloride and mercaptoethanol, amino acid analysis) [3]

Glycoprotein/Lipoprotein

–

---

4 ISOLATION/PREPARATION

Source organism

Hyphomicrobium sp. [1]; Pseudomonas oxalaticus [2]; Pseudomonas putida [3]; Pseudomonas ovalis [4, 5]

Source tissue

Localisation in source

Purification

Pseudomonas putida [3]; Pseudomonas sp. [4] Pseudomonas ovalis [5]

Crystallization

[3–5]

Cloned

–

Renaturated

–

5 STABILITY

pH

Temperature (°C)

Oxidation

Organic solvent

General stability information

Storage

2°C, suspension in ammonium sulfate, pH 7.5, 10% loss in 1 month [4, 5]

6 CROSSREFERENCES TO STRUCTURE DATABANKS

PIR/MIPS code

Brookhaven code

7 LITERATURE REFERENCES

[1] van der Drift, C., de Windt, F.E.: Antonie Leeuwenhoek,49,167–172 (1983)
[2] Dijkhuizen, L., Knight, M., Harder, W.: Arch. Microbiol.,116,77–83 (1978)
[3] Kohn, L.: J. Biol. Chem.,243,4426–4433 (1968)
[4] Kornberg, H.L., Gotto, A.M.: Methods Enzymol.,9,240–247 (1966)
[5] Gotto, A.M., Kornberg, H.L.: Biochem. J.,81,273–284 (1961)

1 NOMENCLATURE

EC number
1.1.1.61

Systematic name
4-Hydroxybutanoate:NAD$^+$ oxidoreductase

Recommended name
4-Hydroxybutyrate dehydrogenase

Synonymes
gamma-Hydroxybutyrate dehydrogenase

CAS Reg. No.
9028-60-8

2 REACTION AND SPECIFICITY

Catalysed reaction
4-Hydroxybutanoate + NAD$^+$ →
→ succinate semialdehyde + NADH

Reaction type
Redox reaction

Natural substrates
4-Hydroxybutanoate + NAD$^+$ [2]

Substrate spectrum
1 4-Hydroxybutanoate + NAD$^+$ (r) [1, 2]

Product spectrum
1 Succinate semialdehyde + NADH [1, 2]

Inhibitor(s)

Cofactor(s)/prosthetic group(s)/activating agents
NAD$^+$ [1, 2]; NADH [1, 2]

Metal compounds/salts

Turnover number (min^{-1})

Specific activity (U/mg)
More (4.8, 1 unit is a change of absorbancy of 0.2/min) [2]

K_m-value (mM)

pH-optimum

10.5 (4-hydroxybutanoate + NAD^+) [2]; 7.0 (succinate semialdehyde + NADH) [2]

pH-range

Temperature optimum (°C)

Temperature range (°C)

3 ENZYME STRUCTURE

Molecular weight

Subunits

Glycoprotein/Lipoprotein

–

4 ISOLATION/PREPARATION

Source organism

Unidentified brackish water bacterium [1]; Pseudomonas sp. [2]

Source tissue

Localisation in source

Purification

Pseudomonas sp. [2]

Crystallization

–

Cloned

–

Renaturated

–

5 STABILITY

pH

Temperature (°C)

Oxidation

Organic solvent

General stability information

Storage

6 CROSSREFERENCES TO STRUCTURE DATABANKS

PIR/MIPS code

Brookhaven code

7 LITERATURE REFERENCES

[1] Matthies, C., Mayer, F., Schink, B.: Arch. Microbiol.,151,498–505 (1989)
[2] Nirenberg, M.W., Jakoby, W.B.: J. Biol. Chem.,235,954–960 (1960)

1 NOMENCLATURE

EC number
1.1.1.62

Systematic name
Estradiol-17beta:NAD(P)$^+$ 17-oxidoreductase

Recommended name
Estradiol 17beta-dehydrogenase

Synonymes
20alpha-Hydroxysteroid dehydrogenase
17beta,20alpha-Hydroxysteroid dehydrogenase
Dehydrogenase, estradiol 17beta-
17beta-Estradiol dehydrogenase
Estradiol dehydrogenase
Estrogen 17-oxidoreductase
17beta-HSD [22]
More (cf. EC 1.1.1.149)

CAS Reg. No.
9028-61-9

2 REACTION AND SPECIFICITY

Catalysed reaction
Estradiol-17beta + NAD(P)$^+$ →
→ estrone + NAD(P)H

Reaction type
Redox reaction

Natural substrates

Substrate spectrum

1 17beta-Estradiol + NAD(P)$^+$ (i.e. estra-1,3,5(10)triene-3,17-diol [21], r [3, 5, 7, 8, 14, 15], estradiols substituted at C-1, C-2, C-4, absolutely specific for 17beta-hydroxy-group [15]) [1–8, 14, 15, 18, 21–25]
2 Testosterone + NAD(P)$^+$ (r [5, 8]) [5, 8, 21, 24]
3 Progesterone + NADH [22, 24, 26]
4 3-O-Methyl-17beta-estradiol + NAD$^+$ [14]
5 5beta-Androstan-3alpha,17beta-diol + NADP$^+$ (i.e. etiocholanediol) [21]
6 17beta-Estradiol 3-glucouronide + NADP$^+$ [25]
7 Cortisol + NADH [26]
8 Cortisone + NADH [26]

Product spectrum

1 Estrone + NAD(P)H (i.e. 3-hydroxy-estra-1,3,5(10)-triene-17-one [21]) [3, 6, 8, 14, 18, 23, 24]
2 4-Androsten-3,17-dione + NAD(P)H [8, 24]
3 20alpha-Hydroxy-pregn-4-en-3-one + NAD$^+$
4 3-O-Methylestrone + NADH
5 3alpha-Hydroxy-5beta-androstan-17-one + NADPH (i.e. etiocholanone) [21]
6 ?
7 11beta,17alpha,20alpha,21-Tetrahydroxy-4-pregnen-3-one + NAD$^+$
8 17alpha,20alpha,21-Trihydroxy-4-pregnen-3,11-dione + NAD$^+$

Inhibitor(s)

17alpha-(or 17beta-)Estradiol 17-bromoacetate [1]; Periodate oxidized NAD$^+$ with aldehyde group at 2'- and 3'-position of ribose ring [2]; p-Chloromercuribenzoate [6, 8, 23]; 16alpha-Bromoacetoxyprogesterone [7]; Cu^{2+} [8, 23]; Zn^{2+} [8, 23]; Mg^{2+} [8, 23]; N-Ethylmaleimide [11]; Iodoacetamide [23]; 2'-AMP [17]; 5'-ATP [17]; 2',5'-ATP [17]; 5'-ADP [17]; 5'-ADPribose [17]; 3',5'-ADP [17]; Testosterone (competitive to 17beta-estradiol) [22]; 17beta-Estradiol (competitive to testosterone) [22]; 5alpha-Dihydrotestosterone [22]; 5beta-Dihydrotestosterone [22]; Progesterone [22]; 20alpha-Dihydroprogesterone [22]; Cibacron Blue [21]; Hexestrol [21]

Cofactor(s)/prosthetic group(s)/activating agents

NAD$^+$ (preferred [4, 8, 14, 24], B-specific, i.e. transfer of 4-pro-S-hydrogen to/from NAD$^+$/NADH [26, 27]) [1–4, 6–8, 13–15, 23–27]; NADP$^+$ (A-specific, i.e. transfer of 4-pro-R hydrogen to/from NADP$^+$/NADPH [21], low activity [4, 8, 14, 24], 20 times greater activity with NADP$^+$ than with NAD$^+$ [25]) [4, 8, 14, 21, 24–27]

Metal compounds/salts

Turnover number (min^{-1})

Specific activity (U/mg)
7.2 [3]; 0.392 [6]; 0.136 [14]; More [8–10, 23, 26]

K_m-value (mM)
0.0008 (17beta-estradiol) [22]; 0.0013 (testosterone) [22]; 0.0017 (17beta-estradiol) [4]; 0.002–0.0028 (estrone, NADPH [6], 20alpha-hydroxy-4-pregnen-3-one [22]) [6, 22]; 0.0032–0.134 (17beta-estradiol, value depends on type of endometrial tissue) [8]; 0.0033 (17beta-estradiol) [18]; 0.0059 (17beta-estradiol) [7]; 0.007–0.0077 (17beta-estradiol [5], $NADP^+$ [6]) [5, 6]; 0.0086 (estrone) [7]; 0.0137 (20alpha-hydroxy-5beta-pregnan-3-one) [26]; 0.0173 (20alpha-hydroxy-pregn-4-en-3-one) [26]; 0.024 (17alpha,20alpha-dihydroxy-4-pregnen-3-one) [26]; 0.03 (estradiol) [14]; 0.042 (estradiol, (+ NAD^+)) [23]; 0.044 (estradiol, (+ $NADP^+$)) [23]; 0.124 (androstenedione) [5]; 0.263 (testosterone) [5]; 0.357 (NAD^+) [23]; 0.37 ($NADP^+$) [23]; 0.52 (NAD^+) [14]

pH-optimum
5.5–6.5 (reduction of estrone) [8, 18, 23]; 7.4 (reduction of estrone) [6]; 8.2 [14]; 9.0–9.5 (oxidation of 17beta-estradiol) [4, 6, 8, 18, 21, 23, 24]

pH-range
8.5–10 (oxidation of 17beta-estradiol) [4]

Temperature optimum (°C)
40 [18]; 41 (cytoplasmic enzyme) [24]; 42 (microsomal enzyme) [24]; 45 [8]

Temperature range (°C)

3 ENZYME STRUCTURE

Molecular weight
103400–105000 (female sheep, sucrose density centrifugation) [14]
97000 (chicken, high molecular weight form, gel filtration) [6]
64000–68500 (human, polyacrylamide pore gradient gel electrophoresis [3], human, ultracentrifugation [13], horse, gel filtration [4], rabbit, gel filtration [23]) [3, 4, 13, 23]
46000–50000 (Streptomyces hydrogenans, gradient gel electrophoresis, HPLC gel exclusion chromatography) [26]
43000 (chicken, low molecular weight form, gel filtration) [6]
36000 (mouse) [21]

Subunits

Dimer (2 × 33000–33600, human [3], horse [4], SDS-PAGE) [3, 4]
Monomer (1 × 36000, mouse) [21]

Glycoprotein/Lipoprotein

No glycoprotein [13]

---

4 ISOLATION/PREPARATION

Source organism

Human [1–3, 5, 7–13, 15–20, 22, 24, 27]; Horse [4]; Chicken [6]; Sheep (female) [14]; Rabbit [23, 25]; Mouse [21]; Streptomyces hydrogenans [26]

Source tissue

Placenta [1–5, 7, 9–13, 16, 17, 19, 20, 22, 24, 27]; Liver [6, 21, 25]; Endometrial tissue [8, 18]; Endometrial carcinoma [18]; Ovary [14]; Uterus [23]

Localisation in source

Microsomes [4, 22, 24]; Cytosol [10, 12, 21, 23–26]; Mitochondria (outer membrane) [8]

Purification

Human [3, 7–10, 12, 13]; Horse [4]; Chicken [6]; Sheep (female) [14]; Rabbit [23, 25]; Streptomyces hydrogenans (partial) [26]

Crystallization

[20]

Cloned

–

Renaturated

–

---

5 STABILITY

pH

6.3–7 (7 days) [26]; 6.5–7.0 [14]

Temperature (°C)

Oxidation

Organic solvent

General stability information

Stabilization by glycerol [14, 24, 26]

Storage

–85°C [14]; –20°C, 50% v/v glycerol, at least 3 months [6]; 4°C, 50% glycerol, at least 5 months [3]; –18°C, ammonium sulfate precipitate, Tris buffer, pH 8.2, 20% glycerol, over 1 year [26]; 4°C, phosphate buffer, pH 7.0, 1 mM dithiothreitol, at least 1 month [4]; 4°C, in microsomes, 4 days, 35–75% decrease of activity [22]; 0°C, 1 h, 60% loss of activity of cytoplasmic enzyme, less then 5% loss of activity of microsomal enzyme [24]

6 CROSSREFERENCES TO STRUCTURE DATABANKS

PIR/MIPS code

PIR1:DEHUE7 (human)

Brookhaven code

7 LITERATURE REFERENCES

[1] Murdock, G.L., Warren, J.C., Sweet, F.: Biochemistry,27,4452–4458 (1988)
[2] Mendoza-Hernandez, G., Lopez-Solache, I., Diaz-Zagoya, J.C.: Biochem. Biophys. Res. Commun.,146,645–651 (1987)
[3] Mendoza-Hernandez, G., Rendon, J.L., Diaz-Zagoya, J.C.: Biochem. Biophys. Res. Commun.,126,477–481 (1985)
[4] LaRhee, L.H., Warren, J.C.: Biochemistry,23,486–491 (1984)
[5] Mendoza-Hernandez, G., Calcagno, M., Sanchez-Nuncio, H.R., Diaz-Zagoya, J.C.: Biochem. Biophys. Res. Commun.,119,83–87 (1984)
[6] Renwick, A.G.C., Soon, C.Y., Chambers, S.M., Brown, C.R.: J. Biol. Chem.,256,1881–1887 (1981)
[7] Strickler, R.C., Tobias, B., Covey, D.F.: J. Biol. Chem.,256,316–321 (1981)
[8] Pollow, K., Luebbert, H., Pollow, B.: J. Steroid Biochem.,7,45–50 (1976)
[9] Nicolas, J.C.: Methods Enzymol.,34,552–554 (1974)
[10] Nicolas, J.C.: Methods Enzymol.,34,555–557 (1974)
[11] Nicolas, J.C., Harris, J.I.: FEBS Lett.,29,173–176 (1973)
[12] Nicolas, J.C., Pons, M., Descomps, B., de Paulet, A.C: FEBS Lett.,23,175–179 (1972)

[13] Burns, D.J.W., Engel, L.L., Bethune, J.L.: Biochemistry,11,2699–2703 (1972)
[14] Kautsky, M.P., Hagerman, D.D.: J. Biol. Chem.,245,1978–1984 (1970)
[15] Langer, L.J., Alexander, J.A, Engel, L.L.: J. Biol. Chem.,234,2609–2614 (1959)
[16] Luebbert, H.: Acta Endocrinol.,99,448–453 (1982)
[17] Engel, L.L., Inano, H.: Adv. Enzyme Regul.,17,363–371 (1979)
[18] Pollow, K., Luebbert, H., Jeske, R., Pollow, B.: Acta Endocrinol.,79,146–156 (1975)
[19] Murdock, G.L., Chin, C.-C., Warren, J.C.: Biochemistry,25,641–646 (1986)
[20] Chin, C.-C., Dence, J.B., Warren, J.C.: J. Biol. Chem.,251,3700–3705 (1976)
[21] Hara, A., Nakayama, T., Nakagawa, M., Inoue, Y., Tanabe, H., Sawada, H.: J. Biochem.,102,1585–1592 (1987)
[22] Blomquist, C.H., Lindemann, N.J., Hakanson, E.Y.: Arch. Biochem. Biophys.,239, 206–215 (1985)
[23] Pollow, K., Elger, W., Heßlinger, H., Pollow, B.: Z. Naturforsch.,34c,726–737 (1979)
[24] Pollow, K., Runge, W., Pollow, B.: Z. Naturforsch.,30c,17–24 (1975)
[25] Hasnain, S., Williamson, D.G.: Can. J. Biochem.,52,120–125 (1974)
[26] Rimsay, R.L., Murphy, G.W., Martin, C.J., Orr, J.C.: Eur. J. Biochem.,174,437–442 (1988)
[27] Jarabak, T., Talalay, P.: J. Biol. Chem.,235,2147–2151 (1960)

1 NOMENCLATURE

EC number
1.1.1.63

Systematic name
17beta-Hydroxysteroid:NAD^+ 17-oxidoreductase

Recommended name
Testosterone 17beta-dehydrogenase

Synonymes
Dehydrogenase, testosterone 17beta-
17-Ketoreductase
17beta-HSD [3]

CAS Reg. No.
9028-62-0

2 REACTION AND SPECIFICITY

Catalysed reaction
Testosterone + NAD^+ →
→ androst-4-ene-3,17-dione + NADH

Reaction type
Redox reaction

Natural substrates

Substrate spectrum
1 Testosterone + NAD^+ (r) [1–5, 7, 8]
2 17beta-Hydroxy-5beta-androstan-3-one + NAD^+ [4, 5, 8]
3 17beta-Hydroxy-5alpha-androstan-3-one + NAD^+ [4, 5, 8]
4 17beta-Hydroxy-1,4-androstadien-3-one + NAD^+ [5]
5 17beta-Hydroxy-1alpha-methyl-5alpha-androstan-3-one + NAD^+ [5]
6 3alpha,17beta-Dihydroxy-5alpha-androstane + NAD^+ [5, 8]
7 3beta,17beta-Dihydroxy-5alpha-androstan + NAD^+ [5, 8]
8 17beta-Hydroxy-1-methyl-5alpha-androst-1-ene-3-one + NAD^+ [5]
9 3,17beta-Dihydroxy-1,3,5(10)estratriene + NAD^+ (i.e. estradiol 17beta) [5, 8]

10 3beta-Hydroxy-5-androsten-17-one + NADH [5]
11 6beta,17beta-Dihydroxy-4-androsten-3-one + NAD^+ [8]
12 11alpha,17beta-Dihydroxy-4-androsten-3-one + NAD^+ [8]
13 11beta,17beta-Dihydroxy-4-androsten-3-one + NAD^+ [8]
14 15alpha,17beta-Dihydroxy-4-androsten-3-one + NAD^+ [8]
15 4-Androsten-3beta,17beta-diol + NAD^+ [8]
16 5-Androsten-3beta,17beta-diol + NAD^+ [8]
17 More (no 3beta-hydroxysteroid dehydrogenase activity, no 17alpha-hydroxysteroid dehydrogenase activity) [8]

Product spectrum

1 Androst-4-ene-3,17-dione + NADH (i.e androstenedione) [1, 5, 7]
2 5beta-Androstan-3,17-dione + NADH
3 5alpha-Androstan-3,17-dione + NADH
4 1,4-Androstadien-3,17-dione + NADH
5 1alpha-Methyl-5alpha-androstan-3,17-dione + NADH
6 3alpha-Hydroxy-5alpha-androstan-17-one + NADH
7 3beta-Hydroxy-5alpha-androstan-17-one + NADH
8 1-Methyl-5alpha-androst-1-ene-3,17-dione + NADH
9 3beta-Hydroxy-1,3,5(10)estratriene-17-one + NADH
10 3beta-Hydroxy-5-androstene-17-ol + NAD^+
11 6beta-Hydroxy-4-androsten-3,17-dione + NADH
12 11alpha-Hydroxy-4-androsten-3,17-dione + NADH
13 11beta-Hydroxy-4-androsten-3,17-dione + NADH
14 15alpha-Hydroxy-4-androsten-3,17-dione + NADH
15 3beta-Hydroxy-4-androsten-17-one + NADH
16 3beta-Hydroxy-5-androsten-17-one + NADH
17 ?

Inhibitor(s)

CN^- [2]; p-Chloromercuribenzoate [2]; Zn^{2+} [2]; Fe^{3+} [2]; Ag^+ [2]; Cu^{2+} [2]; Hg^{2+} [2]; Triton X-100 (competitive to testosterone) [3]; 5alpha-Androstan-3,17-dione [5]; 17beta-Methyltestosterone [5]

Cofactor(s)/prosthetic group(s)/activating agents

NAD^+ (hydrogen transfer from testosterone to the 4-pro-S position of NAD^+ [7]) [1–8]; $NADP^+$ (low activity [4], no activity [5, 7, 8]) [4]; NADH [1–8]

Metal compounds/salts

Turnover number (min^{-1})

4318 (testosterone + NAD^+) [7]; 91 (androstenedione + NADH) [7]

Specific activity (U/mg)

335 [8]

K_m-value (mM)

0.0016 (4-androstene-3,7-dione) [8]; 0.002–0.0073 (testosterone) [3]; 0.0036 (testosterone (+ NAD^+)) [8]; 0.0064 (estradiol-17beta (+ NAD^+)) [8]; 0.0068 (NADH) [7]; 0.0099 (testosterone (+ thio-NAD^+)) [8]; 0.0095–0.014 (testosterone) [7]; 0.013–0.014 (testosterone [5], estradiol-17beta (+ thio-NAD^+) [8]) [5, 8]; 0.021 (estradiol-17beta) [5]; 0.024 (androstenedione) [7]; 0.027–0.028 (NAD^+) [1, 8]; 0.033 (testosterone) [1]; 0.1–0.164 (NAD^+) [3, 5]; 0.198–0.215 (NAD^+) [7]; More (increase of K_m in presence of 2-mercaptoethanol) [6]

pH-optimum

6 (reduction of 17-oxosteroid) [8]; 7 (reduction of androstenedione) [7]; 9 [5]; 9.5–9.6 [2, 3]; 9–10 (oxidation of 17-hydroxysteroid) [8]; 10 (oxidation of testosterone) [7]

pH-range

6.0–10 [3]; 7.4–9 [5]; 8.0 (no activity below) [2]

Temperature optimum (°C)

45 [5]

Temperature range (°C)

20–45 [5]

3 ENZYME STRUCTURE

Molecular weight

176000 (guinea pig, calculation from sedimentation constant and Stokes radius) [3]
130000 (Streptomyces hydrogenans, gel filtration) [5]
86000 (Alcaligenes sp., polyacrylamide gel electrophoresis) [8]
68000 (Alcaligenes sp., gel filtration) [8]
58600 (Cylindrocarpon radicicola, gel filtration, polyacrylamide gel electrophoresis) [7]

Subunits

Dimer (2 × 26000, Cylindrocarpon radicicola SDS-PAGE [7], 2 × 32000, Alcaligenes sp., SDS-PAGE [8]) [7, 8]

Glycoprotein/Lipoprotein

–

4 ISOLATION/PREPARATION

Source organism

Guinea pig [1–3, 6]; Rat [4]; Streptomyces hydrogenans [5]; Cylindrocarpon radicicola [7]; Alcaligenes sp. [8]

Source tissue

Liver [1–4, 6]

Localisation in source

Microsomes [1–4, 6]; Cytosol [5, 7, 8]

Purification

Guinea pig [3]; Cylindrocarpon radicicola [7]; Alcaligenes sp. [8]

Crystallization

–

Cloned

–

Renaturated

–

5 STABILITY

pH

9 (denaturation above) [5]

Temperature (°C)

43 (pH 10.0: half-life 10 min, but stable in presence of testosterone, pH 7.0: 30 min, 10% inactivation) [7]; 45 (denaturation above) [5]; 50 (1 h, 20% loss of activity) [8]

Oxidation

Organic solvent

General stability information

Inactivation by 1% deoxycholate solution [3]; Stable to lyophilization [7]; Stabilization by glycerol [8]

Storage

–20°C or 4°C, relatively stable [1]; 4–6°C [2]; 4°C, pH 9.0, 0.2 M sucrose, 10 mM mercaptoethanol, 10–12 days [3]; –80°C, several months [7]; –20°C, 20 mM potassium phosphate buffer, pH 7.5, 50% glycerol, 1 mM dithiothreitol, 1 mM EDTA, at least 1 year [8]; 4°C, 1 week, 10% loss of activity [8]

6 CROSSREFERENCES TO STRUCTURE DATABANKS

PIR/MIPS code

Brookhaven code

7 LITERATURE REFERENCES

[1] Villee, C.A., Spencer, J.M.: J. Biol. Chem.,235,3615–3619 (1960)
[2] Endahl, G.L., Kochakian, C.D., Hamm, D.: J. Biol. Chem.,235,2792–2796 (1960)
[3] Blomquist, C.H., Kotts, C.E., Hakanson, E.Y.: J. Steroid Biochem.,8,193–198 (1977)
[4] Ghraf, R., Raible, M., Schriefers, H.: Hoppe-Seyler's Z. Physiol. Chem.,354,299–305 (1973)
[5] Markert, C., Traeger, L.: Hoppe-Seyler's Z. Physiol. Chem.,365,1843–1852 (1975)
[6] Blomquist, C.H., Kotts, C.E.: Steroids,32,399–419 (1978)
[7] Itagaki, E., Iwaya, T.: J. Biochem.,103,1039–1044 (1988)
[8] Payne, D.W., Talalay, P.: J. Biol. Chem.,260,13648–13655 (1985)

1 NOMENCLATURE

EC number
1.1.1.64

Systematic name
17beta-Hydroxysteroid:$NADP^+$ 17-oxidoreductase

Recommended name
Testosterone 17beta-dehydrogenase ($NADP^+$)

Synonymes
Dehydrogenase, testosterone 17beta- (nicotinamide adenine dinucleotide phosphate)
17-Ketoreductase
NADP-dependent testosterone-17beta-oxidoreductase

CAS Reg. No.
9028-63-1

2 REACTION AND SPECIFICITY

Catalysed reaction
Testosterone + $NADP^+$ →
→ androst-4-ene-3,17-dione + NADPH (rapid equilibrium random bi bi mechanism [5], mechanism [6])

Reaction type
Redox reaction

Natural substrates
Androstenedione + NADPH (role in metabolism of xenobiotics [3], essential role in testosterone formation [8]) [3, 8]

Substrate spectrum
1 Testosterone + $NADP^+$ (r) [1, 2, 4, 6, 8–12]
2 19-Nortestosterone + $NADP^+$ (i.e. 17beta-hydroxy-4-estren-3-one) [4]
3 5alpha-Androstan-3alpha,17beta-diol + $NADP^+$ [4]
4 5alpha-Androstan-3beta,17beta-diol + $NADP^+$ [4]
5 17beta-Hydroxy-5alpha-androstan-3-one + $NADP^+$ [3, 4]
6 5beta-Androstan-3alpha,17beta-diol + $NADP^+$ (r [6]) [4, 6]
7 5beta-Androstan-3beta,17beta-diol + $NADP^+$ [4]
8 17beta-Hydroxy-5beta-androstan-3-one + $NADP^+$ [3, 4]
9 5beta-Androstan-17beta-ol + $NADP^+$ [4]

10 Pyridine-4-aldehyde + $NADP^+$ [3]
11 Benzene dihydrodiol + $NADP^+$ [3]
12 1-Acenaphthenol + $NADP^+$ [3]
13 1-Indanol + $NADP^+$ [3]
14 1,2,3,4-Tetrahydro-1-naphthol + $NADP^+$ [3]
15 Cyclohex-2-en-1-ol + $NADP^+$ [3]
16 Estrone + NADPH (r) [6, 8–10, 12]
17 Dehydroepiandrosterone + NADPH (i.e. 3beta-hydroxyandrost-5-en-17-one) [8–10, 12]

Product spectrum

1 Androst-4-ene-3,17-dione + NADPH (i.e. androstenedione) [1, 10, 12]
2 4-Estren-3,17-dione + NADPH
3 3alpha-Hydroxy-5alpha-androstan-17-one + NADPH
4 3beta-Hydroxy-5alpha-androstan-17-one + NADPH
5 5alpha-Androstan-3,17-dione + NADPH
6 3alpha-Hydroxy-5beta-androstan-17-one + NADPH [6]
7 3beta-Hydroxy-5beta-androstan-17-one + NADPH
8 5beta-Androstan-3,17-dione + NADPH
9 5beta-Androstan-17-one + NADPH
10 ?
11 ?
12 1-Acenaphthenone + NADPH
13 1-Indanone + NADPH
14 ?
15 ?
16 Estradiol-17beta + $NADP^+$ [6, 12]
17 3beta,17beta-Dihydroxyandrost-5-ene + $NADP^+$ [12]

Inhibitor(s)

CN^- [2]; Zn^{2+} [2]; Fe^{3+} [2]; Pb^{2+} [2]; Cu^{2+} [2]; Ag^+ [2]; Hg^{2+} [2]; More (product inhibition) [5, 6]

Cofactor(s)/prosthetic group(s)/activating agents

$NADP^+$ [1–6, 11, 12]; NAD^+ (low activity) [3–5] NADPH (transfer of hydrogen from 4-pro-S position of NADPH to androstendione i.e. B-specific [5, 12], 4-pro-R hydrogen of NADPH is transferred to substrate, i.e. A-specific [6, 11]) [1–6, 11, 12]; NADH (low activity) [3–5]

Metal compounds/salts

Turnover number (min^{-1})

Specific activity (U/mg)

0.03–3.87 [3]

K_m-value (mM)

0.0067 (17beta-hydroxy-5beta-androstan-3-one) [4]; 0.011 (5beta-androstan-3alpha,17beta-diol [4], $NADP^+$ [1], NADPH [5]) [1, 4, 5]; 0.018 ($NADP^+$, isozyme TD2) [3]; 0.023 (5alpha-androstan-3alpha,17beta-diol) [4]; 0.025 ($NADP^+$, isozyme TD1) [3]; 0.027 (testosterone) [4]; 0.029 (5beta-androstan-3beta,17beta-diol) [4]; 0.053 (19-nortestosterone) [4]; 0.077 (17beta-hydroxy-5alpha-androstan-3-one) [4]; 0.177 (NADH) [5]; 0.27 (testosterone) [1]; 0.47 (benzene dihydrodiol, isozyme TD1) [3]; 6.7 (benzene dihydrodiol, isozyme TD2) [3]

pH-optimum

6.0–6.1 (reduction of androstenedione) [1, 4]; 9.0 (oxidation of testosterone) [1]; 9.5 (oxidation of testosterone) [2]; 10.2 (oxidation of benzene dihydrodiol, isozyme TD2) [3]; 10.6 (oxidation of testosterone) [4]; 10.8 (oxidation of benzene dihyrodiol, isozyme TD1) [3]

pH-range

Temperature optimum (°C)

45 [10]; 47–50 [4]; 50 [8–10]

Temperature range (°C)

25–55 (25°C: 30% of activity maximum, 55°C: 60% of activity maximum)[10]; 30–50 (30°C: abaut 15% of activity maximum, 50°C: activity maximum) [9]

---

3 ENZYME STRUCTURE

Molecular weight

39000 (guinea pig, gel filtration) [4]

33200–35500 (porcine testes, sedimentation [5], porcine testes, gel filtration, thin layer chromatography [9]) [5, 9]

Subunits

Monomer (1 × 31000, guinea pig, SDS-PAGE [4], 1 × 36500, pig, SDS-PAGE [7]) [4, 7]

Glycoprotein/Lipoprotein

–

---

4 ISOLATION/PREPARATION

Source organism

Guinea pig (2 isozymes [3]) [1–4, 11]; Pig [5, 7, 9, 12]; Mouse [6]; Human [10]; Rat [8, 10]; More (overview mammals) [8]

Source tissue
Liver [1–4, 6, 11]; Testis (interstitial tissue [8]) [5, 7–10, 12]

Localisation in source
Cytosol [1–4, 6]; Microsomes [5, 8–10]

Purification
Guinea pig [4]; More (overview mammalian testes) [5]

Crystallization
–

Cloned
–

Renaturated
–

5 STABILITY

pH

Temperature (°C)
42 (rapid inactivation above, protection by NADPH and $NADP^+$) [6]

Oxidation

Organic solvent

General stability information
Inactivation by freezing [4]

Storage
–20°C or 4°C, relatively stable [1]; –20°C, purified enzyme, no loss of activity in 6 months [8, 9]; 4–6°C, 5% loss of activity per day [2]; 0–4°C, 0.05% mercaptoethanol, 10 days [4]

6 CROSSREFERENCES TO STRUCTURE DATABANKS

PIR/MIPS code

Brookhaven code

7 LITERATURE REFERENCES

[1] Villee, C.A., Spencer, J.M.: J. Biol. Chem.,235,3615–3619 (1960)
[2] Endahl, G.L., Kochakian, C.D., Hamm, D.: J. Biol. Chem.,235,2792–2796 (1960)
[3] Hara, A., Hayashibara, M., Nakayama, T., Hasebe, K., Usui, S., Sawada, H.: Biochem. J.,225,177–181 (1985)
[4] Kageura, E., Toki, S.: Biochem. J.,163,401–407 (1977)
[5] Inano, H., Tamaoki, B.-i.: Steroids,48,3–26 (1986) (Review)
[6] Hara, A., Nakayama, T., Nakagawa, M., Inoue, Y., Sawada, H.: J. Biochem.,102, 1585–1592 (1987)
[7] Inano, H., Tamaoki, B.-i, Hamana, K., Nakagawa, H.: J. Steroid Biochem.,13, 287–295 (1980)
[8] Tamaoki, B.-i., Inanao, H.: J. Steroid Biochem.,6,361–363 (1975)
[9] Inano, H., Tamaoki, B.-i.: Eur. J. Biochem.,44,13–23 (1974)
[10] Oshima, H., Ochiai, K.: Biochim. Biophys. Acta,306,227–236 (1973)
[11] Nakayama, T., Hara, A., Kariya, K.-i., Sawada, H.: J. Biochem.,98,1131–1133 (1985)
[12] Inano, H., Tamaoki, B.-i.: Eur. J. Biochem.,53,319–326 (1975)

1 NOMENCLATURE

EC number
1.1.1.65

Systematic name
Pyridoxine:$NADP^+$ 4-oxidoreductase

Recommended name
Pyridoxine 4-dehydrogenase

Synonymes
Pyridoxin dehydrogenase
Dehydrogenase, pyridoxol 4-
Pyridoxol dehydrogenase
Pyridoxine dehydrogenase

CAS Reg. No.
9028-64-2

2 REACTION AND SPECIFICITY

Catalysed reaction
Pyridoxine + $NADP^+$ →
→ pyridoxal + NADPH

Reaction type
Redox reaction

Natural substrates
Pyridoxine + $NADP^+$ (enzyme participates in the conversion of vitamin B_6 to the coenzyme pyridoxalphosphate) [1]

Substrate spectrum
1 Pyridoxine + $NADP^+$ (r [1], pyridoxine is identical with pyridoxol, equilibrium lies far to the side of pyridoxol [1])
2 Pyridoxine phosphate + $NADP^+$ [1]

Product spectrum
1 Pyridoxal + NADPH [1]
2 ?

Inhibitor(s)
$MnCl_2$ [1]; $CuSO_4$ [1]; $CoCl_2$ [1]; $ZnCl_2$ (0.1 mM: activation, 10 mM: inhibition) [1]; p-Chloromercuriphenyl sulfonic acid [1]; SO_3^{2-} [1]

Cofactor(s)/prosthetic group(s)/activating agents
$NADP^+$ [1]

Metal compounds/salts
$ZnCl_2$ (0.1 mM: activation, 10 mM: inhibition) [1]

Turnover number (min^{-1})

Specific activity (U/mg)
More [1]

K_m-value (mM)
0.2 ($NADP^+$) [1]; 0.002 (NADPH) [1]; 1.6 (pyridoxal) [1]; 0.7 (pyridoxol) [1]

pH-optimum
6.0 [1]

pH-range
5.0–8.0 (at pH 5.0 and 8.0: about 50% of activity maximum) [1]

Temperature optimum (°C)
22 (assay at room temperature) [1]

Temperature range (°C)

3 ENZYME STRUCTURE

Molecular weight

Subunits

Glycoprotein/Lipoprotein
–

4 ISOLATION/PREPARATION

Source organism
Brewer's yeast [1]

Source tissue
Cell [1]

Localisation in source

Purification
Brewer's yeast [1]

Crystallization
–

Cloned

–

Renaturated

–

5 STABILITY

pH

Temperature (°C)

Oxidation

Organic solvent

General stability information

Storage

6 CROSSREFERENCES TO STRUCTURE DATABANKS

PIR/MIPS code

Brookhaven code

7 LITERATURE REFERENCES

[1] Holzer, H., Schneider, S.: Biochim. Biophys. Acta,48,71–76 (1961)

1 NOMENCLATURE

EC number
1.1.1.66

Systematic name
10-Hydroxydecanoate:NAD+ 10-oxidoreductase

Recommended name
omega-Hydroxydecanoate dehydrogenase

Synonymes
Dehydrogenase, omega-hydroxydecanoate

CAS Reg. No.
9028-65-3

2 REACTION AND SPECIFICITY

Catalysed reaction
10-Hydroxydecanoate + NAD+ →
→ 10-oxodecanoate + NADH

Reaction type
Redox reaction

Natural substrates

Substrate spectrum
1 10-Hydroxydecanoate + NAD+ (best substrate) [1]
2 9-Hydroxynonanoate + NAD+ (oxidation at 51% the rate of 10-hydroxydecanoic acid oxidation) [1]
3 11-Hydroxyundecanoate + NAD+ (oxidation at 89% the rate of 10-hydroxydecanoic acid oxidation) [1]
4 More (C_1- to C_4- or C_{18}-omega-hydroxy acids, primary alcohols with or without functional groups, i.e. amino, carboxy or a second hydroxyl, or secondary fatty acid alcohols are not oxidized) [1]

Product spectrum
1 10-Oxodecanoate + NADH [1]
2 9-Oxononanoate + NADH [1]
3 11-Oxoundecanoate + NADH [1]
4 ?

Inhibitor(s)

Cofactor(s)/prosthetic group(s)/activating agents
NAD+ (cannot be replaced by NADP+) [1]

Metal compounds/salts

Turnover number (min^{-1})

Specific activity (U/mg)
More (223, units defined as change in optical density at 340 nm of 0.001/min, pig) [1]

K_m-value (mM)
0.066 (NAD+, pig) [1]; 0.08 (11-hydroxyundecanoate, pig) [1]; 0.125 (10-hydroxydecanoate, pig) [1]; 40 (9-hydroxynonanoate, pig) [1]

pH-optimum
10.0 (pig) [1]

pH-range
9.0–11.0 (about half-maximal activity at pH 9.0 and 11.0, pig) [1]

Temperature optimum (°C)
37 (assay at) [1]

Temperature range (°C)

---

3 ENZYME STRUCTURE

Molecular weight

Subunits

Glycoprotein/Lipoprotein
–

---

4 ISOLATION/PREPARATION

Source organism
Horse [1]; Pig [1]; Sheep [1]; Rabbit [1]; Rat [1]

Source tissue
Heart [1]; Kidney [1]; Liver [1]

Localisation in source
Cytoplasm [1]

Purification
Pig [1]

Crystallization

–

Cloned

–

Renaturated

–

5 STABILITY

pH

Temperature (°C)

Oxidation

Organic solvent

General stability information

Storage

4°C, crude extract from pig stable for a week [1]

6 CROSSREFERENCES TO STRUCTURE DATABANKS

PIR/MIPS code

Brookhaven code

7 LITERATURE REFERENCES

[1] Mitz, M.A., Heinrikson, R.L.: Biochim. Biophys. Acta,46,45–50 (1961)

1 NOMENCLATURE

EC number
1.1.1.67

Systematic name
D-Mannitol:NAD^+ 2-oxidoreductase

Recommended name
Mannitol 2-dehydrogenase

Synonymes
D-Mannitol dehydrogenase
Mannitol dehydrogenase

CAS Reg. No.
9001-65-4

2 REACTION AND SPECIFICITY

Catalysed reaction
D-Mannitol + NAD^+ →
→ D-fructose + NADH

Reaction type
Redox reaction

Natural substrates
D-Mannitol + NAD^+ [1–22]

Substrate spectrum
1 D-Mannitol + NAD^+ (r) [1–22]
2 Sorbitol + NAD^+ [1, 13, 19]
3 D-Arabinitol + NAD^+ [1, 10, 13, 15]

Product spectrum
1 D-Fructose + NADH [1–22]
2 L-Sorbose + NADH [1, 13, 19]
3 D-Ribulose + NADH [1, 10, 13, 15]

Inhibitor(s)

Mannitol 1-phosphate [14]; Methyl mercurinitrate [15]; p-Hydroxymercuribenzoate [15, 18, 19]; Cu^{2+} [18]; Zn^{2+} [18]

Cofactor(s)/prosthetic group(s)/activating agents

NAD^+ [1–22]; NADH [1–22]

Metal compounds/salts

Turnover number (min^{-1})

Specific activity (U/mg)

60.3 [1]; 10.24 [15]; 393 [16]; 6.9 [18]; 27.3 [21]

K_m-value (mM)

0.18–0.27 (NAD^+) [1, 13, 16, 19, 22]; 31.8 (sorbitol) [1]; 0.29–21.8 (D-mannitol) [1–3, 7, 13, 14, 16, 19]; 60 (D-mannitol) [22]; 1.8–6.5 (D-arabinitol) [1]; 16.3–79.2 (D-fructose) [1, 3, 5, 16, 19, 22]; 0.01–0.079 (NADH) [1, 5, 16, 19]; 0.13 (NADH) [22]

pH-optimum

8.9–10.0 (D-mannitol oxidation) [1, 3–5, 15, 18]; 6.0–7.0 (D-fructose reduction) [1, 3–5, 15, 18]; 8.6 (D-mannitol oxidation) [16, 19, 21, 22]; 5.3 (D-fructose reduction) [16, 19, 21, 22]

pH-range

6.0–6.5 (not active below, D-mannitol oxidation) [15, 18]; 5.0 (not active below, D-mannitol oxidation) [19, 22]; 9.0–9.5 (not active above, D-fructose reduction) [18, 19]; 8.0 (not active above, D-fructose reduction) [22]

Temperature optimum (°C)

Temperature range (°C)

---

3 ENZYME STRUCTURE

Molecular weight

132000–137000 (Leuconostoc mesenteroides, gel filtration, sucrose gradient centrifugation) [11, 16]
133000 (Absidia glauca, gel filtration) [15]
64500 (Pseudomonas cepacia, sucrose gradient centrifugation) [8]
53000 (Saccharomyces cerevisiae) [5]
47200 (Rhodobacter sphaeroides, sucrose gradient centrifugation) [1]

Subunits

Tetramer (4 × 36000, Leuconostoc mesenteroides, SDS-PAGE) [11]
Dimer (Saccharomyces cerevisiae) [5]
Monomer (1 × 52200, Rhodobacter sphaeroides, SDS-PAGE) [1]

Glycoprotein/Lipoprotein

–

4 ISOLATION/PREPARATION

Source organism

Rhodobacter sphaeroides (phototrophic bacterium) [1]; Brevibacterium flavum [4]; Rhizobium japonicum [10, 13]; Rhizobium meliloti [12, 13]; Bradyrhizobium japonicum [6]; Pseudomonas aeruginosa [17]; Pseudomonas fluorescens [17]; Pseudomonas cepacia [8]; Pseudomonas coronafaciens [17]; Pseudomonas sp. [7]; Sarcina marginata [17]; Sarcina aurantiaca [17]; Leuconostoc mesenteroides [11, 16, 17, 19, 20]; Lactobacillus brevis [12, 16, 17, 21, 22]; Lactobacillus gayonii [16, 17]; Lactobacillus pentoaceticus [16, 17]; Acetobacter suboxydans [12]; Acetobacter melanogenum [18]; Actinoplanes missouriensis [12]; Mycobacterium smegmatis [12]; Nocardia erythropolis [12]; Streptomyces lavendulae [12]; Absidia glauca [14, 15]; Saccharomyces cerevisiae (yeast) [2, 5]; Fomes pinicola [9]; Platimonas subcordiformis [3]

Source tissue

Localisation in source

Purification

Rhodobacter sphaeroides [1]; Absidia glauca [15]; Leuconostoc mesenteroides [16, 20]; Acetobacter melanogenum [18]; Lactobacillus brevis [21, 22]

Crystallization

[11, 16, 19, 20]

Cloned

–

Renaturated

–

5 STABILITY

pH

6.0–6.5 (highest stability) [5, 21, 22]; 5.0–9.0 [16, 19]

Temperature (°C)

40 (not stable above) [1]; 20 (not stable above) [15]

Oxidation

Organic solvent

General stability information

Storage

4°C, 1 month [1]; –20°C, 10 mM dithiothreitol, 45 days [15]; –16°C, 1 mM mercaptoethanol several weeks [21, 22]; 3°C, crystalline enzyme in ammonium sulfate, 2 months [21, 22]

6 CROSSREFERENCES TO STRUCTURE DATABANKS

PIR/MIPS code

Brookhaven code

7 LITERATURE REFERENCES

[1] Schneider, K.H., Giffhorn, F.: Eur. J. Biochem.,184,15–19 (1989)
[2] Quain, D.E., Boulton, C.A.: J. Gen. Microbiol.,133,1675–1684 (1987)
[3] Richter, D.F.E., Kirst, G.O.: Planta,170,528–534 (1987)
[4] Mori, M., Shiio, I.: Agric. Biol. Chem.,51,129–138 (1987)
[5] Kulbe, K.D., Schwab, U., Howaldt, M., Kimmerle, K.: GBF Monogr. Ser.,9,189–200 (1986)
[6] Mathis, J.N., Barbour, W.M., Miller, T.B., Israel, D.W., Elkan, G.H.: Appl. Environ. Microbiol.,52,81–85 (1986)·
[7] Davis, C.L., Robb, F.T.: Appl. Environ. Microbiol.,50,743–748 (1985)
[8] Allenza, P., Lee, Y.N., Lessie, T.G.: J. Bacteriol.,150,1348–1356 (1982)
[9] Hult, K., Veide, A., Gatenbeck, S.: Arch. Microbiol.,128,253–255 (1980)
[10] Mulongoy, K., Elkan, G.H.: Curr. Microbiol.,1,335–340 (1978)
[11] Yamanaka, K., Izawa, K., Tenmizu, K.: Agric. Biol. Chem.,41,1695–1699 (1977)
[12] Mehta, R.J., Fare, L.R., Shearer, M.E., Nash, C.H.: Appl. Environ. Microbiol.,33, 1013–1015 (1977)
[13] Kuykendall, L.D., Elkan, G.H.: J. Gen. Microbiol.,98,291–295 (1977)
[14] Ueng, S.T.H., McGuinness, E.T.: Biochemistry,16,107–111 (1977)
[15] Ueng, S.T.H., Hartanowicz, P., Lewandowski, C., Keller, J., M. Holick, E.T. McGuinness: Biochemistry,15,1743–1749 (1976)
[16] Yamanaka, K.: Methods Enzymol.,41 B,138–142 (1975)
[17] Yamanaka, K., Sakai, S.: Can. J. Microbiol.,14,391–396 (1968)
[18] Sasajima, K., Isono, M.: Agric. Biol. Chem.,32,161–169 (1968)
[19] Sakai, S., Yamanaka, K.: Agric. Biol. Chem.,32,894–899 (1968)
[20] Sakai, S., Yamanaka, K.: Biochim. Biophys. Acta,151,686–688 (1968)
[21] Horecker, B.L.: Methods Enzymol.,9,143–146 (1966)
[22] Martinez, G., Barker, H.A., Horecker, B.L.: J. Biol. Chem.,238,1598–1603 (1963)

1 NOMENCLATURE

EC number
1.1.1.69

Systematic name
D-Gluconate:NAD(P)$^+$ 5-oxidoreductase

Recommended name
Gluconate 5-dehydrogenase

Synonymes
Dehydrogenase, gluconate 5-
Reductase, 5-ketogluconate 5-
5-Keto-D-gluconate 5-reductase
5-Ketogluconate 5-reductase
5-Ketogluconate reductase
5-Keto-D-gluconate reductase

CAS Reg. No.
9028-70-0

2 REACTION AND SPECIFICITY

Catalysed reaction
5-Dehydro-D-gluconate + NAD(P)H →
→ D-gluconate + NAD(P)$^+$

Reaction type
Redox reaction

Natural substrates
5-Dehydro-D-gluconate + NAD(P)H (shuttle between gluconate and 5-keto-gluconate, contributes to reduction of 5-keto-D-gluconate to gluconate to supply carbon source via pentose phosphate cycle and for Entner-Doudoroff pathway, in combination with reoxidation of NADP$^+$) [2]

Substrate spectrum
1 5-Keto-D-gluconate + NAD(P)H (r [1–3], rate of D-gluconate oxidation is about 60% of the rate of 5-keto-D-gluconate reduction [1], highly specific for 5-keto-D-gluconate [1, 2]) [1–7]
2 5-Ketofructose + NADPH [2, 4]
3 More (D-glucono-delta-lactone and D-sorbitol are slightly oxidized, D-fructose and 5-ketofructose are slightly reduced) [4]

Product spectrum

1 D-Gluconate + $NAD(P)^+$ [1–7]
2 ?
3 ?

Inhibitor(s)

Sulfhydryl reagents [1]; Divalent metal ions [1]; p-Chloromercuribenzoate [2, 4]; Hg^{2+} [2, 4]; Cd^{2+} [2]; Cs^{2+} [2]; Cu^{2+} [2]; Sn^{2+} [2]; Ni^{2+} [2]; N-Ethylmaleimide [4]; Sodium lauryl sulfate [4]

Cofactor(s)/prosthetic group(s)/activating agents

NADPH (specific for [4], enzyme from Klebsiella sp. and Escherichia sp.: reaction proceeds twice as fast with NADH as with NADPH, enzyme from Gluconobacter suboxydans: reaction only with NADPH [3]) [1–4]; $NADP^+$ (specific for [4]) [1–4]; NADH (enzyme from Klebsiella sp. and Escherichia sp.: reaction proceeds twice as fast with NADH as with NADPH, enzyme from Gluconobacter suboxydans: reaction only with NADPH) [3]; NAD^+ [3]; EDTA (stimulation) [4]

Metal compounds/salts

Mn^{2+} (1 mM, slight stimulation) [2]

Turnover number (min^{-1})

Specific activity (U/mg)

69.69 [1, 2]; More [4]

K_m-value (mM)

0.9 (5-keto-D-gluconate [1, 2, 4], pH 6.0 [1, 2]) [1, 2, 4]; 0.006 (NADPH, pH 6.0) [1, 2]; 20 (D-gluconate, pH 10.0) [1, 2]; 0.020 ($NADP^+$, pH 10.0) [1, 2]; 18 (gluconic acid) [4]; 0.016 ($NADP^+$) [4]; 0.011 (NADPH) [4]

pH-optimum

10.0 (D-gluconate + $NADP^+$) [1, 2, 4]; 5.5 (5-dehydro-D-gluconate + NADPH) [1, 2]; 6–7 (Klebsiella sp., Escherichia sp., reduction of 5-ketogluconate) [3]; 7.3–7.5 (Gluconobacter suboxydans oxydans, reduction of 5-ketogluconate) [3]; 7.5 (Gluconobacter liquefaciens, 5-ketogluconate + $NADP^+$) [4]; 9.3 (Gluconobacter suboxydans oxydans, oxidation of D-gluconate) [3]; 10–11 (Klebsiella sp., Escherichia sp., oxidation of D-gluconate) [3]

pH-range

4–11 [3]

Temperature optimum (°C)

50 (both directions) [1, 2, 4]

Temperature range (°C)

3 ENZYME STRUCTURE

Molecular weight

110000 (Gluconobacter suboxydans, gel filtration [2], Gluconobacter liquefaciens, gel filtration [4]) [2, 4]

Subunits

Tetramer (4 × 25000, Gluconobacter suboxydans, SDS-PAGE) [1, 2]

Glycoprotein/Lipoprotein

–

4 ISOLATION/PREPARATION

Source organism

Gluconobacter suboxydans (IFO 12528 [2]) [1–3]; Klebsiella sp. [3]; Escherichia sp. [3]; Gluconobacter liquefaciens [4, 7]; Penicillium notatum [5]; Acetic acid bacteria (enzyme occurs only in acetic acid bacteria [1], enzyme is unique in acetic acid bacteria among the aerobic bacteria [6], overview [7]) [1, 6, 7]

Source tissue

Cell [1, 2, 4]

Localisation in source

Cytosol [2]

Purification

Gluconobacter suboxydans (IFO 12528 [2]) [1, 2]; Gluconobacter liquefaciens (partial) [4]

Crystallization

[1, 2]

Cloned

–

Renaturated

–

5 STABILITY

pH

Temperature (°C)

40 (complete loss of activity after a few min with and without addition of D-gluconate or 5-keto-D-gluconate) [1]; 50 (10 min, stable) [4]; 55 (5 min, no loss of activity, addition of D-gluconate or 5-keto-D-gluconate) [1]; 60 (10 min, about 20% loss of activity) [4]; 70 (10 min, 75% loss of activity) [4]

Oxidation

Organic solvent

General stability information

D-Gluconate stabilizes during purification, enhances heat stability [1, 2]; 5-Keto-D-gluconate stabilizes during purification, enhances heat stability [1, 2]; Unstable against dialysis in the cold in presence of sulfhydryl compounds, sucrose and/or glycerol [4]

Storage

5°C, 0.01 M potassium phosphate buffer, pH 7.5, 3 days, complete loss of activity [4]

6 CROSSREFERENCES TO STRUCTURE DATABANKS

PIR/MIPS code

Brookhaven code

7 LITERATURE REFERENCES

[1] Ameyama, M., Adachi, O.: Methods Enzymol.,89,198–202 (1982) (Review)
[2] Adachi, O., Shinagawa, E., Matsushita, K., Ameyama, M. : Agric. Biol. Chem.,43,75–83 (1979)
[3] De Ley, J.: Methods Enzymol.,9,200–203 (1966) (Review)
[4] Ameyama, M., Chiyonobu, T., Adachi, O.: Agric. Biol. Chem.,38,1377–1382 (1974)
[5] Pitt, D., Mosley, M.J.: Antonie Leeuwenhoek,51,353–364 (1985)
[6] Shinagawa, E., Chiyonobu, T., Matsushita, K., Adachi, O., Ameyama, M.: Agric. Biol. Chem.,42,1055–1057 (1978)
[7] Chiyonobu, T., Shinagawa, E., Adachi, O., Ameyama, M.: Agric. Biol. Chem.,39,2425–2427 (1975)

1 NOMENCLATURE

EC number
1.1.1.71

Systematic name
Alcohol:NAD(P)$^+$ oxidoreductase

Recommended name
Alcohol dehydrogenase (NAD(P)$^+$)

Synonymes
Retinal reductase
Dehydrogenase, alcohol (nicotinamide adenine dinucleotide (phosphate))
Aldehyde reductase (NADPH/NADH)

CAS Reg. No.
37250-10-5

2 REACTION AND SPECIFICITY

Catalysed reaction
An aldehyde + NAD(P)H →
→ an alcohol + NAD(P)$^+$

Reaction type
Redox reaction

Natural substrates

Substrate spectrum
1 Aldehyde + NAD(P)H (reduces aliphatic aldehydes of carbon chain length 2 to 14 with greatest activity on C_4, C_6 and C_8 aldehydes, unsaturated C-18 fatty aldehydes are reduced at a low rate [1], not: saturated aldehydes of chain length C-16 or greater, stereospecific for 4R-NADH [1], chain length C_2-C_7 [2]) [1, 2]
2 Primary alcohol + NAD(P)$^+$ (chain length C_2-C_6) [2]
3 Secondary alcohol + NAD(P)$^+$ (overview) [2]
4 Retinal + NAD(P)H [1, 2]
5 More (no substrate: benzyl alcohol) [2]

Product spectrum

1 Alcohol + NAD(P)$^+$ [2]
2 Primary aldehyde + NAD(P)H [2]
3 Secondary aldehyde + NAD(P)H [2]
4 Retinol + NAD(P)$^+$ [1]
5 ?

Inhibitor(s)

Sodium glycocholate [1]; Tween 80 [1]; SH-inhibitors [1]; Mercaptoethanol [1]; p-Hydroxymercuribenzoate (strong [2]) [1, 2]; Urea (inactivation) [2]; 8-Hydroxyquinoline [2]; o-Phenanthroline [2]; IAA [2]; Ag^{2+} [2]; Hg^{2+} [2]; Cu^{2+} [2]; Fe^{2+} [2]; AMP/ADP/ATP (NAD$^+$-dependent reaction, not NADP$^+$) [2]; More (no inhibitors: NaN_3, thiourea, oxalic acid, potassium thiocyanate, EDTA) [2]

Cofactor(s)/prosthetic group(s)/activating agents

NADH (at low concentration NADH is more effective, at high concentration the reaction rate is slightly greater with NADPH, reduction of retinal) [1]; NADPH (at low concentration NADH is more effective, at high concentration the reaction rate is slightly greater with NADPH, reduction of retinal) [1]; NAD$^+$ (non-specific native enzyme, NAD$^+$-specific subunit S_1) [2]; NADP$^+$ (non-specific native enzyme and subunit S_2) [2]; Glutathione (stimulates) [1]; Cysteine (stimulates) [1]

Metal compounds/salts

Turnover number (min^{-1})

Specific activity (U/mg)

0.674 (NAD$^+$) [2]; 3.37 (NADP$^+$) [2]

K_m-value (mM)

0.0004 (NADH) [1]; 0.02 (retinal) [1]; 0.04 (NADPH) [1]

pH-optimum

6.3 [1]; 8.6 (NADP$^+$) [2]; 10.2 (NAD$^+$) [2]

pH-range

5.6–6.7 (5.6: about 60% of activity maximum, 6.7: about 65% of activity maximum) [1]

Temperature optimum (°C)

Temperature range (°C)

3 ENZYME STRUCTURE

Molecular weight
240000 (Leuconostoc mesenteroides, gel filtration) [2]

Subunits
Dimer (1 × 80000 + 1 × 160000, Leuconostoc mesenteroides, gel filtration) [2]

Glycoprotein/Lipoprotein
–

4 ISOLATION/PREPARATION

Source organism
Rat [1]; Leuconostoc mesenteroides [2]

Source tissue
Intestinal mucosa [1]; Cell [2]

Localisation in source
Soluble [1]

Purification
Rat [1]; Leuconostoc mesenteroides [2]

Crystallization
–

Cloned
–

Renaturated
–

5 STABILITY

pH
7.8–9.5 (maximal stability, NAD^+) [2]; 8.0–9.8 (maximal stability, $NADP^+$) [2]

Temperature (°C)
30 (at least 15 min stable) [2]; 60 (inactivation after 15 min) [2]

Oxidation

Organic solvent

General stability information
2-Mercaptoethanol or DTT, 1 mM, stabilizes during storage [2]

Storage

4°C, at least 3 months in 10 mM phosphate-buffer, pH 8.2 [2]

6 CROSSREFERENCES TO STRUCTURE DATABANKS

PIR/MIPS code

Brookhaven code

7 LITERATURE REFERENCES

[1] Fridge, N.H., Goodman, D.S.: J. Biol. Chem.,243,4372–4379 (1968)
[2] Hatanaka, A., Kajiwara, T., Tomohiro, S.: Agric. Biol. Chem.,38,1819–1833 (1974)

1 NOMENCLATURE

EC number
1.1.1.72

Systematic name
Glycerol:NADP$^+$ oxidoreductase

Recommended name
Glycerol dehydrogenase (NADP$^+$)

Synonymes

CAS Reg. No.
37250-11-6

2 REACTION AND SPECIFICITY

Catalysed reaction
D-Glyceraldehyde + NADPH →
→ glycerol + NADP$^+$

Reaction type
Redox reaction

Natural substrates
D-Glyceraldehyde + NADPH [5]

Substrate spectrum
1 D-Glyceraldehyde + NADPH (r) [1, 3, 4, 7, 8]
2 L-Glyceraldehyde + NADPH (36% of D-glyceraldehyde activity [3], no reduction [7]) [1, 3]
3 Glycolaldehyde + NADPH [1, 4]
4 Methylglyoxal + NADPH [1, 4]
5 Formaldehyde + NADPH [4]
6 Propionaldehyde + NADPH [1, 4]
7 n-Butyraldehyde + NADPH [1, 4]
8 n-Pentanal + NADPH [1, 4]
9 D-Erythrose + NADPH [1]
10 Dihydroxyacetone + NADPH (r [8], 43% of D-glyceraldehyde activity [3]) [3, 4, 7, 8]
11 Acetaldehyde + NADPH$^+$ [4]
12 2-Butenal + NADPH [4]
13 Isobutyraldehyde + NADPH [4]
14 3-Methyl-2-butanal + NADPH [4]

15 n-Hexenal + NADPH [4]
16 n-Hexanal + NADPH [4]
17 2,3-Butanedione + NADPH (i.e. diacetyl) [7]
18 Erythritol + NADP+ (i.e. 1,2,3,4-butanetetrol) [7]
19 D-Xylose + NADPH [1]
20 D-Glucose + NADPH (no reduction [4]) [1]
21 4-Nitrobenzaldehyde + NADPH [1]
22 D-Glucuronate + NADP+ [1]
23 More (no substrates: D-fructose, D-ribose, D-xylose) [3]

Product spectrum

1 Glycerol + NADP+ [1–3, 7, 8]
2 Glycerol + NADP+ [1]
3 Ethyleneglycol + NADP+
4 ?
5 Methanol + NADP+
6 Propanol + NADP+
7 n-Butanol + NADP+
8 n-Pentanol + NADP+
9 Erythritol + NADP+ (i.e. 1,2,3,4-butanetetrol)
10 Glycerol + NADP+
11 Ethanol + NADP+
12 2-Butenol + NADP+
13 Isobutanol + NADP+
14 3-Methyl-2-butanol + NADP+
15 n-Hexenol + NADP+
16 n-Hexanol + NADP+
17 ?
18 ?
19 Xylitol + NADP+ [1]
20 Sorbitol + NADP+ [1]
21 4-Nitrobenzylalcohol + NADP+ [1]
22 ?
23 ?

Inhibitor(s)

2-Mercaptoethanol [1]; Phenobarbitol [1]; Diphenylhydantoin [1]; Quercitin [1]; p-Chloromercuribenzoate [4]; p-Hydroxymercuribenzoate [3]; Urea [4]; F^- [5]; Na^+ [8]

Cofactor(s)/prosthetic group(s)/activating agents

NADP+ [1–4, 7, 8]; NADPH (A-specific, pro-R hydrogen is transferred to substrate [6]) [1, 3–7]; NADH (10% of NADPH-activity [1], no reaction [3, 7]) [1]

Metal compounds/salts

HCO_3^- (activates conversion of glycerol to dihydroxyacetone) [8]; More (no effects of NH_4^+, SO_4^{2-}, no metal ion requirement) [3]

Turnover number (min^{-1})

Specific activity (U/mg)

5.75 [1]; 20.7 [3]; More [4, 5, 7]

K_m-value (mM)

0.015 ($NADP^+$, Aspergillus nidulans, pH 10.0) [7]; 0.016 (NADPH) [4]; 0.017 ($NADP^+$, pH 10.0, NADPH, pH 6.0, Aspergillus niger) [7]; 0.019 (NADPH, Aspergillus nidulans, pH 6.0) [7]; 0.024 (NADPH) [3]; 0.03 (NADPH, Aspergillus niger, pH 5.4) [7]; 0.05 (NADPH, Aspergillus nidulans, pH 4.8) [7]; 0.056 ($NADP^+$) [3]; 0.15 (D-glyceraldehyde) [4]; 0.29 (D-glyceraldehyde, Aspergillus nidulans, pH 6.0) [7]; 0.39 (D-glyceraldehyde, Aspergillus niger, pH 6.0) [7]; 0.41 (D-glyceraldehyde, Aspergillus niger, pH 5.4) [7]; 0.45 (D-glyceraldehyde, Aspergillus nidulans, pH 4.8) [7]; 0.7 (glycerol, Aspergillus niger, pH 10.0) [7]; 1 (glycerol, Aspergillus nidulans, pH 10.0) [7]; 11.5 (D-glyceraldehyde) [3]; 143 (glycerol) [3]

pH-optimum

4.8 (dihydroxyacetone reduction, Aspergillus nidulans) [7]; 5.4 (dihydroxyacetone reduction, Aspergillus niger) [7]; 6.5 (D-glyceraldehyde reduction) [3]; 7.0 (DL-glyceraldehyde reduction) [1]; 9.5 (glycerol oxidation) [3]; 10.0 (glycerol oxidation, Aspergillus nidulans, Aspergillus niger) [7]; More [4]

pH-range

6–8 (D-glyceraldehyde reduction) [3]; 8.5–10.0 (glycerol oxidation) [3]

Temperature optimum (°C)

Temperature range (°C)

3 ENZYME STRUCTURE

Molecular weight

34000 (rabbit, gel filtration) [1]

38000 (Aspergillus nidulans, Aspergillus niger, gel permeation chromatography) [7]

160000 (Neurospora crassa, gel filtration) [3]

Subunits

Monomer (1 × 38000, Aspergillus nidulans, Aspergillus niger, SDS-PAGE) [7]

? (x × 43000, Neurospora crassa, SDS-PAGE [3], x × 41500, rabbit, SDS-PAGE [1]) [1, 3]

Glycoprotein/Lipoprotein

–

4 ISOLATION/PREPARATION

Source organism

Rabbit [1, 4, 6]; Aspergillus nidulans [3, 7]; Aspergillus niger [3, 7, 8]; Neurospora crassa [3]; Aspergillus oryzae [2]; Penicillium notatum [2]; Penicillium expansum [2]; Fusarium sp. [2]; Verticillium sp. [2]; Rat [5]

Source tissue

Skeletal muscle [1, 4–6]

Localisation in source

Soluble [3]

Purification

Rabbit [1, 4]; Neurospora crassa [3]; Aspergillus nidulans [7]; Aspergillus niger [7]

Crystallization

–

Cloned

–

Renaturated

–

5 STABILITY

pH

4–9 (denaturation above and below) [4]

Temperature (°C)

37 (30 min, 13% loss of activity, 60 min, 21% loss of activity, 120 min, 35% loss of activity, 180 min, 45% loss of activity) [4]; 45 (the half-life of 25 min increases to 45 min upon addition of 400 mM glycerol) [3]; 50 (5 min, 98% loss of activity) [3]

Oxidation

Organic solvent

General stability information

Glycerol stabilizes [3]

Storage

–20°C, dialyzed against distilled H_2O with 30 mg/ml polyethyleneglycol, or lyophilized, stable [1]; –20°C, 1 year [4]

6 CROSSREFERENCES TO STRUCTURE DATABANKS

PIR/MIPS code

Brookhaven code

7 LITERATURE REFERENCES

[1] Flynn, T.G., Cromlish, J.A.: Methods Enzymol.,89,237–242 (1982)
[2] Yamada, H., Nagao, A., Nishise, H., Tani, Y.: Agric. Biol. Chem.,46,2325–2331 (1982)
[3] Viswanath-Reddy, M., Pyle, J.E., Branch Howe, E.: J. Gen. Microbiol.,107,289–296 (1978)
[4] Korman, A.W., Hurst, R.O., Flynn, T.G.: Biochim. Biophys. Acta,258,40–55 (1972)
[5] Toews, C.J.: Biochem. J.,105,1067–1073 (1967)
[6] Walton, D.J.: Biochemistry,12,3472–3478 (1973)
[7] Schuurink, R., Busink, R., Hondmann, D.H.A., Witteveen, C.F.B., Visser, J.: J. Gen. Microbiol.,136,1043–1050 (1990)
[8] Inayat, M.S., Mattey, M.: Biochem. Soc. Trans.,16,976–977 (1988)

1 NOMENCLATURE

EC number
1.1.1.73

Systematic name
Octanol:NAD^+ oxidoreductase

Recommended name
Octanol dehydrogenase

Synonymes
1-Octanol dehydrogenase

CAS Reg. No.
9031-31-6

2 REACTION AND SPECIFICITY

Catalysed reaction
1-Octanol + NAD^+ →
→ 1-octanal + NADH

Reaction type
Redox reaction

Natural substrates
1-Octanol + NAD^+ [5, 6]

Substrate spectrum
1 1-Octanol + NAD^+ (r [6]) [5, 6]
2 Ethanol + NAD^+ (no substrate [6]) [5]
3 n-Propanol + NAD^+ [5]
4 n-Butanol + NAD^+ [5]
5 n-Pentanol + NAD^+ [5]
6 n-Hexanol + NAD^+ [5]
7 n-Heptanol + NAD^+ [5]
8 Benzylalcohol + NAD^+ [5]
9 Farnesol + NAD^+ [5]
10 More (no substrates: methanol, isopropanol, 2-butanol, 3-butanol, isobutanol, 2-pentanol, 3-pentanol, 2-hexanol, 3-hexanol, 2-heptanol, 3-heptanol, 2-octanol, cyclohexanol) [5]

Product spectrum

1 1-Octanal + NADH [6]
2 Ethanal + NADH
3 n-Propanal + NADH
4 n-Butanal + NADH
5 n-Pentanal + NADH
6 n-Hexanal + NADH
7 n-Heptanal + NADH
8 Benzaldehyde + NADH
9 3,7,11-Trimethyl-2,6,10-dodecatrien-1-al + NADH
10 ?

Inhibitor(s)

$HgCl_2$ [6]; Iodoacetate [6]; 1,10-Phenanthroline [6]; More (not inhibitory: KCN, NaN_3) [6]

Cofactor(s)/prosthetic group(s)/activating agents

NAD^+ (not $NADP^+$) [6]; NADH [6]

Metal compounds/salts

Turnover number (min^{-1})

Specific activity (U/mg)

0.57 [5]

K_m-value (mM)

0.4 (1-octanol) [6]; 0.2 (NAD^+) [6]

pH-optimum

10.2 [5]; 8.3 [6]

pH-range

Temperature optimum (°C)

Temperature range (°C)

3 ENZYME STRUCTURE

Molecular weight

101500 (Drosophila pseudoobscura, Drosophila persimilis, gel electrophoresis) [3]
109000 (Drosophila sp., gel filtration) [5]

Subunits

Glycoprotein/Lipoprotein

–

4 ISOLATION/PREPARATION

Source organism

Nomuraea rileyi [1]; Saccharomyces cerevisiae [6]; Drosophila grimshawi [2]; Drosophila orthofascia [2]; Drosophila formella [2]; Drosophila pseudo-obscura [3]; Drosophila persimilis [3]; Drosophila silvestris [4]; Drosophila heteroneura [4]; Drosophila sp. [5]

Source tissue

Malpighian tubules [2]; Fat body [2]; Abdomen [4]; Whole animals [5]

Localisation in source

Soluble [6]

Purification

Drosophila sp. (partially) [5]

Crystallization

–

Cloned

–

Renaturated

–

5 STABILITY

pH

Temperature (°C)

40 (15 min, 60% loss of activity) [6]; 60 (15 min, 80% loss of activity) [6]

Oxidation

Organic solvent

General stability information

Storage

0°C, 24 h, complete inactivation [6]

6 CROSSREFERENCES TO STRUCTURE DATABANKS

PIR/MIPS code

Brookhaven code

7 LITERATURE REFERENCES

[1] Joslyn, D.J., Boucias, D.G.: Can. J. Microbiol.,27,364–366 (1981)
[2] Dickinson, W.J.: Science,207,995–997 (1980)
[3] Coyne, J.A., Felton, A.A.: Genetics,87,285–304 (1977)
[4] Sene, F.M., Carson, H.L.: Genetics,86,187–198 (1977)
[5] Sieber, F., Fox, D.J., Ursprung, H.: FEBS Lett.,26,274–276 (1972)
[6] Roche, B., Azoulay, E.: Eur. J. Biochem.,8,426–434 (1969)

1 NOMENCLATURE

EC number
1.1.1.75

Systematic name
(R)-1-Aminopropan-2-ol:NAD$^+$ oxidoreductase

Recommended name
(R)-Aminopropanol dehydrogenase

Synonymes
Dehydrogenase, L-aminopropanol
L-Aminopropanol dehydrogenase
1-Aminopropan-2-ol-NAD$^+$ dehydrogenase [1]
L(+)-1-Aminopropan-2-ol:NAD$^+$ oxidoreductase [1]
1-Aminopropan-2-ol-dehydrogenase [3]
DL-1-Aminopropan-2-ol: NAD$^+$ dehydrogenase [5]
L(+)1-Aminopropan-2-ol-NAD/NADP oxidoreductase [8]

CAS Reg. No.
37250-13-8

2 REACTION AND SPECIFICITY

Catalysed reaction
(R)-1-Aminopropan-2-ol + NAD$^+$ →
→ aminoacetone + NADH

Reaction type
Redox reaction

Natural substrates
Aminoacetone + NADH (hypothesis: urinary aminopropanol is formed from L-threonine via aminoacetone) [2]

Substrate spectrum

1 L-1-Aminopropan-2-ol + NAD^+ (r [7, 8], at high aminopropanol concentrations (e.g. 0.2 M) activities towards L- and D-isomers are the same, at lower concentrations activity towards L-aminopropanol becomes predominant [1, 5], in fresh cell-free extracts L(+)-1-aminopropan-2-ol preparations are oxidized more rapidly than racemic or laevo-rotatory material, in appropriately treated extracts activity towards D-enantiomer is detectable and relatively higher than towards the L-enantiomer [4], DL-aminopropan-2-ol [2–4]) [1–5, 7, 8]
2 D-1-Aminopropan-2-ol + NAD^+ [1, 5]
3 Propan-1,2-diol + NAD^+ [5]
4 Propan-1,3-diol + NAD^+ [5]
5 Butan-2,3-diol + NAD^+ [5]
6 Hydroxyacetone + NAD^+ [5]
7 1-Aminopropan-2,3-diol + NAD^+ [5]
8 1,3-Diaminopropan-2-ol + NAD^+ [5]
9 1-Amino-3-diethylaminopropan-2-ol + NAD^+ [5]
10 DL-Phenylserine + NAD^+ [4, 7]
11 DL-1-Aminobutan-2-ol + NAD^+ [4]
12 DL-3-Hydroxybutyrate + NAD^+ (slowly) [7]
13 DL-2-Hydroxy-2-phenylethylamine + NAD^+ [7]
14 5-Aminolaevulate + NADH (little activity) [7]

Product spectrum

1 Aminoacetone + NADH [7, 8]
2 Aminoacetone + NADH
3 ?
4 ?
5 ?
6 ?
7 ?
8 1,3-Diaminoacetone + NADH
9 1-Amino-3-diethylaminoacetone + NADH
10 ?
11 1-Aminobutan-2-one + NADH
12 3-Ketobutyrate + NADH
13 ?
14 ?

Inhibitor(s)

DL-Propan-1,2-diol [1]; DL-1-Chloropropan-2-ol [1]; DL-1-Bromopropan-2-ol [1]; DL-3-Hydroxybutyrate [2, 3]; Mg^{2+} (above 3 mM) [2]; L-Cysteine (5–20 mM) [2]; 2-Mercaptoethanol (5–20 mM) [2]; EDTA (slight) [2]; Salicylate (slight) [2]; Oxalate (slight) [2]; Thiourea (slight) [2]; Cu^{2+} [3]; DL-2-Hydroxy-2-phenylethylamine [3, 4]; DL-Serine [4]; DL-1-Aminopropan-2,3-diol [4]; ATP (3 mM, pH 7: stimulation of NADPH-dependent activity, inhibition of NADH-dependent activity) [6]; D(-)-1-Aminopropan-2-ol [7]; K^+ (200–400 mM [7], 10–80 mM [2]) [2, 7]; Na^+ (200–400 mM) [7]; NH_4^+ (200–400 mM) [7]; Phosphate [7, 8]

Cofactor(s)/prosthetic group(s)/activating agents

Glutathione (reduced, activation at 1–2 mM) [2]; NAD^+ (activity with alpha-NAD^+ (5 mM) is 51% that of an equivalent concentration of the commercially available (alpha + beta)NAD^+ mixture [5], inactive with alpha-NAD^+ [7]) [1–5, 7, 8]; NADH (not [5], at pH 7: 4 times more active with NADPH than with NADH [6]); [1, 6, 7]; NADPH (50% of the activity with NADH [1], at pH 7: 4 times more active with NADPH than with NADH) [6]; ATP (3 mM, pH 7, stimulation of NADPH-dependent activity, inhibition of NADH-dependent activity) [6, 7]; $NADP^+$ [7, 8]

Metal compounds/salts

K^+ (stimulation (optimum: 400 mM) [5], requirement [3]) [3, 5]; Mg^{2+} (1.3 mM, 10–20% activation) [2]; Fe^{2+} (activation) [3]

Turnover number (min^{-1})

Specific activity (U/mg)

0.390 [5]; 8.5 [7]

K_m-value (mM)

1.7 (NAD^+ (+ L-1-aminopropan-2-ol)) [5]; 1.8 (NAD^+ (+ D-1-aminopropan-2-ol)) [5]; 31 (L-1-aminopropan-2-ol) [5]; 54 (D-1-aminopropan-2-ol) [5]; 48 (L-1-aminopropan-2-ol) [1]; 110 (D-1-aminopropan-2-ol) [1]; 4 (NAD^+) [1]; 16.4 (NADH) [1]; 83.0 (aminoacetone) [1]; 15 (1-aminopropan-2-ol) [2]; 0.8 (1-aminopropan-2-ol) [3]; 0.4 (NAD^+, pH 10) [4]; 0.05 (NAD^+, pH 7) [4]; 10 (DL-aminopropan-2-ol) [4]; 1.5 (L(+)-1-aminopropan-2-ol) [4]; 0.12 (DL-1-aminopropan-2-ol) [7]; 0.3 (NAD^+) [7]; 0.2 ($NADP^+$) [7]; 0.7 (DL-3-phenylserine, DL-2-hydroxy-2-phenylethylamine) [7]; 0.4 (NADH) [7]

pH-optimum

5.0 (aminoacetone + NADH [7, 8], a second peak at pH 8.4 [7]) [7, 8]; 7.0 (aminopropanol + NAD^+) [3]; 8 (aminoacetone + NADH) [6]; 8.4 (aminoacetone + NADH, a second peak at pH 5.0) [7]; 8.5 (DL-1-aminopropan-2-ol, L-1-aminopropan-2-ol, D-1-aminopropan-2-ol) [5]; 9 (aminoacetone + NADH) [1]; 9.1 (aminopropanol + NAD^+) [2]; 9.5 (aminopropanol + NAD^+) [7]; 9.6 (aminopropanol + NAD^+) [8]; 10 (aminopropanol + NAD^+) [4]; 10 (aminopropanol + NAD^+) [4]

pH-range

8.8–10 (8.8: about 50% of activity maximum, 10: about 55% of activity maximum) [2]

Temperature optimum (°C)

37 (assay at) [1, 2, 5]

Temperature range (°C)

---

3 ENZYME STRUCTURE

Molecular weight

70000–80000 (Pseudomonas sp., gel filtration) [8]

Subunits

Glycoprotein/Lipoprotein

–

---

4 ISOLATION/PREPARATION

Source organism

E. coli [3, 4]; Rat [1, 2]; Bacillus subtilis [5]; Pseudomonas sp. (N.C.I.B. 8858 [6]) [6–8]

Source tissue

Cell [5, 7]; Liver [1, 2]; Kidney [2]; Heart [2]; Spleen [2]; Muscle [2]

Localisation in source

Mitochondria [1, 2]

Purification

Bacillus subtilis (partial) [5]; Pseudomonas sp. [7]

Crystallization

–

Cloned

–

Renaturated

–

5 STABILITY

pH

Temperature (°C)

40–52 (15 min, 48–78% loss of activity) [1]

Oxidation

Organic solvent

General stability information

Labile enzyme [4]; Dialysis at 0°C, overnight causes appreciable loss of activity [4]; Dialysis at 4°C, against 0.1 M potassium phosphate buffer, pH 7, stable overnight [7]

Storage

–20°C, disrupted mitochondrial preparation in 0.1 M phosphate buffer, pH 7, up to 6 days, no effect [1]; 20°C, disrupted mitochondrial preparation in 0.1 M phosphate buffer, pH 7, 6 days, 50% loss of activity [1]; 0°C, 7 days, 0.1 M Tris-HCl buffer, pH 7.2, 90% loss of activity [5]; –15°C, 7 days, 0.1 M Tris-HCl buffer, pH 7, 49% loss of activity, purified enzyme [5]; –15°C, 1 mg protein/ml, phosphate buffer, pH 7, 1 year, less than 10% loss of activity [7]

6 CROSSREFERENCES TO STRUCTURE DATABANKS

PIR/MIPS code

Brookhaven code

7 LITERATURE REFERENCES

[1] Cox, N., Turner, J.M., Willetts, A.J.: Biochim. Biophys. Acta,170,438–439 (1968)
[2] Turner, J.M., Willetts, A.J.: Biochem. J.,102,511–519 (1967)
[3] Turner, J.M.: Biochem. J.,99,427–433 (1966)
[4] Turner, J.M.: Biochem. J.,104,112–121 (1967)
[5] Willetts, A.J., Turner, J.M.: Biochim. Biophys. Acta,252,98–104 (1971)
[6] Faulkner, A., Turner, J.M.: Biochem. J.,138,263–276 (1974)
[7] Pickard, M.A., Higgins, I.J., Turner, J.M.: J. Gen. Microbiol.,54,115–126 (1968)
[8] Pickard, M.A., Higgins, I.J., Turner, J.M.: J. Gen. Microbiol.,45, i-ii (1966)

1 NOMENCLATURE

EC number
1.1.1.76

Systematic name
(S,S)-Butane-2,3-diol:NAD$^+$ oxidoreductase

Recommended name
(S,S)-Butanediol dehydrogenase

Synonymes
Dehydrogenase, L-butanediol
L-BDH [1]
L(+)-2,3-Butanediol dehydrogenase (L-acetoin forming) [1]

CAS Reg. No.
37250-14-9

---

2 REACTION AND SPECIFICITY

Catalysed reaction
(S,S)-Butane-2,3-diol + NAD$^+$ →
→ acetoin + NADH

Reaction type
Redox reaction

Natural substrates

Substrate spectrum
1 (S,S)-Butane-2,3-diol + NAD$^+$ (r, equilibrium favors formation of acetoin [2], racemic acetoin used for reverse reaction, not D-(-)acetoin [4]) [1–5]

Product spectrum
1 Acetoin + NADH (L-acetoin [1], (+)-acetoin [2]) [1, 2, 4]

Inhibitor(s)

Cofactor(s)/prosthetic group(s)/activating agents
NAD$^+$ [2, 4]; NADH [2, 4]

Metal compounds/salts

Turnover number (min^{-1})

Specific activity (U/mg)

2.8 [2]

K_m-value (mM)

pH-optimum

pH-range

Temperature optimum (°C)

Temperature range (°C)

---

3 ENZYME STRUCTURE

Molecular weight

Subunits

Glycoprotein/Lipoprotein

–

---

4 ISOLATION/PREPARATION

Source organism

Klebsiella pneumoniae (isozyme E2 [4]) [1, 4]; Brevibacterium saccharolyticum (C-1012) [2, 3]; Aerobacter aerogenes [5]; Pseudomonas hydrophila [5]; Bacillus subtilis [5]

Source tissue

Localisation in source

Purification

Crystallization

–

Cloned

–

Renaturated

–

5 STABILITY

pH

Temperature (°C)
More (thermolabile) [1]

Oxidation

Organic solvent

General stability information
Unstable to repeated freezing and thawing [2]

Storage
−17°C, pH 8.0, stable [2]

6 CROSSREFERENCES TO STRUCTURE DATABANKS

PIR/MIPS code

Brookhaven code

7 LITERATURE REFERENCES

[1] Ui, S., Matsuyama, N., Masuda, H., Muraki, H.: J. Ferment. Technol.,62,551–559 (1984)
[2] Ui, S., Masuda, H., Muraki, H.: Agric. Biol. Chem.,48,2837–2838 (1984)
[3] Ui, S., Masuda, H., Muraki, H.: J. Ferment. Technol.,62,151–156 (1984)
[4] Voloch, M., Ladisch, M.R., Rodwell, V.W., Tsao, G.T.: Biotechnol. Bioeng.,25,173–183 (1983)
[5] Taylor, M.B., Juni, E.: Biochim. Biophys. Acta,39,448–457 (1960)

1 NOMENCLATURE

EC number
1.1.1.77

Systematic name
[(R) or (S)]-Propane-1,2-diol:NAD^+ oxidoreductase

Recommended name
Lactaldehyde reductase

Synonymes
Reductase, lactaldehyde
Propanediol:nicotinamide adenine dinucleotide (NAD) oxidoreductase [2]
L-Lactaldehyde:propanediol oxidoreductase [3]

CAS Reg. No.
37250-15-0

2 REACTION AND SPECIFICITY

Catalysed reaction
(R) [or (S)]-Propane-1,2-diol + NAD^+ →
→ (R) [or (S)]-lactaldehyde + NADH

Reaction type
Redox reaction

Natural substrates
(S)-Propane-1,2-diol + NAD^+ (fucose metabolism) [5, 9]

Substrate spectrum
1 (R)-Propane-1,2-diol + NAD^+ (r [1], utilizes D- and L-lactaldehyde in the reverse direction equally well [1], acts only on L-isomer [5]) [1–5, 9]
2 (S)-Propane-1,2-diol + NAD^+ (r [1], only L-propane-1,2-diol oxidized [5]) [1–3, 5, 6]
3 Ethanol + NAD^+ (r [1]) [1, 5]
4 DL-Glyceraldehyde + NADH (r) [5]
5 Propionaldehyde + NADH (r) [5]
6 Glycolaldehyde + NADH (r) [5]
7 More (D-1,2-propanediol neither serves as substrate nor as inhibitor) [5]

Product spectrum

1 (R)-Lactaldehyde + NADH [1–4]
2 (S)-Lactaldehyde + NADH [1]
3 Acetaldehyde + NADH [1]
4 Glycerol + NAD^+ [5]
5 Propanol + NAD^+ [5]
6 Ethylene glycol + NAD^+ [5]
7 ?

Inhibitor(s)

Iodoacetate [1]; N-Ethylmaleimide [1]; L-1,2-Propanediol (above 25 mM: substrate inhibition) [5]; p-Chloromercuribenzoate [1]; More (D-1,2-propanediol neither serves as substrate nor as inhibitor) [5]

Cofactor(s)/prosthetic group(s)/activating agents

NADH [1, 5]; NAD^+ [1–6, 9]; $NADP^+$ (not [5], with $NADP^+$ one tenth of the activity as with NAD^+ [1]) [1]

Metal compounds/salts

Turnover number (min^{-1})

Specific activity (U/mg)

More [1]; 36.6 [2]

K_m-value (mM)

0.56 (DL-lactaldehyde) [1]; 0.065 (acetaldehyde) [1]; 11 (DL-1,2-propanediol) [1]; 0.26 (ethanol) [1]; 0.035 (L-lactaldehyde, pH 7.0) [5]; 1.25 (L-1,2-propanediol, pH 9.5) [5]

pH-optimum

6.5 (reduction of lactaldehyde [5]) [1, 5]; 9.5 (dehydrogenation of L-1,2-propanediol) [5]

pH-range

6.0–7.8 (6.0: 50% of activity maximum (DL-lactaldehyde), 60% of activity maximum (acetaldehyde), 7.8: 45% of activity maximum (DL-lactaldehyde), 50% of activity maximum (acetaldehyde)) [1]; 8.0–11 (dehydrogenation of L-1,2-propanediol, 8.0: about 10% of activity maximum, 11: about 70% of activity maximum) [5]; 4.5–10 (reduction of lactaldehyde, 4.5: about 50% of activity maximum, 10: about 10% of activity maximum) [5]

Temperature optimum (°C)

30 (assay at) [4]

Temperature range (°C)

3 ENZYME STRUCTURE

Molecular weight
76000 (E. coli, gel filtration) [5]
155000 (Microcyclus eburneus, gel filtration) [2]

Subunits
Dimer (2 × 39000, E. coli, SDS-PAGE) [5]
Tetramer (4 × 38000, Microcyclus eburneus, SDS-PAGE) [2]

Glycoprotein/Lipoprotein
–

---

4 ISOLATION/PREPARATION

Source organism
Rat [1]; Bacillus macerans [3]; Clostridium sphenoides [4]; Microcyclus eburneus [2]; E. coli (enzyme acts only on L-isomer [5], mutant with constitutive enzyme [8]) [5–9]

Source tissue
Liver [1]; Heart [1]; Spleen [1]; Lung [1]; Brain [1]; Skeletal muscle [1]

Localisation in source

Purification
Rat [1]; Microcyclus eburneus [2]

Crystallization
–

Cloned
–

Renaturated
–

---

5 STABILITY

pH

Temperature (°C)

Oxidation

Organic solvent

General stability information

Storage

6 CROSSREFERENCES TO STRUCTURE DATABANKS

PIR/MIPS code

PIR1:RDECLA (Escherichia coli)

Brookhaven code

7 LITERATURE REFERENCES

[1] Ting, S.-M., Sellinger, O.Z., Miller, O.-N.: Biochim. Biophys. Acta,89,217–225 (1964)
[2] Kawagishi, T., Nishio, N., Matsuno, R., Kamikubo, T.: Agric. Biol. Chem.,44,949–950 (1980)
[3] Weimer, P.J.: Appl. Environ. Microbiol.,47,263–267 (1984)
[4] Tran-Din, K., Gottschalk, G.: Arch. Microbiol.,142,87–92 (1985)
[5] Boronat, A., Aguilar, J.: J. Bacteriol.,140,320–326 (1979)
[6] Hacking, A.J., Lin, E.C.C.: J. Bacteriol.,126,1166–1172 (1976)
[7] Hacking, A.J., Aguilar, J., Lin, E.C.C.: J. Bacteriol.,136,522–530 (1978)
[8] Cocks, G.T., Aguilar, J., Lin, E.C.C.: J. Bacteriol.,118,83–88 (1974)
[9] Boronat, A., Aguilar, J.: J. Bacteriol.,147,181–185 (1981)

1 NOMENCLATURE

EC number
1.1.1.78

Systematic name
(R)-Lactaldehyde:NAD^+ oxidoreductase

Recommended name
D-Lactaldehyde dehydrogenase

Synonymes
Dehydrogenase, D-lactaldehyde
Methylglyoxal reductase

CAS Reg. No.
37250-16-1

2 REACTION AND SPECIFICITY

Catalysed reaction
(R)-Lactaldehyde + NAD^+ →
→ methylglyoxal + NADH (similar enzyme with NADPH-requirement for methylglyoxal reduction, that is inactive with NAD^+, NADH and $NADP^+$ [4–7])

Reaction type
Redox reaction

Natural substrates
Methylglyoxal + NADH (fermentation of glucose via methylglyoxal by-pass) [3]

Substrate spectrum

1 Methylglyoxal + NAD(P)H (i.e. pyruvaldehyde, ir [2, 4], r [1], similar enzyme with NADPH requirement, no reaction with NAD^+, NADH and $NADP^+$ [4–7]) [1, 2, 4–7]
2 DL-Glyceraldehyde + NADH [1, 8]
3 Phenylglyoxal + NADH (NADPH required [4]) [2, 4, 7–9]
4 Glyoxal + NAD(P)H (slight [8]) [4, 7–9]
5 Acetaldehyde + NAD(P)H [7–9]
6 Glutaraldehyde + NADH [8]
7 4,5-Dioxopentanoate + NADPH [7, 9]
8 Propionaldehyde + NADPH [7, 9]
9 3-Deoxyglucosone + NADH [8]
10 2,3-Pentanedione + NADH [8]
11 More (enzyme MGR I: specific for 2-oxoaldehydes (glyoxal phenylglyoxal), enzyme MGR II: active towards 2-oxoaldehydes (glyoxal, methylglyoxal, phenylglyoxal), 4,5-dioxovalerate and some aldehydes (propionaldehyde and acetaldehyde) [7], specific for lactaldehyde [1], enzyme utilizes both: NADH and NADPH [9], NADPH required [4], NADH better substrate than NADPH [2]) [1, 2, 4, 7, 9]

Product spectrum

1 (R)-Lactaldehyde + NAD^+ [1]
2 ?
3 Phenylglycoaldehyde + NAD^+
4 Glycolaldehyde + NAD^+
5 Ethanol + $NAD(P)^+$
6 ?
7 ?
8 Propanol + $NADP^+$
9 ?
10 ?
11 ?

Inhibitor(s)

2-Mercaptoethanol [7]; Dithiothreitol [7]; $NADP^+$ [4, 7]; N-Dodecylsarcosine [4]; Sodium dodecylsulfate [4]; Sodium cholate [4]; Glyoxal (in absence of NADPH) [7]; Methylglyoxal (in absence of NADPH) [7]; Phenylglyoxal (in absence of NADPH) [7]; Mn^{2+} enzymes (MGR I, MGR II) [7]; Zn^{2+} (enzyme MGR I [7]) [7, 9]; Mg^{2+} (enzymes MGR I and MGR II) [7]; Cu^{2+} (enzymes MGR I and MGR II [7]) [7, 9]; Ca^{2+} (enzymes MGR I (slightly) MGR II) [7]; Ni^{2+} (enzyme MGR I [7]) [7, 9]; Co^{2+} (activates enzyme MGR I, slightly inhibits enzyme MGR II [7]) [7, 9]; EDTA (no effect [9]) [7]; p-Chloromercuribenzoate (activation of enzyme MGR I, inhibition of enzyme MGR II [7]) [1, 2, 4, 7–9]; Iodoacetate (activation of enzyme MGR I, inhibition of enzyme MGR II [7]) [1, 4, 7, 9]; N-Ethylmaleimide (activation of enzyme MGR I, inhibition of enzyme MGR II [7]) [1, 4, 7, 9]; 5,5'-Dithiobis(2-nitrobenzoate) [2]; Hg^{2+} [4, 9]

Cofactor(s)/prosthetic group(s)/activating agents

NAD^+ [1]; NADH (utilizes both NADPH and NADH [9], NADH better substrate than NADPH [2]) [1, 2, 8, 9]; NADPH (utilizes both NADH and NADPH [9], NADPH required [4], NADH better substrate than NADPH [2]) [2–7, 9]; Glutathione (activates [4], slight activation [9]) [4, 9]; Dithiothreitol (activates [4], slight activation [9]) [4, 9]; 2-Mercaptoethanol (activates [4], slight activation [9]) [4, 9]; L-Cysteine (activates) [4]; Triton X-100 (activates) [4]; Tween 80 (activates) [4]; p-Chloromercuribenzoate (activation of enzyme MGR I, inhibition of enzyme MGR II) [7]; Iodoacetate (activation of enzyme MGR I, inhibition of enzyme MGR II) [7]; N-Ethymaleimide (activation of enzyme MGR I, inhibition of enzyme MGR II) [7]

Metal compounds/salts

Co^{2+} (activates enzyme MGR I, slightly inhibits enzyme MGR II) [7]

Turnover number (min^{-1})

Specific activity (U/mg)

69.2 [2]; 8.38 [4]; 2.90 [7]; 2.26 [9]

K_m-value (mM)

0.034 (NAD^+) [1]; 0.013 (DL-lactaldehyde) [1]; 0.4 (methylglyoxal) [2]; 0.2 (NADH) [2]; 0.3 (phenylglyoxal) [2]; 0.0002 (NADH) [4]; 5.88 (methylglyoxal) [4]; 1.54 (phenylglyoxal) [4]; 0.054 (NADPH, enzyme MGR I) [7]; 0.0068 (NADPH, enzyme MGR II) [7]; 15.4 (methylglyoxal, enzyme MGR I) [7]; 10 (glyoxal, enzyme MGR II) [7]; 1.43 (methylglyoxal, enzyme MGR I) [7]; 4.35 (phenylglyoxal, enzyme MGR II) [7]; 0.7 (methylglyoxal) [7]; 10 (3-deoxyglucosone) [8]; More [9]

pH-optimum

11 (oxidation of D-lactaldehyde, increase of reaction rate up to pH 11) [1]; 6.5 (reduction of methylglyoxal) [2]; 7.0 (similar enzyme with NADPH-requirement for methylglyoxal reduction, that is inactive with NAD^+, NADH and $NADP^+$) [4]; 6.5 (enzyme MGR I [7]) [7–9]; 9.0 (enzyme MGR II) [7]

pH-range

6.0–8.0 (6.0 and 8.0: about 50% of activity maximum) [4]; 4.5–9.5 (4.5 and 9.5: about 50% of activity maximum) [9]

Temperature optimum (°C)

45 [9]

Temperature range (°C)

25–75 (25°C and 75°C: about 50% of activity maximum) [9]

3 ENZYME STRUCTURE

Molecular weight

37000 (Aspergillus niger, enzyme MGR I, similar enzyme with NADPH-requirement for methylglyoxal reduction, that is inactive with NAD^+, NADH and $NADP^+$, gel filtration) [7]
38000 (Aspergillus niger, enzyme MGR II, similar enzyme with NADPH-requirement for methylglyoxal reduction, that is inactive with NAD^+, NADH and $NADP^+$, gel filtration) [7]
89000 (goat, gel filtration) [2]
69000 (pig, gel filtration) [8]

Subunits

Dimer (2 × 46000, goat, SDS-PAGE [2], 2 × 35000, pig, SDS-PAGE [8]) [2, 8]
Monomer (1 × 43000, E. coli, SDS-PAGE [9], 1 × 43000, Saccharomyces cerevisiae, SDS-PAGE [4], 1 × 36000, enzyme MGR I, 1 × 38000, enzyme MGR II, Aspergillus niger, similar enzyme with NADPH-requirement for methylglyoxal reduction, that is inactive with NAD^+, NADH and $NADP^+$, SDS-PAGE [7]) [4, 7, 9]

Glycoprotein/Lipoprotein

Glycoprotein (6.6% carbohydrate, similar enzyme with NADPH-requirement for methylglyoxal reduction, that is inactive with NAD^+, NADH and $NADP^+$) [4]

4 ISOLATION/PREPARATION

Source organism

Rat [1]; Pig [8]; E. coli [9]; Goat [2]; Clostridium sphenoides [3]; Saccharomyces cerevisiae (similar enzyme with NADPH-requirement for methylglyoxal reduction, that is inactive with NAD^+, NADH and $NADP^+$) [4–6]; Aspergillus niger (2 methylglyoxylate reductases: MGR I and MGR II, similar enzyme with NADPH-requirement for methylglyoxal reduction, that is inactive with NAD^+, NADH and $NADP^+$) [7]

Source tissue

Liver [1, 2]; Skeletal muscle (highest activity) [1]; Brain [1]; Lung [1]; Kidney [1]; Spleen [1]; Heart [1]; Cell [3, 4]

Localisation in source

Soluble [2]

Purification

Goat [1]; Saccharomyces cerevisiae (similar enzyme with NADPH-requirement for methylglyoxal reduction, that is inactive with NAD^+, NADH and $NADP^+$) [4]; Pig [8]; E. coli [9]; Aspergillus niger (2 methylglyoxylate reductases: MGR I and MGR II, similar enzyme with NADPH-requirement for methylglyoxal reduction, that is inactive with NAD^+, NADH and $NADP^+$) [7]

Crystallization

–

Cloned

(similar enzyme with NADPH-requirement for methylglyoxal reduction, that is inactive with NAD^+, NADH and $NADP^+$) [5]

Renaturated

–

5 STABILITY

pH

6.0 (1 min, 40°C, stable) [4]; 8.0 (1 min, 40°C, stable) [4]; 4.5 (1 min, 40°C, 50% loss of activity) [4]; 11 (1 min, 40°C, 50% loss of activity) [4]

Temperature (°C)

25 (stable below) [4]; 39 (3 min, 50% loss of activity) [4]; 44 (80 s, 50% loss of activity) [4]; 50 (40 s, 50% loss of activity) [4]; 60 (35 s, 50% loss of activity) [4]

Oxidation

Organic solvent

General stability information

Storage

–20°C [4]; 4°C [7]; 4°C, 15 days [2]

6 CROSSREFERENCES TO STRUCTURE DATABANKS

PIR/MIPS code

Brookhaven code

7 LITERATURE REFERENCES

[1] Ting, S.-M., Miller, O.N., Sellinger, O.Z.: Biochim. Biophys. Acta,97,407–415 (1965)
[2] Ray, M., Ray, S.: Biochim. Biophys. Acta,802,119–127 (1984)
[3] Tran-Din, K., Gottschalk, G.: Arch. Microbiol.,142,87–92 (1985)
[4] Murata, K., Fukuda, Y., Simosaka, M., Watanabe, K., Saikusa, T., Kimura, A.: Eur. J. Biochem.,151,631–636 (1985)
[5] Murata, K., Fukuda, Y., Shimosaka, M., Watanabe, K., Saikusa, T., Kimura, A.: Appl. Environ. Microbiol.,50,1200–1207 (1985)
[6] Murata, K., Inoue, Y., Saikusa, T., Watanabe, K., Fukuda, Y., Shimosaka, M., Kimura, A.: J. Ferment. Technol.,64,1–4 (1986)
[7] Inoue, Y., Rhee, H., Watanabe, K., Murata, K., Kimura, A.: Eur. J. Biochem.,171, 213–218 (1988)
[8] Kato, H., Miyauchi, Y., Nishimura, T., Liang, Z.-Q.: Agric. Biol. Chem.,52,2641–2642 (1988)
[9] Saikusa, T., Rhee, H., Watanabe, K., Murata, K., Kimura, A.: Agric. Biol. Chem.,51,1893–1899 (1987)

1 NOMENCLATURE

EC number
1.1.1.79

Systematic name
Glycolate: NADP⁺ oxidoreductase

Recommended name
Glyoxylate reductase (NADP⁺)

Synonymes
NADPH-glyoxylate reductase

CAS Reg. No.
37250-17-2

2 REACTION AND SPECIFICITY

Catalysed reaction
Glycolate + NADP⁺ →
→ glyoxylate + NADPH

Reaction type
Redox reaction

Natural substrates

Substrate spectrum
1 Glyoxylate + NAD(P)H (r [6], ir [9], specific for glyoxylate [1 3], isozyme GR-1 specific for glyoxylate [5]) [1–10]
2 Hydroxypyruvate + NADPH (isozyme GR-2 [5], not reduced [3]) [5, 9, 10]
3 Oxaloacetate + NADPH (28.6% of glyoxylate activity [3], not reduced [2]) [3, 9]
4 2-Oxoglutarate + NADPH (13.9% of glyoxylate activity [3], not reduced [2]) [3]
5 Acetaldehyde + NADPH (10.4% of glyoxylate activity) [3]
6 More (no reduction of 2-oxobutyrate, 4-methyl-2-oxopentanoate, 3-methyl-2-oxobutyrate [2], pyruvate, formaldehyde [3]) [2, 3]

Product spectrum

1 Glycolate + NADP$^+$ [1]
2 Glycerate + NADP$^+$
3 Malate + NADP$^+$
4 2-Hydroxyglutarate + NADP$^+$
5 Ethanol + NADP$^+$
6 ?

Inhibitor(s)

Adenine [3]; N-Ethylmaleimide [3]; p-Chloromercuribenzoate (not [5]) [3, 5, 9, 10]; NO_3^- [5]; More (not inhibitory: dithiothreitol, 2-mercaptoethanol, L-cysteine, gluthathione, iodoacetate, iodoacetamide) [5]

Cofactor(s)/prosthetic group(s)/activating agents

NADPH (preferred [1, 2], specific for [8, 9]) [1–10]; NADH (16% of NADPH activity [5], not [3, 8, 9]) [2, 5, 6]

Metal compounds/salts

Turnover number (min^{-1})

Specific activity (U/mg)

210.6 [2]; 108.1 [5]; 17.1 [3]; 3.89 [9]

K_m-value (mM)

0.00023 (NADH) [5]; 0.003–0.004 (NADPH) [2, 3, 5, 9]; 0.012 (NADPH) [7]; 0.045 (glyoxylate) [3]; 0.085 (glyoxylate (+ NADPH)) [2, 9]; 0.1–0.11 (glyoxylate [1, 6], hydroxypyruvate, isozyme GR-3 [5]) [1, 5, 6]; 0.15 (NADH) [2]; 1.1 (glyoxylate (+ NADH)) [2]; 5.1 (glyoxylate, isozyme GR-3) [5]; 5.7 (glyoxylate, isozyme GR-2) [5]; 13 (glyoxylate) [9]; 30 (glyoxylate, isozyme GR-1) [5]

pH-optimum

5.5–7.2 [9]; 5.8–6.2 [6]; 6.2 (hydroxypyruvate) [5]; 6.5 [3]; 6.5–7.4 [2]; 6.7 (glyoxylate) [5]; 7.2 [7]

pH-range

5.5–8.0 [3]

Temperature optimum (°C)

Temperature range (°C)

3 ENZYME STRUCTURE

Molecular weight

25000 (Neurospora crassa, isozyme GR-1, gel filtration) [5]
31000 (Saccharomyces cerevisiae, reductase I, gel filtration) [9]
70000 (Neurospora crassa, isozyme GR-3, gel filtration) [5]
82000 (Euglena gracilis, gel filtration) [3]
110000 (Neurospora crassa, isozyme GR-3, gel filtration) [5]
125000 (Spinacia oleracea, gel chromatography) [2]

Subunits

Tetramer (4 × 33000, Spinacia oleracea, SDS-PAGE) [2]

Glycoprotein/Lipoprotein

–

4 ISOLATION/PREPARATION

Source organism

Chlamydomonas reinhardtii [1]; Spinacia oleracea [2]; Euglena gracilis [3, 4, 7]; Neurospora crassa [5]; Populus gelrica (poplar tree) [6]; Pseudomonas sp. [8]; Saccharomyces cerevisiae [9, 10]; Chlorella vulgaris [1]; Chlorella miniata [1]; Dunaliella tertiolecta [1]

Source tissue

Leaf [2]; Xylem [6]

Localisation in source

Mitochondria [3, 4, 7]; Chloroplast [1–3]

Purification

Chlamydomonas reinhardtii (partial) [1]; Spinacia oleracea [2]; Euglena gracilis [3]; Neurospora crassa [5]; Saccharomyces cerevisiae [9, 10]

Crystallization

–

Cloned

–

Renaturated

–

5 STABILITY

pH

7.0–11.0 [5]

Temperature (°C)

50 (10 min, less than 50% loss of activity) [5]

Oxidation

Organic solvent

General stability information

No stabilizers like glycerol necessary [5]

Storage

–20°C, several months [1]; –20°C [2]; 4°C, pH 7–11, several weeks [5]; 4°C, 0.01 M potassium phosphate buffer, pH 7.0, 1 week, 20% loss of activity [9]; 4°C, 0.01 M potassium phosphate buffer, pH 7.0, 1 week, 80% loss of activity, but stable in presence of 20% v/v glycerol [10]

6 CROSSREFERENCES TO STRUCTURE DATABANKS

PIR/MIPS code

Brookhaven code

7 LITERATURE REFERENCES

[1] Husic, D.W., Tolbert, N.E.: Arch. Biochem. Biophys.,252,396–408 (1987)
[2] Kleczkowski, L.A., Randall, D.D., Blevins, D.G.: Biochem. J.,239,653–659 (1986)
[3] Yokota, A., Haga, S., Kitaoka, S.: Biochem. J.,227,211–216 (1985)
[4] Yokota, A., Kitaoka, S.: Agric. Biol. Chem.,45,15–22 (1981)
[5] Fukuda, H., Moriguchi, M., Kimura, A., Tokichura, T.: Agric. Biol. Chem.,45,1153–1158 (1981)
[6] Sagisaka, S.: Plant Physiol.,65,377–381 (1980)
[7] Yokota, A., Kitaoka, S.: Biochem. J.,184,189–192 (1979)
[8] Hullin, R.P.: Methods Enzymol.,41, Pt. B,343–348 (1975)
[9] Tochikura, T., Fukuda, H., Moriguchi, M.: J. Biochem.,86,105–110 (1979)
[10] Fukuda, H., Moriguchi, M., Tochikura, T.: J. Biochem.,87,841–846 (1980)

1 NOMENCLATURE

EC number
1.1.1.80

Systematic name
Propan-2-ol:$NADP^+$ oxidoreductase

Recommended name
Isopropanol dehydrogenase ($NADP^+$)

Synonymes
Dehydrogenase, isopropanol (nicotinamide adenine dinucleotide phosphate)

CAS Reg. No.
37250-18-3

2 REACTION AND SPECIFICITY

Catalysed reaction
Propan-2-ol + $NADP^+$ →
→ acetone + NADPH

Reaction type
Redox reaction

Natural substrates
More (possibly takes part in the further metabolic reactions of the fragments formed from the side-chain of the steroid hormone precursor in the adrenals or gonads) [1]

Substrate spectrum
1 Propan-2-ol + $NADP^+$ [1]
2 More (also acts on short-chain secondary alcohols and, slowly on primary alcohols) [1]

Product spectrum
1 Acetone + NADPH
2 ?

Inhibitor(s)

Cofactor(s)/prosthetic group(s)/activating agents
$NADP^+$ [1]

Metal compounds/salts

Turnover number (min^{-1})

Specific activity (U/mg)

K_m-value (mM)

pH-optimum

pH-range

Temperature optimum (°C)

Temperature range (°C)

3 ENZYME STRUCTURE

Molecular weight

Subunits

Glycoprotein/Lipoprotein

–

4 ISOLATION/PREPARATION

Source organism

Rat [1]

Source tissue

Adrenal [1]

Localisation in source

Mitochondria [1]

Purification

Crystallization

–

Cloned

–

Renaturated

–

5 STABILITY

pH

Temperature (°C)

Oxidation

Organic solvent

General stability information

Storage

6 CROSSREFERENCES TO STRUCTURE DATABANKS

PIR/MIPS code

Brookhaven code

7 LITERATURE REFERENCES

[1] Ploc, I., Starka, L.: J. Chromatogr.,172,374–378 (1979)

1 NOMENCLATURE

EC number
1.1.1.81

Systematic name
D-Glycerate:$NADP^+$ 2-oxidoreductase

Recommended name
Hydroxypyruvate reductase

Synonymes
Reductase, hydroxypyruvate
beta-Hydroxypyruvate reductase
NADH:hydroxypyruvate reductase [1, 9]
D-Glycerate dehydrogenase [5, 11]

CAS Reg. No.
9059-44-3

---

2 REACTION AND SPECIFICITY

Catalysed reaction
D-Glycerate + $NAD(P)^+$ →
→ hydroxypyruvate + NAD(P)H

Reaction type
Redox reaction

Natural substrates
Hydroxypyruvate + NAD(P)H (discussion of biological role [2, 3, 12], gluco-neogenesis [20]) [2, 3, 12, 20]

Substrate spectrum
1 Hydroxypyruvate + NAD(P)H (r [5, 7, 11, 15–17, 21, 22, 24], ir [3], reverse reaction a small percentage of forward reaction [1], strong specificity [6]) [1–3, 5–12, 15–17, 21, 22, 24]
2 Glyoxylate + NAD(P)H (5–15% the rate of hydroxypyruvate reduction [7], no reduction [6]) [3, 7, 9, 12, 22]
3 Oxaloacetate + NAD(P)H (32% the rate of hydroxypyruvate reduction [7], no reduction [3, 6]) [7]
4 Acetoin + NAD(P)H (13–14% the rate of hydroxypyruvate reduction) [7]
5 Diacetyl + NAD(P)H (12–13% the rate of hydroxypyruvate reduction) [7]
6 Glyceraldehyde + $NAD(P)^+$ [8]
7 Glycoaldehyde + $NAD(P)^+$ [8]

8 More (no reduction of pyruvate, 2-oxobutyrate, 2-oxoglutarate [3], no reaction with NAD(P)H: glyoxal, glycoaldehyde, glyoxylate, glyceraldehyde, glyceraldehyde 3-phosphate, dihydroxyacetone, pyruvate, mercaptopyruvate, bromopyruvate, 2-oxobutyrate, 3-oxobutyrate, 3-oxoglutarate, no reaction with $NAD(P)^+$: glycolate, D(-)-lactate, L(+)-lactate, propionate, 3-hydroxypropionate, L(+)-tartrate, D(-)-tartrate, mesotartrate [8]) [3, 8]

Product spectrum

1 D-Glycerate + $NAD(P)^+$ [5, 7, 11]
2 Glycolate + $NAD(P)^+$
3 ?
4 ?
5 ?
6 ?
7 ?
8 ?

Inhibitor(s)

Hydroxypyruvate [1, 3]; Citrate [7]; Br^- [7]; I^- [7]; Tris-chloride (induced enzyme) [8]; PO_4^{3-} [8]; SO_4^{2-} [8, 15]; Cl^- [8, 15]; NO_3^- [3, 12, 15]; Glyoxylate [8]; Pyruvate [8, 16, 18]; Dihydroxyfumarate [8, 11]; Glycolate [8]; Propionate [8]; L-Tartrate [11]; D-Tartrate [8]; meso-Tartrate [8]; Arsenite [11]; EDTA [11]; Sodium bisulfite [11]; Ammonium sulfate [12]; KBr [12]; KCl [12]; Hg^{2+} [16]; Ag^+ [16]; alpha-D-Fructose 1,6-diphosphate [16]; 3-Phospho-D-glycerate [16, 18]; 2,3-Diphospho-D-glycerate [16]; 2-Phospho-DL-glycerate [16, 18]; ATP [16, 21]; GTP [21]; UTP [21]; CTP [21]; Glycine [24]; Phosphohydroxypyruvate [18, 21]; p-Chloromercuribenzoate [25]; Bromopyruvate [25]; Iodoacetate [25]; Iodoacetamide [25]; Anions (inhibition of D-glycerate oxidation) [19]

Cofactor(s)/prosthetic group(s)/activating agents

NADH (50% of NADPH activity [3], favoured cofactor for induced enzyme [8], ratio of activity NADH/NADPH 13:1 [12], 8 times faster reduction of hydroxypyruvate than with NADPH [22]) [1, 3, 7–15, 17, 18, 20, 22]; NADPH (30% of NADH activity [1, 9], preferred [3], preferred by constitutive enzyme [8], ratio of activity NADH/NADPH 13:1 [12]) [1–3, 7–9, 11–15, 18, 22]; NAD^+ (NAD^+ and $NADP^+$ equally effective [22]) [21, 22, 24]; $NADP^+$ (NAD^+ and $NADP^+$ equally effective [22]) [21, 22]; Glycolate (activation) [6]

Metal compounds/salts

Cl^- (activation [9], inhibition [8, 15]) [9]; PO_4^{3-} (activation [9], inhibition [8]) [9]; $(NH_4)_2SO_4$ (slight activation) [12]; KCl (slight activation) [12]

Turnover number (min^{-1})

Specific activity (U/mg)

525 [10]; 57.4 [15]; 18 [1, 9]; More [5, 13, 16]

K_m**-value** (mM)

0.003–0.0065 (NADPH, constitutive enzyme, D(-)-glycerate, presence of $NADP^+$ [8], NADH at pH range 6.0–7.1 [10], hydroxypyruvate [16, 17], NADH [17]) [8, 10, 16, 17]; 0.015–0.02 (NADPH [5, 7, 11], NADH, induced enzyme [8], NADH [15, 16]) [5, 7, 8, 11, 15, 16]; 0.037 ($NADP^+$) [16]; 0.05–0.065 (hydroxypyruvate) [2, 9, 10]; 0.07–0.09 (hydroxypyruvate [1], NAD^+, induced enzyme [8], D(-)-glycerate, presence of NAD^+ [8], NAD^+ [16]) [1, 8, 16]; 0.1–0.13 (hydroxypyruvate [2, 7], hydroxypyruvate (+ NADPH) [5, 11], NADH [7], NADPH, induced enzyme [8], hydroxypyruvate (+ NADH) [12], NADPH [15]) [2, 5, 7, 8, 11, 12, 15]; 0.22 (NAD^+) [17]; 0.24 (NADH) [5, 11]; 0.52–0.54 ($NADP^+$, induced enzyme [8], D-glycerate [17]) [8, 17]; 0.8 (hydroxypyruvate) [3]; 1.4 (DL-glycerate) [16]; 1.5 (NADH, constitutive enzyme) [8]; 2.1 (hydroxypyruvate (+ NADH)) [5, 11]; 2.63 (hydroxypyruvate (+ NADPH)) [12]; 5.7 (glyoxylate) [10]; 8.0 (hydroxypyruvate) [8, 15]; 9.9 (glyoxylate) [1, 9]; 10–40 (hydroxypyruvate, isozymes LS, HS) [6]; 15.3–35.0 (glyoxylate, value depends on pH) [12]; More (dependence on salt concentration) [17]

pH-optimum

4.5–8.5 (glyoxylate reduction) [9]; 5 [7]; 5.0–6.5 (constitutive enzyme, hydroxypyruvate reduction) [8]; 5.0–7.0 (hydroxypyruvate reduction) [1, 9]; 5.1 (glyoxylate reduction) [12]; 5.3 (hydroxypyruvate reduction) [11]; 5.5–8.0 (induced enzyme, hydroxypyruvate reduction) [8]; 5.8–6.2 (glyoxylate reduction) [10]; 6.2 (hydroxypyruvate reduction) [12]; 6.5 (hydroxypyruvate reduction) [15]; 6.9–7.3 (hydroxypyruvate reduction) [10]; 7.0 (hydroxypyruvate reduction) [16]; 9.3 (D-glycerate oxidation) [16]; 9–11 (induced and constitutive enzyme, glyoxylate reduction) [8]; More (pH-optimum moves towards acid region with increasing salt concentration) [17]

pH-range

5.3–6.9 (less than half-maximal activity below pH 5.3 and above pH 6.9) [5]; 5.5–8.0 (less than half-maximal activity below pH 5.5 and above pH 8.0) [15]; 5.0–7.1 [11]

Temperature optimum (°C)

Temperature range (°C)

3 ENZYME STRUCTURE

Molecular weight

96000 (Chlamydomonas reinhardtii, gel filtration) [1]
91000–95000 (Cucumis sativus, gel filtration, native PAGE) [10]
84500–85500 (Pseudomonas acidovorans, sedimentation and diffusion coefficients [8, 15], meniscus depletion [15]) [8, 15]
72000–75600 (Pseudomonas acidovorans, sedimentation and diffusion coefficients [5, 11], bovine, meniscus depletion [16]) [5, 11, 16]
50000 (Methylobacterium extorquens, gel filtration, PAGE) [6]

Subunits

Monomer (1 × 50000, Methylobacterium extorquens, SDS-PAGE) [6]
Dimer (2 × 43800–45500, Pseudomonas acidovorans, meniscus depletion in presence of 6 M guanidine-HCl, ultracentrifugation of guanidine-HCl treated enzyme, SDS-PAGE [8], 2 × 40500, Cucumis sativus, SDS-PAGE [10], 2 × 36000, bovine, SDS-PAGE [25]) [8, 10, 25]
? (x × 38000–45000, Spinacia oleracea, SDS-PAGE) [3, 13]

Glycoprotein/Lipoprotein

–

4 ISOLATION/PREPARATION

Source organism

Chlamydomonas reinhardtii [1, 9]; Zea mays [2]; Spinacia oleracea (spinach) [2, 3, 12, 13]; Pisum sativum (pea) [4]; Pseudomonas acidovorans (induced by growth on glyoxylate [8]) [5, 8, 11, 15]; Methylobacterium extorquens (AM1) [6]; Paracoccus denitrificans [7]; Chlorella vulgaris [9]; Cucumis sativus [10]; Chlorella miniata [9]; Dunaliella tertiolecta [9]; Hordeum vulgare (barley) [14]; Cucurbita pepo (pumpkin) [13]; Citrullus vulgaris (watermelon) [13]; Bovine [16–19, 21, 25]; Rat [20, 24]; Pig (female) [21, 22]; Euglena gracilis [23]

Source tissue

Leaf [2–4, 12, 13]; Bundle sheath cells [2]; Mesophyll cells [2]; Protoplasts [4]; Cotyledons [10]; Liver [16–20, 25]; Spinal cord [21, 22]; Brain [24]

Localisation in source

Chloroplasts [4]; Peroxisomes (NADH-specific activity mainly in peroxisomes [4], peroxisome-like particles [23]) [4, 10, 12, 13, 23]; Cytosol (NADPH-specific activity mainly in cytosol [4]) [4, 20]

Purification

Chlamydomonas reinhardtii (partial) [1, 9]; Pseudomonas acidovorans [5, 8, 11, 15]; Methylobacterium extorquens [6]; Paracoccus denitrificans [7]; Cucumis sativus [10]; Bovine [16]; Pig [22]

Crystallization
[5, 11]

Cloned
–

Renaturated
[8]

5 STABILITY

pH
5–8 (not dependent on salt concentration) [17]

Temperature (°C)
42 (half-life 35 min) [7]

Oxidation

Organic solvent

General stability information
Slight inactivation by freezing/thawing [7]; Glycerol: stabilization [8]; Mercaptoethanol: stabilization [8]; DTT: stabilization [16]

Storage
–20°C, 50% v/v glycerol [3, 20]; Crystalline suspension 3 months or solution of 1 mg protein/ml in 0.1 M potassium phosphate buffer, pH 7.5, several days [5]; –15°C [7]; 0°C, 0.04 M Tris-HCl, pH 7.4, 25% v/v glycerol, 0.1 M NaCl, 3 mM 2-mercaptoethanol, 1 year [8, 15]; –15°C, 6 months [17]; –70°C, 0.005 M potassium phosphate buffer, pH 7.0, 0.3 M NaCl, 3.0 mM DTT, 1 mg protein/ml, several months [21]; –15°C, 1–2 mg protein/ml, several weeks [22]

6 CROSSREFERENCES TO STRUCTURE DATABANKS

PIR/MIPS code

Brookhaven code

7 LITERATURE REFERENCES

[1] Husic, D.W., Tolbert, N.E.: Arch. Biochem. Biophys.,252,396–408 (1987)
[2] Kleczkowski, L.A., Edwards, G.E.: Plant Physiol.,91,278–286 (1989)
[3] Kleczkowski, L.A., Randall, D.D.: Biochem. J.,250,145–152 (1988)
[4] Kleczkowski, L.A., Givan, C.V., Hodgson, J.M., Randall, D.D.: Plant Physiol.,88, 1182–1185 (1988)
[5] Kohn, L.D., Jakoby, W.B.: Methods Enzymol.,9,229–232 (1966)
[6] Krema, C., Lidstrom, M.E.: Methods Enzymol.,188,373–378 (1990)
[7] Bamforth, C.W., Quayle, J.R.: J. Gen. Microbiol.,101,259–267 (1977)
[8] Utting, J.M., Kohn, L.D.: J. Biol. Chem.,250,5233–5242 (1975)
[9] Husic, D.W., Tolbert, N.E.: Arch. Biochem. Biophys.,252,396–408 (1987)
[10] Titus, D.E., Hondred, D., Becker, W.M.: Plant Physiol.,72,402–408 (1983)
[11] Kohn, L.D., Jakoby, W.B.: J. Biol. Chem.,243,2494–2499 (1968)
[12] Tolbert, N.E., Yamazaki, R.K., Oeser, A.: J. Biol. Chem.,245,5129–5136 (1970)
[13] Sautter, C., Sautter, E., Hock, B.: Planta,176,149–158 (1988)
[14] Murray, A.J.S., Blackwell, R.D., Lea, P.J.: Plant Physiol.,91,395–400 (1989)
[15] Kohn, L.D., Utting, J.M.: Methods Enzymol.,89,341–345 (1982)
[16] Sugimoto, E., Kitagawa, Y., Nakanishi, K., Chiba, H.: J. Biochem.,72,1307–1315 (1972)
[17] Coderch, R., Lluis, C., Bozal, J.: Biochim. Biophys. Acta,566,21–31 (1979)
[18] Kitagawa, Y., Sugimoto, E., Chiba, H.: Agric. Biol. Chem.,39,193–198 (1975)
[19] Sugimoto, E., Kitagawa, Y., Hirose, M., Chiba, H.: J. Biochem.,72,1317–1325 (1972)
[20] Kitagawa, Y., Sugimoto, E.: Biochim. Biophys. Acta,582,276–282 (1979)
[21] Feld, R.D., Sallach, H.J.: Arch. Biochem. Biophys.,166,417–425 (1975)
[22] Feld, R.D., Sallach, H.J.: Methods Enzymol.,41, Pt.B,289–293 (1975)
[23] Collins, N., Merrett, M.J.: Biochem. J.,148,321–328 (1975)
[24] Uhr, M.L., Sneddon, M.K.: FEBS Lett.,17,137–140 (1971)
[25] Kitagawa, Y., Sugimoto, E., Chiba, H.: Agric. Biol. Chem.,39,199–206 (1975)

1 NOMENCLATURE

EC number
1.1.1.82

Systematic name
(S)-Malate:$NADP^+$ oxidoreductase

Recommended name
Malate dehydrogenase ($NADP^+$)

Synonymes
NADP-malic enzyme [7]
NADP-malate dehydrogenase [1]
Dehydrogenase, malate (nicotinamide adenine dinucleotide phosphate)
Malic dehydrogenase (nicotinamide adenine dinucleotide phosphate)
Malate NADP dehydrogenase
NADP malate dehydrogenase
NADP-linked malate dehydrogenase

CAS Reg. No.
37250-19-4

2 REACTION AND SPECIFICITY

Catalysed reaction
Oxaloacetate + NADPH →
→ (S)-malate + $NADP^+$

Reaction type
Redox reaction

Natural substrates
Oxaloacetate + NADPH (key enzyme of C_4-photosynthesis [1–4, 10]) [1–4, 10]

Substrate spectrum
1 Oxaloacetate + NADPH (r [10, 18, 20], low activity in direction of malate oxidation [16]) [4–6, 8–10, 16, 18–20, 22, 25, 28]

Product spectrum
1 (S)-Malate + $NADP^+$ [16]

Inhibitor(s)

Oxaloacetate (substrate inhibition, below pH 8, substrate inhibition of reduced truncated enzyme [9]) [9, 20]; Diethyldicarbonate (partial protection by $NADP^+$, complete protection by $NADP^+$ + Mg^{2+}) [11]; p-Chloromercuribenzoate [16]; Hg^{2+} [16]; Cd^{2+} [16]; Arsenite (added with an equimolar concentration of 2,3-dimercaptopropanol) [16]; N-Ethylmaleimide [16]; Dichlorotriazine dye (added with an equimolar concentration of 2,3-dimercaptopropanol) [16]; NADPH [20]

Cofactor(s)/prosthetic group(s)/activating agents

NADPH [4–6, 8–10, 16, 18–20, 22, 25, 28]; $NADP^+$ [10]; NADH (ratio of utilization of NADH to NADPH in reduction of oxaloacetate is 1:160 [28]) [10, 28]; NAD^+ (40–50% of $NADP^+$-activity) [10]; More (activated by light [5, 8, 13, 15, 16, 19, 21, 24, 29], regulation of light-activation [5], light effect mediator (surface exposed tightly bound protein in thylakoid membrane) [29], noncyclic electron flow required for activation [15], overview: mechanism of light/dark regulation [16], regulation by light-actuated ferredoxin/thioredoxin system of chloroplast [21], $NADP^+$ regulates light activation [24], chemical or photochemical activation with: 2-mercaptoethanol [10], dithiothreitol (enzyme from higher plants is activated by dithiothreitol, Euglena enzyme not [25]) [10, 25, 27, 28], thioredoxin [14, 17, 27, 28], low-molecular weight thiols [17], mechanism of reductive activation by thioredoxin and low-molecular weight thiols [17]) [5, 8, 10, 13–17, 19, 21, 24, 25, 27–29]

Metal compounds/salts

Turnover number (min^{-1})

57300 (oxaloacetate + NADPH, corn) [18]; 49980 (oxaloacetate + NADPH, spinach) [20]; 174 (malate + $NADP^+$, corn) [18]; 360–402 (malate + $NADP^+$, spinach) [18]

Specific activity (U/mg)

181 [4]; 600–1000 [10]; 184 [26]; 160 [28]; More [10]

K_m-value (mM)

0.039 (NADPH, intact reduced enzyme) [9]; 0.089 (NADPH, truncated enzyme (reduced)) [9]; 0.046 (NADPH, truncated enzyme (oxidized)) [9]; 0.048 (oxaloacetate, intact reduced enzyme) [9]; 0.042 (oxaloacetate, truncated enzyme (reduced)) [9]; 1.2 (oxaloacetate, truncated enzyme (oxidized)) [9]; 0.024 (NADPH) [10]; 0.056 (oxaloacetate (+ NADPH)) [10]; 0.073 ($NADP^+$) [10]; 32 (malate (+ $NADP^+$)) [10]; 0.83 (NADH) [10]; 0.061 (oxaloacetate (+ NADH)) [10]; 0.80 (NAD^+) [10]; 29 (malate (+ NAD^+)) [10]; 0.214 (oxaloacetate, corn) [18]; 0.195 (NADPH, corn) [18]; 75 (malate, corn) [18]; 1.26 ($NADP^+$, corn) [18]; 0.11 (oxaloacetate, spinach) [18]; 0.36 (NADPH, spinach) [18]; 18 (malate, spinach) [18]; 2 ($NADP^+$, spinach) [18]; More [16, 20, 22, 25, 26]

pH-optimum

6.3 [25]; 6.5–7.5 (oxidized truncated enzyme, optimum for oxaloacetate reduction) [9]; 7.0–8.5 ($NADP^+$ + malate) [28]; 8 [22]; 8–8.5 (intact reduced enzyme) [9]; 8.5 (NADPH-dependent oxaloacetate reduction) [10]; 9 (NAD^+ + malate) [28]

pH-range

5–9.5 (5: about 25% of activity maximum, 9.5: about 50% of activity maximum, $NADP^+$ + malate) [28]

Temperature optimum (°C)

40 (at pH 9.0) [12]

Temperature range (°C)

3 ENZYME STRUCTURE

Molecular weight

56000–57000 (Spinacia oleracea, gel filtration [27], sedimentation equilibrium analysis [20]) [20, 27]
61000 (Zea mays, sedimentation equilibrium analysis) [18]
81000 (Zea mays, gradipore gel electrophoresis) [18]
83500 (Echinochloa crus-galli (population Quebec), PAGE) [4]
85450–86700 (Echinochloa crus-galli (population Mississippi), PAGE) [4]
87000 (Zea mays, gel filtration, sedimentation velocity, gradient pore-PAGE) [10]
More (enzyme exists in at least 2 molecular weight forms [1], overview [16], enzyme exists in an (oxidized) active dithiol form and a (reduced) inactive disulfide-containing form [6]) [1, 6, 16]

Subunits

Monomer (1 × 38500, Pisum sativum, SDS-PAGE, enzyme can be interconverted between monomer (MW 40000 by gel filtration), dimer and tetramer by varying pH and ionic strength) [26]
Dimer (2 × 28500, Spinacia oleracea, SDS-PAGE [27], 2 × 43000, Zea mays, SDS-PAGE [10], 2 × 39000, Zea mays, SDS-PAGE [18], 2 × 40000, Pisum sativum, SDS-PAGE, gel filtration, enzyme can be interconverted between monomer (MW 38500), dimer and tetramer by varying pH and ionic strength [26], Pisum sativum, in oxidized state the enzyme undergoes guanidine-dependent dissociation from dimer (MW 75000 by analytical ultracentrifugation) to monomer [6]) [6, 10, 18, 26, 27]
Tetramer (4 × 38000, Pisum sativum, SDS-PAGE [9], 4 × 40000, Pisum sativum, SDS-PAGE, gel filtration, enzyme can be interconverted between monomer (MW 38500), dimer and tetramer by varying pH and ionic strength [26]) [9, 26]
? (x × 42207, Sorghum vulgare, amino acid sequence deduced from cDNA sequence [3], x × 40740, Echinochloa crus-galli, SDS-PAGE [4]) [3, 4]
More (overview [16], enzyme can be interconverted between monomer (MW 38500), dimer and tetramer by varying pH and ionic strength [26]) [16, 26]

Glycoprotein/Lipoprotein

–

4 ISOLATION/PREPARATION

Source organism

Echinochloa crus-galli (populations: Quebec and Mississippi) [4]; Flaveria brownii [7]; Pisum sativum [6, 9, 17, 22, 24, 26, 29]; Streptomyces aureofaciens NRRL-B 1286 [12]; Vicia faba L. [19]; Euglena gracilis SM-ZK [25]; Spinacia oleracea (spinach) [1, 5, 8, 13, 18, 20, 21, 27]; Hordeum vulgare (barley) [1]; Zea mays (corn) [1, 10, 11, 14–16, 18, 23, 28]; Sorghum vulgare [2, 3]

Source tissue

Leaf [1, 6, 11, 14, 16, 20, 22, 23, 27]; Mesophyll [7]; Bundle sheath [7]; Shoots [9]; Mycelium [12]; Guard cells [19]

Localisation in source

Mitochondria (matrix, inner membrane space, inner membrane (slight activity)) [25]; Chloroplast (located exclusively in chloroplasts of mesophyll cells [23]) [5, 6, 8, 9, 13, 16, 17, 23, 24]; Intracellular [12]

Purification

Echinochloa crus-galli [4]; Pisum sativum [26]; Spinacia oleracea (partial, isoenzymes) [1, 20]; Zea mays [10, 28]

Crystallization

–

Cloned

–

Renaturated

(renaturation of reduced enzyme is more rapid and occurs at higher yields than of oxidized enzyme) [6]

---

5 STABILITY

pH

Temperature (°C)

60 (30 min, stable) [12]

Oxidation

Organic solvent

General stability information

Stable to thawing and freezing [28]

Storage

–15°C, 37.5 mM sodium-acetate solution, pH 5.5, 0.37 mM EDTA, 25% glycerol, no loss of activity after 3 months [28]

---

6 CROSSREFERENCES TO STRUCTURE DATABANKS

PIR/MIPS code

PIR3:S33066 (common ice plant); PIR3:S28176 (garden pea); PIR3:S38346 (garden pea); PIR3:S38347 (garden pea); PIR3:S32512 (maize); PIR3:S20743 (sorghum); PIR3:S32513 (spinach); PIR3:S17781 (II sorghum); PIR1:DEMZMC (precursor chloroplast maize); PIR2:JH0151 (precursor chloroplast sorghum); PIR3:S13588 (precursor chloroplast sorghum); PIR2:S13472 (chloroplast garden pea); PIR3:S36733 (chloroplast garden pea)

Brookhaven code

7 LITERATURE REFERENCES

[1] Johnson, H.S.: Biochem. Biophys. Res. Commun.,43,703–709 (1971)
[2] Luchetta, P., Cretin, C., Gadal, P.: Gene,89,171–177 (1990)
[3] Cretin, C., Luchetta, P., Joly, C., Decottignies, P., Lepiniec, L., Gadal, P., Sallantin, M., Huet, J.-C., Pernollet, J.-C.: Eur. J. Biochem.,192,299–303 (1990)
[4] Vairinhos, F., Simon, J.-P.: Plant Sci.,71,173–177 (1990)
[5] Miginiac-Maslow, M., Decottignies, P., Jacquot, J.-P., Gadal, P.: Biochim. Biophys. Acta,1017,273–279 (1990)
[6] Scheibe, R., Rudolph, R., Reng, W., Jaenicke, R.: Eur. J. Biochem.,189,581–587 (1990)
[7] Cheng, S.-H., Moore, B.D., Edwards, G.E., Ku, M.S.B.: Plant Physiol.,87,867–873 (1988)
[8] Migniac-Maslow, M., Cornic, G., Jacquot, J.-P.: Planta,173,468–473 (1988)
[9] Fickenscher, K., Scheibe, R.: Arch. Biochem. Biophys.,260,771–779 (1988)
[10] Kagawa, T., Bruno, P.L.: Arch. Biochem. Biophys.,260,674–695 (1988)
[11] Jawali, N., Bhagwat, A.S.: Phytochemistry,26,1859–1862 (1987)
[12] Laluce, C., Ernandes, J.R., Molinari, R.: Appl. Environ. Microbiol.,53,1913–1917 (1987)
[13] Scheibe, R., Wagenpfeil, D., Fischer, J.: J. Plant Physiol.,124,103–110 (1986)
[14] Jacquot, J.-P., Decottignies, P.: FEBS Lett.,209,87–91 (1986)
[15] Nakamoto, H., Edwards, G.E.: Plant Physiol.,82,312–315 (1986)
[16] Edwards, G.E., Nakamoto, H., Burnell, J.N., Hatch, M. D.: Annu. Rev. Plant Physiol.,36,255–286 (1985) (Review)
[17] Scheibe, R., Fickenscher, K., Ashton, A.R.: Biochim. Biophys. Acta,870,191–197 (1986)
[18] Ferte, N., Jaquot, J.-P., Meunier, J.-C.: Eur. J. Biochem.,154,587–595 (1986)
[19] Gotow, K., Tanaka, K., Kondo, N., Kobayashi, K., Syono, K.: Plant Physiol.,79, 829–832 (1985)
[20] Ferte, N., Meunier, J.-C.: Plant Sci. Lett.,37,115–121 (1984)
[21] Wolosiuk, R.A., Buchanan, B.B., Crawford, N.A.: FEBS Lett.,81,253–258 (1977)
[22] Scheibe, R., Fickenscher, K.: FEBS Lett.,180,317–320 (1985)
[23] Perrot-Rechenmann, C., Jacquot, J.P., Gadal, P., Weeden, N.F., Cseke, C., Buchanan, B.B.: Plant Sci. Lett.,30,219–226 (1983)
[24] Scheibe, R., Jacquot, J.-P.: Planta,157,548–553 (1983)
[25] Isegawa, Y., Nakano, Y., Kitaoka, S.: Agric. Biol. Chem.,48,549–552 (1984)
[26] Fickenscher, K., Scheibe, R.: Biochim. Biophys. Acta,749,249–254 (1983)
[27] Ferte, N., Meunier, J.-C., Ricard, J., Buc, J., Sauve, P.: FEBS Lett.,146,133–138 (1982)
[28] Jacquot, J.-P. P., Buchanan, B.B., Martin, F., Vidal, J.: Plant Physiol.,68,300–304 (1981)
[29] Mohamed, A.H., Anderson, L.E.: Arch. Biochem. Biophys.,209,606–612 (1981)

1 NOMENCLATURE

EC number
1.1.1.83

Systematic name
(R)-Malate:NAD^+ oxidoreductase (decarboxylating)

Recommended name
D-Malate dehydrogenase (decarboxylating)

Synonymes
Dehydrogenase, D-malate (decarboxylating)
D-Malate dehydrogenase
D-Malic enzyme [3]
Bifunctional L(+)-tartrate dehydrogenase-D(+)-malate dehydrogenase (decarboxylating, EC 1.1.1.93 and EC 1.1.1.83) [1, 4]

CAS Reg. No.
37250-20-7

2 REACTION AND SPECIFICITY

Catalysed reaction
(R)-Malate + NAD^+ →
→ pyruvate + CO_2 + NADH

Reaction type
Redox reaction
Oxidative decarboxylation

Natural substrates

Substrate spectrum
1 (R)-Malate + NAD^+ [2, 3, 5, 6]
2 More (the bifunctional L(+)-tartrate dehydrogenase (decarboxylating) EC 1.1.1.9 and 1.1.1.83 also catalyzes nonoxidative decarboxylation of meso-tartrate to D(-)glycerate) [1, 4]

Product spectrum
1 Pyruvate + CO_2 + NADH
2 ?

Inhibitor(s)

1,2-Butanedione [2]; N-Ethylmaleimide (slight effect) [2]; p-Chloromercuribenzoate (slight effect) [2, 6]; meso-Tartrate [4, 5]; Oxaloacetate [4, 5]; Dihydroxyfumarate [4]; Iodoacetate [4]; D-Lactic acid [5]; ATP [5]; More (Ca^{2+}: no inhibition) [4]

Cofactor(s)/prosthetic group(s)/activating agents

NAD^+ (no activity with $NADP^+$ [4]) [2–6]

Metal compounds/salts

Mn^{2+} (required, optimal activity depends on the presence of both: Mn^{2+} and NH_4^+ [4], Mn^{2+} or Mg^{2+} required [5], K_m: 0.016 mM [4]) [4, 5]; NH_4^+ (required, optimal activity depends on the presence of both: Mn^{2+} and NH_4^+) [4]; Mg^{2+} (required [3], Mn^{2+} or Mg^{2+} required [5], can partially substitute for Mn^{2+} [4]) [3–5]; Co^{2+} (can partially substitute for Mn^{2+}) [4]; Zn^{2+} (can partially substitute for Mn^{2+}) [4]; KCl (can partially substitute for $(NH_4)_2SO_4$) [4]; K_2SO_4 (can partially substitute for $(NH_4)_2SO_4$) [4]; NaCl (can partially substitute for $(NH_4)_2SO_4$) [4]

Turnover number (min^{-1})

Specific activity (U/mg)

13.02 [1]; More [4]

K_m-value (mM)

0.49 (NAD^+) [2]; 2.2 (D-malate) [2]; 0.17 (D-(+)-malate) [4]; 0.13 (NAD^+) [4]; 0.3 (malate, Pseudomonas fluorescens) [5]; 0.08 (NAD^+, Pseudomonas fluorescens) [5]

pH-optimum

8.4 [2]; 8.4–9.0 [4]; 8.1–8.8 (Pseudomonas fluorescens) [5]

pH-range

7.0–10.6 (at pH 7.0 and 10.6: about 50% of activity maximum) [4]

Temperature optimum (°C)

50 [4]; 57–60 [5]

Temperature range (°C)

---

3 ENZYME STRUCTURE

Molecular weight

140000 (Pseudomonas fluorescens, gel filtration) [6]

158000–162000 (Rhodopseudomonas sphaeroides, gel filtration, ultracentrifugation) [4]

175000 (Pseudomonas fluorescens, gel filtration, gradient gel electrophoresis) [5]

Subunits

Tetramer (4 × 38500, Rhodopseudomonas sphaeroides, SDS-PAGE [4], 4 × 34000, Pseudomonas fluorescens, SDS-PAGE [5]) [4, 5]

Glycoprotein/Lipoprotein

–

4 ISOLATION/PREPARATION

Source organism

Rhodopseudomonas sphaeroides [1, 4]; Pseudomonas fluorescens UK-1 [2, 5, 6]; Salmonella typhimurium [3]; Aerobacter aerogenes [3]; Pseudomonas putida [5]

Source tissue

Localisation in source

Purification

Pseudomonas fluorescens (partial [5]) [2, 5, 6]; Rhodopseudomonas sphaeroides (bifunctional L-(+)-tartrate-D-(+)-malate dehydrogenase) [4]

Crystallization

–

Cloned

–

Renaturated

–

5 STABILITY

pH

8.5 (30°C, 4 h, stable) [4]

Temperature (°C)

30 (half-life: 30 h) [4]; 50 (half-life: 120 min) [4]; 55 (half-life: 30 min) [4]; 60 (half-life: 9 min) [4]; 65 (2 min, 64% loss of activity) [2]

Oxidation

Organic solvent

General stability information

NAD^+ protects [2]; NAD^+ protects against heat inactivation and trypsinization but not against protein denaturants [6]; 10–40% loss of activity during repeated freezing and thawing [4]

Storage

–70°C, pH 7.2, 1 mM EDTA [2, 6]; –20°C, stable for months [4]; 0°C, stable for 30 h, even at protein concentration below 1 mg/ml [4]

6 CROSSREFERENCES TO STRUCTURE DATABANKS

PIR/MIPS code

Brookhaven code

7 LITERATURE REFERENCES

[1] Ebbighausen, H., Gifforn, F.: Arch. Microbiol.,138,338–344 (1984)
[2] Lähdesmäki, M., Mäntsälä, P.: Biochim. Biophys. Acta,613,266–274 (1980)
[3] Stern, J.R., O'Brien, R.W.: J. Bacteriol.,98,147–151 (1969)
[4] Gifforn, F., Kuhn, A.: J. Bacteriol.,155,281–290 (1983)
[5] Knichel, W., Radler, F.: Eur. J. Biochem.,123,547–552 (1982)
[6] Lähdesmäki, M., Mäntsälä, P.: Acta Chem. Scand., B34,423–427 (1980)

1 NOMENCLATURE

EC number
1.1.1.84

Systematic name
(R)-3,3-Dimethylmalate:NAD+ oxidoreductase (decarboxylating)

Recommended name
Dimethylmalate dehydrogenase

Synonymes
beta,beta-Dimethylmalate dehydrogenase

CAS Reg. No.
37250-21-8

2 REACTION AND SPECIFICITY

Catalysed reaction
(R)-3,3-Dimethylmalate + NAD^+ →
→ 3-methyl-2-oxobutanoate + CO_2 + NADH

Reaction type
Redox reaction
Decarboxylation

Natural substrates
(R)-3,3-Dimethylmalate + NAD^+ [1–3]

Substrate spectrum
1 (R)-3,3-Dimethylmalate + NAD^+ (r) [1–3]
2 D-Malate + NAD^+ [1, 3]
3 3-(n-Propyl)malate + NAD^+ [3]

Product spectrum
1 3-Methyl-2-oxobutanoate + CO_2 + NADH [1–3]
2 Pyruvate + CO_2 + NADH [1, 3]
3 2-Oxopentanoate + CO_2 + NADH [3]

Inhibitor(s)
1,2-Butanedione [1]

Cofactor(s)/prosthetic group(s)/activating agents
NAD^+ [1–3]

Metal compounds/salts
K^+ (K^+ or NH_4^+ required) [3]; NH_4^+ (K^+ or NH_4^+ required) [3]; Mn^{2+} (Mn^{2+} or Co^{2+} required) [3]; Co^{2+} (Mn^{2+} or Co^{2+} required) [3]

Turnover number (min^{-1})

Specific activity (U/mg)
12.54 [1]; 4.1 [3]

K_m-value (mM)
0.27 (NAD^+) [1, 3]; 0.15–0.2 (3,3-dimethyl-D-malate) [1, 3]; 8.9–11.0 (D-malate) [1, 3]; 0.33 (3-(n-propyl)malate) [3]

pH-optimum
8.4 (3,3-dimethylmalate + NAD^+) [1]; 8.0–8.5 (3,3-dimethylmalate + NAD^+) [3]

pH-range

Temperature optimum (°C)

Temperature range (°C)

---

3 ENZYME STRUCTURE

Molecular weight
142000 (Pseudomonas fluorescens, gel filtration) [1, 2]
136000 (Pseudomonas fluorescens, ultracentrifugation) [2]

Subunits
Tetramer (4 × 34000, Pseudomonas fluorescens, SDS-PAGE) [1, 2]

Glycoprotein/Lipoprotein
–

---

4 ISOLATION/PREPARATION

Source organism
Pseudomonas sp. P-2 [1, 3]; Pseudomonas fluorescens [1, 2]

Source tissue

Localisation in source

Purification
Pseudomonas fluorescens [2]; Pseudomonas sp. P-2 [3]

Crystallization

–

Cloned

–

Renaturated

–

5 STABILITY

pH

Temperature (°C)

60 (not stable above) [3]

Oxidation

Organic solvent

General stability information

Storage

50 days, 0°C [3]

6 CROSSREFERENCES TO STRUCTURE DATABANKS

PIR/MIPS code

Brookhaven code

7 LITERATURE REFERENCES

[1] Lähdesmäki, M., Mäntsälsä, P.: Biochim. Biophys. Acta,613,266–274 (1980)
[2] Mäntsälä, P.: Biochim. Biophys. Acta,526,25–33 (1978)
[3] Magee, P.T., Snell, E.E.: Biochemistry,5,409–416 (1966)

1 NOMENCLATURE

EC number
1.1.1.85

Systematic name
3-Carboxy-2-hydroxy-4-methylpentanoate:NAD^+ oxidoreductase

Recommended name
3-Isopropylmalate dehydrogenase

Synonymes
beta-Isopropylmalic enzyme
beta-Isopropylmalate dehydrogenase
threo-Ds-3-Isopropylmalate dehydrogenase

CAS Reg. No.
9030-97-1

2 REACTION AND SPECIFICITY

Catalysed reaction
3-Carboxy-2-hydroxy-4-methylpentanoate + NAD^+ →
→ 3-carboxy-4-methyl-2-oxopentanoate + NADH

Reaction type
Redox reaction

Natural substrates
threo-Ds-3-Isopropylmalate + NAD^+ [2, 4, 5, 14, 19–22]

Substrate spectrum
1 threo-Ds-3-Isopropylmalate + NAD^+ [2, 4, 5, 14, 19–22]

Product spectrum
1 alpha-Ketoisocaproate (decarboxylates to 2-oxo-4-methylpentanoate) + NADH + CO_2 [2, 4, 5, 14, 19–22]

Inhibitor(s)
1,2-Cyclohexane-diamine-N,N,N',N'-tetraacetate [2]; Cu^{2+} [5, 14, 18]; Ni^{2+} [5, 18]; p-Chloromercuribenzoate [14, 19, 21]; Ba^{2+} [19, 22]; Li^+ [19]; Ca^{2+} [19, 22]; N-Ethylmaleimide [19, 21]; Iodoacetate [19]; Iodoacetamide [19]; Zn^{2+} [22]

Cofactor(s)/prosthetic group(s)/activating agents
NAD^+ [2, 4, 5, 14, 19–22]

Metal compounds/salts

Mn^{2+} (essential) [2, 5, 14, 18, 19, 22]; Mg^{2+} (can replace Mn^{2+}) [2, 5, 14, 18, 19, 22]; Cd^{2+} (can replace Mn^{2+}) [5, 14, 18]; Co^{2+} (can replace Mn^{2+}) [5, 14, 18, 19, 22]; K^+ (activation) [2, 19, 22]; Rb^+ (activation) [2, 19, 22]

Turnover number (min^{-1})

Specific activity (U/mg)

18.4–19.3 [5, 18]; 10.5 [14]; 54 [19, 21]; 32.4 [20]

K_m-value (mM)

0.0038–0.08 (threo-Ds-3-isopropylmalate) [2, 5, 18–22]; 0.054–0.28 (NAD^+) [2, 5, 18–22]

pH-optimum

7.2 [2]; 8.9–10.1 [5]; 9.5 [14, 22]; 9.0–9.5 [19, 21]

pH-range

Temperature optimum (°C)

Temperature range (°C)

---

3 ENZYME STRUCTURE

Molecular weight

90000 (Bacillus coagulans, gel electrophoresis) [14]
72000 (Saccharomyces cerevisiae, gel filtration) [18]
70000–75000 (Salmonella typhimurium, sedimentation equilibrium centrifugation) [19, 21]
68000 (Bacillus coagulans, gel filtration) [14]
50000 (Pseudomonas aeruginosa, gel filtration) [20]

Subunits

Dimer (2 × 39000–45000, Saccharomyces cerevisiae, SDS-PAGE, nucleotide sequence analysis [5, 18], 2 × 44000, Bacillus coagulans, SDS-PAGE [14], 2 × 40000, Thermus thermophilus, SDS-PAGE [2], 2 × 39808, Bacillus coagulans, nucleotide sequence analysis [11], 2 × 38700, Candida utilis, nucleotide sequence analysis [9], 2 × 35000, Salmonella typhimurium, sedimentation equilibrium centrifugation after treatment with mercaptoethanol [19, 21]) [2, 5, 9, 11, 14, 19, 21]

Glycoprotein/Lipoprotein

–

4 ISOLATION/PREPARATION

Source organism

Bacillus subtilis [1, 10, 11]; Bacillus caldotenax [10]; Bacillus caldolyticus [10]; Bacillus coagulans [11, 14]; Bacillus sp. [15]; Thermus thermophilus [1, 2, 4, 7, 9–11, 17]; Thiobacillus ferrooxidans [3]; Salmonella typhimurium [5, 19, 21, 22]; Acetobacter aceti [6]; E. coli [10]; Citrobacter freundii [12, 13]; Clostridium butyricum [16, 18]; Pseudomonas aeruginosa [20]; Saccharomyces cerevisiae [5, 9–11, 18]; Candida albicans [8]; Candida utilis [9]

Source tissue

Localisation in source

Cytoplasm [11]

Purification

Thermus thermophilus (gene cloned in E. coli) [2, 17]; Bacillus coagulans (gene cloned in E. coli) [14]; Saccharomyces cerevisiae [5, 18]; Salmonella typhimurium [19, 21, 22]; Pseudomonas aeruginosa [20]

Crystallization

[1, 2, 7]

Cloned

[1–3, 5–10, 12–17]

Renaturated

[2]

---

5 STABILITY

pH

Temperature (°C)

87 (not stable above) [2]

Oxidation

Organic solvent

General stability information

Thermostabilization by monovalent cations [14]

Storage

–20°C, 6 months [2]; 4°C, 50 mM potassium phosphate buffer, pH 7.6, 0.5 mM EDTA 6 months [2]; 23°C, pH 6.9, 1.5 mM ammonium sulfate, 20% glycerol, 0.03% sodium azide, 2 weeks [18]

6 CROSSREFERENCES TO STRUCTURE DATABANKS

PIR/MIPS code

PIR2:A26447 (Bacillus caldotenax); PIR1:DEBSIC (Bacillus coagulans); PIR2:A26522 (Bacillus subtilis); PIR3:S32710 (imperfect fungus (Candida maltosa)); PIR3:S32711 (imperfect fungus (Candida maltosa)); PIR3:S32712 (imperfect fungus (Candida maltosa)); PIR3:S32713 (imperfect fungus (Candida maltosa)); PIR2:A44851 (Spirulina platensis); PIR1:DETWIT (Thermus aquaticus); PIR2:JX0173 (Thermus aquaticus); PIR2:JX0286 (Thiobacillus ferrooxidans); PIR2:A43324 (yeast (Candida boidinii)); PIR2:S06257 (yeast (Candida maltosa)); PIR2:S25369 (yeast (Kluyveromyces marxianus var. lactis)); PIR3:S32969 (yeast (Kluyveromyces marxianus)); PIR2:A47620 (yeast (Pichia jadinii)); PIR1:DEBYI (yeast (Saccharomyces cerevisiae)); PIR2:S25670 (precursor potato); PIR2:S20510 (precursor rape); PIR2:S35133 (Lactococcus lactis subsp. lactis)

Brookhaven code

1IPD (Thermus thermophilus, strain hb8)

7 LITERATURE REFERENCES

[1] Onodera, K., Moriyama, H., Takenaka, A., Tanaka, N., Akutsu, N., Muro, M., Oshima, T., Imada, K., Sato, M., Katsube, Y.: J. Biochem.,109,1–2 (1991)

[2] Yamada, T., Akutsu, N., Miyazaki, K., Kakinuma, K., Yoshida, M., Oshima, T.: J. Biochem.,108,449–456 (1990)

[3] Inagaki, K., Kawaguchi, H., Kuwata, Y., Sugio, T., Tanaka, H., Tano, T.: J. Ferment. Bioeng.,70,71–74 (1990)

[4] Kakinuma, K., Ozawa, K., Fujimoto, Y., Akutsu, N., Oshima, T.: J. Chem. Soc. Chem. Commun.,1190–1192 (1989)

[5] Kohlhaw, G.B.: Methods Enzymol.,166,429–435 (1988)

[6] Okumura, H., Tagami, H., Fukaya, M., Masai, H., Kawamura, Y., Horinouchi, S., Beppu, T.: Agric. Biol. Chem.,52,3125–3129 (1988)

[7] Katsube, Y., Tanaka, N., Takenaka, A., Yamada, T., Oshima, T.: J. Biochem.,104, 679–680 (1988)

[8] Jenkinson, H.F., Shep, G.P., Shepherd, M.G.: FEMS Microbiol. Lett.,49,285–288 (1988)

[9] Hamasawa, K., Kobayashi, Y., Harada, S., Yoda, K., Yamasaki, M., Tamura, G.: J. Gen. Microbiol.,133,1089–1097 (1987)

[10] Sekiguchi, T., Suda, M., Sekiguchi, T., Nosoh, Y.: FEMS Microbiol. Lett.,36,41–45 (1986)

[11] Sekiguchi, T., Ortega-Cesena, J., Nosoh, Y., Ohashi, S., Tsuda, K., Kanaya, S.: Biochim. Biophys. Acta,867,36–44 (1986)

[12] Urano, N., Kanayama, H., Karube, I.: J. Biotechnol.,2,39–45 (1985)
[13] Karube, I., Urano, N., Kanayama, H.: Appl. Microbiol. Biotechnol.,20,340–343 (1984)
[14] Sekiguchi, T., Harada, Y., Shishido, K., Nosoh, Y.: Biochim. Biophys. Acta,788,267–273 (1984)
[15] Kato, C., Honda, H., Kudo, T., Horikoshi, K.: J. Ferment. Technol.,62,77–80 (1984)
[16] Ishi, K., Kudo, T., Honda, H., Horikoshi, K.: Agric. Biol. Chem.,47,2313–2317 (1983)
[17] Tanaka, T., Kawano, N., Oshima, T.: Biochemistry,89,677–682 (1981)
[18] Hsu, Y.P., Kohlhaw, G.B.: J. Biol. Chem.,255,7255–7260 (1980)
[19] Parsons, S.J., Burns, R.O.: Methods Enzymol.,17 A,793–799 (1970)
[20] Rabin, R., Urbano, C., Ajl, S.J.: Arch. Biochem. Biophys.,134,259–261 (1969)
[21] Parsons, S.J., Burns, R.O.: J. Biol. Chem.,244,996–1003 (1969)
[22] Burns, R.O., Umbarger, H.E., Gross, S.R.: Biochemistry,2,1053–1058 (1963)

1 NOMENCLATURE

EC number
1.1.1.86

Systematic name
(R)-2,3-Dihydroxy-3-methylbutanoate:NADP+ oxidoreductase (isomerizing)

Recommended name
Ketol-acid reductoisomerase

Synonymes
Dihydroxyisovalerate dehydrogenase (isomerizing)
Acetohydroxy acid isomeroreductase
Isomerase, ketol acid reducto-
alpha-Keto-beta-hydroxylacyl reductoisomerase
2-Hydroxy-3-keto acid reductoisomerase
Acetohydroxy acid reductoisomerase
Acetolactate reductoisomerase
Dehydrogenase, dihydroxyisovalerate (isomerizing)
Isomeroreductase
Reductoisomerase

CAS Reg. No.
9075-02-9

2 REACTION AND SPECIFICITY

Catalysed reaction
(S)-2-Hydroxy-2-methyl-3-oxobutanoate + NADPH →
→ (R)-2,3-dihydroxy-3-methylbutanoate + NADP+ (stereospecific for the B side of NADPH [2], kinetic mechanism [5], steady-state kinetics [8])

Reaction type
Redox reaction
Acetoin rearrangement [8]

Natural substrates
More (enzyme catalyzes both the second step specific to valine biosynthesis and the third step specific to isoleucine biosynthesis) [2]

Substrate spectrum

1 2-Hydroxy-2-ethyl-3-oxobutanoate + NADPH (i.e. 2-aceto-2-hydroxybutyrate) [1–3, 5–9]
2 2-Acetolactate + NADPH (i.e. 2-hydroxy-2-methyl-3-oxobutanoate, r [2], at pH 7.5, which is the optimum of forward reaction, reversibility cannot be shown, at higher pH values the reaction is reversible [2]) [1–3, 6, 7, 9]
3 More (base-catalyzed proton shuttle mechanism for the alkyl migration reaction followed by an acid-assisted ketone reduction by NADPH) [10]

Product spectrum

1 2,3-Dihydroxy-3-methylpentanoate + $NADP^+$ [1, 2]
2 2,3-Dihydroxy-3-methylbutanoate + $NADP^+$ [1, 2]
3 ?

Inhibitor(s)

More (not: p-hydroxybenzoate [1], cysteine [1], 2-mercaptoethanol [1], compounds which inhibit less than 10% at concentration of 0.01 M: 3-hydroxybutyrate, isovalerate, 3-oxopentanoate, acetoin, 3-hydroxy-3-methyl-2-butanone [2]) [1, 2]; 2-Methyllactate [2]; 2-Oxo-3-hydroxyisovalerate [2]; 2-Hydroxybutyrate [2]; 2-Hydroxy-2-methylbutyrate [2]; 2-Hydroxyisovalerate [2]; 2-Ketoisovalerate [2]; Fe^{2+} [3]; Mn^{2+} [3]; Zn^{2+} [3]; Ca^{2+} [3]; $NADP^+$ (competitive to NADPH [7]) [5, 7]; 2,3-Dihydroxyisovalerate [5]; N-Hydroxy-N-isopropyloxamate [6]; Arsenite (slight) [1]

Cofactor(s)/prosthetic group(s)/activating agents

NADPH [1–9]

Metal compounds/salts

L-Ascorbic acid (enhances reaction rate when 2-aceto-2-hydroxybutyrate is substrate, no effect or slightly inhibitory with 2-acetolactate as substrate [1], no stimulation when present in addition to the pyridine nucleotides [2]) [1, 2]; Mn^{2+} (supports catalysis with 3-hydroxy-3-methyl-2-oxobutyrate) [8]; More (contains no cobalt and cobalamine) [9]

Turnover number (min^{-1})

108 (2-acetolactate + NADPH) [6]; 1100 (2-acetolactate) [9]; 4700 (2-aceto-2-hydroxybutyrate) [9]

Specific activity (U/mg)

1.91 [2]; 18.5 [3]; 4.94 [9]; More [1, 6–9]

K_m-value (mM)

0.0015 (NADPH) [9]; 0.002 (2-aceto-2-hydroxybutyrate) [6]; 0.0057 (NADPH) [7]; 0.014 (2-acetolactate) [6]; 0.025 (2-acetolactate) [7]; 0.037 (2-aceto-2-hydroxybutyrate) [1]; 0.12 (2-aceto-2-hydroxybutyrate) [1]; 0.16 (2-aceto-2-hydroxybutyrate) [3]; 0.29 (2-acetolactate) [9]; 0.32 (2-acetolactate) [3]; 0.78 (2-aceto-2-hydroxybutyrate) [9]

pH-optimum
7.5 (2-acetolactate) [2, 3]; 7.5–8.5 (2-acetolactate) [7]; 8.6 (2-aceto-2-hydroxybutyrate) [1]; More (no definite pH optimum observed with 2-acetolactate as substrate) [1]

pH-range
6.5–9.0 (6.5: about 30% of activity maximum, 9.0: about 90% of activity maximum, 2-aceto-2-hydroxybutyrate) [1]

Temperature optimum (°C)
25 (assay at) [1]; 37 (assay at) [2]

Temperature range (°C)

3 ENZYME STRUCTURE

Molecular weight
205000 (spinach, enzyme iso 3, nondenaturing PAGE) [7]
220000 (Salmonella typhimurium, sedimentation equilibrium ultracentrifugation [2, 9], spinach, enzyme iso 2, nondenaturing PAGE [7]) [2, 7, 9]
235000 (spinach, enzyme iso 1, nondenaturing PAGE) [7]

Subunits
Tetramer (4 × 59000, spinach, enzyme iso 1, 2 and 3, SDS-PAGE [7], 4 × 57000, Salmonella typhimurium, sedimentation equilibrium in 6 M guanidinium hydrochloride [9]) [7, 9]

Glycoprotein/Lipoprotein
Lipoprotein (39–46% lipid) [3]; Contains no carbohydrate [9]

4 ISOLATION/PREPARATION

Source organism
Spinach (multiple enzyme forms: iso 1, 2 and 3) [7]; E. coli (recombinant [8]) [6, 8, 10]; Phaseolus radiatus [1]; Salmonella typhimurium [2, 5, 9]; Neurospora crassa [3]; Saccharomyces cerevisiae [4]

Source tissue
Cell [2]; Leaf [7]

Localisation in source
Mitochondria (associated with) [4]; Chloroplast (stroma, not intrinsically membrane bound) [7]

Purification
Spinach [7]; Phaseolus radiatus (partial) [1]; Salmonella typhimurium [2, 9]; Neurospora crassa [3]; E. coli (recombinant [8]) [6, 8]

Crystallization
[2]

Cloned
–

Renaturated
–

5 STABILITY

pH
6.2 (40°C, 5 min, 88% loss of activity) [3]; 6.5 (40°C, 5 min, 55% loss of activity) [3]; 7.0 (40°C, 5 min, 35% loss of activity) [3]; 7.5 (40°C, 5 min, 30% loss of activity) [3]; 8.0 (40°C, 5 min, 85% loss of activity) [3]

Temperature (°C)
40 (5 min, pH 7.5, about 30% loss of activity, pH 8.0, 85% loss of activity, pH 7.0, 35% loss of activity, pH 6.5, 55% loss of activity, pH 6.2, 88% loss of activity) [3]

Oxidation

Organic solvent

General stability information
NADPH: enzyme is unstable, except in presence of NADPH [3]; Freezing and thawing inactivates [1]; Dialysis has no effect [1]

Storage
–80°C, 10 mM potassium phosphate buffer, pH 7.5, 0.5 mM dithiothreitol [7]

6 CROSSREFERENCES TO STRUCTURE DATABANKS

PIR/MIPS code
PIR3:S36884 (Arabidopsis thaliana); PIR2:C48648 (Corynebacterium glutamicum); PIR1:ISECKR (Escherichia coli); PIR3:A47037 (Synechocystis sp.); PIR2:A24709 (ILV5 yeast (Saccharomyces cerevisiae)); PIR2:S30145 (precursor Arabidopsis thaliana); PIR2:S17180 (precursor spinach); PIR2:S35140 (Lactococcus lactis subsp. lactis)

Brookhaven code

7 LITERATURE REFERENCES

[1] Satyanarayana, T., Radhakrishnan, A.N.: Biochim. Biophys. Acta,110,380–388 (1965)
[2] Arfin, S.M., Umbarger, H.E.: J. Biol. Chem.,244,1118–1127 (1969)
[3] Kiritani, K., Narise, S., Wagner, R.P.: J. Biol. Chem.,241,2047–2051 (1966)
[4] Ryan, E.D., Kohlhaw, G.B.: J. Bacteriol.,120,631–637 (1974)
[5] Shematek, E.M., Arfin, S.M., Diven, W.F.: Arch. Biochem. Biophys.,158,132–138 (1973)
[6] Aulabaugh, A., Schloss, J.V.: Biochemistry,29,2824–2830 (1990)
[7] Dumas, R., Joyard, J., Douce, R.: Biochem. J.,262,971–976 (1989)
[8] Chunduru, S.K., Mrachko, G.T., Calvo, K.C.: Biochemistry,28,486–493 (1989)
[9] Hofler, J.G., Decedue, C.J., Luginbuhl, G.H., Reynolds, J.A., Burns, R.O.: J. Biol. Chem.,250,877–882 (1975)
[10] Mrachko, G.T., Chunduru, S.K., Calvo, K.C.: Arch. Biochem. Biophys.,294,446–453 (1992)

1 NOMENCLATURE

EC number
1.1.1.87

Systematic name
3-Carboxy-2-hydroxyadipate:NAD$^+$ oxidoreductase (decarboxylating)

Recommended name
3-Carboxy-2-hydroxyadipate dehydrogenase

Synonymes
2-Hydroxy-3-carboxyadipate dehydrogenase

CAS Reg. No.
37250-23-0

---

2 REACTION AND SPECIFICITY

Catalysed reaction
3-Carboxy-2-hydroxyadipate + NAD$^+$ →
→ 2-oxoadipate + CO_2 + NADH

Reaction type
Redox reaction
Decarboxylation

Natural substrates
3-Carboxy-2-hydroxyadipate + NAD$^+$ [1]

Substrate spectrum
1 3-Carboxy-2-hydroxyadipate + NAD$^+$ [1]

Product spectrum
1 2-Oxoadipate + CO_2 + NADH [1]

Inhibitor(s)

Cofactor(s)/prosthetic group(s)/activating agents
NAD$^+$ [1]

Metal compounds/salts
Mg^{2+} (required) [1]

Turnover number (min^{-1})

Specific activity (U/mg)

K_m-value (mM)
1.4 (3-carboxy-2-hydroxyadipate) [1]

pH-optimum
7.5 [1]

pH-range

Temperature optimum (°C)

Temperature range (°C)

3 ENZYME STRUCTURE

Molecular weight

Subunits

Glycoprotein/Lipoprotein
–

4 ISOLATION/PREPARATION

Source organism
Saccharomyces cerevisiae [1]

Source tissue

Localisation in source

Purification
Saccharomyces cerevisiae (partial) [1]

Crystallization
–

Cloned
–

Renaturated
–

5 STABILITY

pH

Temperature (°C)

Oxidation

Organic solvent

General stability information

Storage

6 CROSSREFERENCES TO STRUCTURE DATABANKS

PIR/MIPS code

Brookhaven code

7 LITERATURE REFERENCES

[1] Strassman, M., Ceci, L.N.: J. Biol. Chem.,240,4357–4361 (1965)

1 NOMENCLATURE

EC number
1.1.1.88

Systematic name
(S)-Mevalonate:NAD^+ oxidoreductase (CoA-acylating)

Recommended name
Hydroxymethylglutaryl-CoA reductase

Synonymes
Reductase, hydroxymethylglutaryl coenzyme A
beta-Hydroxy-beta-methylglutaryl coenzyme A reductase
beta-Hydroxy-beta-methylglutaryl CoA-reductase
3-Hydroxy-3-methylglutaryl coenzyme A reductase (EC 1.1.1.88)
3-Hydroxy-3-methylglutaryl-CoA reductase
Hydroxymethylglutaryl coenzyme A reductase

CAS Reg. No.
37250-24-1

2 REACTION AND SPECIFICITY

Catalysed reaction
(S)-Mevalonate + 2 NAD^+ + CoA →
→ 3-hydroxy-3-methylglutaryl-CoA + 2 NADH

Reaction type
Redox reaction
Oxidative acylation
Reductive deacylation

Natural substrates
(S)-Mevalonate + NAD^+ + CoA (first committed intermediate in polyisoprenoid biosynthesis [3, 5], catabolism proceeds via reversal of the biosynthetic pathway [3, 5], with mevalonate as sole C-source during cell growth, mevalonate oxidation is the favored reaction [4], in all other instances the reverse reaction is favored [4]) [3–5]

Substrate spectrum

1 (S)-Mevalonate + NAD^+ + CoA ((r) [1–7], both directions of reaction proceed at the same rate under optimal conditions [6], mevalonate formation strongly favored [3]) [1–7]
2 Mevaldate + NADH (i.e. mevaldehyde, reaction proceeds by 1/10 the rate of the main reactions [6]) [3, 4, 6, 7]
3 Mevaldate + CoA + NAD^+ (i.e. mevaldehyde, reaction proceeds by 1/10 the rate of the main reactions [6]) [3, 4, 6, 7]
4 More (glyceraldehyde, glutarate semialdehyde (i.e. 5-oxopentanoate) are no substrates [3], mevalonate oxidation: the carboxy-, 3-methyl-, and 3-hydroxy groups are essential for substrate binding to the enzyme [1, 4]) [1, 3, 4]

Product spectrum

1 3-Hydroxy-3-methylglutaryl-CoA + NADH (r) [1–7]
2 (S)-Mevalonate + NAD^+ [3, 4]
3 3-Hydroxy-3-methylglutaryl-CoA + NADH [3, 4]
4 ?

Inhibitor(s)

3-Hydroxy-3-methylglutarate (competitive, the oxidative acylation of mevalonate, mevaldate reduction and oxidative acylation of mevaldate are each 100-fold more sensitive to hydroxymethylglutarate inhibition than reductive deacylation of hydroxymethylglutaryl-CoA [3], the latter reaction is only inhibited when the hydroxymethylglutaryl-CoA concentration is low [4]) [1, 3, 4]; Deoxycholate (competitive) [1]; 3-Hydroxy-3-methylcarboxylic acid analogues (competitive to 3-hydroxy-3-methylglutarate) [1]; DL-3-Hydroxybutyrate (competitive) [1]; 3-Hydroxy–3-methylbutyrate (competitive) [1]; Diethyl dicarbonate (inactivation, partially restorable by hydroxylamine, His^{381}-mutant enzyme is not inactivated) [2]; 4-Vinylpyridine [5]; DTNB (enzyme mutant C296A is twice as sensitive to DTNB as wild-type enzyme, no inhibition of C156A and C156/296A [7]) [5, 7]; p-Hydroxymercuribenzoate [5]; N-Ethylmaleimide (NAD^+ [5, 6] or mevalonate retard inactivation [5], mevalonate does not protect, inhibition of enzyme mutant C296A from 62% to 81%, not enzyme mutants C156A or C156/296A [7]) [5–7]; Acetoacetate (competitive to mevalonate, non-competitive to NAD^+) [5]; 3-Methylmethanethiosulfonate (complete, irreversible [6], enzyme mutant 296A is inactivated to the same extent as wild-type enzyme [7]) [6, 7]; More (no inhibition by 3-methylbutan-1-ol, DL-1,3-butanediol, 3-methylbutyrate, decanediodate, DL-2-phenylbutyrate) [1]

Cofactor(s)/prosthetic group(s)/activating agents

NAD^+ (four-electron pyridine oxidoreductase [7]) [1–7]; Coenzyme A (remains bound to enzyme throughout the whole catalytic cycle [6]) [1–7]; NADH (reverse reaction) [1–7]; Pantetheine (replaces CoA, 50% as effective) [6]; More (DTT, pantothenate or desulfo-CoA cannot replace CoASH) [6]

Metal compounds/salts

Mg^{2+} (Mycobacterium sp. S4, not Pseudomonas sp. M1) [1]; More (titration with DTNB indicates two SH-groups per subunit, both accessible to DTNB in the presence of mevalonate and/or NAD^+, only one group accessible with hydroxymethylglutaryl-CoA) [6]

Turnover number (min^{-1})

Specific activity (U/mg)

More (specific activities of 4 enzyme mutants under rate limiting conditions) [7]; 0.085 (mevalonate, Mycobacterium sp.) [1]; 0.17 (Mycobacterium sp. S4) [1]; 1.63 (mevalonate, Pseudomonas sp. M1) [1]; 3.26 (NAD^+, Pseudomonas sp. M1) [1]; 7.07 (reduction of mevaldate) [3]; 28.5 (reductive deacylation) [3]; 38.9 (oxidative acylation) [3, 4]; 47.5 [6]; 54.2 (enzyme C156/296A) [7]; 60.5 [5]; 65.6 (enzyme C156A) [7]; 70.5 (wild-type) [7]; 77.7 (enzyme C296A) [7]

K_m-value (mM)

0.03 (coenzyme A, enzyme H381A) [2]; 0.032 (NADH, reductive deacylation of hydroxymethylglutaryl-CoA) [3, 4]; 0.039 (coenzyme A [3], coenzyme A, oxidative acylation of mevalonate [3, 4]) [3, 4]; 0.046 (coenzyme A, oxidative acylation of mevaldate) [3, 4]; 0.05 (coenzyme A, wild-type [2], coenzyme A, oxidative acylation [6], hydroxymethylglutaryl-CoA, reductive deacylation [4]) [2, 4, 6]; 0.060 (NAD^+, oxidative acylation of mevaldate) [3, 4]; 0.1 (coenzyme A) [1]; 0.11 (coenzyme A, enzyme H381Q) [2]; 0.14 (NAD^+, enzyme H381A) [2]; 0.15 (NAD^+, pH 11, wild-type [2], mevalonate [5]) [2, 5]; 0.17 (NAD^+, oxidative acylation of mevaldate, NADH, reductive deacylation of hydroxymethylglutaryl-CoA) [6]; 0.20 (coenzyme A, enzyme H381N) [2]; 0.22 (NAD^+, wild-type [2], mevalonate, oxidative acylation [6]) [2, 6]; 0.24 (coenzyme A, pH 11, wild-type) [2]; 0.27 (R,S-mevalonate, enzyme H381A [2], NAD^+, mevalonate, oxidative acylation of mevalonate [3, 4]) [2–4]; 0.30 (mevalonate, oxidative acylation) [3, 4, 6]; 0.32 (mevalonate, pH 11, wild-type) [2]; 0.35 (NAD^+, DL-mevalonate, Pseudomonas sp. M1) [1]; 0.36 (NADH, reduction of mevaldate) [3, 4]; 0.37 (NAD^+, oxidative acylation of mevalonate) [6]; 0.45 (R,S-mevalonate, wild-type) [2]; 0.49 (mevalonate, Mycobacterium sp. S4) [1]; 0.5 (coenzyme A, pH 11, enzyme H381K) [2]; 0.6 (NAD^+, Mycobacterium sp. S4 [1], coenzyme A, oxidative acylation of mevaldate [6]) [1, 6]; 0.61 (mevaldate, oxidative acylation) [3, 4]; 0.62 (NAD^+, enzyme H381Q) [2]; 0.71 (R,S-mevalonate, enzyme H381Q) [2]; 0.75 (mevalonate, pH 11, enzyme H381K) [2]; 0.81 (NAD^+, enzyme H381N) [2]; 1.23 (NAD^+, pH 11, enzyme H381K) [2]; 1.36 (mevaldate, oxidative acylation) [6]; 3.7 (R,S-mevalonate, enzyme H381M) [2]; 6.0–12.0 (mevaldate, reduction) [4]; 9.0 (mevaldate, reduction) [3]; 36.0 (mevaldate, reduction) [6]

pH-optimum

8.1 [5]; 8.7–9.2 (wild-type) [2]; 9.2–9.6 (mevalonate oxidation [4]) [3, 4]; 9.6 (mevalonate oxidation at enhanced coenzyme A concentration) [4]; 9.8 (H381N and H381Q) [2]; 10.0 (H381A) [2]; 10.8 (H381K) [2]

pH-range

7.3–10.5 (about half-maximal activity at pH 7.3 and 10.5, high coenzyme A concentration enhances activity at pH values above 10.5) [3]

Temperature optimum (°C)

40 [5]

Temperature range (°C)

3 ENZYME STRUCTURE

Molecular weight

178000 (Pseudomonas sp. M1, PAGE) [5]
260000 (Pseudomonas mevalonii, gel electrophoresis) [3]
270000 (Pseudomonas sp. M1, PAGE) [4]

Subunits

Tetramer (4 × 43000, Pseudomonas sp. M1, SDS-PAGE [5], 4 × 45000, Pseudomonas mevalonii, SDS-PAGE [6], 4 × 45538, Pseudomonas mevalonii, amino acid analysis [6]) [5, 6]

Glycoprotein/Lipoprotein

–

4 ISOLATION/PREPARATION

Source organism

Mycobacterium sp. S4 [1]; Pseudomonas mevalonii (formerly: Pseudomonas sp. M1 [1, 4, 5], wild-type and various H381-mutants [2], wild-type and enzyme mutants C156A, C296A and C156/296A (cysteine-free) [7]) [1–8]

Source tissue

Cell [1–8]

Localisation in source

Soluble [1–8]

Purification

Mycobacterium sp. S4 [1]; Pseudomonas mevalonii [1–7]

Crystallization

(Pseudomonas mevalonii) [2]

Cloned

(Pseudomonas plasmid DNA behind the tac promoter and over-expressed in E. coli [2, 6], wild-type and mutant DNA overexpressed in E. coli JM101 [7], mvaA-gene overexpressed in E. coli JM103 [8]) [2, 6–8]

Renaturated

–

5 STABILITY

pH

8.1 (optimal pH for stability) [5]

Temperature (°C)

Oxidation

Organic solvent

General stability information

EDTA stabilizes [4]; DTT stabilizes [4]; Sucrose 300 mM, stabilizes during purification [5]; Mevalonate stabilizes during purification [5]; Phenylmethylsulfonyl fluoride stabilizes during purification [5]; Thawing: purified preparation unstable to [5]; Bovine serum albumin enhances storage stability at 0°C for up to 4 h [5]; KCl, stabilizes [5]; No thiol required for stabilization [6]; Glycerol, 10% v/v, stabilizes [5]

Storage

–180°C, stable for at least a year [6]; –80°C, half-life about 3 months [5]; Frozen, ammonium sulfate precipitate stable for weeks [3]; 0°C, stable for 4 h in phosphate buffer, pH 7.2, with the addition of 0.1% (w/v) bovine serum albumin [5]; 0°C, stable for 10 h in Tris/HCl buffer with 0.4 M KCl, pH 8.1, with 5% loss of activity [5]

6 CROSSREFERENCES TO STRUCTURE DATABANKS

PIR/MIPS code

PIR2:A32107 (Arabidopsis thaliana); PIR2:A23586 (golden hamster); PIR2:A44756 (Pseudomonas sp.); PIR2:B43714 (Pseudomonas sp. (fragment)); PIR2:A30239 (I yeast (Saccharomyces cerevisiae)); PIR2:B30239 (II yeast (Saccharomyces cerevisiae))

Brookhaven code

7 LITERATURE REFERENCES

[1] Fimognari, G.M., Rodwell, V.W.: Biochemistry,4,2086–2090 (1965)
[2] Darnay, B.G., Wang, Y., Rodwell, V.W.: J. Biol. Chem.,267,15064–15070 (1992)
[3] Bensch, W.R., Rodwell, V.W.: J. Biol. Chem.,245,3755–3762 (1970)
[4] Rodwell, V.W., Bensch, W.R.: Methods Enzymol.,71,480–486 (1981)
[5] Gill, J.F., Beach, M.J., Rodwell, V.W.: J. Biol. Chem.,260,9393–9398 (1985)
[6] Jordan-Starck, T.C., Rodwell, V.W.: J. Biol. Chem.,264,17913–17918 (1989)
[7] Jordan-Starck, T.C., Rodwell, V.W.: J. Biol. Chem.,264,17919–17923 (1989)
[8] Beach, M.J., Rodwell, V.W.: J. Bacteriol.,171,2994–3001 (1989)

1 NOMENCLATURE

EC number
1.1.1.90

Systematic name
Aryl-alcohol:NAD^+ oxidoreductase

Recommended name
Aryl-alcohol dehydrogenase

Synonymes
Dehydrogenase, aryl alcohol
Benzyl alcohol dehydrogenase
Coniferyl alcohol dehydrogenase
p-Hydroxybenzyl alcohol dehydrogenase

CAS Reg. No.
37250-26-3

2 REACTION AND SPECIFICITY

Catalysed reaction
Aromatic alcohol + NAD^+ →
→ aromatic aldehyde + NADH

Reaction type
Redox reaction

Natural substrates
Benzyl alcohol + NAD^+ (constitutive enzyme, degradation of benzyl alcohol when it is the sole carbon and energy source) [10]

Substrate spectrum
1 Aromatic alcohol + NAD^+ (r [1, 6]) [1–10]
2 2-Methylbenzyl alcohol + NAD^+ [1–3]
3 4-Methylbenzyl alcohol + NAD^+ [1–3]
4 3-Methylbenzyl alcohol + NAD^+ [1–4]
5 2-Aminobenzyl alcohol + NAD^+ [1]
6 3-Aminobenzyl alcohol + NAD^+ [1]
7 4-Aminobenzyl alcohol + NAD^+ [1]
8 3-Methoxybenzyl alcohol + NAD^+ [1]
9 3-Hydroxybenzyl alcohol + NAD^+ [1, 3]
10 3-Chlorobenzyl alcohol + NAD^+ [1, 3]
11 3-Nitrobenzyl alcohol + NAD^+ [1]

12 3,4-Dimethylbenzyl alcohol + NAD^+ [1]
13 Furfuryl alcohol + NAD^+ [1]
14 3,5-Dichlorobenzyl alcohol + NAD^+ [3]
15 2-Hydroxymethylbenzyl alcohol + NAD^+ [1]
16 4-Ethylbenzyl alcohol + NAD^+ [3]
17 4-Chlorobenzyl alcohol + NAD^+ [3]
18 4-Hydroxybenzyl alcohol + NAD^+ [7]
19 Cinnamyl alcohol + NAD^+ [7]
20 4-Isopropylbenzyl alcohol + NAD^+ [7]
21 4-Hydroxy-3-methoxybenzyl alcohol + NAD^+ [7]
22 Coniferyl alcohol + NAD^+ [7]
23 3,4-Dimethoxybenzyl alcohol + NAD^+ [7]
24 2-Hydroxybenzyl alcohol + NAD^+ [7]
25 Thiophene-2-methanol + NAD^+ [7]
26 2-Methoxybenzyl alcohol + NAD^+ [7]
27 Furan-2-methanol + NAD^+ [7]
28 Perillyl alcohol + NAD^+ [7]
29 Benzyl alcohol + NAD^+ (r [1, 6]) [1–3, 6, 7]
30 More (a group of enzymes with broad specificity towards alcohols with an aromatic or cyclohex-1-ene ring, but with low or no activity towards short chain alipathic alcohols [1], enzyme also capable of oxidizing aliphatic primary alcohols with NAD^+ [11], not: 4-nitrophenyl acetate [1], alcohols that are not oxidized [7], enzyme from Rhodopseudomonas acidophila has broad specificity for aromatic alcohols and is also capable of oxidizing aliphatic alcohols with NAD^+ [11]) [1, 7, 11]

Product spectrum

1 Aromatic aldehyde + NADH (r [1, 6]) [1–10]
2 2-Methylbenzaldehyde + NADH
3 4-Methylbenzaldehyde + NADH
4 3-Methylbenzaldehyde + NADH
5 2-Aminobenzaldehyde + NADH
6 3-Aminobenzaldehyde + NADH
7 4-Aminobenzaldehyde + NADH
8 3-Methoxybenzaldehyde + NADH
9 3-Hydroxybenzaldehyde + NADH
10 3-Chlorobenzaldehyde + NADH
11 3-Nitrobenzaldehyde + NADH
12 3,4-Dimethylbenzaldehyde + NADH
13 Furfural + NADH
14 3,5-Dichlorobenzaldehyde + NADH
15 2-Hydroxymethylbenzaldehyde + NADH
16 4-Ethylbenzaldehyde + NADH
17 4-Chlorobenzaldehyde + NADH
18 4-Hydroxybenzaldehyde + NADH

19 Cinnamic aldehyde + NADH
20 4-Isopropylbenzaldehyde + NADH
21 4-Hydroxy-3-methoxybenzaldehyde + NADH
22 Coniferyl aldehyde + NADH
23 3,4-Dimethoxybenzaldehyde + NADH
24 2-Hydroxybenzaldehyde + NADH
25 2-Formylthiophene + NADH
26 2-Methoxybenzaldehyde + NADH
27 2-Formylfuran + NADH
28 Perillaldehyde + NADH
29 Benzaldehyde + NADH (r [1, 6]) [1–3, 6, 7]
30 ?

Inhibitor(s)

Non-substrate aldehydes (e.g. 2-bromobenzaldehyde, 2-chlorobenzaldehyde, 2-methylbenzaldehyde) [7]; SH-blocking agents [1]; N-Ethylmaleimide [1, 7]; Dithiothreitol (above 2.5 mM) [1]; p-Chloromercuribenzoate [1, 7, 9, 12]; Iodoacetamide (slight) [1, 7]; NADH (noncompetitive) [1]; Iodoacetate [7]; Silver nitrate [12]

Cofactor(s)/prosthetic group(s)/activating agents

NAD^+ (specific for [2, 9], enzyme transfers hydride to the pro-R side of the prochiral C_4 to the pyridine ring of NAD^+ [1]) [1–10]; More (contains no prosthetic group such as FAD or FMN) [1]

Metal compounds/salts

KCl (activation) [1]; $MgCl_2$ (slight activation) [1]

Turnover number (min^{-1})

13680 (benzyl alcohol) [1]; 37320 (benzaldehyde) [1]; 7860 (benzyl alcohol) [7]; 4980 (perillyl alcohol) [7]; 6960 (4-methoxybenzyl alcohol) [7]; 4800 (4-hydroxybenzyl alcohol) [7]; 8460 (cinnamyl alcohol) [7]; 8640 (4-isopropylbenzyl alcohol) [7]; 9360 (3-methoxybenzyl alcohol) [7]

Specific activity (U/mg)

230.3 [1]; 376 [2]; 238 [6]; 20.31–58.7 [9]; More [11]

K_m-value (mM)

More [6, 7, 9, 11]; 0.2194 (benzyl alcohol) [1]; 0.0476 (NAD^+) [1]; 0.0257 (benzaldehyde) [1]; 0.047 (NADH) [1]; 0.219 (NAD^+, Pseudomonas putida, plasmid-encoded enzyme) [2]; 0.233 (benzyl alcohol, Pseudomonas putida, plasmid-encoded enzyme) [2]; 0.605 (2-methylbenzyl alcohol, Pseudomonas putida, plasmid-encoded enzyme) [2]; 0.081 (3-methylbenzyl alcohol, Pseudomonas putida, plasmid-encoded enzyme) [2]; 0.106 (4-methylbenzyl alcohol, plasmid-encoded enzyme) [2]; 0.0406 (NAD^+, Acinetobacter calcoaceticus, chromosomally encoded enzyme) [2]; 0.121 (benzyl alcohol, Acinetobacter calcoaceticus, chromosomally encoded enzyme) [2]; 0.992 (2-methylbenzyl alcohol, Acinetobacter calcoaceticus, chromosomally encoded enzyme) [2]; 0.146 (3-methylbenzyl alcohol, Acinetobacter calcoaceticus, chromosomally encoded enzyme) [2]; 0.118 (4-methylbenzyl alcohol, Acinetobacter calcoaceticus, chromosomally encoded enzyme) [2]; 0.018 (perillyl alcohol) [7]; 0.057 (4-methoxybenzyl alcohol) [7]; 0.046 (4-hydroxybenzyl alcohol) [7]; 0.093 (cinnamyl alcohol) [7]; 0.111 (4-isopropylbenzyl alcohol) [7]; 0.044 (coniferyl alcohol) [7]

pH-optimum

8.9 (reduction of benzaldehyde) [6]; 9.2 (oxidation of benzyl alcohol) [6]; 9.4 [1, 2]; 9.5 [9]

pH-range

7.5–10.5 (7.5: about 15% of activity maximum, 10.5: about 20% of activity maximum) [1]

Temperature optimum (°C)

25 (assay at) [1]; 30 (assay at) [9]

Temperature range (°C)

3 ENZYME STRUCTURE

Molecular weight

27000 (Rhodopseudomonas acidophila, gel filtration) [11]
75000 (Pseudomonas putida, gel filtration, p-cresol grown cells) [9]
122000 (Pseudomonas putida, gel filtration, 3,5-xylenol grown cells, activity 2) [9]
145000 (Pseudomonas putida, gel filtration, 3,5-xylenol grown cells, activity 1) [9]

Subunits

Monomer (Secale cereale) [12]
Dimer (2 × 42000, Pseudomonas putida, SDS-PAGE) [1]
Tetramer (4 × 43000, Pseudomonas putida, SDS-PAGE [2], 4 × 39700, Acinetobacter calcoaceticus, SDS-PAGE, cross-linking experiments in vitro using dimethylsuberimidate [6]) [2, 6]

Glycoprotein/Lipoprotein

–

4 ISOLATION/PREPARATION

Source organism

Pseudomonas putida (enzyme encoded by the TOL plasmid [1–4], 2 distinct NAD^+-dependent m-hydroxybenzyl alcohol dehydrogenases from cells grown on 3,5-xylenol differing in relative rates with the various substrates, another NAD^+-dependent alcohol dehydrogenase, not identical with the former ones from p-cresol-grown cells [9], gene cloned in E. coli [5]) [1–5, 9, 10]; Acinetobacter calcoaceticus (chromosomally encoded) [2, 6, 7]; Pseudomonas alcaligenes (plasmid-encoded) [8]; Rhodopseudomonas acidophila (enzyme also capable of oxidizing aliphatic primary alcohols with NAD^+) [11]; Secale cereale (rye) [12]; Secale montanum [12]; Secale sylvestre [12]; Dasypyrum villosum [12]; Triticum baeoticum (wheat) [12]; Triticum turgidum [12]; Triticale turgidocereale [12]; Elytrigia repens (wheatgrass) [12]

Source tissue

Cell [1, 2, 6]; Seedlings [12]

Localisation in source

Purification

Pseudomonas putida (enzyme encoded by plasmid TOL [1, 2], 2 distinct NAD^+-dependent m-hydroxybenzyl alcohol dehydrogenases from cells grown on 3,5-xylenol, differing in relative rates with the various substrates, another NAD^+-dependent alcohol dehydrogenase, not identical with the former ones from p-cresol-grown cells [9]) [1, 2, 9]; Acinetobacter calcoaceticus [6]; Rhodopseudomonas acidophila (enzyme also capable of oxidizing aliphatic primary alcohols with NAD^+) [11]

Crystallization

–

Cloned

[4, 5]

Renaturated

–

5 STABILITY

pH

Temperature (°C)

37 (21 h, stable) [1]; 56 (10 min, complete loss of activity) [1]

Oxidation

Organic solvent

General stability information

$MgCl_2$ stabilizes slightly [1]; 2-Mercaptoethanol stabilizes [1]; Dithiothreitol stabilizes [1, 9]; Benzyl alcohol stabilizes slightly [1]

Storage

–70°C, enzyme dialyzed against buffer containing 40% glycerol [1]; –20°C, concentrated, mixed with ethanediol [2]

6 CROSSREFERENCES TO STRUCTURE DATABANKS

PIR/MIPS code

PIR2:A46704 (Pseudomonas putida plasmid pWW0)

Brookhaven code

7 LITERATURE REFERENCES

[1] Shaw, J.P., Harayama, S.: Eur. J. Biochem.,191,705–714 (1990)
[2] Chalmers, R.M., Scott, A.J., Fewson, C.A.: J. Gen. Microbiol.,136,637–643 (1990)
[3] Abril, M.-A., Michan, C., Timmis, K.N., Ramos, J.L.: J. Bacteriol.,171,6782–6790 (1989)
[4] Harayama, S., Rekik, M., Wubbolts, M., Rose, K., Leppik, R.A., Timmis, K.N.: J. Bacteriol.,171,5048–5055 (1989)
[5] Inouye, S., Nakazawa, A., Nakazawa, T.: J. Bacteriol.,145,1137–1143 (1981)
[6] MacKintosh, R.W., Fewson, C.A.: Biochem. J.,250,743–751 (1988)
[7] MacKintosh, R.W., Fewson, C.A.: Biochem. J.,255,653–661 (1988)
[8] Poh, C.L., Bayly, R.C.: J. Bacteriol.,143,59–69 (1980)
[9] Keat, M.J., Hopper, D.J.: Biochem. J.,175,659–667 (1978)
[10] Collins, J., Hegeman, G.: Arch. Microbiol.,138,153–160 (1984)
[11] Yamanaka, K., Minoshima, R.: Agric. Biol. Chem.,48,1161–1171 (1984)
[12] Jaaska, V., Jaaska, V.: Biochem. Physiol. Pflanz.,179,21–30 (1984)

1 NOMENCLATURE

EC number
1.1.1.91

Systematic name
Aryl-alcohol:NADP⁺ oxidoreductase

Recommended name
Aryl-alcohol dehydrogenase (NADP⁺)

Synonymes
Dehydrogenase, aryl alcohol (nicotinamide adenine dinucleotide phosphate)
Aryl alcohol dehydrogenase (nicotinamide adenine dinucleotide phosphate)
Coniferyl alcohol dehydrogenase
NADPH-linked benzaldehyde reductase [1]

CAS Reg. No.
37250-27-4

2 REACTION AND SPECIFICITY

Catalysed reaction
An aromatic alcohol + NADP⁺ →
→ an aromatic aldehyde + NADPH

Reaction type
Redox reaction

Natural substrates
Benzyl alcohol + NADP⁺ (enzyme participates in benzoate reduction) [1]

Substrate spectrum
1 Aromatic alcohol + NADP⁺ (r [2, 3]) [1–4]
2 Benzyl alcohol + NADP⁺ (r) [3]
3 Salicyl alcohol + NADP⁺ (r, i.e. p-hydroxybenzyl alcohol) [3]
4 Anisic alcohol + NADP⁺ (r, i.e. p-methoxybenzyl alcohol) [3]
5 Veratryl alcohol + NADP⁺ (r, i.e. 3,4-dimethoxybenzyl alcohol) [3]
6 Cinnamic alcohol + NADP⁺ (r) [3]
7 n-Butanol + NADP⁺ (r) [3]
8 n-Pentanol + NADP⁺ (r) [3]
9 n-Propanol + NADP⁺ (r) [3]
10 Coniferyl alcohol + NADP⁺ (r, i.e. 4-hydroxy-3-methoxycinnamic alcohol) [3]
11 More (broad substrate specificity towards both aromatic and to a lesser extent aliphatic aldehydes [3], low activity with aliphatic aldehydes [2]) [2, 3]

Product spectrum

1 Aromatic aldehyde + NADPH (r [2, 3])
2 Benzaldehyde + NADPH
3 p-Hydroxybenzaldehyde + NADPH
4 p-Methoxybenzaldehyde + NADPH (i.e. p-anisaldehyde)
5 Veratraldehyde + NADPH (i.e. 3,4-dimethoxybenzaldehyde)
6 Cinnamic aldehyde + NADPH
7 n-Butanal + NADPH
8 n-Pentanal + NADPH
9 n-Propanal + NADPH
10 Coniferylaldehyde + NADPH
11 ?

Inhibitor(s)

Iodosobenzoate [3]; p-Chloromercuribenzoate [3, 4]; $AgNO_3$ [4]

Cofactor(s)/prosthetic group(s)/activating agents

$NADP^+$ [2, 3]; NADPH (needs NADPH as cofactor, with NADH the reaction is less than 1% of the rate with NADPH [3]) [2, 3]

Metal compounds/salts

Turnover number (min^{-1})

Specific activity (U/mg)

More [3]

K_m-value (mM)

More [2, 3]; 0.1 (NADPH, aldehyde reduction) [3]; 0.5 (p-methoxybenzaldehyde) [3]; 1.7 (p-methoxybenzaldehyde) [3]; 0.024 ($NADP^+$) [3]; 0.34 (benzaldehyde) [2]; 0.22 (cinnamic aldehyde) [2]; 0.55 (3-phenylpropionaldehyde) [2]; 4.35 (benzyl alcohol) [2]; 0.18 (NADPH) [2]; 0.30 (p-methylbenzaldehyde) [2]; 1.1 (m-hydroxybenzaldehyde) [2]

pH-optimum

6.8 (aldehyde reduction) [3]; 8.8 (alcohol oxidation) [3]

pH-range

Temperature optimum (°C)

25 (assay at) [1]; 30 (assay at) [3]

Temperature range (°C)

3 ENZYME STRUCTURE

Molecular weight

75000 (Neurospora crassa, gel filtration) [3]

Subunits

Dimer (Secale cereale) [4]

Glycoprotein/Lipoprotein

–

4 ISOLATION/PREPARATION

Source organism

Nocardia asteroides [1]; Neurospora crassa [3]; Solanum tuberosum [2]; Brassica napobrassica [2]; Secale cereale (rye) [4]; Secale montanum [4]; Secale sylvestre [4]; Dasypyrum villosum [4]; Triticum baeoticum (wheat) [4]; Triticum turgidum [4]; Triticale turgidocereale [4]; Elytrigia repens (wheatgrass) [4]

Source tissue

Tuber [2]; Mycelium [3]; Seedlings [4]

Localisation in source

Purification

Neurospora crassa [3]; Solanum tuberosum (partial) [2]

Crystallization

–

Cloned

–

Renaturated

–

5 STABILITY

pH

Temperature (°C)

Oxidation

Organic solvent

General stability information

Storage

6 CROSSREFERENCES TO STRUCTURE DATABANKS

PIR/MIPS code

Brookhaven code

7 LITERATURE REFERENCES

[1] Kato, N., Konishi, H., Masuda, M., Joung, E.-H., Shimao, M., Sakazawa, C.: J. Ferment. Bioeng.,69,220–223 (1990)
[2] Davies, D.D., Ugochukwu, E.N., Patil, K.D., Towers, G. H.N.: Phytochemistry,12, 531–536 (1973)
[3] Gross, G.G., Zenk, M.H.: Eur. J. Biochem.,8,420–425 (1969)
[4] Jaaska, V., Jaaska, V.: Biochem. Physiol. Pflanz.,179,21–30 (1984)

1 NOMENCLATURE

EC number
1.1.1.92

Systematic name
D-Glycerate:NAD(P)$^+$ oxidoreductase (carboxylating)

Recommended name
Oxaloglycolate reductase (decarboxylating)

Synonymes

CAS Reg. No.
37250-28-5

2 REACTION AND SPECIFICITY

Catalysed reaction
D-Glycerate + NAD(P)$^+$ + CO_2 →
→ 2-hydroxy-3-oxosuccinate + NAD(P)H

Reaction type
Redox reaction

Natural substrates
2-Hydroxy-3-oxosuccinate + NAD(P)H [1, 2]

Substrate spectrum
1 2-Hydroxy-3-oxosuccinate + NAD(P)H (i.e. oxaloglycolate) [1, 2]
2 Hydroxypyruvate + NAD(P)H (r) [2]
3 Glyoxylate + NAD(P)H [2]

Product spectrum
1 D-Glycerate + NAD(P)$^+$ + CO_2 [1, 2]
2 D-Glycerate + NAD(P)$^+$ [2]
3 Glycolate + NAD(P)$^+$ [2]

Inhibitor(s)

Cofactor(s)/prosthetic group(s)/activating agents
NADH [1, 2]; NADPH [1, 2]; NADP$^+$ [2]; NAD$^+$ [2]

Metal compounds/salts

Turnover number (min^{-1})

Specific activity (U/mg)

2.6 [2]

K_m-value (mM)

1.2 (2-hydroxy-3-oxosuccinate (+ NADH)) [2]; 2.5 (2-hydroxy-3-oxosuccinate, hydroxypyruvate (+ NADPH)) [2]; 1.0 (hydroxypyruvate (+ NADH)) [2]; 3.3 (glyoxylate (+ NADH)) [2]; 8.0 (glyoxylate (+ NADPH)) [2]; 0.14 (NADH) [2]; 0.024 (NADPH) [2]

pH-optimum

6.0 [2]

pH-range

4.4 (not active below) [2]

Temperature optimum (°C)

Temperature range (°C)

3 ENZYME STRUCTURE

Molecular weight

63000 (Pseudomonas putida, sedimentation velocity) [2]

Subunits

Glycoprotein/Lipoprotein

–

4 ISOLATION/PREPARATION

Source organism

Pseudomonas putida [1, 2]

Source tissue

Localisation in source

Purification

Pseudomonas putida [2]

Crystallization

[2]

Cloned

–

Renaturated

–

5 STABILITY

pH

Temperature (°C)

Oxidation

Organic solvent

General stability information

Storage

–15°C, 0.04 M Tris-chloride, 30% glycerol, 5 mM mercaptoethanol, 6 months [2]

6 CROSSREFERENCES TO STRUCTURE DATABANKS

PIR/MIPS code

Brookhaven code

7 LITERATURE REFERENCES

[1] Do Nascimento, K.H., Davies, D.D.: Biochem. J.,149,553–557 (1975)
[2] Kohn, L.D., Jakoby, W.B.: J. Biol. Chem.,243,2486–2493 (1968)

1 NOMENCLATURE

EC number
1.1.1.93

Systematic name
Tartrate:NAD^+ oxidoreductase

Recommended name
Tartrate dehydrogenase

Synonymes
Dehydrogenase, tartrate
Mesotartrate dehydrogenase [4]
More (bifunctional L-(+)-tartrate dehydrogenase/D-(+)-malate dehydrogenase (decarboxylating) (EC 1.1.1.93/EC 1.1.1.83)) [1, 2]

CAS Reg. No.
37250-29-6

2 REACTION AND SPECIFICITY

Catalysed reaction
Tartrate + NAD^+ →
→ oxaloglycolate + NADH

Reaction type
Redox reaction

Natural substrates
L-(+)-Tartrate + NAD^+ (bifunctional L-(+)-tartrate dehydrogenase/D-(+)-malate dehydrogenase (decarboxylating) (EC 1.1.1.93/EC 1.1.1.83), accounts for ability of Rhodobacter sphaeroides to grow on L-(+)-tartrate and D-(+)-malate) [2]

Substrate spectrum
1 Tartrate + NAD^+ (r [3], L-(+)-tartrate [2], not: D-(-)-tartrate [2, 3], mesotartrate [2], L-(+)-tartrate and mesotartrate [3], reduction of oxaloglycolate or its enol-form dihydroxyfumarate is 1% of the rate of meso-tartrate oxidation [3], $NADP^+$ cannot substitute NAD^+ [2], enzyme is A-specific [4]) [1–4]
2 More (not: L-(-)-malate [3], D-(+)-malate [3], D,L-glycerate [3], tartronate [3], 3-hydroxybutyrate, 2-hydroxybutyrate [3], hydroxypyruvate [3], glycolate [3], tartronic semialdehyde [3], citrate [3], glycolaldehyde [3]) [3]

Product spectrum

1 Oxaloglycolate + NADH (r [3], oxaloglycolate is in tautomeric equilibrium with dihydroxyfumarate [2], both meso- and L-(+)-tartrate formed in the reverse direction [3]) [1–4]

2 ?

Inhibitor(s)

ATP (bifunctional L-(+)-tartrate dehydrogenase/D-(+)-malate dehydrogenase (decarboxylating) (EC 1.1.1.93/EC 1.1.1.83)) [2]; meso-Tartrate (competitive) [2]; Oxaloacetate (competitive) [2]; Dihydroxyfumarate (noncompetitive) [2]

Cofactor(s)/prosthetic group(s)/activating agents

NAD^+ [1–4]; NADH [3]

Metal compounds/salts

Mg^{2+} (slight stimulation) [3]; Mn^{2+} (requires Mn^{2+} and one of several monovalent cations: oxidation of meso-tartrate (K^+, Rb^+, Cs^+, NH_4^+ or Na^+), oxidation of L-(+)-tartrate (K^+ or Rb^+) [3], requires Mn^{2+} and NH_4^+ [2], K_m: 0.016 mM (reduction of L-(+)-tartrate, cofactor NAD^+) [2]) [2, 3]; Monovalent cations (requires Mn^{2+} and one of several monovalent cations: oxidation of meso-tartrate (K^+, Rb^+, Cs^+, NH_4^+ or Na^+), oxidation of L-(+)-tartrate (K^+ or Rb^+)) [3]

Turnover number (min^{-1})

Specific activity (U/mg)

1.88 [2]

K_m-value (mM)

2.3 (L-(+)-tartrate) [2]; 0.28 (NAD (+ L-(+)-tartrate)) [2]; 0.83 (mesotartrate) [3]; 55 (L-(+)-tartrate) [3]; 0.09 (NAD^+) [3]; 1.4 (NADH (in presence of 1.5 mM dihydroxyfumarate)) [3]; 11 (dihydroxyfumarate (in presence of 0.19 mM NADH)) [3]

pH-optimum

7 (dihydroxyfumarate + NADH) [3]; 8.4–9.0 (bifunctional L-(+)-tartrate dehydrogenase/D-(+)-malate dehydrogenase (decarboxylating) (EC 1.1.1.93/EC 1.1.1.83)) [2]; 8.5 (meso-tartrate + NAD^+) [3]

pH-range

7.0–10.6 (at pH 7.0 and 10.6: 50% of activity maximum) [2]; 7–9 (at pH 7 and 9: about 50% of activity maximum, mesotartrate + NAD^+) [3]; 6–7.8 (6: about 40% of activity maximum, 7.8: about 80% of activity maximum, dihydroxyfumarate + NADH) [3]

Temperature optimum (°C)

50 [2]

Temperature range (°C)

3 ENZYME STRUCTURE

Molecular weight

162000 (Rhodobacter sphaeroides, ultracentrifugation) [2]
158000 (Rhodobacter sphaeroides, gel filtration) [2]
145000 (Pseudomonas putida, diffusion and sedimentation data) [3]

Subunits

Tetramer (4 × 38500, Rhodobacter sphaeroides, SDS-PAGE [2], 4 × 36800, Pseudomonas putida, sedimentation study after dialysis with guanidine-mercaptoethanol solution [3]) [2, 3]

Glycoprotein/Lipoprotein

–

4 ISOLATION/PREPARATION

Source organism

Rhodobacter sphaeroides (bifunctional L-(+)-tartrate dehydrogenase/D-(+)-malate dehydrogenase (decarboxylating) (EC 1.1.1.93/EC 1.1.1.83)) [1, 2]; Pseudomonas putida [3, 4]

Source tissue

Cell [2, 3]

Localisation in source

Purification

Rhodobacter sphaeroides (bifunctional L-(+)-tartrate dehydrogenase/D-(+)-malate dehydrogenase (decarboxylating) (EC 1.1.1.93/EC 1.1.1.83)) [2]; Pseudomonas putida [3]

Crystallization

(Pseudomonas putida) [3]

Cloned

–

Renaturated

–

5 STABILITY

pH

Temperature (°C)

30 (30 h: 50% loss of activity, 4 h stable) [2]

Oxidation

Organic solvent

General stability information

Repeated freezing and thawing of purified enzyme, 10–40% loss of activity [2]

Storage

–20°C, months, partially purified enzyme [2]; 0°C, 30 h, 1 mg/ml protein concentration [2]

6 CROSSREFERENCES TO STRUCTURE DATABANKS

PIR/MIPS code

Brookhaven code

7 LITERATURE REFERENCES

[1] Ebbighausen, H., Giffhorn, F.: Arch. Microbiol.,138,338–344 (1984)
[2] Giffhorn, F., Kuhn, A.: J. Bacteriol.,155,281–290 (1983)
[3] Kohn, L.D., Packman, P.M., Allen, R.H., Jakoby, W.B.: J. Biol. Chem.,243,2479–2485 (1968)
[4] Do Nascimento, K.H., Davies, D.D.: Biochem. J.,149,553–557 (1975)

1 NOMENCLATURE

EC number
1.1.1.94

Systematic name
sn-Glycerol-3-phosphate:NAD(P)$^+$ 2-oxidoreductase

Recommended name
Glycerol-3-phosphate dehydrogenase (NAD(P)$^+$)

Synonymes
L-Glycerol-3-phosphate:NAD(P) oxidoreductase
Dehydrogenase, glycerol phosphate (nicotinamide adenine dinucleotide (phosphate))
Glycerol phosphate dehydrogenase (nicotinamide adenine dinucleotide (phosphate))
Glycerol 3-phosphate dehydrogenase (NADP)

CAS Reg. No.
37250-30-9

2 REACTION AND SPECIFICITY

Catalysed reaction
Glycerone phosphate + NAD(P)H →
→ sn-glycerol 3-phosphate + NAD(P)$^+$

Reaction type
Redox reaction

Natural substrates
Dihydroxyacetone phosphate + NADPH (likely function is to provide glycerol 3-phosphate to lipid synthesis) [1]

Substrate spectrum
1 Glycerone phosphate + NAD(P)H (r, glycerone phosphate is dihydroxyacetone phosphate) [1]
2 DL-Glyceraldehyde 3-phosphate + NAD(P)H [1]

Product spectrum
1 sn-Glycerol 3-phosphate + NAD(P)$^+$ [1]
2 Glycerol 3-phosphate + NAD(P)$^+$

Inhibitor(s)

Palmitoyl-CoA [1]; L-Glycerol 3-phosphate [1]; Bovine serum albumin [1]; NH_4Cl (100 mM) [1]; NaCl (100 mM) [1]; KCl (100 mM) [1]; K_2HPO_4 (100 mM) [1]; Na_2HPO_4 (100 mM) [1]; $(NH_4)_2SO_4$ (100 mM) [1]; Na_2SO_4 (100 mM) [1]; K_2SO_4 (100 mM) [1]

Cofactor(s)/prosthetic group(s)/activating agents

NADPH (deamino-NADH and NADPH have twice the activity of NADH) [1]; $NADP^+$ [1]; Deamino-NADH (deamino-NADH and NADPH have twice the activity of NADH) [1]; NADH (deamino-NADH and NADPH have twice the activity of NADH) [1]; More (no activity with acetylpyridine-NADH) [1]

Metal compounds/salts

Turnover number (min^{-1})

Specific activity (U/mg)

4.8 (dihydroxyacetone phosphate + NADPH) [1]

K_m-value (mM)

0.170 (dihydroxyacetone phosphate) [1]; 0.210 (glycerol 3-phosphate) [1]; 0.010 (NADPH) [1]

pH-optimum

7.3 [1]

pH-range

6.2–8 (at pH 6.2 and 8.0 about 50% of activity maximum) [1]

Temperature optimum (°C)

25 (assay at) [1]

Temperature range (°C)

---

3 ENZYME STRUCTURE

Molecular weight

Subunits

Glycoprotein/Lipoprotein

–

---

4 ISOLATION/PREPARATION

Source organism

E. coli [1]

Source tissue
Cell [1]

Localisation in source

Purification
E. coli [1]

Crystallization
–

Cloned
–

Renaturated
–

5 STABILITY

pH

Temperature (°C)

Oxidation

Organic solvent

General stability information
Salts, 0.1–1.0 mM, stabilize [1]

Storage

6 CROSSREFERENCES TO STRUCTURE DATABANKS

PIR/MIPS code

Brookhaven code

7 LITERATURE REFERENCES

[1] Kito, M., Pizer, L.I.: J. Biol. Chem.,244,3316–3323 (1969)

1 NOMENCLATURE

EC number
1.1.1.95

Systematic name
3-Phosphoglycerate:NAD^+ 2-oxidoreductase

Recommended name
Phosphoglycerate dehydrogenase

Synonymes
D-3-Phosphoglycerate:NAD oxidoreductase [2]
Dehydrogenase, phosphoglycerate
alpha-Phosphoglycerate dehydrogenase
3-Phosphoglycerate dehydrogenase
3-Phosphoglyceric acid dehydrogenase
D-3-Phosphoglycerate dehydrogenase
Glycerate 3-phosphate dehydrogenase
Glycerate-1,3-phosphate dehydrogenase
Phosphoglycerate oxidoreductase
Phosphoglyceric acid dehydrogenase

CAS Reg. No.
9075-29-0

2 REACTION AND SPECIFICITY

Catalysed reaction
3-Phosphoglycerate + NAD^+ →
→ 3-phosphohydroxypyruvate + NADH

Reaction type
Redox reaction

Natural substrates
3-Phosphoglycerate + NAD^+ (first enzyme in metabolic sequence of synthesis of serine from 3-phosphoglycerate) [3, 14, 16]

Substrate spectrum
1 3-Phosphoglycerate + NAD^+ (r [1–20], reduction of hydroxypyruvate-phosphate is faster than oxidation of phosphoglycerate [1]) [1–20]

Product spectrum
1 3-Phosphohydroxypyruvate + NADH [1–20]

Inhibitor(s)

N-Alkylmaleimides (N-ethyl-, N-butyl-, N-pentyl-, N-heptyl-, N-phenyl-) [17]; Pyridoxal 5'-phosphate [18]; Pyridoxamine 5'-phosphate [18]; L-Alanine [1]; L-Threonine (slight) [1]; L-Homoserine (slight) [1]; p-Hydroxymercuribenzoate [3, 12]; Iodoacetate [1]; N-Ethylmaleimide [1]; NAD^+ (product inhibition, competitive to NADH) [16]; Phosphoglycerate (noncompetitive to phosphohydroxypyruvate) [16]; ATP (free ATP is more effective than the magnesium complex [12]) [1, 12]; GTP [1]; ADP (free ADP is more effective than the magnesium complex [12]) [1, 12]: GDP [1]; Sulfhydryl reagents [5]; Mercurials [6]; Ag^+ [6]: L-Serine (0.001–0.010 mM [1], inhibition of enzyme from E. coli, Salmonella typhimurium and Haemophilus influenzae, not of mammalian enzyme [6], not [16]) [1, 3, 6, 11–13]; NADH (inhibition of phosphoglycerate oxidation) [1]; Glycine [1, 11]; L-Allothreonine [1]; Amino acids (exept serine at concentrations of 1 mM and higher) [1, 6]

Cofactor(s)/prosthetic group(s)/activating agents

$NADP^+$ (ineffective [2–5], 8% of the activity with NAD^+ [6]) [6]; NAD^+ [1–20]; NADH (A-stereospecific for NADH [3, 10]) [1–20]; Deamino-NAD^+ (220% [4], 81% of the activity with NAD^+ [5]) [4, 5, 12]; 3-Acetylpyridine-NAD^+ (125% [4], 199% of the activity with NAD^+ [5]) [4, 5, 12]; Thio-NAD^+ (17.5% [4], 66% [5] of the activity with NAD^+) [4, 5]; Deamino-NADH (as effective as NADH) [6]; 3-Acetylpyridine-NADH (40% as effective as NADH [6]) [6, 12]

Metal compounds/salts

NaCl (stimulates) [2]; $MgCl_2$ (stimulates) [16]; Cl^- (stimulates) [16]; SO_4^{2-} (stimulates) [16]

Turnover number (min^{-1})

Specific activity (U/mg)

6.65 [1, 6]; 11.8 [3]; 15.6 [13]; 21.3 [16]; 81.3 [15]; More [4, 5]

K_m-value (mM)

40.2 (phosphohydroxypyruvate) [5]; 12.0 (NADH) [5]; 7.8 (NAD^+) [6]; 1.35 (3-phosphoglycerate) [5]; 1.3 (hydroxypyruvate phosphate) [6]; 1.1 (phosphoglycerate) [1, 6]; 0.5 (NAD^+) [5]; 0.125 (NAD^+) [3]; 0.07 (hydroxypyruvate phosphate, 37°C) [13]; 0.05 (3-phosphoglycerate, phosphohydroxypyruvate) [3]; 0.045 (hydroxypyruvate phosphate, 25°C) [13]; 0.010 (NADH) [3]; 0.008 (NAD^+) [1, 6]; 0.0078 (NAD^+) [12]; 0.0054 (3-acetylpyridine-NAD^+, 37°C) [12]; 0.0023 (NADH, 25°C) [13]; 0.005 (NADH, 37°C) [13]; 0.0005 (NADH, 25°C) [12]; 0.00025 (3-acetylpyridine-NAD^+, 25°C) [12]; 0.00016 (3-acetylpyridine-NADH, 25°C) [12]; 0.00013 (hydroxypyruvate phosphate, 25°C) [12]; More [12, 16]

pH-optimum

5.5 (NADH oxidation) [3]; 6.1 (NADH oxidation) [16]; 8.5 (initial rate maximum) [1, 6]; 9.4 (NAD^+ reduction) [16]; 9.5 (NAD^+ reduction) [3]

pH-range

8.5–10.3 (about 50% of activity maximum at pH 8.5 and 10.3, NAD^+ reduction) [16]; 5–7.8 (about 50% of activity maximum at pH 5 and 7.8, NADH oxidation) [16]

Temperature optimum (°C)

25 (assay at) [1, 2, 5, 6, 13]

Temperature range (°C)

3 ENZYME STRUCTURE

Molecular weight

163000 (E. coli) [6]
165000 (chicken, sedimentation equilibrium analysis) [20]
166000 (Bacillus subtilis, gel filtration) [13]
247000 (rabbit, HPLC gel filtration) [15]
250000 (Glycine max, gel filtration) [16]

Subunits

Tetramer (4 × 40000–50000, E. coli, sedimentation equilibrium in guanidinium-HCl [6], 4 × 60000, rabbit, SDS-PAGE [15], 4 × 40000–43000, chicken, SDS-PAGE, sedimentation equilibrium studies in guanidine hydrochloride [20]) [6, 15, 20]

Glycoprotein/Lipoprotein

–

4 ISOLATION/PREPARATION

Source organism

More (overview: occurence in animal tissues) [2]; Neurospora crassa [7]; Salmonella typhimurium [8]; Haemophilus influenzae [9]; Bacillus subtilis [13]; Glycine max [16]; Rabbit [2, 15]; Bovine [2]; Frog [2]; E. coli [1, 6, 10, 12, 14]; Chicken [2, 17, 19, 20]; Pigeon [2]; Pig [2, 4]; Dog [2]; Pisum sativum [3, 12, 18]; Wheat [5]; Rat [6, 11]

Source tissue

Adipocyte [4]; Ovary [4]; Prostate [4]; Testis [4]; Etiolated epicotyl [12]; Brain [2, 4, 11]; Liver [2, 4, 15, 17, 19, 20]; Kidney [2, 4]; Seedlings [3]; Spinal cord [4]; Adrenals [4]; Germ [5]; Nodules [16]

Localisation in source

Purification

E. coli [1, 6]; Glycine max [16]; Chicken (partial [2]) [2, 19]; Pisum sativum [3]; Pig [4]; Wheat [5]; Bacillus subtilis [13]; Rabbit [15]

Crystallization
[1, 6, 17]

Cloned
[14]

Renaturated
–

5 STABILITY

pH
5.5 (stable) [1]; 7.5 (stable) [1]; 8.5 (unstable) [1]; 9. 5 (unstable) [1]

Temperature (°C)

Oxidation

Organic solvent

General stability information
Dilution inactivates [5, 16]; Dithiothreitol increases stability [5, 6]; Inorganic ions essential for stability [16]

Storage
2°C, 7 days, 40% loss of activity [3]; –15°C, stable for several weeks [4]; –80°C, 20 mM sodium phosphate buffer, pH 7.2 [15]

6 CROSSREFERENCES TO STRUCTURE DATABANKS

PIR/MIPS code
PIR1:DEECPG (Escherichia coli); PIR2:B38156 (Escherichia coli (fragment)); PIR3:S22096 (Escherichia coli (fragment))

Brookhaven code

7 LITERATURE REFERENCES

[1] Sugimoto, E., Pizer, L.I.: J. Biol. Chem.,243,2081–2089 (1968)
[2] Willis, J.E., Sallach, H.J.: Biochim. Biophys. Acta,81,39–54 (1964)
[3] Slaughter, J.C., Davies, D.D.: Methods Enzymol.,41,278–281 (1975) (Review)
[4] Feld, R.D., Sallach, H.J.: Methods Enzymol.,41,282–285 (1975) (Review)
[5] Rosenblum, I.Y., Sallach, H.J.: Methods Enzymol.,41,285–289 (1975) (Review)
[6] Pizer, L.I., Sugimoto, E.: Methods Enzymol.,17B,325–331 (1971) (Review)
[7] Sojka, G.A., Garner, H.R.: Biochim. Biophys. Acta,148,42–47 (1967)
[8] Umbarger, H.E., Umbarger, M.A.: Biochim. Biophys. Acta,62,193–195 (1962)
[9] Pizer, L.I., Ponce-de-Leon, M., Michalka, J.: J. Bacteriol.,97,1357–1361 (1969)
[10] Winicov, I.: Biochim. Biophys. Acta,397,288–293 (1975)
[11] Uhr, M.L., Sneddon, M.K.: FEBS Lett.,17,137–140 (1971)
[12] Slaughter, J.C.: Biochem. J.,135,563–565 (1973)
[13] Saski, R., Pizer, L.I.: Eur. J. Biochem.,51,415–427 (1975)
[14] Tobey, K.L., Grant, G.A.: J. Biol. Chem.,261,12179–12183 (1986)
[15] Lund, K., Merrill, D.K., Guynn, R.W.: Biochem. J.,238,919–922 (1986)
[16] Boland, M.J., Schubert, K.R.: Plant Physiol.,71,658–661 (1983)
[17] Anderson, B.M., Dubler, R.E.: Arch. Biochem. Biophys.,200,583–589 (1980)
[18] Slaughter, J.C.: Biochem. Soc. Trans.,3,1058–1061 (1975)
[19] Grant, G.A., Keefer, L.M., Bradshaw, R.A.: J. Biol. Chem.,253,2724–2726 (1978)
[20] Grant, G.A., Bradshaw, R.A.: J. Biol. Chem.,253,2727–2731 (1978)

1 NOMENCLATURE

EC number
1.1.1.96

Systematic name
3-(3,5-Diiodo-4-hydroxyphenyl)lactate:NAD$^+$ oxidoreductase

Recommended name
Diiodophenylpyruvate reductase

Synonymes
Reductase, diiodophenylpyruvate
Aromatic alpha-keto acid reductase [1]
KAR [1]
2-Oxo acid reductase [3]

CAS Reg. No.
37250-31-0

2 REACTION AND SPECIFICITY

Catalysed reaction
3-(3,5-Diiodo-4-hydroxyphenyl)pyruvate + NADH →
→ 3-(3,5-diiodo-4-hydroxyphenyl)lactate + NAD$^+$

Reaction type
Redox reaction

Natural substrates

Substrate spectrum
1 3-(3,5-Diiodo-4-hydroxyphenyl)pyruvate + NADH [2, 3]
2 p-Hydroxyphenylpyruvate + NADH (low activity) [2]
3 Phenylpyruvate + NADH (low activity [2]) [2, 3]
4 3,5-Dibromo-p-hydroxyphenylpyruvate + NADH [2]
5 3-Iodo-4-hydroxyphenylpyruvate + NADH [2]
6 Oxaloacetate + NADH [3]
7 Indolepyruvate + NADH [3]
8 More (substrates contain an aromatic ring with a pyruvate side chain, the most active substrates are halogenated derivatives, compounds with hydroxyl or amino groups in the 3 or 5 position are inactive)

Product spectrum

1 3-(3,5-Diiodo-4-hydroxyphenyl)lactate + NAD^+
2 p-Hydroxyphenyllactate + NAD^+
3 Phenyllactate + NAD^+
4 3,5-Dibromo-4-hydroxyphenyllactate + NAD^+
5 3-Iodo-4-hydroxyphenyllactate + NAD^+
6 Malate + NAD^+
7 Indolelactate + NAD^+
8 ?

Inhibitor(s)

3,5-Diiodotyrosine (slight) [2]; 3,5-Diiodo-4-hydroxyphenyllactic acid (slight) [2]; 2,5-Dihydroxybenzoic acid [2]; 2,5-Dihydroxyphenylacetic acid (slight) [2]

Cofactor(s)/prosthetic group(s)/activating agents

NADH [1–3]; NADPH (15% of the activity with NADH [2], no activity [1, 3]) [2]

Metal compounds/salts

Turnover number (min^{-1})

Specific activity (U/mg)

41.75 [2]; More [3]

K_m-value (mM)

0.043 (3-(3,5-diiodo-4-hydroxyphenyl)pyruvate) [2]; 0.052 (NADH) [2]; 1.9 (4-hydroxyphenylpyruvate) [2]; 4.8 (phenylpyruvate) [2]; 0.14 (oxaloacetate (+ 3-(3,5-diiodo-4-hydroxyphenyl)pyruvate)) [3]; 0.04 (3-(3,5-diiodo-4-hydroxyphenyl)pyruvate) [3]; 0.2 (indolepyruvate) [3]; 1.6 (phenylpyruvate) [3]

pH-optimum

6.0–7.0 (rapid decrease of activity outside the optimum range) [2]

pH-range

Temperature optimum (°C)

Temperature range (°C)

---

3 ENZYME STRUCTURE

Molecular weight

75000–80000 (dog, gel filtration, sucrose-density gradient centrifugation) [3]

Subunits

Dimer (human [1], 2 × 40000, dog, SDS-PAGE [3]) [1, 3]

Glycoprotein/Lipoprotein

–

4 ISOLATION/PREPARATION

Source organism
Human [1]; Bovine [2]; Dog [2, 3]

Source tissue
Red cell lysates [1]; Fibroblasts [1]; Spleen [1]; Muscle [1, 2]; Lung [1]; Kidney [1–3]; Heart [1–3]; Adrenal [1, 2]; Brain [1, 3]; Liver [1–3]; Placenta [1]; Thyroid [2]; Blood [2]

Localisation in source
Cytosol (strictly localized in) [3]

Purification
Dog [2, 3]

Crystallization
–

Cloned
–

Renaturated
–

5 STABILITY

pH

Temperature (°C)
62 (2 min, no complete inactivation) [2]; 72 (2 min, complete inactivation) [2]

Oxidation

Organic solvent

General stability information
No inactivation by repeated freezing and thawing [2]

Storage
5°C, maintains over 80% of the activity for at least 1 month, crude and purified preparations [2]; –20°C, purified enzyme, for at least 2 months without loss of activity [3]; 4°C, 2 weeks, little or no loss of activity [3]

6 CROSSREFERENCES TO STRUCTURE DATABANKS

PIR/MIPS code

Brookhaven code

7 LITERATURE REFERENCES

[1] Donald, L.J.: Ann. Hum. Genet.,46,299–306 (1982)
[2] Zannoni, V.G., Weber, W.W.: J. Biol. Chem.,241,1340–1344 (1966)
[3] Takada, Y., Noguchi, T., Kido, R.: Life Sci.,20,609–616 (1977)

1 NOMENCLATURE

EC number
1.1.1.97

Systematic name
3-Hydroxybenzyl-alcohol:NADP$^+$ oxidoreductase

Recommended name
3-Hydroxybenzyl-alcohol dehydrogenase

Synonymes
Dehydrogenase, 3-hydroxybenzyl alcohol
m-Hydroxybenzyl alcohol dehydrogenase
m-Hydroxybenzyl alcohol (NADP) dehydrogenase [2]
m-Hydroxybenzylalcohol dehydrogenase [3]

CAS Reg. No.
9075-73-4

2 REACTION AND SPECIFICITY

Catalysed reaction
3-Hydroxybenzyl alcohol + NADP$^+$ →
→ 3-hydroxybenzaldehyde + NADPH

Reaction type
Redox reaction

Natural substrates
3-Hydroxybenzyl alcohol + NADP$^+$ (rate-limiting step of patulin biosynthesis, key enzyme in conversion of 6-methyl-salicylic acid to patulin [2]) [1, 2]

Substrate spectrum
1 3-Hydroxybenzyl alcohol + NADP$^+$ (r, reduction of 3-hydroxybenzaldehyde is favored) [1, 2]
2 3-Methoxybenzaldehyde + NADPH [1, 2]
3 Acetaldehyde + NADPH (29% of the activity with 3-hydroxybenzyl alcohol) [1, 2]
4 Benzaldehyde + NADPH (20% of the activity with 3-hydroxybenzaldehyde) [1, 2]
5 4-Hydroxybenzaldehyde + NADPH (10% of the activity with 3-hydroxybenzyl alcohol) [1, 2]

Product spectrum

1 3-Hydroxybenzaldehyde + NADPH [1, 2]
2 3-Methoxybenzyl alcohol + $NADP^+$
3 Ethanol + $NADP^+$
4 Benzyl alcohol + $NADP^+$
5 4-Hydroxybenzyl alcohol + $NADP^+$

Inhibitor(s)

Iodoacetate [1, 2]; Diethyldicarbonate [1, 2]; Acetaldehyde (slight) [2]; 4-Hydroxybenzaldehyde [2]; 4-Hydroxyacetophenone [2]; Gentisate [2]

Cofactor(s)/prosthetic group(s)/activating agents

$NADP^+$ (specific for [1]) [1–3]; NADPH [1, 2]

Metal compounds/salts

Turnover number (min^{-1})

Specific activity (U/mg)

0.655 [1, 2]; 522 [3]

K_m-value (mM)

pH-optimum

7.6 [1, 2]

pH-range

6–9 (6: about 15% of activity maximum, 9: about 25% of activity maximum) [2]

Temperature optimum (°C)

30 (assay at) [1, 3]

Temperature range (°C)

3 ENZYME STRUCTURE

Molecular weight

120000 (Penicillium urticae, gel filtration) [1, 2]

Subunits

More (only one major band by SDS-PAGE) [1, 2]

Glycoprotein/Lipoprotein

–

4 ISOLATION/PREPARATION

Source organism

Penicillium urticae [1–3]; Pseudomonas putida [4]; Penicillium patulinum [2]; More (occurence is restricted to closely related fungi which possess all or part of the patulin biosynthetic pathway, no constitutive enzyme) [2]

Source tissue

Cells [1, 2]

Localisation in source

Intracellular [1, 2]; Soluble [1, 2]

Purification

Penicillium urticae (partial [1, 2]) [1–3]

Crystallization

–

Cloned

–

Renaturated

–

5 STABILITY

pH

Temperature (°C)

30 (buffer increases half-life from 14 to over 800 min) [3]

Oxidation

Organic solvent

General stability information

Bovine serum albumin and glycerol enhance in vitro half-life at 30°C by 32 min [3]; $MnCl_2$, 2 mM, enhances in vitro half-life at 30°C by 77 min [3]; $FeCl_3$, 2 mM, enhances in vitro half-life at 30°C by 80 min [3]; $CaCl_2$, 2 mM, enhances in vitro half-life at 30°C by 20 min [3]; $MoCl_2$, 2 mM, enhances in vitro half-life at 30°C by 18 min [3]; NADPH, enhances in vitro half-life at 30°C [3]; Dithiothreitol stabilizes [3]

Storage

6 CROSSREFERENCES TO STRUCTURE DATABANKS

PIR/MIPS code

Brookhaven code

7 LITERATURE REFERENCES

[1] Gaucher, G.M.: Methods Enzymol.,43,540–548 (1975) (Review)
[2] Forrester, P.I., Gaucher, G.M.: Biochemistry,11,1108–1114 (1972)
[3] Scott, R.E., Lam, K.S., Gaucher, G.M.: Can. J. Microbiol.,32,167–175 (1986)
[4] Hopper, D.J., Kemp, P.D.: J. Bacteriol.,142,21–26 (1980)

1 NOMENCLATURE

EC number
1.1.1.98

Systematic name
(R)-2-Hydroxystearate:NAD^+ oxidoreductase

Recommended name
(R)-2-Hydroxy-fatty-acid dehydrogenase

Synonymes
Dehydrogenase, D-2-hydroxy fatty acid
D-2-Hydroxy fatty acid dehydrogenase
2-Hydroxy fatty acid oxidase [1]

CAS Reg. No.
37250-32-1

2 REACTION AND SPECIFICITY

Catalysed reaction
(R)-2-Hydroxystearate + NAD^+ →
→ 2-oxostearate + NADH

Reaction type
Redox reaction

Natural substrates
D-2-Hydroxystearate + NAD^+ [1]

Substrate spectrum
1 D-2-Hydroxystearate + NAD^+ [1]

Product spectrum
1 2-Oxostearate + NADH [1]

Inhibitor(s)

Cofactor(s)/prosthetic group(s)/activating agents
NAD^+

Metal compounds/salts

Turnover number (min^{-1})

Specific activity (U/mg)

K_m-value (mM)

pH-optimum
7.5 (assay at) [1]

pH-range

Temperature optimum (°C)
37 (assay at) [1]

Temperature range (°C)

---

3 ENZYME STRUCTURE

Molecular weight

Subunits

Glycoprotein/Lipoprotein
–

---

4 ISOLATION/PREPARATION

Source organism
Rat [1]

Source tissue
Kidney [1]

Localisation in source
Cytoplasm [1]; Microsomes [1]

Purification
Rat (partial) [1]

Crystallization
–

Cloned
–

Renaturated
–

5 STABILITY

pH

Temperature (°C)

Oxidation

Organic solvent

General stability information

Storage

6 CROSSREFERENCES TO STRUCTURE DATABANKS

PIR/MIPS code

Brookhaven code

7 LITERATURE REFERENCES

[1] Levis, G.M.: Biochem. Biophys. Res. Commun.,38,470–477 (1970)

1 NOMENCLATURE

EC number
1.1.1.99

Systematic name
(S)-2-Hydroxystearate:NAD^+ oxidoreductase

Recommended name
(S)-2-Hydroxy-fatty-acid dehydrogenase

Synonymes
Dehydrogenase, L-2-hydroxy fatty acid
L-2-Hydroxy fatty acid dehydrogenase
2-Hydroxy fatty acid oxidase [1]

CAS Reg. No.
37250-33-2

2 REACTION AND SPECIFICITY

Catalysed reaction
(S)-2-Hydroxystearate + NAD^+ →
→ 2-oxostearate + NADH

Reaction type
Redox reaction

Natural substrates
L-2-Hydroxystearate + NAD^+ (may participate in the alpha-oxidation system involved in the degradation of normal and/or branched long-chain fatty acids) [1]

Substrate spectrum
1 L-2-Hydroxystearate + NAD^+ [1]

Product spectrum
1 2-Oxostearate + NADH [1]

Inhibitor(s)

Cofactor(s)/prosthetic group(s)/activating agents
NAD^+

Metal compounds/salts

Turnover number (min^{-1})

Specific activity (U/mg)

K_m-value (mM)

pH-optimum
7.5 (assay at) [1]

pH-range

Temperature optimum (°C)
37 (assay at) [1]

Temperature range (°C)

3 ENZYME STRUCTURE

Molecular weight

Subunits

Glycoprotein/Lipoprotein
–

4 ISOLATION/PREPARATION

Source organism
Rat [1]

Source tissue
Kidney [1]

Localisation in source
Cytoplasm [1]; Microsomes [1]

Purification
Rat (partial) [1]

Crystallization
–

Cloned
–

Renaturated
–

5 STABILITY

pH

Temperature (°C)

Oxidation

Organic solvent

General stability information

Storage

6 CROSSREFERENCES TO STRUCTURE DATABANKS

PIR/MIPS code

Brookhaven code

7 LITERATURE REFERENCES

[1] Levis, G.M.: Biochem. Biophys. Res. Commun.,38,470–477 (1970)

1 NOMENCLATURE

EC number

1.1.1.100

Systematic name

(3R)-3-Hydroxyacyl-[acyl-carrier-protein]:$NADP^+$ oxidoreductase

Recommended name

3-Oxoacyl-[acyl-carrier-protein] reductase

Synonymes

Reductase, 3-oxoacyl-[acyl carrier protein]
beta-Ketoacyl-[acyl-carrier protein](ACP) reductase [1]
beta-Ketoacyl acyl carrier protein (ACP) reductase [5]
beta-Ketoacyl reductase
beta-Ketoacyl thioester reductase
beta-Ketoacyl-ACP reductase
beta-Ketoacyl-acyl carrier protein reductase
3-Ketoacyl acyl carrier protein reductase
NADPH-specific 3-oxoacyl-[acylcarrier protein]reductase
3-Oxoacyl-[ACP]reductase [6]

CAS Reg. No.

37250-34-3

2 REACTION AND SPECIFICITY

Catalysed reaction

3-Oxoacyl-[acyl-carrier-protein] + NADPH →
→ (3R)-3-hydroxyacyl-[acyl-carrier-protein] + $NADP^+$

Reaction type

Redox reaction

Natural substrates

3-Oxoacyl-[acyl-carrier-protein] + NADPH (enzyme in fatty acid synthesis [2, 10], first reduction step of fatty acid synthase [3, 4]) [2–4, 10]

Substrate spectrum

1 3-Oxoacyl-[acyl-carrier-protein] + NADPH (r [1], pH 6.0–7.0, equilibrium almost completely favors formation of the beta-hydroxyacyl ACP derivatives [5]) [1–11]
2 Acetoacetyl-[acyl-carrier-protein] + NADPH (r [1], much greater specificity for acetoacetyl-ACP than for either of the model substrates acetoacetyl-CoA or acetoacetyl-N-acetylcysteamine [3]) [1, 3]
3 Acetoacetyl-CoA + NADPH [1, 3]
4 Acetoacetyl-N-cysteamine + NADPH [3, 6, 7]
5 More (beta-ketoacyl thioesters of CoA and pantetheine metabolized at slower rates [4], only D(-) isomer is oxidized [4], broad specificity for chain length of substrates [4, 5], equally active on beta-ketoacyl-ACP derivatives of C_4 to C_{16} [4], animals, yeast and Mycobacteria have fatty acid synthetase containing all the individual activities on one or two multifunctional polypeptide chains, plants and most bacterial systems (including E. coli) possess individual monofunctional enzymes and a separate acyl carrier protein [3]) [3–5]

Product spectrum

1 (3R)-3-Hydroxyacyl-[acyl-carrier-protein] + $NADP^+$ (r [1], product is the D(-)-stereoisomer [5]) [1, 5]
2 D-beta-Hydroxybutyryl-[acyl-carrier-protein] + $NADP^+$
3 ?
4 ?
5 ?

Inhibitor(s)

Phenylglyoxal [7]; p-Chloromercuribenzoate [1]; More (not: N-ethylmaleimide, arsenite) [1]

Cofactor(s)/prosthetic group(s)/activating agents

NADPH (absolute specificity for NADPH [3, 5], much more effective than NADH [1, 10]) [1–11]; NADH (NADPH is much more effective than NADH) [1, 10]; $NADP^+$ [1]

Metal compounds/salts

Turnover number (min^{-1})

Specific activity (U/mg)

7.95 [1]; More [6, 7]

K_m-value (mM)

0.003 (acetoacetyl-[acyl-carrier-protein]) [6]; 0.0037 (acetoacetyl-[acyl-carrier-protein]) [1]; 0.0079 (acetoacetyl-[acyl-carrier-protein]) [7]; 0.009 (acetoacetyl-[acyl-carrier-protein]) [2]; 0.0093 (NADPH) [7]; 0.016 (NADPH) [2]; 0.023 (NADPH (+ acetoacetyl-[acyl-carrier-protein])) [6]; 0.025 (NADPH) [1]; 0.035 (N-acetoacetyl-N-cysteamine) [6]; 0.250 (acetoacetyl-CoA) [1]; 0.261 (acetoacetyl-CoA) [6]; 0.470 (acetoacetyl-CoA) [7]; 48.0 (acetoacetyl-N-acetylcysteamine) [7]; More [10, 11]

pH-optimum

5.7–6.2 [10]; 6.0–7.0 [5]; 6.5 [11]

pH-range

5.5–8.5 (about 50% of activity maximum at pH 5.5 and 8.5) [11]

Temperature optimum (°C)

Temperature range (°C)

3 ENZYME STRUCTURE

Molecular weight

40000–46000 (Persea americana, gel filtration, sucrose density gradient centrifugation) [11]
44000 (Euglena gracilis, gel filtration) [9]
97000 (Spinacia oleracea, gel filtration) [1]
120000 (Brassica napus, gel filtration) [6]
130000 (Persea americana, gel filtration) [7]

Subunits

Tetramer (4 × 24200, Spinacia oloracea, SDS-PAGE [1], 4 × 20000–30000, Brassica napus, SDS-PAGE [3, 6], 4 × 28000, Persea americana, SDS-PAGE [7]) [1, 3, 6, 7]
More (animals, yeast and Mycobacteria have fatty acid synthetases containing all the individual activities on one or two multifunctional polypeptide chains, plants and most bacterial systems (including E. coli) possess individual monofunctional enzymes and a separate acyl carrier protein) [3]

Glycoprotein/Lipoprotein

–

4 ISOLATION/PREPARATION

Source organism

E. coli [4, 5]; Spinacia oleracea [1, 8, 10]; Cuphea lutea [8]; Pisum sativum [8]; Brassica juncea (rapeseed) [8]; Carthamus tinctorius (safflower) [2, 8]; Persea americana (avocado [3]) [3, 7, 8, 11]; Brassica napus (oilseed [3]) [3, 6]; Euglena gracilis [9]

Source tissue

Leaves [1, 8]; Cell [5]; Seed [2, 8]; Rape [3, 6]; Fruit mesocarp [3, 7, 11]

Localisation in source

Plastids [6, 7, 11]

Purification

E. coli [5]; Spinacia oleracea [1]; Persea americana [3, 7, 11]; Brassica napus [3, 6]

Crystallization

–

Cloned

–

Renaturated

–

5 STABILITY

pH

Temperature (°C)

Oxidation

Organic solvent

General stability information

Inactivation by dilution can be partly prevented by inclusion of NADPH [6, 7]

Storage

–20°C, stable for 2 months [1]; 4°C, 100 mM potassium phosphate, 52% loss of activity after 24 h [6]; 27°C, 100 mM potassium phosphate buffer, 13% loss of activity after 24 h [6]; 27°C, 2 M NaCl, 24 h, stable [7]

6 CROSSREFERENCES TO STRUCTURE DATABANKS

PIR/MIPS code

PIR3:S22416 (Arabidopsis thaliana); PIR3:S13055 (avocado); PIR2:S19832 (Cuphea lanceolata); PIR3:S21742 (rape)

Brookhaven code

7 LITERATURE REFERENCES

[1] Shimakata, T., Stumpf, P.K.: Arch. Biochem. Biophys.,218,77–91 (1982)
[2] Shimakata, T., Stumpf, P.K.: Arch. Biochem. Biophys.,217,144–154 (1982)
[3] Sheldon, P.S., Safford, R., Slabas, A.R., Kekwick, R. G.O.: Biochem. Soc. Trans.,16,392–393 (1988) (Review)
[4] Volpe, J.J., Vagelos, P.R.: Annu. Rev. Biochem.,42,21–60 (1973) (Review)
[5] Toomey, R.E., Wakil, S.J.: Biochim. Biophys. Acta,116,189–197 (1966)
[6] Sheldon, P.S., Kekwick, R.G.O., Smith, C.G., Sidebottom, C., Slabas, A.R.: Biochim. Biophys. Acta,1130,151–159 (1992)
[7] Sheldon, P.S., Kekwick, R.G.O., Sidebottom, C., Smith, C.G., Slabas, A.R.: Biochem. J.,271,713–720 (1990)
[8] Shimakata, T., Stumpf, P.K.: J. Biol. Chem.,258,3592–3598 (1983)
[9] Hendren, R.W., Bloch, K.: J. Biol. Chem.,255,1504–1508 (1980)
[10] Shimakata, T., Stumpf, P.K.: Plant Physiol.,69,1257–1262 (1982)
[11] Gaughey, I., Kekwick, R.G.O.: Eur. J. Biochem.,123,553–561 (1982)

1 NOMENCLATURE

EC number
1.1.1.101

Systematic name
1-Palmitoylglycerol-3-phosphate:$NADP^+$ oxidoreductase

Recommended name
Acylglycerone-phosphate reductase

Synonymes
Reductase, palmitoyl dihydroxyacetone phosphate
Palmitoyl-dihydroxyacetone-phosphate reductase
Acyldihydroxyacetone phosphate reductase
Palmitoyl dihydroxyacetone phosphate reductase

CAS Reg. No.
37250-35-4

2 REACTION AND SPECIFICITY

Catalysed reaction
1-Palmitoylglycerol 3-phosphate + $NADP^+$ →
→ palmitoylglycerone phosphate + NADPH

Reaction type
Redox reaction

Natural substrates
Acyldihydroxyacetone phosphate + NADPH (reaction in glycerolipid anabolism [1, 2, 5], acyldihydroxyacetone phosphate pathway [6], enzyme links acyldihydroxyacetone phosphate pathway intermediates to cellular glycerolipid biosynthesis [7]) [1, 2, 5–7]

Substrate spectrum
1 Palmitoylglycerone phosphate + NADPH (i.e. palmitoyldihydroxyacetone phosphate) [1–5, 7]
2 Hexadecyldihydroxyacetone phosphate + NADPH [1–8]
3 Acyldihydroxyacetone phosphate + NADPH [1–6, 8]
4 Alkyldihydroxyacetone phosphate + NADPH (ether analog of acyl dihydroxyacetone phosphate [5]) [1–6, 8]

Product spectrum

1 1-Palmitoylglycerol 3-phosphate + $NADP^+$ [1–5, 7]
2 1-Hexadecyl-sn-glycerol-3-phosphate + $NADP^+$ [1–8]
3 1-Acyl-sn-glycerol-3-phosphate + $NADP^+$ [1–6, 8]
4 1-Alkyl-sn-glycerol-3-phosphate + $NADP^+$ [1–6, 8]

Inhibitor(s)

$NADP^+$ (competitive to NADPH) [5, 7]; Alkyldihydroxyacetone phosphate (at high concentrations) [5]; Palmitoyldihydroxyacetone phosphate (competitive to hexadecyldihydroxyacetone phosphate) [7]; Palmitoyl-CoA (competitive to NADPH, not to palmitoyldihydroxyacetone phosphate [7]) [5, 7]; PCMB [7]; Trypsin (inactivation) [5]; Phosphatidylcholine (at high concentrations) [7]; Phosphatidylglycerol (at high concentrations) [7]; N-Ethylmaleimide (strong inhibitor) [8]; Divalent cations (i.e. Cu^{2+}, Co^{2+}, Ca^{2+}, Mg^{2+}, Mn^{2+}, Ni^{2+}, Zn^{2+}, strong inhibition) [8]; F^- (weak inhibition) [8]; EDTA (weak inhibition) [8]; More (stable to N-ethylmaleimide [7], DTNB and thiols) [5, 7]

Cofactor(s)/prosthetic group(s)/activating agents

NADPH (specific for NADPH [1, 8], hydrogen from B-side of nicotinamide ring is transferred to reduce keto substrates) [1–8]; NADH (can substitute for NADPH [7], at high concentrations, rat liver [1]) [1, 5, 7]; EDTA (activation, acyldihydroxyacetone phosphate as substrate) [1]; Triton X-100 (activation [3], destabilizes during solubilization [7]) [3]; Phosphatidylcholine micelles (activation, in presence of Triton X-100, slight stimulation at low concentrations) [7]; Phosphatidylglycerol micelles (activation, in presence of Triton X-100, slight stimulation at low concentrations) [7]

Metal compounds/salts

Turnover number (min^{-1})

6000 (NADPH) [7]

Specific activity (U/mg)

0.00158 [3]; 0.00546 (light mitochondrial fraction) [2]; 62.5 [7]

K_m-value (mM)

0.005 (NADPH (+ hexadecyldihydroxyacetone phosphate), rat brain) [1]; 0.0051 (NADPH, human skin fibroblasts, normal cells) [6]; 0.00540 (NADPH, human skin Zellwenger syndrome fibroblasts) [6]; 0.007 (NADPH (+ palmitoyldihydroxyacetone phosphate), rat brain) [1]; 0.01 (NADPH, ascites microsomes [1], K_m-value below: hexadecyldihydroxyacetone phosphate (+ NADPH), microsomal enzyme [5]) [1, 5]; 0.0150 (hexadecyldihydroxyacetone, microsomal enzyme [5]) [1, 5]; 0.0150 (hexadecyldihydroxyacetone phosphate) [8]; 0.0154 (palmitoyldihydroxyacetone phosphate) [7]; 0.02 (NADPH) [7, 8]; 0.021 (hexadecyldihydroxyacetone phosphate) [7]; 0.05 (NADPH, guinea pig liver mitochondria [1], hexadecyldihydroxyacetone phosphate, peroxisomal enzyme [5]) [1, 5]; 0.06 (NADPH (+ palmitoyldihydroxyacetone phosphate), peroxisomal enzyme) [5]; 0.07 (NADPH, in presence of $NADP^+$, microsomal enzyme) [5]; 0.08 (NADH (+ hexadecyldihydroxyacetone phosphate), microsomal enzyme, NADPH (+ hexadecyldihydroxyacetone phosphate), peroxisomal enzyme) [5]; 0.13 (NADPH, in presence of $NADP^+$, peroxisomal enzyme) [5]; 0.8 (NADH (+ hexadecyldihydroxyacetone phosphate), peroxisomal enzyme) [5]; 1.7 (NADH) [7]

pH-optimum

6.5 [5]; 6.5–7.5 (broad optimum) [1]; 6.7–7.2 (broad optimum) [8]

pH-range

5.5–8.2 (about half-maximal activity at pH 5.5 and 8.2) [5]

Temperature optimum (°C)

40 [8]

Temperature range (°C)

28–43 (about half-maximal activity at 28°C and 43°C) [8]

3 ENZYME STRUCTURE

Molecular weight

70000–75000 (guinea pig, size exclusion chromatography) [7]

Subunits

Monomer (1 × 60000, guinea pig, SDS-PAGE) [7]
More (27% hydrophobic amino acids by amino acid analysis) [7]

Glycoprotein/Lipoprotein

–

4 ISOLATION/PREPARATION

Source organism

Guinea pig [1, 5, 7]; Rat (Sprague-Dawley strain [2], Charles River CD strain [3]) [1–3, 5]; Trypanosoma brucei (stock 427, causative agent of nagana in domestic animals) [4]; Human [6]; Saccharomyces cerevisiae (wild-type) [8]

Source tissue

Brain [1]; Liver [1–3, 5, 7]; Kidney [1]; Adipose tissue [1]; Spleen [1]; Testis [1]; Lung [1]; Ascites tumor cells (grown in the abdomen of swiss white mice) [1]; Cell homogenate [4]; Skin fibroblasts (Zellweger syndrome cell lines GM0228 and GM4340 and normal control cells) [6]; Cell [8]

Localisation in source

Peroxisomes (microbodies [2]) [2, 5, 7]; Microsomes (membrane–bound, located on cytoplasmic surface of microsomal vesicles [3]) [1–3, 5]; Mitochondria (light mitochondrial fraction [2]) [1, 2, 5]; Lysosomes [2]; Glycosomes [4]; Membranes (twice the activity of anaerobically grown cells in aerobically grown yeast) [8]; More (distribution in guinea pig liver cells: 25% in mitochondria, 35% in peroxisomes and 35% in microsomes, in rat predominantly in micro- and peroxisomes) [5]

Purification

Guinea pig (affinity chromatography on NADPH-agarose gel matrix [7], partial, solubilization [5]) [5, 7]

Crystallization

–

Cloned

–

Renaturated

–

5 STABILITY

pH

Temperature (°C)

30 (10 min stable below) [8]; 45 ($t_{1/2}$: 10 min) [8]; 50 ($t_{1/2}$: 15 min, control fibroblasts, $t_{1/2}$: 20 min, Zellweger syndrome fibroblasts) [6]; 55 (10% loss of activity after 15 min, NADPH enhances thermal stability) [5]; 100 (complete inactivation after 10 min) [1]

Oxidation

Organic solvent

Glycerol stabilizes [7]

General stability information

Freeze-thawing, stable after several cycles [1]; Freeze-thawing leads to rapid loss of activity [7]; NADPH enhances thermal stability [5]; Triton X-100, stable to solubilization in the presence of high salt concentrations [5]; Trypsin inactivates [5]; Thiols stabilize solubilized enzyme [5]; NADPH stabilizes during storage and solubilization [7]; DTT stabilizes during storage [7]; High ionic strength facilitates solubilization with neutral or anionic detergents [7]; Triton X-100, enzyme becomes unstable during solubilization unless NADPH and 1 M KCl are present [7]

Storage

-80°C, membrane fraction stable [8]; -20°C, membrane fraction stable [8]; 4°C, stable for at least a month in the presence of DTT and NADPH [7]

6 CROSSREFERENCES TO STRUCTURE DATABANKS

PIR/MIPS code

Brookhaven code

7 LITERATURE REFERENCES

[1] LaBelle, E.F., Hajra, A.K.: J. Biol. Chem.,247,5825–5834 (1972)
[2] Hajra, A.K., Burke, C.L., Jones, C.L.: J. Biol. Chem.,254,10896–10900 (1979)
[3] Ballas, L.M., Bell, R.M.: Biochim. Biophys. Acta,665,586–595 (1981)
[4] Opperdoes, F.R.: FEBS Lett.,169,35–39 (1984)
[5] Ghosh, M.K., Hajra, A.K.: Arch. Biochem. Biophys.,245,523–530 (1986)
[6] Webber, K.O., Datta, N.S., Hajra, A.K.: Arch. Biochem. Biophys.,254,611–620 (1987)
[7] Datta, S.C., Ghosh, M.K., Hajra, A.K.: J. Biol. Chem.,265,8268–8274 (1990)
[8] Racenis, P.V., Lai, J.L., Das, A.K., Mullick, P.C., Hajra, A.K., Greenberg, M.L.: J. Bacteriol.,174,5704–5710 (1992)

1 NOMENCLATURE

EC number
1.1.1.102

Systematic name
D-erythro-Dihydrosphingosine:$NADP^+$ 3-oxidoreductase

Recommended name
3-Dehydrosphinganine reductase

Synonymes
Reductase, 3-dehydrosphinganine
D-3-Dehydrosphinganine reductase
D-3-Oxosphinganine reductase
Reductase, D-3-oxosphinganine
DSR [3]
3-Oxosphinganine reductase [1]
3-Oxosphinganine:NADPH oxidoreductase [1]
D-3-Oxosphinganine:B-NADPH oxidoreductase [2]

CAS Reg. No.
37250-36-5

2 REACTION AND SPECIFICITY

Catalysed reaction
3-Dehydrosphinganine + NADPH →
→ sphinganine + $NADP^+$

Reaction type
Redox reaction

Natural substrates
3-Dehydrosphinganine + NADPH (biosynthesis of dihydrosphingosine [2], one of the first steps in sphingolipid biosynthesis [3]) [2, 3]

Substrate spectrum
1 3-Dehydrosphinganine + NADPH (absolute requirement for NADPH [1], only the D-isomer of 3-oxodihydrosphingosine is reduced by the B-specific NADPH dependent reductase [2]) [1–3]

Product spectrum
1 Sphinganine + $NADP^+$ (i.e. D-erythro-dihydrosphingosine, (2S,3R)-2-amino-1,3-dihydroxyoctadecane [2]) [1–3]

Inhibitor(s)

Cofactor(s)/prosthetic group(s)/activating agents

NADPH (absolute requirement [1], B-specific, i.e. pro-S-hydrogen of NADPH is transferred to carbonyl of 3-oxodihydrosphingosine [2]) [1, 2]; More (NADH completely ineffective) [1]

Metal compounds/salts

Turnover number (min^{-1})

Specific activity (U/mg)

25 [2]

K_m-value (mM)

0.015 (C_{18}–3-oxodihydrosphingosine) [2]; 0.03 (C_{20}–3-oxodihydrosphingosine) [2]

pH-optimum

6.0–7.0 [1]; 6.5–7.0 [2]

pH-range

5.3–8 (5.3: 32% of activity maximum, 8: about 90% of activity maximum) [1]; 6.2–7.4 (6.2: about 50% of activity maximum, 7.4: 47% of activity maximum) [2]

Temperature optimum (°C)

Temperature range (°C)

---

3 ENZYME STRUCTURE

Molecular weight

Subunits

Glycoprotein/Lipoprotein

Lipoprotein [2]

---

4 ISOLATION/PREPARATION

Source organism

Hansenula ciferrii [1]; Rat [1, 2]; Bovine [2]; Mouse [3]

Source tissue

Spleen [2]; Liver [1–3]; Brain [2]; Heart muscle [2]; Skeletal muscle [2]; Kidney [2]

Localisation in source

Endoplasmic reticulum [1, 3]; Microsomes [2]

Purification
Bovine [2]

Crystallization
–

Cloned
–

Renaturated
–

5 STABILITY

pH

Temperature (°C)

Oxidation

Organic solvent

General stability information
Triton X-100: stable even in presence of high concentrations [2]

Storage
In frozen state stable for many weeks [1]

6 CROSSREFERENCES TO STRUCTURE DATABANKS

PIR/MIPS code

Brookhaven code

7 LITERATURE REFERENCES

[1] Stoffel, W., LeKim, D., Sticht, G.: Hoppe-Seyler's Z. Physiol. Chem.,349,664–670 (1968)
[2] Stoffel, W., LeKim, D., Sticht, G.: Hoppe-Seyler's Z. Physiol. Chem.,349,1637–1644 (1968)
[3] Mandon, E.C., Ehses, I., Rother, J., van Echten, G., Sandhoff, K.: J. Biol. Chem.,267,11144–11148 (1992)

1 NOMENCLATURE

EC number
1.1.1.103

Systematic name
L-Threonine:NAD^+ oxidoreductase

Recommended name
L-Threonine 3-dehydrogenase

Synonymes
Dehydrogenase, threonine 3-
L-Threonine dehydrogenase
Threonine 3-dehydrogenase
Threonine dehydrogenase

CAS Reg. No.
9067-99-6

2 REACTION AND SPECIFICITY

Catalysed reaction
L-Threonine + NAD^+ →
→ L-2-amino-3-oxobutanoate + NADH (ordered bi-bi mechanism: NAD^+ is added first [2, 7], the enzyme forms a stable, active threonine cleavage complex with aminoacetone synthetase with an apparent stoichiometry of 2 synthetase dimers/1 dehydrogenase tetramer [14], ordered sequential bi-ter mechanism: NAD^+ is the leading substrate, followed by threonine, products are released in the order CO_2, aminoacetone, NADH [14])

Reaction type
Redox reaction
More (oxidative decarboxylation of L-threonine presumed) [14]

Natural substrates
L-Threonine + NAD^+ (reaction in the main of several catabolic pathways of L-threonine by pro- and eukaryotes) [4, 14]

Substrate spectrum

1 L-Threonine + NAD^+ (highly specific, (r) [7, 13]) [1–14]
2 L-Allothreonine + NAD^+ (oxidation at 10% the oxidation rate of L-threonine, no substrate [1, 2]) [14]
3 D-Allothreonine + NAD^+ (no substrate [1, 2]) [4, 7, 12]
4 L-Threonine amide + NAD^+ (as good as L-threonine [12]) [4, 7, 12]
5 DL-alpha-Amino-beta-hydroxyvaleric acid + NAD^+ (71% of the relative V_{max} of L-threonine oxidation [12]) [4, 12]
6 DL-Threonine methylester + NAD^+ (27% of the relative V_{max} of L-threonine oxidation [12]) [4, 12]
7 DL-Threonine hydroxamate + NAD^+ (25% of the relative V_{max} of L-threonine oxidation [12])
8 L-Serine + NAD^+ (very low activity) [12]
9 threo-DL-beta-Phenylserine + NAD^+ (very low activity) [12]
10 More (no substrates are D-threonine [1, 2, 14], L-serine [1, 2, 7, 14], DL-homoserine, hydroxyproline, DL-4-amino-3-hydroxybutyrate, DL-3-hydroxybutyrate, DL-glycerate, DL-malate, DL-lactate, DL-1-amino-propan-2-ol [1, 4], ethanolamine, ethanol, formaldehyde [1], beta-hydroxybutyrate [2, 4], alpha-aminobutyrate [4, 7], no reverse reaction with aminoacetone plus NADH or the methyl ester of 2-amino-3-oxobutyrate plus NADH [14]) [1, 2, 4, 7, 14]

Product spectrum

1 L-2-Amino-3-oxobutanoate + NADH (spontaneous decarboxylation to aminoacetone [1–7, 13], enzymatic decarboxylation rather than spontaneous, L-2-amino-3-oxobutanoate is not detectable, it presumably does not leave the enzyme [14]) [1–14]
2 ?
3 ?
4 ?
5 ?
6 ?
7 ?
8 ?
9 ?
10 ?

Inhibitor(s)

L-Allothreonine (Staphylococcus aureus [1], competitive to L-threonine [7]) [1, 7]; NADH (competitive to NAD^+ [7, 8, 13, 14], 2 mM yield 65% inhibition [7, 13], non–competitive to L-threonine [7]) [7, 8, 13, 14]; ADP (competitive to NAD^+, non-competitive to L-threonine [7], not [13]) [7]; DTT (inhibits Cd^{2+} activation more than 2-mercaptoethanol does) [11]; Hg^{2+} (bull frog [2]) [2, 3, 9]; Ag^+ [3, 9]; Cu^{2+} [4, 7]; Ni^{2+} (at higher concentrations) [4, 7]; Zn^{2+} [7]; Cd^{2+} [7]; Mn^{2+} (at high concentration) [7]; PCMB (bull frog, partially reversible by GSH [2], rapid inactivation, enzyme loses ability to bind Mn^{2+} [8]) [2, 8]; Mercuribenzoate (p-substituted, rapid inactivation, enzyme loses ability to bind Mn^{2+} [8]) [3, 4, 8]; Mitochondrial inhibitory factor [10]; EDTA (very potent when preincubated in the absence of substrate [3], 8-hydroxyquinoline increases EDTA-inhibition [4]) [3–5]; DTNB (gradual, nearly complete loss of activity under non-denaturating conditions [8]) [4, 8]; 8-Hydroxyquinoline (increases EDTA inhibition) [4]; Iodoacetate [5, 8]; Iodoacetamide (slight inhibition) [8]; KCl (rat enzyme) [1]; N-Ethylmaleimide (partial, bull frog [2]) [2, 4]; 1,10-Phenanthroline (not reversible by dialysis) [5]; Dipicolinic acid [5]; Methyl methanethiosulfonate (complete inactivation) [8]; Methylglyoxal (allosteric process, trypsine treatment of the enzyme reduces inhibition, possible physiological role: feed-back inhibition) [13]; HCO_3^- (non–competitive) [14]; Aminoacetone (uncompetitive) [14]; L-Amino-n-butyric acid (at equimolar concentration 50% inhibition with respect to L-threonine) [14]; Ammonium sulfate (inhibits during purification) [13]; More (glycine, acetyl-CoA, pyruvate, D- or L-lactate, glucose, fructose 1,6-bisphosphate, phosphoenolpyruvate, ATP, ADP or AMP [13], Mg^{2+} [9], D-threonine, DL-serine, L-valine, D-amino-n-butyric acid do not inhibit L-threonine oxidation [14]) [9, 13, 14]

Cofactor(s)/prosthetic group(s)/activating agents

NAD^+ (essential, cannot be replaced by $NADP^+$ [2, 4] or NAD^+-analogs [7]) [1–14]; NADH (reverse reaction, NADH is not oxidized [2]) [7]; 3-Acetyl-pyridine-NAD^+ (can replace NAD^+, 1/10 as effective as NAD^+ [14]) [4, 14]; 3-Pyridinealdehyde-NAD^+ (can replace NAD^+) [4]; Thionicotinamide-NAD^+ (can replace NAD^+, 1/10 as effective as NAD^+ [14]) [4, 14]; Deamino-NAD^+ (can replace NAD^+ with the highest oxidation rate of all pyridines) [4]; DTT (activation) [4, 9]; 2-Mercaptoethanol (activation) [9]; Thiols (requirement, irreversible inactivation without [7]) [7, 9, 11]; 1,10-Phenanthroline (activation) [7]; L-Cysteine (activation) [9]; alpha-Thioglycerol (activation) [9]; Mercaptoacetic acid (activation) [9]; GSH (activation) [9]; More ($NADP^+$ is completely inert to reduction, no reverse reaction with aminoacetone or methyl ester of 2-amino-3-oxobutyrate and NADH) [2, 4, 14]

Metal compounds/salts

Zn^{2+} (requirement, zinc-containing dehydrogenase [6], 1 mol Zn^{2+} per mol of enzyme subunit, can be substituted by Co^{2+} or Cd^{2+} [5], EDTA-treated enzyme is reactivated by Zn^{2+}, less well by Co^{2+} and Mn^{2+} [3]) [3, 5, 6]; K^+ (activation, at high concentration, Staphylococcus aureus) [1]; Mn^{2+} (activation, thiol dependent [8, 9], added to native enzyme or to Zn^{2+}-, Co^{2+}- or Cd^{2+}-substituted form of the enzyme, stimulates to the same extent [5], carboxymethylation reduces Mn-dependent velocity [8]) [4, 5, 8, 9, 11]; Cd^{2+} (concentration dependent activation [11], not thiol dependent [8, 11], full activity after incubation at 37°C for 5 min in the absence of thiol [11]) [5, 8, 9, 11]; More (no activation by Li^+, Na^+, K^+, Rb^+, Cs^+, Mg^{2+}, Ca^{2+}, Sr^{2+}, Ba^{2+}, Co^{2+}, Zn^{2+}, Fe^{2+}, Fe^{3+}, Al^{3+}, Sn^{2+}, V^{4+}, Mo^{6+} [9], no metal requirement, bull frog [2]) [2, 9]

Turnover number (min^{-1})

Specific activity (U/mg)

0.0001 (Streptococcus faecalis) [1]; 0.0007 (Bacillus subtilis) [1]; 0.021 (E. coli) [1]; 0.024 (Corynebacterium erythrogenes) [1]; 0.25 (Staphylococcus aureus) [1]; 6.05 [3]; 8.0–16.0 (demetalized enzyme) [9]; 8.45 [13]; 19.0 [7]; 20–30 (Mn^{2+}-saturated enzyme) [9]; 21 (Zn^{2+}-saturated enzyme, specific activity remains unchanged when Zn^{2+} is replaced by Cd^{2+} or Co^{2+}) [5]; 30.3 [14]; 34.0 [4, 12]

K_m-value (mM)

0.047 (NAD^+, Staphylococcus aureus) [1]; 0.1 (NAD^+) [14]; 0.11 (NAD^+, demetalized enzyme) [9]; 0.14 (NAD^+, pH 7.5) [3]; 0.18 (NAD^+, Cd^{2+}-activated enzyme) [11]; 0.19 (NAD^+) [4, 12]; 0.24 (NAD^+, bull frog) [2]; 0.6 (NAD^+, Mn^{2+}-saturated enzyme) [9]; 0.98 (NAD^+) [7]; 1.0 (NAD^+) [13]; 1.1 (L-threonine, demetalized enzyme) [9]; 1.43 (L-threonine) [4, 12]; 5.0 (L-threonine) [14]; 5.5 (L-threonine) [13]; 7.5 (L-threonine, bull frog) [2]; 8.2 (L-threonine, Cd^{2+}-activated enzyme) [11]; 8.4 (L-threonine) [4, 7]; 8.7 (L-threonine, Staphylococcus aureus) [1]; 10.5 (L-threonine, pH 7.5) [3]; 16.0 (L-allothreonine) [14]; 221.0 (L-threonine, Mn-saturated enzyme) [9]

pH-optimum

7.8 (native enzyme [10], bull frog [2]) [2, 10]; 8.2–8.8 (0.1 M Tris/HCl buffer) [13]; 8.4 [14]; 8.4–8.5 (Staphylococcus aureus) [1]; 8.5–8.9 [3]; 8.6–8.7 [7]; 9.0 (rat enzyme after freezing and storage) [10]; 10.3 [4, 12]; 10.8 (demetalized enzyme) [9]

pH-range

6.0–11.5 (demetalized enzyme, steady increase of activity from 6.0 to 10.8, precipitous decline thereafter) [9]; 7.4–8.2 (about 80% of maximal activity at pH 7.4 and 8.2, bull frog) [2]; 7.4–8.8 (33% of maximal activity at pH 7.4 and 100% at 8.8) [13]; 7.5–9.8 (about half-maximal activity at pH 7.5 and 9.8, Staphylococcus aureus [1], about 30% of maximal activity at pH 7.5 and about 80% of maximal activity at pH 9.8 [7]) [1, 7]; 8.0–8.4 (over 95% of maximal activity at pH 8.0 and 8.4) [14]; 9.0–11.0 (about 35% of maximal activity at pH 9.0 and 6% at pH 11.0) [4]; 9.5–10.8 (about half-maximal activity at pH 9.5 and 10.8) [12]

Temperature optimum (°C)

25 (assay at) [2]; 37 (assay at) [1, 5, 8–13]

Temperature range (°C)

3 ENZYME STRUCTURE

Molecular weight

More (together with aminoacetone synthetase, MW 56000, threonine dehydrogenase coelutes with an apparent molecular weight of 150000) [14]
78000 (chicken, gel filtration) [7]
86000 (goat, gel filtration) [13]
88000 (chicken, gel electrophoresis) [7]
100000 (pig, gel filtration) [14]
140000 (E. coli K-12, sucrose density gradient centrifugation, sedimentation equilibrium centrifugation) [4, 12]
160000 (E. coli K-12, gel filtration) [4]

Subunits

Monomer (1 × 88000, chicken, SDS-PAGE [7], 1 × 89000, goat, SDS-PAGE [13]) [7, 13]
Tetramer (4 × 24903, pig, amino acid analysis [14], 4 × 25000, pig, SDS-PAGE [14], 4 × 35000, E. coli K-12, SDS-PAGE of enzyme cross-linked with dimethylsuberimidate, equilibrium centrifugation in 6 M guanidinium-chloride [4, 12]) [4, 12, 14]
More (35400, minimal molecular weight of E. coli K-12 enzyme by amino acid analysis) [12]

Glycoprotein/Lipoprotein

No monosaccharides [4]

4 ISOLATION/PREPARATION

Source organism

Staphylococcus aureus (strain Duncan) [1]; E. coli (B2 [1], K-12 [4–6, 8, 9, 11, 12]) [1, 4–6, 8, 9, 11, 12]; Bacillus subtilis [1]; Corynebacterium erythrogenes [1]; Streptococcus faecalis [1]; Rat (Wistar [10]) [1, 2, 10]; Bull frog [2]; Frog [2]; Toad [2]; Mouse [2]; Arthrobacter sp. [3, 12]; Chicken [7]; Bovine [10]; Goat [13]; Pig [14]; Guinea pig [2]

Source tissue

Cell [1, 3–6, 8, 9, 11, 12]; Liver [1, 2, 7, 10, 13, 14]

Localisation in source

Cytoplasm (bacterial enzyme) [1, 3–6, 8, 9, 11, 12]; Mitochondria (inner mitochondrial membrane matrix [7]) [1, 2, 7, 10, 13, 14]

Purification

Bull frog [2]; Staphylococcus aureus [1]; Arthrobacter sp. [3]; Chicken [7]; E. coli (affinity chromatography [4, 9], with added Mn^{2+} to loading and buffer [9]) [4, 9, 12]; Goat [13]; Pig [14]

Crystallization

–

Cloned

(E. coli K-12) [6]

Renaturated

–

---

5 STABILITY

pH

6.0–7.0 (near neutral pH, stable in the presence of 2-mercaptoethanol) [2]; 8.0 (most stable in Tris/HCl buffer) [3]; 8.8 (preincubation at 37°C at pH-values above, leads to sharp loss of activity of demetalized enzyme) [9]

Temperature (°C)

37 (complete inactivation after 2 min, NAD^+ and KCl increase thermal stability) [1]

Oxidation

Organic solvent

General stability information

Freezing inactivates [2]; Lyophilization inactivates [2]; Purification leads to loss of activity overnight [3]; 2-Mercaptoethanol stabilizes [4, 7]; Exhaustive dialysis leads to 15% loss of activity, restorable by DTT or 2-mercaptoethanol [4]; Dialysis after storage increases activity [10]; Phenylmethylsulfonyl fluoride recovers activity during early purification [7]; L-Threonine, 30% protection against iodoacetate inactivation [8]; NADH stabilizes against iodoacetate inactivation [8]; Dialysis, stable to [13, 14]; Glycerol, 10% vlv stabilizes during purification [7]

Storage

-20°C, increase of activity of rat enzyme to a maximum value after 2 weeks [10]; -20°C, 50% loss of maximum activity in 4 weeks, bovine [10]; -20°C, stable for at least a month in the presence of 2-mercaptoethanol plus glycerol [7]; -15°C, crude enzyme preparation stable [3]; -15°C, purified preparation loses activity within a week [3]; -10°C, stable [1]; 4°C, purified preparation stable for 2 months [4, 9, 12]

6 CROSSREFERENCES TO STRUCTURE DATABANKS

PIR/MIPS code

PIR1:DEECTH (Escherichia coli)

Brookhaven code

7 LITERATURE REFERENCES

[1] Green, M.L., Elliott, W.H.: Biochem. J.,92,537–549 (1964)
[2] Hartshorne, D., Greenberg, D.M.: Arch. Biochem. Biophys.,105,173–178 (1964)
[3] McGilvray, D., Morris, J.G.: Methods Enzymol.,17B,580–584 (1971) (Review)
[4] Boylan, S.A., Dekker, E.E.: J. Biol. Chem.,256,1809–1815 (1981)
[5] Epperly, B.R., Dekker, E.E: J. Biol. Chem.,266,6086–6092 (1991)
[6] Aronson, B.D., Somerville, R.L., Epperly, B.R., Dekker, E.E.: J. Biol. Chem.,264,5226–5232 (1989)
[7] Aoyama, Y., Motokawa,Y.: J. Biol. Chem.,256,12367–12373 (1981)
[8] Craig, P.A., Dekker, E.E.: Biochim. Biophys. Acta,1037,30–38 (1990)
[9] Craig, P.A., Dekker, E.E.: Biochemistry,25,1870–1876 (1986)
[10] Pagani, R., Guerranti, R., Righi, S., Leoncini, R., Vannoni, D., Marinello, E.: Biochem. Soc. Trans.,20,24 (1991)
[11] Craig, P.A., Dekker, E.E.: Biochim. Biophys. Acta,957,222–229 (1988)
[12] Boylan, S.A., Dekker, E.E.: Biochem. Biophys. Res. Commun.,85,190–197 (1978)
[13] Ray, M., Ray, S.: J. Biol. Chem.,260,5913–5918 (1985)
[14] Tressel, T., Thompson, R., Zieske, L.R., Menendez, M.I.T.S., Davis, L.: J. Biol. Chem.,261,16428–16437 (1986)

1 NOMENCLATURE

EC number
1.1.1.104

Systematic name
4-Hydroxy-L-proline:NAD^+ oxidoreductase

Recommended name
4-Oxoproline reductase

Synonymes
Hydroxy-L-proline oxidase

CAS Reg. No.
37250-37-6

---

2 REACTION AND SPECIFICITY

Catalysed reaction
4-Oxoproline + NADH →
→ 4-hydroxy-L-proline + NAD^+

Reaction type
Redox reaction

Natural substrates
4-Oxoproline + NADH [1]

Substrate spectrum
1 4-Oxoproline + NADH (ir) [1]

Product spectrum
1 4-Hydroxy-L-proline + NAD^+ [1]

Inhibitor(s)

Cofactor(s)/prosthetic group(s)/activating agents
NADH [1]

Metal compounds/salts

Turnover number (min^{-1})

Specific activity (U/mg)
6.2 [1]

K_m-value (mM)
0.84 (NADH) [1]; 0.60 (4-oxoproline) [1]

pH-optimum
6.0–6.5 [1]

pH-range

Temperature optimum (°C)

Temperature range (°C)

3 ENZYME STRUCTURE

Molecular weight

Subunits

Glycoprotein/Lipoprotein
–

4 ISOLATION/PREPARATION

Source organism
Rat [1]; Rabbit [1]

Source tissue
Kidney [1]

Localisation in source

Purification
Rabbit (partial) [1]

Crystallization
–

Cloned
–

Renaturated
–

5 STABILITY

pH

Temperature (°C)

Oxidation

Organic solvent

General stability information

Storage

−10°C, 3 months [1]

6 CROSSREFERENCES TO STRUCTURE DATABANKS

PIR/MIPS code

Brookhaven code

7 LITERATURE REFERENCES

[1] Smith, T.E., Mitoma, C.: J. Biol. Chem.,237,1177–1180 (1962)

1 NOMENCLATURE

EC number
1.1.1.105

Systematic name
Retinol:NAD^+ oxidoreductase

Recommended name
Retinol dehydrogenase

Synonymes
Retinol (vitamin A1) dehydrogenase
MDR [3]
Microsomal retinol dehydrogenase [3]
Dehydrogenase, retinol
All-trans retinol dehydrogenase
Retinal reductase
Retinene reductase

CAS Reg. No.
9033-53-8

---

2 REACTION AND SPECIFICITY

Catalysed reaction
Retinol + NAD^+ →
→ retinal + NADH

Reaction type
Redox reaction

Natural substrates

Substrate spectrum
1 Retinol + NAD^+ [1–5]
2 More (a similar bovine enzyme catalyzes the reaction with $NADP^+$ as cofactor, in some references only $NADP^+$ as cofactor is mentioned, or much lower activity is observed with NAD^+ than with $NADP^+$) [6–9]

Product spectrum
1 Retinal + NADH
2 ?

Inhibitor(s)
Acetaldehyde [2]; Formaldehyde [2]; More (not: 4-methylpyrazole, CO, ethanol) [3, 4]

Cofactor(s)/prosthetic group(s)/activating agents
NAD^+ [1–5]; $NADP^+$ [3–5]; More (a similar bovine enzyme catalyzes the reaction with $NADP^+$ as cofactor, in some references only $NADP^+$ as cofactor is mentioned, or much lower activity is observed with NAD^+ than with $NADP^+$) [6–9]

Metal compounds/salts

Turnover number (min^{-1})

Specific activity (U/mg)

K_m-value (mM)
0.12 (retinol) [3, 4]; 0.163 (retinal) [5]

pH-optimum
8.5 [5]; More (optimal activity at physiological pH) [3, 4]

pH-range

Temperature optimum (°C)

Temperature range (°C)

---

3 ENZYME STRUCTURE

Molecular weight

Subunits

Glycoprotein/Lipoprotein
–

---

4 ISOLATION/PREPARATION

Source organism
Rat [1–4]; Mouse [5]; More (a similar bovine enzyme catalyzes the reaction with $NADP^+$ as cofactor, in some references only $NADP^+$ as cofactor is mentioned, or much lower activity is observed with NAD^+ than with $NADP^+$) [6–9]

Source tissue
Retina [1]; Liver [1–4]; Epidermis [5]

Localisation in source
Microsomes [3, 4]; Cytosol [5]

Purification

Crystallization

–

Cloned

–

Renaturated

–

5 STABILITY

pH

Temperature (°C)

Oxidation

Organic solvent

General stability information

Storage

6 CROSSREFERENCES TO STRUCTURE DATABANKS

PIR/MIPS code

Brookhaven code

7 LITERATURE REFERENCES

[1] Koen, A.L., Shaw, C.R.: Biochim. Biophys. Acta,128,48–54 (1966)
[2] Grisolia, S., Guerri, C., Godfrey, W.: Biochem. Biophys. Res. Commun.,66,1112–1117 (1975)
[3] Leo, M.A., Lieber, C.S.: Methods Enzymol.,189,520–524 (1990) (Review)
[4] Leo, M.A., Kim., C.-I., Lieber, C.S.: Arch. Biochem. Biophys.,259,241–249 (1987)
[5] Kishore, G.S., Boutwell, R.K.: Biochem. Biophys. Res. Commun.,94,1381–1386 (1980)
[6] Blaner, W.S., Churchich, J.E.: Biochem. Biophys. Res. Commun.,94,820–826 (1980)
[7] Ishiguro, S., Suzuki, Y., Tamai, M., Mizuno, K.: J. Biol. Chem.,266,15520–15524 (1991)
[8] Lion, F., Rotmans, J.P., Daemen, F.J.M., Bonting, S.L. : Biochim. Biophys. Acta,384,283–292 (1975)
[9] Nicotra, C., Livrea, M.A.: J. Biol. Chem.,257,11836–11841 (1982)

1 NOMENCLATURE

EC number
1.1.1.106

Systematic name
(R)-Pantoate:NAD^+ 4-oxidoreductase

Recommended name
Pantoate 4-dehydrogenase

Synonymes
Dehydrogenase, pantoate
Pantothenase [1]
D-Pantoate:NAD^+ 4-oxidoreductase [2]

CAS Reg. No.
37250-38-7

2 REACTION AND SPECIFICITY

Catalysed reaction
(R)-Pantoate + NAD^+ →
→ (R)-4-dehydropantoate + NADH

Reaction type
Redox reaction

Natural substrates
(R)-Pantoate + NAD^+ (pantoate degradation) [1, 3]

Substrate spectrum
1 (R)-Pantoate + NAD^+ (r [3], the equilibrium of the reaction favors pantoate formation at all pH values [3]) [1–3]
2 Ketopantoate + NAD^+ [3]
3 More (not: 2-hydroxy-DL-isovalerate, DL-lactate, 4-hydroxybutyrate) [3]

Product spectrum
1 (R)-4-Dehydropantoate + NADH [1–3]
2 ?
3 ?

Inhibitor(s)

p-Chloromercuribenzoate (potassium D-pantoate and NAD^+ protect) [2]; 5,5'-Dithiobis(2-nitrobenzoic acid) (potassium D-pantoate and NAD^+ protect) [2]; Iodoacetic acid (potassium D-pantoate and NAD^+ protect) [2]; Phenylglyoxal (potassium D-pantoate and NAD^+ protect) [2]

Cofactor(s)/prosthetic group(s)/activating agents

NAD^+ [1–3]; NADH [3]; More (not $NADP^+$) [3]

Metal compounds/salts

Turnover number (min^{-1})

Specific activity (U/mg)

19.34 [1]; 21.0 [2]; 6.0 [3]

K_m-value (mM)

0.33 (D-pantoate) [3]; 0.5 (about, ketopantoate) [3]

pH-optimum

10.0 (no true optimum found, increase of activity throughout the range tested (pH 7.0–10.0)) [3]

pH-range

7.0–10.0 (increase of activity throughout the range) [3]

Temperature optimum (°C)

25 [2]

Temperature range (°C)

---

3 ENZYME STRUCTURE

Molecular weight

80000 (Pseudomonas fluorescens, ultracentrifugation) [1]
86000 (Pseudomonas fluorescens, gel filtration) [1]

Subunits

Tetramer (4 × 24000, identical, Pseudomonas fluorescens, SDS-PAGE) [1]

Glycoprotein/Lipoprotein

–

---

4 ISOLATION/PREPARATION

Source organism

Pseudomonas fluorescens UK-1 [1, 2]; Pseudomonas sp. P-2 (inducible) [3]

Source tissue

Cell [3]

Localisation in source

Purification

Pseudomonas fluorescens UK-1 [1, 2]; Pseudomonas sp. P-2 [3]

Crystallization

–

Cloned

–

Renaturated

–

5 STABILITY

pH

4.0 (22°C, 30 min: complete loss of activity) [3]; 5.5–10 (22°C, purified enzyme stable for 7 h) [3]; 5.5 (below pH 5.5, slow loss of activity) [3]

Temperature (°C)

22 (7 h, pH 5.5–10, purified enzyme stable) [3]; 50 (5 min: crude enzyme extract stable, purified enzyme not) [3]

Oxidation

Organic solvent

General stability information

NAD^+ protects against heat inactivation [3]

Storage

–70°C, after dialysis for 4 h against 50 mM potassium phosphate buffer, pH 7.2, 1 mM EDTA [1]

6 CROSSREFERENCES TO STRUCTURE DATABANKS

PIR/MIPS code

Brookhaven code

7 LITERATURE REFERENCES

[1] Mäntsälä, P.: Biochim. Biophys. Acta,526,25–33 (1978)
[2] Myöhänen, T., Mäntsälä, P.: Biochim. Biophys. Acta,614,266–273 (1980)
[3] Goodhue, C.T., Snell, E.E.: Biochemistry,5,403–408 (1966)

1 NOMENCLATURE

EC number
1.1.1.107

Systematic name
Pyridoxal:NAD^+ 4-oxidoreductase

Recommended name
Pyridoxal 4-dehydrogenase

Synonymes
Pyridoxal dehydrogenase

CAS Reg. No.
37250-39-8

2 REACTION AND SPECIFICITY

Catalysed reaction
Pyridoxal + NAD^+ →
→ 4-pyridoxolactone + NADH

Reaction type
Redox reaction

Natural substrates
Pyridoxal + NAD^+ [1]

Substrate spectrum
1 Pyridoxal + NAD^+ (ir) [1]
2 2-Demethyl-2-ethylpyridoxal + NAD^+ [1]

Product spectrum
1 4-Pyridoxolactone + NADH [1]
2 ? + NADH [1]

Inhibitor(s)
o-Phenanthroline [1]; m-Phenanthroline [1]; p-Phenanthroline [1]; 4-Pyridoxolactone [1]; 4-Pyridoxate [1]; Pyridoxine [1]; Pyridoxamine [1]; $HgCl_2$ [1]; 8-Hydroxyquinoline [1]; 2,2'-Dipyridyl [1]; Quinoline [1]

Cofactor(s)/prosthetic group(s)/activating agents
NAD^+ [1]; 5-Deoxypyridoxal (stimulation) [1]

Metal compounds/salts

Turnover number (min^{-1})

Specific activity (U/mg)

K_m-value (mM)
0.076 (pyridoxal) [1]; 0.29 (NAD^+) [1]

pH-optimum
9.3–9.5 [1]

pH-range
6.1 (not active below) [1]

Temperature optimum (°C)

Temperature range (°C)

---

3 ENZYME STRUCTURE

Molecular weight

Subunits

Glycoprotein/Lipoprotein
–

---

4 ISOLATION/PREPARATION

Source organism
Pseudomonas sp. [1]

Source tissue

Localisation in source

Purification
Pseudomonas sp. (partial) [1]

Crystallization
–

Cloned
–

Renaturated
–

5 STABILITY

pH

Temperature (°C)

Oxidation

Organic solvent

General stability information

Storage

6 CROSSREFERENCES TO STRUCTURE DATABANKS

PIR/MIPS code

Brookhaven code

7 LITERATURE REFERENCES

[1] Burg, R.W., Snell, E.E.: J. Biol. Chem.,244,2585–2589 (1969)

1 NOMENCLATURE

EC number
1.1.1.108

Systematic name
Carnitine:NAD^+ 3-oxidoreductase

Recommended name
Carnitine 3-dehydrogenase

Synonymes

CAS Reg. No.
9045-45-8

---

2 REACTION AND SPECIFICITY

Catalysed reaction
Carnitine + NAD^+ →
→ 3-dehydrocarnitine + NADH

Reaction type
Redox reaction

Natural substrates
Carnitine + NAD^+ [1–7]

Substrate spectrum
1 Carnitine + NAD^+ (r) [1–7]

Product spectrum
1 3-Dehydrocarnitine + NADH [1–7]

Inhibitor(s)
Ag^+ [3]; Ni^+ [3]; Hg^+ [3]; Li^+ [3]; Ca^{2+} [3]; Mn^{2+} [3]; Co^{2+} [3]; Cu^{2+} [3]; Zn^{2+} [3]; p-Chloromercuribenzoate [3, 7]; D-Carnitine [6]; Glycine betaine [6]; Choline [6]; Hydroxylamine [7]; Borate [7]; Hydrazine hydrate [7]

Cofactor(s)/prosthetic group(s)/activating agents
NAD^+ [1–7]; NADH [1–7]

Metal compounds/salts

Turnover number (min^{-1})

Specific activity (U/mg)
50.0–55.3 [1, 2]; 21 [3]; 600 [5]; 199 [6]

K_m-value (mM)
5.0–16.0 (L-carnitine) [1–3, 5–7]; 0.042–0.25 (NAD^+) [1–3, 5–7]; 1.25–1.71 (3-dehydrocarnitine) [2–6]; 0.022–0.067 (NADH) [2–6]

pH-optimum
9.5 (carnitine + NAD^+) [1, 3]; 9.0 (carnitine + NAD^+) [2, 7]; 9.0–9.5 (carnitine + NAD^+) [6]; 7.0–7.5 (3-dehydrocarnitine + NADH) [2]; 6.5 (3-dehydrocarnitine + NADH) [3]; 7.0 (3-dehydrocarnitine + NADH) [3]

pH-range
5.5–11.0 (carnitine + NAD^+) [2, 6]; 4.0–10.0 (3-dehydrocarnitine + NADH) [2]; 11 (not active above, 3-dehydrocarnitine + NADH) [6]

Temperature optimum (°C)
30 (carnitine + NAD^+) [2]; 35 [6]

Temperature range (°C)
45 (not active above, carnitine + NAD^+) [2]

---

3 ENZYME STRUCTURE

Molecular weight
74000 (Xanthomonas translucens, gel filtration) [3]
60000 (Pseudomonas putida, gel filtration) [2]

Subunits
Dimer (2 × 32000, identical, Pseudomonas putida, SDS-PAGE [2], 2 × 37000, Xanthomonas translucens, SDS-PAGE [3]) [2, 3]

Glycoprotein/Lipoprotein
–

---

4 ISOLATION/PREPARATION

Source organism
Pseudomonas aeruginosa [1, 4–7]; Pseudomonas putida [2, 4]; Pseudomonas ovalis [4]; Pseudomonas fluorescens [4]; Xanthomonas translucens [3]

Source tissue

Localisation in source

Purification
Pseudomonas putida [2]; Xanthomonas translucens [3]; Pseudomonas aeruginosa [6]

Crystallization

–

Cloned

–

Renaturated

–

5 STABILITY

pH

Temperature (°C)

Oxidation

Organic solvent

General stability information

Storage

4°C, lyophilized, several months [1]

6 CROSSREFERENCES TO STRUCTURE DATABANKS

PIR/MIPS code

Brookhaven code

7 LITERATURE REFERENCES

[1] Matsumoto, K., Yamada, Y., Takahashi, M., Todoroki, T., Mizoguchi, K., Misaki, H., Yuki, H.: Clin. Chem.,36,2072–2076 (1990)
[2] Goulas, P.: Biochim. Biophys. Acta,957,335–339 (1988)
[3] Mori, N., Kasugai, T., Kitamoto, Y., Ichikawa, Y.: Agric. Biol. Chem.,52,249–250 (1988)
[4] Vandecasteele, J.P.: Appl. Environ. Microbiol.,39,327–334 (1980)
[5] Schöpp, W., Sorger, H., Kleber, H.P., Aurich, H.: Eur. J. Biochem.,10,56–60 (1969)
[6] Aurich, H., Kleber, H.P., Sorger, H., Tauchert, H.: Eur. J. Biochem.,6,196–201 (1968)
[7] Aurich, H., Kleber, H.P., Schöpp, W.D.: Biochim. Biophys. Acta,139,505–507 (1967)

1 NOMENCLATURE

EC number
1.1.1.110

Systematic name
Indolelactate:NAD^+ oxidoreductase

Recommended name
Indolelactate dehydrogenase

Synonymes

CAS Reg. No.
37250-41-2

2 REACTION AND SPECIFICITY

Catalysed reaction
Indolelactate + NAD^+ →
→ indolepyruvate + NADH

Reaction type
Redox reaction

Natural substrates
Indolelactate + NAD^+ [1]

Substrate spectrum
1 Indolelactate + NAD^+ (r) [1]
2 Phenyllactate + NAD^+ (r) [1]
3 p-Hydroxyphenyllactate + NAD^+ (r) [1]

Product spectrum
1 Indolepyruvate + NADH [1]
2 Phenylpyruvate + NADH [1]
3 p-Hydroxyphenylpyruvate + NADH [1]

Inhibitor(s)

Cofactor(s)/prosthetic group(s)/activating agents
NAD^+ [1]; NADH [1]

Metal compounds/salts

Turnover number (min^{-1})

Specific activity (U/mg)
1.89 [1]

K_m-value (mM)
1.03 (indolelactate) [1]; 1.77 (p-hydroxyphenyllactate) [1]; 2.21 (phenyllactate) [1]; 0.075 (p-hydroxyphenylpyruvate) [1]; 0.106 (phenylpyruvate) [1]; 0.195 (NAD^+) [1]; 0.0142 (NADH) [1]

pH-optimum
8.5 (indolelactate + NAD^+) [1]; 7.0 (indolepyruvate + NADH) [1]

pH-range
10.0 (not active above, indolelactate + NAD^+) [1]; 3.5–10.5 (indolepyruvate + NADH) [1]

Temperature optimum (°C)

Temperature range (°C)

3 ENZYME STRUCTURE

Molecular weight

Subunits

Glycoprotein/Lipoprotein
–

4 ISOLATION/PREPARATION

Source organism
Clostridium sporogenes [1]

Source tissue

Localisation in source

Purification
Clostridium sporogenes [1]

Crystallization
–

Cloned
–

Renaturated
–

5 STABILITY

pH

Temperature (°C)

Oxidation

Organic solvent

General stability information

Storage
-20°C, 4 months [1]

6 CROSSREFERENCES TO STRUCTURE DATABANKS

PIR/MIPS code

Brookhaven code

7 LITERATURE REFERENCES

[1] Jean, M., DeMoss, R.D.: Can. J. Microbiol.,14,429–435 (1968)

1 NOMENCLATURE

EC number
1.1.1.111

Systematic name
(S)-3-(Imidazol-5-yl)lactate:NAD(P)$^+$ oxidoreductase

Recommended name
3-(Imidazol-5-yl)lactate dehydrogenase

Synonymes
Imidazol-5-yl lactate dehydrogenase

CAS Reg. No.
37250-42-3

---

2 REACTION AND SPECIFICITY

Catalysed reaction
(S)-3-(Imidazol-5-yl)lactate + NAD(P)$^+$ →
→ 3-(imidazol-5-yl)pyruvate + NAD(P)H

Reaction type
Redox reaction

Natural substrates
(S)-3-(Imidazol-5-yl)lactate + NADP$^+$ [1]

Substrate spectrum
1 (S)-3-(Imidazol-5-yl)lactate + NAD(P)$^+$ (r, NAD$^+$: 58% activity of NADP$^+$) [1]

Product spectrum
1 3-(Imidazol-5-yl)pyruvate + NAD(P)H [1]

Inhibitor(s)

Cofactor(s)/prosthetic group(s)/activating agents
NADPH [1]; NAD$^+$ (58% activity of NADP$^+$) [1]; NADP$^+$ [1]; NADH [1]

Metal compounds/salts

Turnover number (min^{-1})

Specific activity (U/mg)

K_m-value (mM)

pH-optimum

pH-range

Temperature optimum (°C)

Temperature range (°C)

3 ENZYME STRUCTURE

Molecular weight

Subunits

Glycoprotein/Lipoprotein

–

4 ISOLATION/PREPARATION

Source organism

Pseudomonas acidovorans [1]; E. coli [1]

Source tissue

Localisation in source

Purification

Crystallization

–

Cloned

–

Renaturated

–

5 STABILITY

pH

Temperature (°C)

Oxidation

Organic solvent

General stability information

Storage

6 CROSSREFERENCES TO STRUCTURE DATABANKS

PIR/MIPS code

Brookhaven code

7 LITERATURE REFERENCES

[1] Coote, J.G., Hasall, H.: Biochem. J.,111,237–239 (1969)

1 NOMENCLATURE

EC number
1.1.1.112

Systematic name
Indan-1-ol:NAD(P)$^+$ 1-oxidoreductase

Recommended name
Indanol dehydrogenase

Synonymes

CAS Reg. No.
37250-43-4

---

2 REACTION AND SPECIFICITY

Catalysed reaction
Indan-1-ol + NAD(P)$^+$ →
→ indanone + NAD(P)H

Reaction type
Redox reaction

Natural substrates
1-Indanol + NAD(P)$^+$ [1–5]
Progesterone + NAD(P)H [2]
Bile acids + NAD(P)H [2]

Substrate spectrum
1 1-Indanol + NAD(P)$^+$ (r [1–3], ir [4], 5 times lower activity with NAD$^+$ than with NADP$^+$ [1, 5], higher activity with NAD$^+$ than with NADP$^+$ [4]) [1–5]
2 Alicyclic alcohols + NAD(P)$^+$ (r) [1–3, 5]
3 3-Hydroxysteroids + NAD(P)$^+$ (r) [2, 3]
4 20-Hydroxysteroids + NAD(P)$^+$ (r) [2, 3]

Product spectrum
1 Indanone + NAD(P)H [1–5]
2 Corresponding alicyclic ketones + NAD(P)H [1–3, 5]
3 3-Ketosteroids + NAD(P)H [2, 3]
4 20-Ketosteroids + NAD(P)H [2, 3]

Inhibitor(s)

1,10-Phenanthroline [1, 2]; 2,2'-Bipyridine [1]; 4,4'-Bipyridine [1]; Cibacron blue [1]; Hexestrol [1]; Medroxyprogesterone acetate [1]; Progesterone [1]; 8-Hydroxyquinoline [5]

Cofactor(s)/prosthetic group(s)/activating agents

NAD^+ [1–5]; $NADP^+$ [1–5]; NADH [1, 3]; NADPH [1, 3]

Metal compounds/salts

Turnover number (min^{-1})

Specific activity (U/mg)

11.1 [1]; 7.77 [3]

K_m-value (mM)

0.002–0.097 ($NADP^+$) [1, 3, 4]; 0.009–0.250 (NAD^+) [1, 3, 4]; 0.001 (1-indanol (+ NAD^+)) [4]; 0.0017 (1-indanol (+ $NADP^+$)) [4]; 0.601 ((S)-(+)-1-indanol) [1]; 0.240 ((RS)-1-indanol) [1]; 0.028 mM (1,2,3,4-tetrahydro-1-naphtol) [1]; 0.0061 mM (progesterone) [2]; More (alicyclic alcohols and aldehydes [1–3, 5], 1-indanol, various isozymes [3]) [1–3, 5]

pH-optimum

9.3–10.0 [1, 2]; 9.0–11.0 (depending on isoenzyme) [3]; 9.8 [4]; 10.0–10.5 (1-indanol + $NADP^+$) [5]; 9.0–10.0 (1-indanol + NAD^+) [5]

pH-range

Temperature optimum (°C)

Temperature range (°C)

---

3 ENZYME STRUCTURE

Molecular weight

38000 (Japanese monkey, high pressure exclusion chromatography) [1]
30000–36000 (rabbit, gel filtration, isoenzymes) [3]
30000 (bovine, gel filtration) [5]

Subunits

Monomer (1 × 36000, Japanese monkey, SDS-PAGE [1], 1 × 33000–37000, rabbit, SDS-PAGE, isoenzymes [3]) [1, 3]

Glycoprotein/Lipoprotein

No glycoprotein [1]

4 ISOLATION/PREPARATION

Source organism
Japanese monkey [1, 2]; Rabbit [3]; Human [4]; Bovine [5]

Source tissue
Liver [1–3, 5]; Kidney [1, 5]; Brain [1]; Spleen [1]; Lung [1]; Heart [1, 5]; Placenta [4]; Adrenal [5]

Localisation in source
Cytosol [1, 3]; Microsomes [1, 4, 5]

Purification
Monkey [1]; Rabbit [3]; Bovine (partially) [5]

Crystallization
–

Cloned
–

Renaturated
–

5 STABILITY

pH
5.2–10.5 [1]; 5.5 (not stable below) [3]

Temperature (°C)
48 (10 min, 60% loss of activity) [1]; 42 (not stable above) [3]

Oxidation

Organic solvent

General stability information

Storage
4°C, 0.15 M KCl, 1 month [1]

6 CROSSREFERENCES TO STRUCTURE DATABANKS

PIR/MIPS code

Brookhaven code

7 LITERATURE REFERENCES

[1] Hara, A., Mouri, K., Nakagawa, M., Nakamura, M., Nakayama, T., Matsuura, K., Sawada, H.: J. Biochem.,106,126–132 (1989)
[2] Hara, A., Nakagawa, M., Taniguchi, H., Sawada, H.: J. Biochem.,106,900–903 (1989)
[3] Hara, A., Kariya, K., Nakamura, M., Nakayama, T., Sawada, H.: Arch. Biochem. Biophys.,249,225–236 (1986)
[4] Kulkarni, A.P., Strohm, B.H., Houser, W.H.: Xenobiotica,15,513–519 (1985)
[5] Billings, R.E., Sullivan, H.R., McMaron, R.E.: J. Biol. Chem.,246,3512–3517 (1971)

1 NOMENCLATURE

EC number
1.1.1.113

Systematic name
L-Xylose:$NADP^+$ 1-oxidoreductase

Recommended name
L-Xylose 1-dehydrogenase

Synonymes
Dehydrogenase, L-xylose
NADPH-xylose reductase

CAS Reg. No.
37250-44-5

---

2 REACTION AND SPECIFICITY

Catalysed reaction
L-Xylose + $NADP^+$ →
→ L-xylono-1,4-lactone + NADPH

Reaction type
Redox reaction

Natural substrates

Substrate spectrum
1 L-Xylose + $NADP^+$ [1]
2 D-Arabinose + $NADP^+$ [1]
3 D-Lyxose + $NADP^+$ (very slowly) [1]

Product spectrum
1 L-Xylono-1,4-lactone + NADPH
2 D-Arabono-gamma-lactone + NADPH
3 D-Lyxono-gamma-lactone + NADPH [1]

Inhibitor(s)

Cofactor(s)/prosthetic group(s)/activating agents
$NADP^+$ [1]; More (not: NAD^+) [1]

Metal compounds/salts

Turnover number (min^{-1})

Specific activity (U/mg)
More [1]

K_m-value (mM)

pH-optimum
9.5 [1]

pH-range
7.6–10.5 (7.6: about 30% of activity maximum, 10.5: about 70% of activity maximum) [1]

Temperature optimum (°C)
20 (assay at) [1]

Temperature range (°C)

---

3 ENZYME STRUCTURE

Molecular weight

Subunits

Glycoprotein/Lipoprotein
–

---

4 ISOLATION/PREPARATION

Source organism
Saccharomyces cerevisiae [1]

Source tissue

Localisation in source

Purification
Saccharomyces cerevisiae [1]

Crystallization
–

Cloned
–

Renaturated
–

5 STABILITY

pH

Temperature (°C)

Oxidation

Organic solvent

General stability information

Storage

6 CROSSREFERENCES TO STRUCTURE DATABANKS

PIR/MIPS code

Brookhaven code

7 LITERATURE REFERENCES

[1] Uehara, K., Takeda, M.: J. Biochem.,52,461–463 (1962)

1 NOMENCLATURE

EC number
1.1.1.114

Systematic name
D-Apiitol:NAD^+ 1-oxidoreductase

Recommended name
Apiose 1-reductase

Synonymes
Reductase, D-apiose
D-Apiose reductase
D-Apiitol reductase

CAS Reg. No.
37250-45-6

2 REACTION AND SPECIFICITY

Catalysed reaction
D-Apiitol + NAD^+ →
→ D-apiose + NADH

Reaction type
Redox reaction

Natural substrates
D-Apiose + NADH (bacterium utilizes this sugar as a sole source of carbon for growth) [2, 3]

Substrate spectrum
1 D-Apiitol + NAD^+ (r) [1–3]
2 More (highly specific for apiose [1, 2], specific for D-apiitol with a few exceptions (myo-inositol (38%), meso-inositol (13%) of the rate of the activity obtained with D-apiitol) [1, 2], low activity with D-ribulose (8.5% of the activity obtained with apiose) [3], D-erythrose (1.5% of the activity obtained with apiose) [3]) [1–3]

Product spectrum
1 D-Apiose + NADH [1–3]
2 ?

Inhibitor(s)

NADH (above 0.1 mM) [1, 2]; More (activity not affected by EDTA) [3]

Cofactor(s)/prosthetic group(s)/activating agents

NADH (specificity for NADH as electron acceptor) [1–3]; NAD^+ (specificity for NAD^+ as electron donor) [1–3]; More (FAD, $FADH_2$, cytochrome c and ferricyanide are inactive) [1, 2]

Metal compounds/salts

More (activity not affected by EDTA) [3]

Turnover number (min^{-1})

Specific activity (U/mg)

0.102 [1, 2]

K_m-value (mM)

71.4 (D-apiose) [1, 2]; 0.035 (NAD^+) [1, 2]; 11.6 (D-apiitol) [1, 2]; 0.015 (NADH) [1, 2]; 20 (D-apiose) [3]; 10 (D-apiitol) [3]

pH-optimum

7.5 (apiose reduction, Tris-HCl or glycine buffer) [1, 2]; 10 (apiitol oxidation, glycine-NaOH buffer) [1, 2]; 10.5 (apiitol oxidation, glycine-NaOH buffer) [3]

pH-range

5.5–9.5 (5.5: about 85% of activity maximum, 9.5: 45% of activity maximum, apiose reduction) [3]

Temperature optimum (°C)

Temperature range (°C)

---

3 ENZYME STRUCTURE

Molecular weight

110000–115000 (bacterium (gram-negative coccus, isolated from surface of germinating parsley), sucrose density gradient centrifugation, gel filtration) [1, 2]

Subunits

Glycoprotein/Lipoprotein

–

---

4 ISOLATION/PREPARATION

Source organism

Bacterium (gram-negative coccus, isolated from surface of germinating parsley) [1, 2]; Aerobacter aerogenes PRL-R3 [3]

Source tissue
Cell [1]

Localisation in source

Purification
Bacterium (gram-negative coccus isolated from surface of germinating parsley) [1]

Crystallization
–

Cloned
–

Renaturated
–

5 STABILITY

pH

Temperature (°C)

Oxidation

Organic solvent

General stability information
Mercaptoethanol increases stability in dilute solution at 0–4°C [3]; Two cycles of freezing/thawing inactivate [3]

Storage
–20°C, several weeks [1, 2]; 0 4°C, 50% loss of activity after 48 h [3]; Frozen, 2 months [3]

6 CROSSREFERENCES TO STRUCTURE DATABANKS

PIR/MIPS code

Brookhaven code

7 LITERATURE REFERENCES

[1] Chandrasekaran, E.V., Davila, M., Mendicino, J.: Methods Enzymol.,89,228–232 (1982) (Review)
[2] Hanna, R., Picken, M., Mendicino, J.: Biochim. Biophys. Acta,315,259–271 (1973)
[3] Neal, D.L., Kindel, P.K.: J. Bacteriol.,101,910–915 (1970)

1 NOMENCLATURE

EC number
1.1.1.115

Systematic name
D-Ribose:NADP$^+$ 1-oxidoreductase

Recommended name
Ribose 1-dehydrogenase (NADP$^+$)

Synonymes
Dehydrogenase, ribose (nicotinamide adenine dinucleotide phosphate)
D-Ribose dehydrogenase (NADP$^+$)
NADP-pentose-dehydrogenase [1]

CAS Reg. No.
37250-46-7

2 REACTION AND SPECIFICITY

Catalysed reaction
D-Ribose + NADP$^+$ + H_2O →
→ D-ribonate + NADPH

Reaction type
Redox reaction

Natural substrates

Substrate spectrum
1 D-Ribose + NADP$^+$ [1]
2 D-Xylose + NADP$^+$ [1]
3 L-Arabinose + NADP$^+$ [1]
4 D-Arabinose + NADP$^+$ [1]
5 Digitoxose + NADP$^+$ [1]
6 2-Deoxy-D-ribose + NADP$^+$ [1]

Product spectrum
1 D-Ribonate + NADPH [1]
2 ?
3 ?
4 ?
5 ?
6 ?

Inhibitor(s)
N-Ethylmaleimide [1]; Iodoacetamide [1]; p-Chloromercuribenzoate (weak); $HgCl_2$ (weak) [1]

Cofactor(s)/prosthetic group(s)/activating agents
$NADP^+$ [1]; NAD^+ (2% of the activity with $NADP^+$) [1]

Metal compounds/salts

Turnover number (min^{-1})

Specific activity (U/mg)

K_m-value (mM)
0.05 ($NADP^+$) [1]; 1.4 (D-xylose) [1]; 7.7 (D-arabinose) [1]; 8.0 (D-ribose) [1]

pH-optimum
9.8 [1]

pH-range
8.9–10.5 (about 50% of maximal activity at pH 8.9 and 10.5) [1]

Temperature optimum (°C)

Temperature range (°C)

---

3 ENZYME STRUCTURE

Molecular weight

Subunits

Glycoprotein/Lipoprotein
–

---

4 ISOLATION/PREPARATION

Source organism
Pig [1]

Source tissue
Liver [1]

Localisation in source
Cytoplasm [1]

Purification
Pig (partial) [1]

Crystallization
–

Cloned

–

Renaturated

–

5 STABILITY

pH

Temperature (°C)

Oxidation

Organic solvent

General stability information

Storage

6 CROSSREFERENCES TO STRUCTURE DATABANKS

PIR/MIPS code

Brookhaven code

7 LITERATURE REFERENCES

[1] Schiwara, H.W., Domschke, W., Domagk, G.F.: Hoppe-Seyler's Z. Physiol. Chem.,349,1575–1581 (1968)

1 NOMENCLATURE

EC number
1.1.1.116

Systematic name
D-Arabinose:NAD^+ 1-oxidoreductase

Recommended name
D-Arabinose 1-dehydrogenase

Synonymes
Dehydrogenase, D-arabinose
NAD-pentose-dehydrogenase [6]
Arabinose(fucose)dehydrogenase [2]

CAS Reg. No.
37250-47-8

2 REACTION AND SPECIFICITY

Catalysed reaction
D-Arabinose + NAD^+ →
→ D-arabinono-1,4-lactone + NADH (ordered sequential mechanism)

Reaction type
Redox reaction

Natural substrates
D-Arabinose + NAD^+ (one of a variety of NAD^+ or $NADP^+$ requiring arabinose dehydrogenases, see also 1.1.1.115) [2]

Substrate spectrum
1 D-Arabinose + NAD^+ (irreversible reaction at pH 10 [3], by far the best substrate) [1–6]
2 L-Fucose + NAD^+ [1–3, 5]
3 L-Arabinose + NAD^+ (poor [6], not [5]) [1, 6]
4 D-Ribose + NAD^+ (not [5]) [1, 6]
5 D-Glucose + NAD^+ [1]
6 D-Xylose + NAD^+ (poor substrate [6], no substrate for bacterial enzyme [5]) [1, 5, 6]
7 2-Deoxy-D-ribose + NAD^+ (no substrate for fungal enzyme [1]) [1, 6]
8 Digitoxose + NAD^+ [6]

9 D-Lyxose + NAD^+ [6]
10 More (no substrates for the bacterial enzyme are D-lyxose, D-fucose, L-rhamnose, 2-deoxy-D-glucose, D-fructose, sucrose, xylitol, L-arabitol, D-arabitol, ribitol, D-sorbitol, D-mannitol, D-galactose, D-mannose [5], no substrates for the fungal enzyme are D-fucose, L-xylose, 6-phosphogluconate [1]) [1, 5]

Product spectrum
1 Arabinonolactone + NADH (D-arabinonolactone [2, 3]) [1–3]
2 L-Fuconolactone + NADH [3]
3 Arabinonolactone + NADH [1]
4 ?
5 ?
6 ?
7 ?
8 ?
9 ?
10 ?

Inhibitor(s)
L-Fucose (at high concentrations) [3]; D-Arabinose (at high concentrations) [3]; NADH (non competitive inhibition at non-saturating NAD^+-level) [3]; D-Glucose (excess concentration) [5]; L-Xylose (excess concentration) [5]; p-Chloromercuribenzoate [6]; $HgCl_2$ [6]; N-Ethylmaleimide [6]; Iodoacetamide [6]

Cofactor(s)/prosthetic group(s)/activating agents
NAD^+ [1–6]; $NADP^+$ (66% [5], 6% [6] as effective as NAD^+ [5]) [5, 6]

Metal compounds/salts
Tris-buffer (stimulates) [2]

Turnover number (min^{-1})

Specific activity (U/mg)
0.432 (D-arabinose) [2]; 0.605 (L-fucose) [2]; 1.11 [4]

K_m-value (mM)
0.017 (NAD^+) [6]; 0.024 (NAD^+ (+ L-fucose)) [2]; 0.048 ($NADP^+$) [5]; 0.087 (NAD^+) [5]; 0.18 (L-fucose) [5]; 0.3 (NAD^+) [1, 4]; 0.5 (D-arabinose) [5]; 0.67 (D-arabinose) [1, 4]; 5.3 (L-fucose) [2]; 19 (D-arabinose) [2]; 20 (D-arabinose) [6]; 24 (2-deoxy-D-ribose) [6]; 49 (digitoxose) [6]; 670 (D-ribose) [6]

pH-optimum
8.6 (Tris/HCl buffer) [5]; 9.0 [4]; 10.0 (glycine/NaOH buffer) [2, 3]; 10.6 [6]

pH-range

8.5–10.5 (about 75% of maximal activity at pH 8.5 and 10.5) [2]; 9.4–11.8 (about 60% of maximal activity at pH 9.4 and 11.8) [6]

Temperature optimum (°C)

25 (assay at) [6]; 30 (assay at) [2–5]

Temperature range (°C)

---

3 ENZYME STRUCTURE

Molecular weight

60000 (Neurospora crassa, PAGE) [4]
245000 (pig, gel filtration) [2]

Subunits

Monomer (1 × 52000, Neurospora crassa, SDS-PAGE, PAGE with 6.2 M urea) [4]

Glycoprotein/Lipoprotein

–

---

4 ISOLATION/PREPARATION

Source organism

Neurospora crassa (wild type strain 74A, mutants col-15a and col-16a) [1, 4]; Pseudomonas sp. [5]; Pig [2, 3, 6]

Source tissue

Cell [1, 4, 5]; Liver [2, 3, 6]

Localisation in source

Cytoplasm [1–5]

Purification

Neurospora crassa (partial [1], mutants col-15a [4], col-16a [4]) [1, 4]; Pseudomonas sp. [5]; Pig (partial [6]) [2, 6]

Crystallization

–

Cloned

–

Renaturated

–

5 STABILITY

pH

7–8 [5]

Temperature (°C)

50 (half-life of partially purified enzyme 2.25 to 2.4 min) [4]

Oxidation

Organic solvent

General stability information

Ammonium sulfate precipitation, almost complete inactivation [1]; Freeze-drying: inactivation [1]; Redissolution of $(NH_4)_2SO_4$-precipitates, complete inactivation [2]

Storage

6 CROSSREFERENCES TO STRUCTURE DATABANKS

PIR/MIPS code

Brookhaven code

7 LITERATURE REFERENCES

[1] Pincheira G., Leon G., Ureta, T.: FEBS Lett.,30,111–114 (1973)
[2] Maijub, A.G., Pecht, M.A., Miller, G.R., Carper, W.R.: Biochim. Biophys. Acta,315,37–42 (1973)
[3] Carper, W.R., Chang, K.W., Thorpe, W.G., Carper, M.A., Buess, C.M.: Biochim. Biophys. Acta,358,49–56 (1974)
[4] Carrasco, A., Pincheira, G., Ureta, T.: J. Bacteriol.,145,164–170 (1981)
[5] Yamanaka, K.: Agric. Biol. Chem.,39,2227–2234 (1975)
[6] Schiwara, H.W., Domschke, W., Domagk, G.F.: Hoppe-Seyler's Z. Physiol. Chem.,349,1575–1581 (1968)

1 NOMENCLATURE

EC number
1.1.1.117

Systematic name
D-Arabinose:NAD(P)$^+$ 1-oxidoreductase

Recommended name
D-Arabinose 1-dehydrogenase (NAD(P)$^+$)

Synonymes
Dehydrogenase, D-arabinose (nicotinamide adenine dinucleotide (phosphate))

CAS Reg. No.
37250-48-9

2 REACTION AND SPECIFICITY

Catalysed reaction
D-Arabinose + NAD(P)$^+$ →
→ D-arabinono-1,4-lactone + NAD(P)H

Reaction type
Redox reaction

Natural substrates
D-Arabinose + NAD(P)$^+$ (specific for sugars of the optical configuration of L-galactose) [2]

Substrate spectrum
1 D-Arabinose + NAD(P)$^+$ (alpha-anomer preferred [2]) [1, 2]
2 L-Galactose + NAD(P)$^+$ (beta-anomer preferred) [2]
3 L-Fucose + NAD(P)$^+$ (i.e. 6-deoxy-L-galactose) [2]
4 L-Colitose + NAD(P)$^+$ (i.e. 3,6-dideoxy-L-galactose) [2]

Product spectrum
1 Arabinonolactone + NAD(P)H [1, 2]
2 ?
3 ?
4 ?

Inhibitor(s)

Cofactor(s)/prosthetic group(s)/activating agents
NADP+ [1–3]; NAD+ (the affinity for NADP+ is considerably higher [2]) [1–3]

Metal compounds/salts

Turnover number (min^{-1})

Specific activity (U/mg)

K_m-value (mM)
0.82 (D-arabinose) [2]; 2.2 (NADP+) [3]; 2.3 (NADP+) [2]; 22.0 (NAD+) [2]

pH-optimum
8.0–8.5 [2]

pH-range
7.3–9.2 (about 80% of maximal activity at pH 7.3 and 9.2) [2]

Temperature optimum (°C)
23 (assay at) [2]; 24 (assay at) [1]

Temperature range (°C)

---

3 ENZYME STRUCTURE

Molecular weight
50000 (Pseudomonas sp. G6, gel filtration) [3]
104000 (Pseudomonas sp. G6, sedimentation analysis in EDTA/phosphate buffer) [3]

Subunits
? (x × 20000, Pseudomonas sp. G6, sedimentation analysis in 4.2 M guanidinium chloride, presence of subunits of unequal size might be possible) [3]

Glycoprotein/Lipoprotein
–

---

4 ISOLATION/PREPARATION

Source organism
Pseudomonas sp. G6 [1–3]

Source tissue
Cell [1–3]

Localisation in source
Cytoplasm [1–3]

Purification

Pseudomonas sp. G6 [1–3]

Crystallization

–

Cloned

–

Renaturated

–

5 STABILITY

pH

Temperature (°C)

50 (after 10 min 90% of maximal activity retained) [1]; 58 (half-life of 20 s is increased to 80 s in presence of cofactor, in both cases biphasic inactivation curve) [3]

Oxidation

Organic solvent

General stability information

Storage

Frozen, purified enzyme stable for many months [1]; Lyophilized, purified enzyme stable for many months [1]; –20°C, acetone-dried cells retain full enzyme activity for extended periods of time [1]; 4°C, stable as ammonium sulfate precipitate [1]

6 CROSSREFERENCES TO STRUCTURE DATABANKS

PIR/MIPS code

Brookhaven code

7 LITERATURE REFERENCES

[1] Cline, A.L., Hu, A.S.L.: J. Biol. Chem.,240,4488–4492 (1965)
[2] Cline, A.L., Hu, A.S.L.: J. Biol. Chem.,240,4493–4497 (1965)
[3] Cline, A.L., Hu, A.S.L.: J. Biol. Chem.,240,4498–4502 (1965)

1 NOMENCLATURE

EC number
1.1.1.118

Systematic name
D-Glucose:NAD^+ 1-oxidoreductase

Recommended name
Glucose 1-dehydrogenase (NAD^+)

Synonymes
D-Glucose:NAD oxidoreductase [1]
D-Aldohexose dehydrogenase [3]
Dehydrogenase, glucose (nicotinamide adenine dinucleotide)

CAS Reg. No.
37250-49-0

2 REACTION AND SPECIFICITY

Catalysed reaction
D-Glucose + NAD^+ →
→ D-glucono-1,5-lactone + NADH

Reaction type
Redox reaction

Natural substrates

Substrate spectrum
1 D-Glucose + NAD^+ (i.e. beta-D-glucopyranose [3]) [1–3]
2 D-Galactose + NAD^+ [3]
3 D-Mannose + NAD^+ [3]
4 2-Deoxy-D-glucose + NAD^+ [3]
5 6-Deoxy-D-galactose + NAD^+ (i. e. beta-D-fucopyranose) [3]
6 2-Deoxy-D-galactose + NAD^+ [3]
7 D-Altrose + NAD^+ [3]
8 6-Deoxy-D-glucose + NAD^+ (i.e. D-quinose) [3]
9 D-Allose + NAD^+ [3]
10 3,6-Dideoxy-D-galactose + NAD^+ (i.e. abequose) [3]
11 More (not: D-glucuronic acid, D-galacturonic acid, D-galactose 6-phosphate, D-glucose 6-phosphate, D-glucosamine, N-acetyl-D-glucosamine, 6-deoxy-D-allose, 6-iodo-6-deoxy-D-galactose, 2-acetamide-6-deoxy-D-allose, 2-acetamido-6-deoxy-D-altrose) [3]

Product spectrum

1 D-Glucono-1,5-lactone + NADH
2 ?
3 ?
4 ?
5 ?
6 ?
7 ?
8 ?
9 ?
10 ?
11 ?

Inhibitor(s)

Cofactor(s)/prosthetic group(s)/activating agents

NAD^+ [1–3]; More ($NADP^+$ not used) [3]

Metal compounds/salts

Turnover number (min^{-1})

Specific activity (U/mg)

More [3]

K_m-value (mM)

0.86 (D-glucose) [3]; 5.8 (D-fucose) [3]; 80 (NAD^+) [3]

pH-optimum

8.0–8.5 (Tris-HCl buffer) [3]; 9–10 (glycine-NaOH buffer) [3]

pH-range

Temperature optimum (°C)

Temperature range (°C)

3 ENZYME STRUCTURE

Molecular weight

Subunits

Glycoprotein/Lipoprotein

–

4 ISOLATION/PREPARATION

Source organism

Pseudomonas sp. (MSU-1 [3]) [1, 3]; Corynebacterium sp. (No. 93–1) [2]

Source tissue

Localisation in source

Purification

Pseudomonas sp. (MSU-1) [3]

Crystallization

–

Cloned

–

Renaturated

–

5 STABILITY

pH

Temperature (°C)

Oxidation

Organic solvent

General stability information

Storage

Stable to freezing for at least 6 months, pH 7.0 [3]

6 CROSSREFERENCES TO STRUCTURE DATABANKS

PIR/MIPS code

PIR2:A17150 (bovine (fragment))

Brookhaven code

7 LITERATURE REFERENCES

[1] Hu, A.S.L., Cline, A.L.: Biochim. Biophys. Acta,93,237–245 (1964)
[2] Kobayashi, Y., Horikoshi, K.: Agric. Biol. Chem.,44,41–47 (1980)
[3] Anderson, R.L., Dahms, A.S.: Methods Enzymol.,41,147–150 (1975) (Review)

1 NOMENCLATURE

EC number
1.1.1.119

Systematic name
D-Glucose:NADP+ 1-oxidoreductase

Recommended name
Glucose 1-dehydrogenase (NADP+)

Synonymes
Nicotinamide adenine dinucleotide phosphate-linked aldohexose dehydrogenase [1]
NADP-linked aldohexose dehydrogenase
Dehydrogenase, glucose (nicotinamide adenine dinucleotide phosphate)
NADP-dependent glucose dehydrogenase

CAS Reg. No.
37250-50-3

2 REACTION AND SPECIFICITY

Catalysed reaction
D-Glucose + NADP+ →
→ D-glucono-1,5-lactone + NADPH

Reaction type
Redox reaction

Natural substrates

Substrate spectrum
1 D-Glucose + NADP+ (r [1, 3], 4% of the reverse reaction when assayed with D-glucono-1,5-lactone and NADPH at neutral or weak alkaline pH [3]) [1–3]
2 D-Mannose + NADP+ [1–3]
3 2-Amino-2-deoxy-D-mannose + NADP+ [1]
4 2-Deoxy-D-glucose + NADP+ [1]
5 More (overview: sugars being neither substrates nor inhibitors) [1, 2]

Product spectrum

1 D-Glucono-1,5-lactone + NADPH (D-glucono-delta-lactone [1–3])
2 ?
3 ?
4 ?
5 ?

Inhibitor(s)

NADPH (competitive to $NADP^+$) [1]; $ZnSO_4$ [1, 2]; $CuSO_4$ [1, 2]; Hg^{2+} [3]; Cu^{2+} [3]; Ni^{2+} [3]; p-Chloromercuribenzoate [3]; Sulfhydryl reagents [3]; More (overview: sugars found to be neither substrates nor inhibitors [1, 2], not: iodoacetate, N-ethylmaleimide, p-hydroxymercuribenzoate, 5,5'-dithio-bis(2-nitrobenzate)) [1, 2]

Cofactor(s)/prosthetic group(s)/activating agents

$NADP^+$ [1–4]; NADPH [1, 3]; More (inactive with NAD^+ [1–3], acceptance of adenine-modified $NADP^+$ [4]) [1–4]

Metal compounds/salts

Turnover number (min^{-1})

Specific activity (U/mg)

10.1 [1, 2]; 140 [3]

K_m-value (mM)

0.01 ($NADP^+$) [3]; 0.012 ($NADP^+$) [1]; 0.02 ($NADP^+$) [4]; 0.03 (N^6-(2-aminoethyl)-$NADP^+$) [4]; 0.038 (NADPH) [1]; 0.07 (N^1-(2-aminoethyl)-$NADP^+$, N^6-ethenoadenine-$NADP^+$) [4]; 0.09 (polyethylene glycol(MW 20000)-N^6-(2-aminoethyl)-$NADP^+$) [4]; 1.8 (D-mannose) [1, 2]; 5 (D-glucose) [3]; 5.3 (D-glucose) [1, 2]; 29 (2-amino-2-deoxy-D-mannose) [1, 2]; 39 (2-deoxy-D-glucose) [1, 2]

pH-optimum

7.5–9.0 [1–3]

pH-range

Temperature optimum (°C)

50 [3]

Temperature range (°C)

3 ENZYME STRUCTURE

Molecular weight

150000 (Gluconobacter suboxydans, gel filtration) [3]

Subunits

Tetramer (4 × 40000, Gluconobacter suboxydans, SDS-PAGE) [3]

Glycoprotein/Lipoprotein

–

4 ISOLATION/PREPARATION

Source organism

Gluconobacter cerinus [1, 2]; Gluconobacter suboxydans (IFO 12528) [3]; Cryptococcus uniguttulatus [4]

Source tissue

Cell [1]

Localisation in source

Cytosol [3]

Purification

Gluconobacter suboxydans (IFO 12528) [3]; Gluconobacter cerinus (partial) [1, 2]

Crystallization

[3]

Cloned

–

Renaturated

–

5 STABILITY

pH

Temperature (°C)

Oxidation

Organic solvent

General stability information

Lyophilization of dilute enzyme solutions, below 0.5 ng/ml results in significant loss of activity [1, 2]; Freezing and thawing: significant loss of activity [1, 2]

Storage

10% loss of activity after 1 year, lyophilized enzyme in deep freezer [1]; 4°C, pH 5.4–8.5, 0.5 mg/ml protein concentration, stable for 8 weeks [1, 2]; 5°C, crystalline preparation as suspension in ammonium sulfate is quite stable for more than 6 months [3]

6 CROSSREFERENCES TO STRUCTURE DATABANKS

PIR/MIPS code

Brookhaven code

7 LITERATURE REFERENCES

[1] Avigad, G., Alroy, Y., Englard, S.: J. Biol. Chem.,243,1936–1941 (1968)
[2] Avigad, G., Englard, S.: Methods Enzymol.,41,142–147 (1975) (Review)
[3] Adachi, O., Matsushita, K., Shinagawa, E., Ameyama, M. : Agric. Biol. Chem.,44,301–308 (1980)
[4] Bückmann, A.F., Schmid, R.: Biotechnol. Appl. Biochem.,14,104–113 (1991)

1 NOMENCLATURE

EC number
1.1.1.120

Systematic name
D-Galactose:$NADP^+$ 1-oxidoreductase

Recommended name
Galactose 1-dehydrogenase ($NADP^+$)

Synonymes
Dehydrogenase, galactose (nicotinamide adenine dinucleotide phosphate)
D-Galactose dehydrogenase ($NADP^+$)

CAS Reg. No.
37250-51-4

2 REACTION AND SPECIFICITY

Catalysed reaction
D-Galactose + $NADP^+$ →
→ D-galactonolactone + NADPH

Reaction type
Redox reaction

Natural substrates
D-Galactose + $NADP^+$ (beta-anomer, the enzyme is specific for sugars of the optical configuration of D-galactose) [2]

Substrate spectrum
1 D-Galactose + $NADP^+$ (beta-anomer preferred [2]) [1–3]
2 L-Arabinose + $NADP^+$ (alpha-anomer preferred) [2]
3 D-Fucose + $NADP^+$ (i.e. 6-deoxy-D-galactose) [2]
4 2-Deoxy-D-galactose + $NADP^+$ [2]

Product spectrum
1 D-Galactonolactone + NADPH [1–3]
2 ?
3 ?
4 ?

Inhibitor(s)

Cofactor(s)/prosthetic group(s)/activating agents
$NADP^+$ [1–3]

Metal compounds/salts

Turnover number (min^{-1})

Specific activity (U/mg)

K_m**-value** (mM)

pH-optimum
9.7 [2]

pH-range
8.3–10.2 (about 80% of maximal activity at pH 8.3 and 10.2) [2]

Temperature optimum (°C)
23 (assay at) [2]; 24 (assay at) [1]

Temperature range (°C)

---

3 ENZYME STRUCTURE

Molecular weight
64000 (Pseudomonas sp. G6, sedimentation analysis in EDTA/phosphate-buffer) [3]

Subunits
? (x × 30000, Pseudomonas sp. G6, gel filtration, sedimentation analysis in 4.2 M guanidinium chloride) [3]

Glycoprotein/Lipoprotein
–

---

4 ISOLATION/PREPARATION

Source organism
Pseudomonas sp. G6 [1–3]

Source tissue
Cell [1–3]

Localisation in source
Cytoplasm [1–3]

Purification
Pseudomonas sp. G6 [1–3]

Crystallization

(Pseudomonas sp. G6) [1, 2]

Cloned

–

Renaturated

–

5 STABILITY

pH

Temperature (°C)

50 (after 10 min more than 90% of the maximal activity retained) [1]

Oxidation

Organic solvent

General stability information

Storage

4°C, stable as ammonium sulfate precipitate without loss of activity [1]; Frozen, purified enzyme stable for many months [1]; Lyophilized, purified enzyme stable for many months [1]

6 CROSSREFERENCES TO STRUCTURE DATABANKS

PIR/MIPS code

Brookhaven code

7 LITERATURE REFERENCES

[1] Cline, A.L., Hu, A.S.L.: J. Biol. Chem.,240,4488–4492 (1965)
[2] Cline, A.L., Hu, A.S.L.: J. Biol. Chem.,240,4493–4497 (1965)
[3] Cline, A.L., Hu, A.S.L.: J. Biol. Chem.,240,4498–4502 (1965)

1 NOMENCLATURE

EC number
1.1.1.121

Systematic name
D-Aldose:NAD^+ 1-oxidoreductase

Recommended name
Aldose 1-dehydrogenase

Synonymes
Dehydrogenase, aldose
Dehydrogenase, D-aldohexose

CAS Reg. No.
9076-61-3

2 REACTION AND SPECIFICITY

Catalysed reaction
D-Aldose + NAD^+ →
→ D-aldonolactone + NADH

Reaction type
Redox reaction

Natural substrates
D-Aldose + NAD^+ (D-aldoses of pyranose ring size [2], broad substrate specificity [1, 2, 4], the enzyme is coordinately regulated with EC 1.1.1.120 and EC 1.1.1.117 [4]) [1–4]

Substrate spectrum
1 D-Glucose + NAD^+ (r [2], beta-anomer preferred to alpha-anomer [4], the most probable produced aldonolactones from this and the following substrates have not been isolated so far [2, 4]) [1, 2, 4]
2 D-Galactose + NAD^+ (beta-anomer preferred to alpha-anomer [4]) [1–4]
3 L-Arabinose + NAD^+ (alpha-anomer preferred to beta-anomer [4]) [1, 4]
4 D-Fucose + NAD^+ [1, 2, 4]
5 D-Xylose + NAD^+ [1, 4]
6 D-Mannose + NAD^+ [1, 2]
7 D-Altrose + NAD^+ [2]
8 D-Allose + NAD^+ [2]
9 2-Deoxy-D-glucose + NAD^+ [2, 4]
10 2-Deoxy-D-galactose + NAD^+ [2, 4]
11 D-Quinovose + NAD^+ [4]

12 Chalcose + NAD^+ (demethylated) [4]
13 Viosamine + NAD^+ (i.e. 4-amino-4,6-dideoxy-D-glucose) [4]
14 5-Hydroxypentanal + NAD^+ [4]
15 2,3,4-Trideoxyaldopyranose + NAD^+ [4]
16 More (substrates are D-aldoses of pyranose ring size [2], broad substrate specificity [1, 2, 4], L-aldohexoses, ketohexoses, pentoses, trioses, polyols, di- and trisaccharides are no substrates [2]) [1, 2, 4]

Product spectrum

1 D-Gluconolactone + NADH [1, 2, 4]
2 D-Galactonolactone + NADH [1–4]
3 ?
4 D-Fucono-delta-lactone + NADH (hydrolyzes spontaneously to D-fuconate) [1, 2, 4]
5 ?
6 ?
7 ?
8 ?
9 ?
10 ?
11 ?
12 ?
13 ?
14 ?
15 ?
16 ?

Inhibitor(s)

Zn^{2+} [2]; Fe^{2+} [2]; Cu^{2+} [2]; Ni^{2+} [2]; Ca^{2+} [2]; Co^{2+} [2]; NH_4^+ [2]

Cofactor(s)/prosthetic group(s)/activating agents

NAD^+ [1–5]; $NADP^+$ (active at concentrations above 0.02 M [2, 4]) [1, 2, 4]

Metal compounds/salts

Turnover number (min^{-1})

Specific activity (U/mg)

13.8 [2]

K_m-value (mM)

0.08 (NAD^+) [2]; 0.16 (NAD^+) [4]; 0.25 (NAD^+) [5]; 0.78 (D-galactose) [4]; 0.86 (D-glucose) [2]; 1.6 (D-galactose, 2-deoxy-D-glucose) [2]; 2.4 (D-altrose) [2]; 4.5 (D-mannose) [2]; 5.8 (D-fucose) [2]; 6.3 (2-deoxy-D-galactose) [2]; 13.0 (D-allose) [2]

pH-optimum

8–8.5 (Tris/HCl buffer) [2]; 9.0–10.0 (glycine/NaOH buffer) [2]; 9.2 (Tris/acetate buffer) [4]

pH-range

7.3–9.3 (about 66% of maximal activity at pH 7.2 and 9.3 in Tris/HCl buffer) [2]; 8.0–9.8 (about 80% of maximal activity at pH 8.0 and 9.8 in Tris/acetate buffer) [4]; 8.5–10.4 (about 80% of maximal activity at pH 8.5 and 10.4 in glycine/NaOH buffer) [2]

Temperature optimum (°C)

23 (assay at) [3, 4]; 25 (assay at) [2]

Temperature range (°C)

3 ENZYME STRUCTURE

Molecular weight

40000 (Pseudomonas sp., gel filtration) [5]

140000 (Pseudomonas sp., sedimentation analysis in EDTA/phosphate buffer) [5]

Subunits

? (x × 38000, Pseudomonas sp., sedimentation analysis in 4.2 M guanidinium chloride, presence of subunits of unequal size might be possible) [5]

Glycoprotein/Lipoprotein

–

4 ISOLATION/PREPARATION

Source organism

Bacterium strain 58 [1]; Pseudomonas sp. (MSU-1 [2], G6 [3–5]) [2–5]

Source tissue

Cell [1–5]

Localisation in source

Cytoplasm [1–5]

Purification

Pseudomonas sp. (MSU-1 [2], G6 [3–5]) [2–5]

Crystallization

–

Cloned

–

Renaturated

–

5 STABILITY

pH

Temperature (°C)

46 (half-life: 1 min, biphasic inactivation curve) [5]; 50 (rapidly inactivated) [3]; 55 (half-life: 40 s in sodium phosphate buffer, pH 7.0, with D-fucose, D-galactose, D-glucose or D-mannose as substrate) [2]

Oxidation

Organic solvent

General stability information

Storage

–20°C, Sephadex G-200 fractions are stable for at least 6 months [2]; –20°C, acetone-dried cells retain full activity over a long period [3]; Frozen, stable for many months [3]; Lyophilized, stable for many months [3]; 5°C, in presence of 2-mercaptoethanol 50% loss of activity after several months [3]

6 CROSSREFERENCES TO STRUCTURE DATABANKS

PIR/MIPS code

Brookhaven code

7 LITERATURE REFERENCES

[1] Green, P.N., Gibson, D.M.: FEMS Microbiol. Lett.,23,31–34 (1984)
[2] Dahms, A.S., Anderson, R.L.: J. Biol. Chem.,247,2222–2227 (1972)
[3] Cline, A.L., Hu, A.S.L.: J. Biol. Chem.,240,4488–4492 (1965)
[4] Cline, A.L., Hu, A.S.L.: J. Biol. Chem.,240,4493–4497 (1965)
[5] Cline, A.L., Hu, A.S.L.: J. Biol. Chem.,240,4498–4502 (1965)

1 NOMENCLATURE

EC number
1.1.1.122

Systematic name
D-threo-Aldose:NAD^+ 1-oxidoreductase

Recommended name
D-threo-Aldose 1-dehydrogenase

Synonymes
(2S,3R)-Aldose dehydrogenase
Dehydrogenase, L-fucose
L-Fucose (D-arabinose) dehydrogenase
2S,3R-Aldose-dehydrogenase [3]

CAS Reg. No.
9082-70-6

2 REACTION AND SPECIFICITY

Catalysed reaction
A D-threo aldose + NAD^+ →
→ a D-threo-aldono-1,5-lactone + NADH

Reaction type
Redox reaction

Natural substrates
L-Fucose + NAD^+ (irreversible reaction, initial step of L-fucose catabolism) [6]

Substrate spectrum
1 L-Fucose + NAD^+ (steric requirement is the configuration of hydroxy-groups from C-2 to C-4 [6]) [1–6]
2 D-Arabinose + NAD^+ [1–6]
3 3-Amino-D-arabinose + NAD^+ [2]
4 L-Galactose + NAD^+ [1, 2, 4]
5 D-Ribose + NAD^+ (poor substrate, no substrate for sheep enzyme [4]) [2–4]
6 L-Xylose + NAD^+ (poor substrate) [2, 3]
7 D-Xylose + NAD^+ (poor substrate [3]) [3, 6]
8 D-Lyxose + NAD^+ (poor substrate) [2]
9 2-Deoxy-D-ribose + NAD^+ [5]

10 L-Glucose + NAD^+ [3]
11 More (enzyme acts on pyranose form of sugars [2], catalyzes reactions with sugars of the same hydroxyl-group configuration from C-2 to C-4 [2], no substrates are D-glucose [3, 5, 6], D-galactose [6], 2-deoxy-D-glucose [5], D-arabitol, L-arabinose, D-fucose, D-mannose, L-rhamnose, D-fructose, lactose, melibiose, D-galactono-1,4-lactone, D-glucono-1,4-lactone, D-glucosamine, D-galactosamine, N-acetylglucosamine, N-acetylgalactosamine [2, 3, 5]) [2, 3, 5, 6]

Product spectrum

1 L-Fucono-1,5-lactone + NADH (L-fucono-delta-lactone [5], hydrolyzes spontaneously to L-fuconate [1, 2, 5]) [1–6]
2 D-Arabinono-1,5-lactone + NADPH (hydrolyzes spontaneously to D-arabinate [5]) [1–6]
3 ?
4 ?
5 ?
6 ?
7 ?
8 ?
9 ?
10 ?
11 ?

Inhibitor(s)

Thiol reagents [1]; Heavy metal ions (e.g. Cd^{2+} [6], Hg^{2+} [6]) [1, 6]; D-Arabinose (competitive inhibition of L-fucose oxidation) [2]; 3-Amino-3-deoxy-D-arabinose (competitive inhibition of L-fucose oxidation) [2]; PCMB (20% loss of activity with saturated inhibitor [2]) [2, 6]; Iodoacetate (at higher concentration [2]) [2, 6]; D-Glucose (competitive inhibition of L-glucose oxidation) [3]; D-Galactose [3]; L-Arabinose [3]; L-Rhamnose [3]; NADH [4]; ATP (slight inhibition) [4]; ADP (slight inhibition) [4]

Cofactor(s)/prosthetic group(s)/activating agents

NAD^+ [1–6]; $NADP^+$ (10 to 20% as effective as NAD^+ [3], $NADP^+$ is no cofactor for mammalian enzyme [2, 5, 6]) [3]; Increased ionic strength (activation) [4]

Metal compounds/salts

Turnover number (min^{-1})

Specific activity (U/mg)

0.433 [2]; 1.03 [5]; 3.66 [3]

K_m-value (mM)

0.01 (NAD^+, pH 10.4) [5]; 0.02 (NAD^+) [2]; 0.04 (NAD^+) [5]; 0.07 (L-fucose, pH 8.6 [5], NAD^+ [6]) [5, 6]; 0.15 (L-fucose) [6]; 0.20 (L-galactose) [4]; 0.32 (L-fucose) [2]; 0.33 (NAD^+) [1]; 0.40 (D-arabinose) [5]; 1.4 (D-arabinose) [6]; 1.5 (L-fucose, pH 10.4) [5]; 1.96 (L-fucose) [1]; 2.1 (D-arabinose [2], L-fucose [3]) [2, 3]; 2.8 (D-arabinose) [3]; 4.5 (L-xylose) [3]; 7.3 (D-arabinose, pH 10.4) [5]; 8.0 (L-galactose [2], 3-amino-3-deoxy-D-arabinose [2]) [2]; 15 (L-glucose) [3]; 40.0 (D-arabinose) [1]; 44 (L-xylose) [2]; 50 (L-galactose) [1]

pH-optimum

8.7 [2]; 9.0–9.5 (high salt concentration) [4, 5]; 10.0 [3, 6]; 10.4 (low salt concentration) [4, 5]; 10.5 [1]

pH-range

7.2–10.4 (about 50% of maximal activity at pH 7.2 and 10.4) [2]; 9.3–12.0 (about 85% of maximal activity at pH 9.3 and 12.0) [2]

Temperature optimum (°C)

33 [1]; 50 [6]

Temperature range (°C)

3 ENZYME STRUCTURE

Molecular weight

40000 (Pullularia pullulans, gel filtration) [1]
70000 (Pseudomonas caryophylli, gel filtration) [3]
82000 (sheep, disc gel electrophoresis) [4]
92000 (rabbit, gel filtration) [6]
96000 (sheep, gel filtration) [5]
123000 (sheep, equilibrium sedimentation) [4]

Subunits

Monomer (1 × 40000, Pullularia pullulans, SDS-PAGE) [1]
Tetramer (4 × 30000, sheep, SDS-PAGE [4], 4 × 35000, sheep, gel electrophoresis after treatment with 6 M guanidinium-HCl [4]) [4]

Glycoprotein/Lipoprotein

–

4 ISOLATION/PREPARATION

Source organism

Pullularia pullulans (fungus) [1]; Pseudomonas caryophylli [3]; Pig [2]; Rabbit [6]; Sheep [4, 5]

Source tissue
Cell [1, 3]; Liver [2, 4–6]

Localisation in source
Cytoplasm [1–6]

Purification
Pullularia pullulans [1]; Pseudomonas caryophylli [3]; Pig [2]; Rabbit [6]; Sheep [4, 5]

Crystallization
–

Cloned
–

Renaturated
–

5 STABILITY

pH
5.0–12.0 (stable) [6]; 7.2 (half-life: 1.8 days at 25°C) [4]; 8.5 (and above, half-life: 6 min at 60°C) [4]

Temperature (°C)
4 (half-life: 39 days) [4]; 25 (half-life: 1.8 days) [4]; 50 (more than 50% of maximal activity retained for 30 min) [6]; 60 (pH 7.2, without NAD^+, half-life: 25 min, with NAD^+, half-life: 35 min [4], stable for 10 min [6]) [4, 6]

Oxidation

Organic solvent

General stability information
NAD^+ stabilizes [4]

Storage
–20°C, stable for a year [6]; –20°C, stable for 2 months [1]; –16°C, higher than 5-fold purified enzyme preparations stable for 1 month [2, 5]; 4°C, rapid loss of activity [1]; 4°C, stable in presence of NAD^+ for 39 days [4, 6]; 4°C, purified enzyme stable for a month [2, 3]; 5°C, stable for a week [5]

6 CROSSREFERENCES TO STRUCTURE DATABANKS

PIR/MIPS code

Brookhaven code

7 LITERATURE REFERENCES

[1] Guimaraes, M.F., Veiga, L.A.: Arq. Biol. Tecnol. (Curitiba) ,32,575–587 (1989)
[2] Schachter, H., Sarney, J., McGuire, E.J., Roseman, S: J. Biol. Chem.,244,4785–4792 (1969)
[3] Sasajima, K.-I., Sinskey, A.J.: Biochim. Biophys. Acta,571,120–126 (1979)
[4] Mobley, P.W., Metzger, R.: Arch. Biochem. Biophys.,186,184–188 (1978)
[5] Mobley, P.W., Metzger, R., Wick, A.N.: Methods Enzymol.,41 (B) ,173–177 (1975)
[6] Endo, M., Hiyama, N.: J. Biochem.,86,1559–1565 (1979)

1 NOMENCLATURE

EC number
1.1.1.123

Systematic name
L-Sorbose:$NADP^+$ 5-oxidoreductase

Recommended name
Sorbose 5-dehydrogenase ($NADP^+$)

Synonymes
5-Keto-D-fructose reductase
Dehydrogenase, sorbose (nicotinamide adenine dinucleotide phosphate)
Reduced nicotinamide adenine dinucleotide phosphate-linked reductase [3]
5-Ketofructose reductase

CAS Reg. No.
37250-52-5

2 REACTION AND SPECIFICITY

Catalysed reaction
5-Dehydro-D-fructose + NADPH →
→ L-sorbose + $NADP^+$

Reaction type
Redox reaction

Natural substrates
5-Keto-D-fructose + NADPH [1]

Substrate spectrum
1 5-Keto-D-fructose + NADPH (reverse reaction not observed [1, 2]) [1–3]
2 5-Keto-D-fructose 1-phosphate + NADPH (low activity: 1.8% of the rate observed with 5-keto-D-fructose) [2, 3]
3 More (relatively inactive on 2,5-diketo-D-gluconate + NADPH) [1]

Product spectrum
1 L-Sorbose + $NADP^+$ [2, 3]
2 L-Sorbose phosphate + $NADP^+$
3 ?

Inhibitor(s)

$NADP^+$ [2, 3]; p-Hydroxymercuribenzoate [2, 3]

Cofactor(s)/prosthetic group(s)/activating agents

NADPH (stereospecific transfer of hydrogen from the A side (i.e. pro-R) of reduced pyridine to the ketohexose substrate [3], not: NADH [1]) [1–3]

Metal compounds/salts

Turnover number (min^{-1})

Specific activity (U/mg)

28.4 [2, 3]

K_m-value (mM)

5.9 (5-keto-D-fructose) [1]; 1.0 (5-keto-D-fructose) [2, 3]; 0.046 (NADPH) [2, 3]; 0.29 (5-keto-D-fructose 2-phosphate) [2, 3]

pH-optimum

6.0–7.0 [1]; 7.4–8.5 [2, 3]

pH-range

5–9 (5: about 20% of activity maximum, 9: about 35% of activity maximum) [1]; 5.5–10.2 (pH 5.5 and 10.2: about 30% of activity maximum) [3]

Temperature optimum (°C)

27 (activity increases gradually with increasing temperature from 0°C to 27°C, falls rapidly above 27°C and is completely lost at 38°C) [1]

Temperature range (°C)

0–38 (activity increases gradually with increasing temperature from 0°C to 27°C, falls rapidly above 27°C and is completely lost at 38°C) [1]

---

3 ENZYME STRUCTURE

Molecular weight

33000 (Corynebacterium sp., gel filtration, SDS-PAGE) [1]

Subunits

Monomer (1 × 33000, Corynebacterium sp., SDS-PAGE) [1]

Glycoprotein/Lipoprotein

–

4 ISOLATION/PREPARATION

Source organism

Corynebacterium sp. (mutant strain) [1]; Torula utilis [2, 3]; Brewer's yeast [2, 3]; Saccharomyces cerevisiae [2, 3]; Bacillus subtilis var. atereum (low activity) [2, 3]; Aerobacter cloacae (low activity) [2, 3]; Bacillus subtilis 168–2 (low activity) [2, 3]; Proteus vulgaris (low activity) [2, 3]; Aerobacter aerogenes (low activity) [2, 3]; Bacillus megaterium (low activity) [2, 3]; Bacillus brevis (low activity) [2, 3]; Acetobacter melanogenum (low activity) [2]; Gluconobacter cerinus [3]; More (in presence of NADH, rat and sheep liver sorbitol dehydrogenase (EC 1.1.1.14) catalyze the reduction of 5-keto-D-fructose specifically to sorbose [2], no activity: E. coli K-12, Bacillus cereus, Candida tropicalis [3]) [2, 3]

Source tissue

Cell [1–3]

Localisation in source

Purification

Corynebacterium sp. [1]; Saccharomyces cerevisiae [2, 3]

Crystallization

–

Cloned

–

Renaturated

–

5 STABILITY

pH

Temperature (°C)

Oxidation

Organic solvent

General stability information

Storage

6 CROSSREFERENCES TO STRUCTURE DATABANKS

PIR/MIPS code

Brookhaven code

7 LITERATURE REFERENCES

[1] Yagi, S., Kobayahi, K., Sonoyama, T.: J. Ferment. Bioeng.,67,212–214 (1989)
[2] Englard, S., Avigad, G.: Methods Enzymol.,41B,132–138 (1975) (Review)
[3] Englard, S., Kaysen, G., Avigad, G.: J. Biol. Chem.,245,1311–1318 (1970)

1 NOMENCLATURE

EC number

1.1.1.124

Systematic name

D-Fructose:NADP⁺ 5-oxidoreductase

Recommended name

Fructose 5-dehydrogenase (NADP⁺)

Synonymes

5-Keto-D-fructose reductase (NADP⁺)
Dehydrogenase, fructose 5-(nicotinamide adenine dinucleotide phosphate)
D-(-)Fructose:(NADP⁺) 5-oxidoreductase [1]

CAS Reg. No.

37250-53-6

2 REACTION AND SPECIFICITY

Catalysed reaction

5-Dehydro-D-fructose + NADPH →
→ D-fructose + NADP⁺ (ordered steady-state kinetic mechanism [1])

Reaction type

Redox reaction

Natural substrates

5-Dehydro-D-fructose + NADPH [4]

Substrate spectrum

1 5-Dehydro-D-fructose + NADPH (r [1, 2], enzyme reaction predominant in reduction of 5-keto-D-fructose to fructose, reverse reaction occurs at an extremely low rate [4]) [1–4]

Product spectrum

1 D-Fructose + NADP⁺ (r [1, 2])

Inhibitor(s)

$NADP^+$ (product inhibition) [1, 3]; NADPH (product inhibition) [1]; Fructose (product inhibition) [1, 3]; Sulfhydryl reagents [4]

Cofactor(s)/prosthetic group(s)/activating agents

NADPH (strictly specific for [4]) [1–4]; NADH (Erwinia citreus: NADH can substitute for NADPH [1], Gluconobacter cerinus: NADH cannot substitute for NADPH [3]) [1]

Metal compounds/salts

Turnover number (min^{-1})

Specific activity (U/mg)

2450 [1]; 627 [2, 3]; 620 [3]

K_m-value (mM)

0.0075 (NADPH) [4]; 0.053 ($NADP^+$) [1]; 0.071 (NADPH) [1]; 12.7 (5-ketofructose) [1]; 9.1 (D-fructose) [1]; 1.97 ($NADP^+$) [1]; 0.32 (NADPH) [1]; 4.5 (5-keto-D-fructose) [2–4]; 1.8 (NADPH) [2, 3]; 70 (D-fructose) [2, 3]; 1.3 ($NADP^+$) [2, 3]

pH-optimum

6.5 (reduction of 5-ketofructose [2–4], phosphate buffer [2, 3]) [2–4]; 7.0 (reduction of 5-ketofructose) [1]; 7.4 (reduction of 5-ketofructose, Tris-HCl buffer) [2, 3]; 8.5–10.0 (D-fructose oxidation) [3]; 9.5 (oxidation of D-fructose) [1]

pH-range

5.1–7.9 (half-maximal activity at pH 5.1 and 7.9) [3]

Temperature optimum (°C)

30 (assay at) [1–3]

Temperature range (°C)

---

3 ENZYME STRUCTURE

Molecular weight

38000–40000 (Erwinia citreus, SDS-PAGE, isoelectric focusing) [1]
50000 (Erwinia citreus, sedimentation equilibrium, association of monomers) [1]
100000 (Gluconobacter industrius, gel filtration) [4]

Subunits

Tetramer (4 × 25000, Gluconobacter industrius, SDS-PAGE) [4]

Glycoprotein/Lipoprotein

–

4 ISOLATION/PREPARATION

Source organism

Erwinia citreus [1]; Gluconobacter cerinus [2, 3]; Gluconobacter industrius IFO 3260 [4]

Source tissue

Cell [1–3]

Localisation in source

Purification

Erwinia citreus [1]; Gluconobacter cerinus [2, 3]; Gluconobacter industrius IFO 3260 [4]

Crystallization

[4]

Cloned

–

Renaturated

–

5 STABILITY

pH

4.0–8.0 (stable) [1]

Temperature (°C)

Oxidation

Organic solvent

General stability information

Storage

4°C, pH 7.8, 30 mM Tris buffer, 0.1–1 mg/ml protein, 4 weeks [1]; –20°C, pH 7.8, 30 mM Tris buffer, 0.1–1 mg/ml, 3 months [1]; 4°C, 3–4 weeks [2];Dilute solution, protein concentration: 0.001–0.010 mg/ml, 20% loss of activity per week at 4°C [1]

6 CROSSREFERENCES TO STRUCTURE DATABANKS

PIR/MIPS code

PIR2:S01003 (Erwinia sp. (fragment))

Brookhaven code

7 LITERATURE REFERENCES

[1] Schrimsher, J.L., Wingfield, P.T., Bernard, A., Mattaliano, R., Payton, M.A.: Biochem. J.,253,511–516 (1988)
[2] Englard, S., Avigad, G.: Methods Enzymol.,41B,127–131 (1975)
[3] Avigad, G., Englard, S., Pifko, S.: J. Biol. Chem.,241,373–378 (1966)
[4] Ameyama, M., Matsushita, K., Shinagawa, E.: Agric. Biol. Chem.,45,863–869 (1981)

1 NOMENCLATURE

EC number
1.1.1.125

Systematic name
2-Deoxy-D-gluconate:NAD+ 3-oxidoreductase

Recommended name
2-Deoxy-D-gluconate 3-dehydrogenase

Synonymes
Dehydrogenase, 2-deoxygluconate
2-Deoxygluconate dehydrogenase

CAS Reg. No.
37254-50-7

2 REACTION AND SPECIFICITY

Catalysed reaction
2-Deoxy-D-gluconate + NAD+ →
→ 2-deoxy-3-dehydro-D-gluconate + NADH

Reaction type
Redox reaction

Natural substrates

Substrate spectrum
1 2-Deoxy-D-gluconate + NAD+ [1]

Product spectrum
1 2-Deoxy-3-dehydro-D-gluconate + NADH [1]

Inhibitor(s)

Cofactor(s)/prosthetic group(s)/activating agents
NAD+ [1]

Metal compounds/salts

Turnover number (min^{-1})

Specific activity (U/mg)

K_m-value (mM)
15.4–50 (2-deoxygluconate) [1]

pH-optimum
7.9–8.9 [1]

pH-range
7.0–9.5 (7.0: about 30% of activity maximum, 9.5: about 45% of activity maximum) [1]

Temperature optimum (°C)
30 (assay at) [1]

Temperature range (°C)

---

3 ENZYME STRUCTURE

Molecular weight

Subunits

Glycoprotein/Lipoprotein
–

---

4 ISOLATION/PREPARATION

Source organism
Pseudomonas sp. DG [1]

Source tissue
Cell [1]

Localisation in source

Purification
Pseudomonas sp. DG [1]

Crystallization
–

Cloned
–

Renaturated
–

5 STABILITY

pH

Temperature (°C)

Oxidation

Organic solvent

General stability information

Cysteine, 2-mercaptoethanol, bovine serum albumin or NAD^+ does not stabilize [1]

Storage

–5°C, partially purified enzyme, remaining activity of 60% after purification is stable for 10 days [1]

6 CROSSREFERENCES TO STRUCTURE DATABANKS

PIR/MIPS code

Brookhaven code

7 LITERATURE REFERENCES

[1] Eichhorn, M.M., Cynkin, M.A.: Biochemistry,4,159–165 (1965)

1 NOMENCLATURE

EC number
1.1.1.126

Systematic name
2-Dehydro-3-deoxy-D-gluconate: $NADP^+$ 6-oxidoreductase

Recommended name
2-Dehydro-3-deoxy-D-gluconate 6-dehydrogenase

Synonymes
2-Keto-3-deoxy-D-gluconate oxidoreductase
Dehydrogenase, 2-keto-3-deoxygluconate

CAS Reg. No.
37250-55-8

2 REACTION AND SPECIFICITY

Catalysed reaction
(4S,5S)-4,5-Dihydroxy-2,6-dioxohexanoate + NADPH →
→ 2-dehydro-3-deoxy-D-gluconate + $NADP^+$

Reaction type
Redox reaction

Natural substrates
(4S,5S)-4,5-Dihydroxy-2,6-dioxohexanoate + NADPH (ir, enzyme of metabolism of alginic acid) [1]

Substrate spectrum
1 (4S,5S)-4,5-Dihydroxy-2,6-dioxohexanoate + NADPH (ir, i.e. ketodeoxygluconaldehyde) [1]

Product spectrum
1 2-Dehydro-3-deoxy-D-gluconate + $NADP^+$ [1]

Inhibitor(s)

Cofactor(s)/prosthetic group(s)/activating agents
NADPH [1]

Metal compounds/salts

Turnover number (min^{-1})

Specific activity (U/mg)
15.9 [1]

K_m-value (mM)
0.36 (4,5-dihydroxy-2,6-dioxohexanoate) [1]; 0.032 (NADPH) [1]

pH-optimum
7.0–7.5 [1]

pH-range
6–8.5 (6: about 80% of activity maximum, 8.5: about 65% of activity maximum) [1]

Temperature optimum (°C)

Temperature range (°C)

3 ENZYME STRUCTURE

Molecular weight

Subunits

Glycoprotein/Lipoprotein
–

4 ISOLATION/PREPARATION

Source organism
Pseudomonas sp. (alginінic acid adapted) [1]

Source tissue

Localisation in source

Purification
Pseudomonas sp. (partial) [1]

Crystallization
–

Cloned
–

Renaturated
–

5 STABILITY

pH

Temperature (°C)

Oxidation

Organic solvent

General stability information

Storage
0–3°C, $NADP^+$, 3 weeks, 90% of initial activity retained [1]

6 CROSSREFERENCES TO STRUCTURE DATABANKS

PIR/MIPS code

Brookhaven code

7 LITERATURE REFERENCES

[1] Preiss, J., Ashwell, G.: J. Biol. Chem.,237,317–321 (1962)

1 NOMENCLATURE

EC number
1.1.1.127

Systematic name
2-Dehydro-3-deoxy-D-gluconate:NAD+ 5-oxidoreductase

Recommended name
2-Dehydro-3-deoxy-D-gluconate 5-dehydrogenase

Synonymes
2-Keto-3-deoxy-D-gluconate dehydrogenase
Dehydrogenase, 2-keto-3-deoxygluconate (nicotinamide adenine dinucleotide (phosphate))
2-Keto-3-deoxy-D-gluconate (3-deoxy-D-glycero-2,5-hexodiulosonic acid) dehydrogenase
2-Keto-3-deoxygluconate-5-dehydrogenase

CAS Reg. No.
37250-56-9

2 REACTION AND SPECIFICITY

Catalysed reaction
(4S)-4,6-Dihydroxy-2,5-dioxohexanoate + NADH →
→ 2-dehydro-3-deoxy-D-gluconate + NAD+

Reaction type
Redox reaction

Natural substrates
(4S)-4,6-Dihydroxy-2,5-dioxohexanoate + NADH (bacterial metabolism of polygalacturonic acid [1], enzyme used in pectinolysis [2]) [1, 2]

Substrate spectrum
1 (4S)-4,6-Dihydroxy-2,5-dioxohexanoate + NADH (r (pH 8.8 [1]) [1, 2], ir (pH 7.0) [1], in the reverse reaction the enzyme from Pseudomonas acts equally well with NAD+ or NADP+, while those from Erwinia chrysanthemi and E. coli are more specific for NAD+ [2]) [1, 2]
2 Pyruvate + NAD+ (oxidized at 10% of the rate of 2-keto-3-deoxy-D-gluconate) [2]

Product spectrum

1 2-Dehydro-3-deoxy-D-gluconate + NAD^+ (i.e. 2-keto-3-deoxy-D-gluconate) [1, 2]
2 ?

Inhibitor(s)

ATP [2]; Ag^{2+} [2]; p-Chloromercuribenzoate [2]; NADH (above 0.2 mM) [2]

Cofactor(s)/prosthetic group(s)/activating agents

NADPH (62% of the activity with NADH [2]) [1, 2]; NADH [1, 2]; NAD^+ (53% of the activity with $NADP^+$, the enzyme from Pseudomonas acts equally well with NAD^+ or $NADP^+$, while those from Erwinia chrysanthemi and E. coli are more specific for NAD^+ [2]) [1, 2]; $NADP^+$ (the enzyme from Pseudomonas acts equally well with NAD^+ or $NADP^+$, while those from Erwinia chrysanthemi and E. coli are more specific for NAD^+) [2]

Metal compounds/salts

Turnover number (min^{-1})

Specific activity (U/mg)

80 [1]; 39.0 [2]

K_m-value (mM)

0.03 (NADH) [2]; 7.7 (2-keto-3-deoxygluconate) [2]; 0.4 (NAD^+) [2]; 0.34 (2,5-diketo-3-deoxygluconate) [2]

pH-optimum

7–7.5 ((4S)-4,6-dihydroxy-2,5-dioxohexanoate + NADH) [2]; 7.0–8.0 [1]; 10 (2-dehydro-3-deoxy-D-gluconate + NAD^+) [2]

pH-range

5.5–9 (at pH 5.5 and 9.0 about 50% of activity maximum, (4S)-4,6-dihydroxy-2,5-dioxohexanoate + NADH) [2]; 6–8.8 (6: about 95% of activity maximum, 8.8: about 45% of activity maximum) [1]; 8–10.5 (8: about 50% of activity maximum, 10.5: about 90% of activity maximum, 2-dehydro-3-deoxy-D-gluconate + NAD^+) [2]

Temperature optimum (°C)

45 [2]

Temperature range (°C)

30–50 (30°C: about 45% of activity maximum, 50°C: about 60% of activity maximum) [2]

3 ENZYME STRUCTURE

Molecular weight

Subunits

? (x × 23000, Erwinia chrysanthemi, SDS-PAGE, formation of large non-specific aggregates of at least 10 monomers (each 23000)) [2]

Glycoprotein/Lipoprotein

–

4 ISOLATION/PREPARATION

Source organism

Pseudomonas sp. [1]; Erwinia chrysanthemi [2]

Source tissue

Cell [1, 2]

Localisation in source

Purification

Pseudomonas sp. [1]; Erwinia chrysanthemi [2]

Crystallization

–

Cloned

–

Renaturated

–

5 STABILITY

pH

Temperature (°C)

4 (3 weeks, 10% loss of activity in absence of dithiothreitol, 5% loss of activity in presence of 2 mM dithiothreitol) [2]; 20 (15 h, stable) [2]; 43 (half-life: 45 min, without protector) [2]; 46 (half-life: 10 min (without protector), 14 min (+ NAD^+), 27 min (+ NADH), 12 min (+ 2-keto-3-deoxygluconate), 20 min (+ 2,5-diketo-3-deoxygluconate)) [2]; 49 (half-life: 2.5 min, without protector) [2]

Oxidation

Organic solvent

General stability information

NAD^+ stabilizes, unstable in absence [1]; $NADP^+$ stabilizes, unstable in absence [1]

Storage

0–3°C, 0.001 M NAD^+ or $NADP^+$, 33–50% loss of activity after 5 weeks [1]; –20°C, 6 months, stable [2]; 4°C, 3 weeks, 2 mM dithiothreitol, 5% loss of activity [2]

6 CROSSREFERENCES TO STRUCTURE DATABANKS

PIR/MIPS code

Brookhaven code

7 LITERATURE REFERENCES

[1] Preiss, J., Ashwell, G.: J. Biol. Chem.,238,1577–1583 (1963)
[2] Condemine, G., Hugouvieux-Cotte-Pattat, N., Robert-Baudouy, J.: J. Gen. Microbiol., 130,2839–2844 (1984)

1 NOMENCLATURE

EC number
1.1.1.128

Systematic name
L-Idonate:$NADP^+$ 2-oxidoreductase

Recommended name
L-Idonate 2-dehydrogenase

Synonymes
5-Keto-D-gluconate 2-reductase
Dehydrogenase, L-idonate
5-Ketogluconate 2-reductase
Reductase, 5-ketogluconate 5- (L-idonate-forming)
5KGR [1]
5-Ketoglucono-idono-reductase [2]

CAS Reg. No.
37250-57-0

2 REACTION AND SPECIFICITY

Catalysed reaction
L-Idonate + $NADP^+$ →
→ 5-dehydro-D-gluconate + NADPH

Reaction type
Redox reaction

Natural substrates
L-Idonate + $NADP^+$ (significant role in control of intracellular concentration of gluconate) [1]

Substrate spectrum
1 L-Idonate + $NADP^+$ (i.e. 5-ketogluconate, r, specific for, no reaction with: 2-keto-D-gluconic acid, 2-keto-L-gulonic acid, fructose, sorbose, acetone [2]) [1, 2]

Product spectrum
1 5-Dehydro-D-gluconate + NADPH

Inhibitor(s)
p-Chloromercuribenzoate [2]; Cu^{2+} [2]

Cofactor(s)/prosthetic group(s)/activating agents
$NADP^+$ [1, 2]; NADPH [2]; More (no activity with NADH) [2]

Metal compounds/salts

Turnover number (min^{-1})

Specific activity (U/mg)

K_m-value (mM)
1.1 (5-ketogluconate) [2]

pH-optimum
6.5–7.4 (L-idonate formation) [2]; 9.5 (5-ketogluconate formation) [2]

pH-range
5.5–10 [2]

Temperature optimum (°C)

Temperature range (°C)

---

3 ENZYME STRUCTURE

Molecular weight

Subunits

Glycoprotein/Lipoprotein
–

---

4 ISOLATION/PREPARATION

Source organism
Escherichia sp. (inducible) [1]; Klebsiella sp. (inducible) [1]; Pseudomonas sp. (inducible) [1]; Gluconobacter sp. (several strains, noninducible enzyme) [1]; Acetobacter aceti [1]; Acetobacter aurantium [1]; Fusarium sp. [2]; More (overview: distribution in acetic acid bacteria) [1]

Source tissue

Localisation in source

Purification

Crystallization
–

Cloned

–

Renaturated

–

5 STABILITY

pH

Temperature (°C)

50 (5 min, complete loss of activity) [2]

Oxidation

Organic solvent

General stability information

Storage

–20°C or 0°C, 1 week, loss of activity [2]

6 CROSSREFERENCES TO STRUCTURE DATABANKS

PIR/MIPS code

Brookhaven code

7 LITERATURE REFERENCES

[1] Chiyonobu, T., Shinagawa, E., Adachi, O., Ameyama, M.: Agric. Biol. Chem.,39,2425–2427 (1975)
[2] Takagi, Y.: Agric. Biol. Chem.,26,719–720 (1962)

1 NOMENCLATURE

EC number
1.1.1.129

Systematic name
L-Threonate:NAD^+ 3-oxidoreductase

Recommended name
L-Threonate 3-dehydrogenase

Synonymes
Dehydrogenase, threonate
L-Threonic acid dehydrogenase [1]

CAS Reg. No.
37250-59-2

2 REACTION AND SPECIFICITY

Catalysed reaction
L-Threonate + NAD^+ →
→ 3-dehydro-L-threonate + NADH

Reaction type
Redox reaction

Natural substrates
L-Threonate + NAD^+ [1]

Substrate spectrum
1 L-Threonate + NAD^+ (specific for straight chain form, inactive with lactone of threonic acid) [1]

Product spectrum
1 3-Dehydro-L-threonate + NADH [1]

Inhibitor(s)
Hydrazine [1]; Hydroxylamine [1]

Cofactor(s)/prosthetic group(s)/activating agents
NAD^+ [1]; More (not: $NADP^+$) [1]

Metal compounds/salts
No evidence for metal ion requirement [1]

Turnover number (min^{-1})

Specific activity (U/mg)
124 [1]

K_m-value (mM)

pH-optimum
10.8 [1]

pH-range
8.9–11.5 (at pH 8.9 and 11.5 about 50% of activity maximum) [1]

Temperature optimum (°C)

Temperature range (°C)

3 ENZYME STRUCTURE

Molecular weight

Subunits

Glycoprotein/Lipoprotein
–

4 ISOLATION/PREPARATION

Source organism
Pseudomonas sp. [1]

Source tissue
Cell [1]

Localisation in source

Purification
Pseudomonas sp. [1]

Crystallization
–

Cloned
–

Renaturated
–

5 STABILITY

pH

Temperature (°C)

Oxidation

Organic solvent

General stability information

Diluted enzyme labile, loss of activity after overnight storage at –15°C [1]

Storage

6 CROSSREFERENCES TO STRUCTURE DATABANKS

PIR/MIPS code

Brookhaven code

7 LITERATURE REFERENCES

[1] Aspen, A.J., Jakoby, W.B.: J. Biol. Chem.,239,710–713 (1964)

1 NOMENCLATURE

EC number
1.1.1.130

Systematic name
3-Dehydro-L-gulonate:NAD(P)$^+$ 2-oxidoreductase

Recommended name
3-Dehydro-L-gulonate 2-dehydrogenase

Synonymes
Dehydrogenase, 3-ketogulonate
3-Keto-L-gulonate dehydrogenase
3-Ketogulonate dehydrogenase

CAS Reg. No.
37250-61-6

---

2 REACTION AND SPECIFICITY

Catalysed reaction
(4R,5S)-4,5,6-Trihydroxy-2,3-dioxohexanoate + NAD(P)H →
→ 3-dehydro-L-gulonate + NAD(P)$^+$

Reaction type
Redox reaction

Natural substrates
(4R,5S)-4,5,6-Trihydroxy-2,3-dioxohexanoate + NAD(P)H (breakdown of ascorbic acid) [1]

Substrate spectrum
1 (4R,5S)-4,5,6-Trihydroxy-2,3-dioxohexanoate + NAD(P)H [1]
2 More (not: L-xylulose, D-ribulose, L-ribulose, D-fructose, 2-keto-D-gluconate, 2-keto-L-gluconate) [1]

Product spectrum
1 3-Dehydro-L-gulonate + NAD(P)$^+$
2 ?

Inhibitor(s)
p-Chloromercuribenzoate [1]

Cofactor(s)/prosthetic group(s)/activating agents
NADPH (50% of the activity with NADH) [1]; NADH [1]

Metal compounds/salts

Turnover number (min^{-1})

Specific activity (U/mg)
More [1]

K_m**-value** (mM)
1.16 (2,3-diketo-L-gulonic acid) [1]; 0.0162 (NADH) [1]; 0.0216 (NADPH) [1]

pH-optimum
6.5 [1]

pH-range
5.2–8.4 (at pH 5.2 and 8.4: about 50% of activity maximum) [1]

Temperature optimum (°C)

Temperature range (°C)

---

3 ENZYME STRUCTURE

Molecular weight

Subunits

Glycoprotein/Lipoprotein
–

---

4 ISOLATION/PREPARATION

Source organism
Aerobacter aerogenes [1]

Source tissue
Cell [1]

Localisation in source

Purification
Aerobacter aerogenes [1]

Crystallization
–

Cloned

–

Renaturated

–

5 STABILITY

pH

7 (unstable above) [1]

Temperature (°C)

Oxidation

Organic solvent

General stability information

Quite stable even in dilute solutions at 0°C [1]

Storage

0°C, 0.03 M Tris, pH 7.5, 0.62 mg of protein per ml, 15% loss of activity after 4 weeks, 64% loss of activity after 9 weeks [1]

6 CROSSREFERENCES TO STRUCTURE DATABANKS

PIR/MIPS code

Brookhaven code

7 LITERATURE REFERENCES

[1] Volk, W.A., Larsen, J.L.: J. Biol. Chem.,237,2454–2457 (1962)

1 NOMENCLATURE

EC number
1.1.1.131

Systematic name
D-Mannonate:NAD(P)$^+$ 6-oxidoreductase

Recommended name
Mannuronate reductase

Synonymes
Dehydrogenase, mannonate (nicotinamide adenine dinucleotide (phosphate))
Mannonate dehydrogenase
Reductase, mannuronate
Mannonate dehydrogenase (NAD(P)$^+$)
D-Mannonate:nicotinamide adenine dinucleotide (phosphate oxidoreductase (D-mannuronate-forming))
EC 1.1.1.180 (formerly)
EC 1.2.1.34 (formerly)

CAS Reg. No.
37250-62-7; 37250-97-8

2 REACTION AND SPECIFICITY

Catalysed reaction
D-Mannuronate + NAD(P)H →
→ D-mannonate + NAD(P)$^+$

Reaction type
Redox reaction

Natural substrates
D-Mannuronate + NAD(P)H (utilization of mannuronic acid as energy source) [1]

Substrate spectrum
1 D-Mannuronate + NAD(P)H [1]

Product spectrum
1 D-Mannonate + NAD(P)$^+$ [1]

Inhibitor(s)

Cofactor(s)/prosthetic group(s)/activating agents

Metal compounds/salts

Turnover number (min^{-1})

Specific activity (U/mg)

K_m-value (mM)

pH-optimum
7.0 (assay at) [1]

pH-range

Temperature optimum (°C)
30 (assay at) [1]

Temperature range (°C)

3 ENZYME STRUCTURE

Molecular weight

Subunits

Glycoprotein/Lipoprotein
–

4 ISOLATION/PREPARATION

Source organism
Aeromonas sp. C11–2B [1]

Source tissue
Cell [1]

Localisation in source

Purification

Crystallization
–

Cloned
–

Renaturated
–

5 STABILITY

pH

Temperature (°C)
45 (20 min: 36% loss of activity) [1]

Oxidation

Organic solvent

General stability information

Storage

6 CROSSREFERENCES TO STRUCTURE DATABANKS

PIR/MIPS code

Brookhaven code

7 LITERATURE REFERENCES

[1] Farmer, J.J., Eagon, R.G.: J. Bacteriol.,97,97–106 (1969)

1 NOMENCLATURE

EC number
1.1.1.132

Systematic name
GDP-D-mannose:NAD^+ 6-oxidoreductase

Recommended name
GDPmannose 6-dehydrogenase

Synonymes
Dehydrogenase, guanosine diphosphomannose
GDP mannose dehydrogenase
Guanosine diphosphomannose dehydrogenase
Guanosine diphospho-D-mannose dehydrogenase

CAS Reg. No.
37250-63-8

2 REACTION AND SPECIFICITY

Catalysed reaction
GDP-D-mannose + 2 NAD^+ + H_2O →
→ GDP-D-mannuronate + 2 NADH

Reaction type
Redox reaction

Natural substrates
GDPmannose + NAD^+ + H_2O (key enzyme in biosynthesis of alginate [2, 5], enzyme involved in biosynthesis of alginate-like polysaccharide by Pseudomonas aeruginosa [4]) [2, 4, 5]

Substrate spectrum
1 GDPmannose + NAD^+ + H_2O [1–7]
2 Deoxy-GDPmannose + NAD^+ + H_2O [1, 3]

Product spectrum
1 GDPmannuronate + NADH [1–7]
2 Deoxy-GDPmannuronate + NADH [3]

Inhibitor(s)

Iodoacetamide [2]; p-Hydroxymercuribenzoate [2]; GMP [2]; ATP [2]; GDP-D-glucose [2]; Maltose [2]

Cofactor(s)/prosthetic group(s)/activating agents

NAD^+ [1–7]; More ($NADP^+$ has no cofactor activity) [2, 3]

Metal compounds/salts

Turnover number (min^{-1})

Specific activity (U/mg)

0.39 [1]; 14.7 [3]; 1.068 [2]

K_m-value (mM)

0.017 (GDPmannose) [1]; 0.26 (NAD^+ (+ GDPmannose)) [1]; 0.084 (deoxy-GDPmannose) [1]

pH-optimum

8.2 (Tris buffer) [1, 3]; 7.7 [2]

pH-range

Temperature optimum (°C)

50 [2]

Temperature range (°C)

---

3 ENZYME STRUCTURE

Molecular weight

290000 (Pseudomonas aeruginosa, gel filtration) [2]

Subunits

Hexamer (6 × 48000, Pseudomonas aeruginosa, SDS-PAGE) [2]

Glycoprotein/Lipoprotein

–

---

4 ISOLATION/PREPARATION

Source organism

Arthrobacter sp. (NRRL B1973 [1]) [1, 3]; Pseudomonas aeruginosa (enzyme overproduced using a plasmid vector containing algD under control of the tac promotor [2]) [2, 4, 5]; Azotobacter vinelandii [6, 7]

Source tissue

Cells [1, 2, 7]

Localisation in source

Purification

Arthrobacter sp. (NRRL B1973 [1]) [1, 3]; Pseudomonas aeruginosa [2]

Crystallization

–

Cloned

[5]

Renaturated

–

5 STABILITY

pH

Temperature (°C)

50 (above, 2–3 min, loss of activity) [2]

Oxidation

Organic solvent

General stability information

Storage

–15°C, protein concentration above 1 mg/ml, stable for at least 1 month [1]

6 CROSSREFERENCES TO STRUCTURE DATABANKS

PIR/MIPS code

PIR1:DEPSGD (Pseudomonas aeruginosa)

Brookhaven code

7 LITERATURE REFERENCES

[1] Preiss, J.: Methods Enzymol.,8,285–287 (1966) (Review)
[2] Roychoudhury, S., Mmay, T.B., Gill, J.F., Singh, S.K., Feingold, D.S., Chakrabarty, A.M.: J. Biol. Chem.,264,9380–9385 (1989)
[3] Preiss, J.: J. Biol. Chem.,239,3127–3132 (1964)
[4] Pugashetti, B.K., Vadas, L., Prihar, H.S., Feingold, D.S.: J. Bacteriol.,153,1107–1110 (1983)
[5] Deretic, V., Gill, J.F., Chakrabarty, A.M.: J. Bacteriol.,169,351–358 (1987)
[6] Horan, N.J., Jarman, T.R., Dawes, E.A.: J. Gen. Microbiol.,129,2985–2990 (1983)
[7] Couperwhite, I., McCallum, M.F.: Antonie Leeuwenhoek,41,25–32 (1975)

1 NOMENCLATURE

EC number
1.1.1.133

Systematic name
dTDP-6-deoxy-L-mannose:NADP+ 4-oxidoreductase

Recommended name
dTDP-4-dehydrorhamnose reductase

Synonymes
dTDP-4-keto-L-rhamnose reductase
Reductase, thymidine diphospho-4-ketorhamnose
dTDP-4-ketorhamnose reductase
TDP-4-keto-rhamnose reductase
Thymidine diphospho-4-ketorhamnose reductase

CAS Reg. No.
37250-64-9

2 REACTION AND SPECIFICITY

Catalysed reaction
dTDP-6-deoxy-L-mannose + NADP+ →
→ dTDP-4-dehydro-6-deoxy-L-mannose + NADPH (in the reverse direction reduction at the 4-position of the hexose moiety takes place only while the substrate is bound to another enzyme which catalyzes epimerization at C-3 and C-5, the complex has been referred to as dTDP-L-rhamnose synthase)

Reaction type
Redox reaction

Natural substrates

Substrate spectrum
1 dTDP-4-keto-6-deoxy-D-glucose + NADPH [1]

Product spectrum
1 dTDP-L-rhamnose + NADP+ (possible mechanism via dTDP-4-dehydro-6-deoxy-L-mannose, bound to the epimerizing enzyme 2, dissociates from the enzyme only after reduction to dTDP-L-rhamnose or reversion to dTDP-4-keto-6-deoxy-D-glucose) [1]

Inhibitor(s)

Cofactor(s)/prosthetic group(s)/activating agents
NADPH [1]

Metal compounds/salts

Turnover number (min^{-1})

Specific activity (U/mg)

K_m-value (mM)

pH-optimum
8.0 (assay at) [1]

pH-range

Temperature optimum (°C)
25 (assay at) [1]

Temperature range (°C)

---

3 ENZYME STRUCTURE

Molecular weight

Subunits

Glycoprotein/Lipoprotein
–

---

4 ISOLATION/PREPARATION

Source organism
Pseudomonas aeruginosa [1]

Source tissue
Cell [1]

Localisation in source

Purification

Crystallization
–

Cloned
–

Renaturated
–

5 STABILITY

pH

Temperature (°C)

50 (after 5 min 50% loss of activity, NADPH prevents denaturation [1], after 8 min 74% loss of activity, NADPH, and to a lesser degree $NADP^+$, prevent denaturation, NAD^+ and NADH are less effective [1]) [1]

Oxidation

Organic solvent

General stability information

NADPH, and to a lesser degree $NADP^+$, prevent denaturation at 50°C, NAD^+ and NADH are less effective [1]

Storage

6 CROSSREFERENCES TO STRUCTURE DATABANKS

PIR/MIPS code

Brookhaven code

7 LITERATURE REFERENCES

[1] Melo, A., Glaser, L.: J. Biol. Chem.,243,1475–1478 (1968)

1 NOMENCLATURE

EC number
1.1.1.134

Systematic name
dTDP-6-deoxy-L-talose:$NADP^+$ 4-oxidoreductase

Recommended name
dTDP-6-deoxy-L-talose 4-dehydrogenase

Synonymes
Dehydrogenase, thymidine diphospho-6-deoxy-L-talose
TDP-6-deoxy-L-talose dehydrogenase
Thymidine diphospho-6-deoxy-L-talose dehydrogenase
dTDP-6-deoxy-L-talose dehydrogenase (4-reductase) [1]

CAS Reg. No.
37250-65-0

2 REACTION AND SPECIFICITY

Catalysed reaction
dTDP-6-deoxy-L-talose + $NADP^+$ →
→ dTDP-4-dehydro-6-deoxy-L-mannose + NADPH (oxidation at the 4-position of the hexose moiety takes place only while the substrate is bound to another enzyme which catalyzes epimerization at C-3 and C-5, the latter enzyme being interchangeable against an epimerizing enzyme from Pseudomonas aeruginosa [1])

Reaction type
Redox reaction

Natural substrates

Substrate spectrum
1 TDP-6-deoxy-D-xylo-4-hexosulose + NADPH [1]

Product spectrum
1 TDP-6-deoxy-L-talose + $NADP^+$ (via TDP-6-deoxy-L-lyxo-4-hexosulose, enzyme-bound intermediate) [1]

Inhibitor(s)

Cofactor(s)/prosthetic group(s)/activating agents
NADPH [1]

Metal compounds/salts

Turnover number (min^{-1})

Specific activity (U/mg)

K_m-value (mM)

pH-optimum
8.0 (assay at) [1]

pH-range

Temperature optimum (°C)
37 (assay at) [1]

Temperature range (°C)

---

3 ENZYME STRUCTURE

Molecular weight

Subunits

Glycoprotein/Lipoprotein
–

---

4 ISOLATION/PREPARATION

Source organism
E. coli 045 [1]

Source tissue
Cell [1]

Localisation in source

Purification

Crystallization
–

Cloned
–

Renaturated
–

5 STABILITY

pH

Temperature (°C)

Oxidation

Organic solvent

General stability information

Storage

6 CROSSREFERENCES TO STRUCTURE DATABANKS

PIR/MIPS code

Brookhaven code

7 LITERATURE REFERENCES

[1] Gaugler, R.W., Gabriel, O.: J.Biol.Chem.,248,6041–6049 (1973)

1 NOMENCLATURE

EC number
1.1.1.135

Systematic name
GDP-6-deoxy-D-talose:NAD(P)$^+$ 4-oxidoreductase

Recommended name
GDP-6-deoxy-D-talose 4-dehydrogenase

Synonymes
Dehydrogenase, guanosine diphospho-6-deoxy-D-talose
Guanosine diphospho-6-deoxy-D-talose dehydrogenase

CAS Reg. No.
37250-66-1

2 REACTION AND SPECIFICITY

Catalysed reaction
GDP-6-deoxy-D-talose + NAD(P)$^+$ →
→ GDP-4-dehydro-6-deoxy-D-talose + NAD(P)H

Reaction type
Redox reaction

Natural substrates

Substrate spectrum
1 GDP-6-deoxy-D-talose + NAD(P)$^+$ [1]

Product spectrum
1 GDP-4-dehydro-6-deoxy-D-talose + NAD(P)H [1]

Inhibitor(s)

Cofactor(s)/prosthetic group(s)/activating agents

Metal compounds/salts

Turnover number (min^{-1})

Specific activity (U/mg)

K_m-value (mM)

pH-optimum

pH-range

Temperature optimum (°C)

Temperature range (°C)

---

3 ENZYME STRUCTURE

Molecular weight

Subunits

Glycoprotein/Lipoprotein

–

---

4 ISOLATION/PREPARATION

Source organism

Bacterium strain GS (gram-negative motile bacterium isolated from soil) [1]

Source tissue

Localisation in source

Purification

Crystallization

–

Cloned

–

Renaturated

–

5 STABILITY

pH

Temperature (°C)

Oxidation

Organic solvent

General stability information

Storage

6 CROSSREFERENCES TO STRUCTURE DATABANKS

PIR/MIPS code

Brookhaven code

7 LITERATURE REFERENCES

[1] Markovitz, A.: J. Biol. Chem.,239,2091–2098 (1964)

1 NOMENCLATURE

EC number
1.1.1.136

Systematic name
UDP-N-acetyl-D-glucosamine:NAD^+ 6-oxidoreductase

Recommended name
UDP-N-acetylglucosamine 6-dehydrogenase

Synonymes
Dehydrogenase, uridine diphosphoacetylglucosamine
UDPacetylglucosamine dehydrogenase [1]
UDP-2-acetamido-2-deoxy-D-glucose:NAD oxidoreductase [1]
UDP-GlcNAc dehydrogenase [2]

CAS Reg. No.
9054-83-5

2 REACTION AND SPECIFICITY

Catalysed reaction
UDP-N-acetyl-D-glucosamine + 2 NAD^+ + H_2O →
→ UDP-N-acetyl-2-amino-2-deoxy-D-glucuronate + 2 NADH

Reaction type
Redox reaction

Natural substrates

Substrate spectrum
1 UDP-N-acetyl-D-glucosamine + NAD^+ (UDP-D-glucose and $NADP^+$ are also utilized at 0.1 the rate of UDP-N-acetyl-D-glucosamine and NAD [1])

Product spectrum
1 UDP-N-acetyl-2-amino-2-deoxy-D-glucuronate + NADH

Inhibitor(s)
NADH (competitive to NAD^+) [1]; p-Chloromercuribenzoate [2]; Ethanol [2]

Cofactor(s)/prosthetic group(s)/activating agents
NAD^+ [1, 2]; $NADP^+$ (reduced at 0.1 the rate of NAD^+) [1]; More (dithiothreitol and the particulate fraction required for full activity) [2]

Metal compounds/salts
More (no activation by Mg^{2+}, Mn^{2+}, Co^{2+} or Ca^{2+}) [2]

Turnover number (min^{-1})

Specific activity (U/mg)
41 [2]

K_m-value (mM)
0.5 (UDP-N-acetyl-D-glucosamine) [1]; 1.5 (NAD^+) [1]; 0.28 (UDP-N-acetyl-D-glucosamine) [2]; 1.43 (NAD^+) [2]

pH-optimum
8.8 (above, Tris buffer) [2]; 9.0 (glycine buffer, in Tris buffer half as active as in glycine buffer, no activity in sodium borate buffer) [1]; 10.0 (above, glycine-NaOH buffer) [2]

pH-range

Temperature optimum (°C)
37 (assay at) [2]

Temperature range (°C)

3 ENZYME STRUCTURE

Molecular weight

Subunits

Glycoprotein/Lipoprotein
–

4 ISOLATION/PREPARATION

Source organism
Achromobacter georgiopolitanum (wild-type and capsuleless mutant) [1]; Micrococcus luteus ATCC 4698 [2]; Citrobacter freundii ATCC 10053 [3]

Source tissue
Cell [1]

Localisation in source
Soluble [3]; Cytoplasm [3]

Purification
Micrococus luteus ATCC 4698 (partial) [2]

Crystallization

–

Cloned

–

Renaturated

–

5 STABILITY

pH

Temperature (°C)

Oxidation

Organic solvent

General stability information

Ammonium sulfate protects against inactivation during storage and incubation [2]; Monovalent cations protect against inactivation during incubation [2]

Storage

4°C, 1 M $(NH_4)_2SO_4$, more than 3 weeks [2]

6 CROSSREFERENCES TO STRUCTURE DATABANKS

PIR/MIPS code

Brookhaven code

7 LITERATURE REFERENCES

[1] Der-Fong Fan, John, C.E., Zalitis, J., Feingold, D. S.: Arch. Biochem. Biophys.,135, 45–49 (1969)
[2] Kawamura, T., Ichihara, N., Sugiyama, S., Yokota, H., Ishimoto, N., Ito, E.: J. Biochem.,98,105–116 (1985)
[3] Wheat, R.W., Barrow, R.O., Land, G.A.: J. Bacteriol.,127,1032–1035 (1976)

1 NOMENCLATURE

EC number
1.1.1.137

Systematic name
D-Ribitol-5-phosphate:NAD(P)$^+$ 2-oxidoreductase

Recommended name
Ribitol-5-phosphate 2-dehydrogenase

Synonymes
Dehydrogenase, ribitol 5-phosphate

CAS Reg. No.
37250-67-2

2 REACTION AND SPECIFICITY

Catalysed reaction
D-Ribitol 5-phosphate + NAD(P)$^+$ →
→ D-ribulose 5-phosphate + NAD(P)H

Reaction type
Redox reaction

Natural substrates

Substrate spectrum
1 D-Ribitol 5-phosphate + NAD(P)$^+$ (r [1], high specificity for ribitol-5-phosphate [2], absolutely specific for NAD$^+$ [2], NADP$^+$ or NAD$^+$ [1, 3], not: glucose-6-phosphate, fructose-6-phosphate, ribose, ribitol, arabitol [1]) [1–3]

Product spectrum
1 D-Ribulose 5-phosphate + NAD(P)H [1–3]

Inhibitor(s)
ATP [2]

Cofactor(s)/prosthetic group(s)/activating agents
NAD$^+$ [1–3]; NADP$^+$ (not [2]) [1, 3]; NADPH [1]

Metal compounds/salts
More (no stimulation by Mg^{2+} [1], no metal ion requirement [2]) [1, 2]

Turnover number (min^{-1})

Specific activity (U/mg)
49.3 [3]; More [1]

K_m-value (mM)
0.15 (ribulose 5-phosphate) [1, 3]; 0.009 ($NADP^+$) [1, 3]; 0.66 (ribitol-5-phosphate) [2]; 0.025 (NAD^+) [2]

pH-optimum
5.5–6.5 (D-ribitol 5-phosphate + NAD^+) [2]; 8.0–9.0 (D-ribulose 5-phosphate + NADH) [2]; 7.5 [1, 3]

pH-range

Temperature optimum (°C)
25 (assay at) [1, 3]

Temperature range (°C)

3 ENZYME STRUCTURE

Molecular weight
115000 (Lactobacillus casei, molecular sieve chromatography) [2]

Subunits
Dimer (2 × 49000, Lactobacillus casei, SDS-PAGE) [2]

Glycoprotein/Lipoprotein
–

4 ISOLATION/PREPARATION

Source organism
Lactobacillus plantarum (ATCC 8014) [1, 3]; Lactobacillus casei [2]; More (not: Bacillus subtilis) [3]

Source tissue
Cell [1]

Localisation in source

Purification
Lactobacillus plantarum (ATCC 8014, partial) [1, 3]; Lactobacillus casei [2]

Crystallization

–

Cloned

–

Renaturated

–

5 STABILITY

pH

Temperature (°C)

Oxidation

Organic solvent

General stability information

Storage

Frozen, for at least a week [1]; –20°C, lyophilized powder, 3 weeks stable [3]

6 CROSSREFERENCES TO STRUCTURE DATABANKS

PIR/MIPS code

Brookhaven code

7 LITERATURE REFERENCES

[1] Glaser, L.: Biochim. Biophys. Acta,67,252–530 (1963)
[2] Hausman, S.Z., London, J.: J. Bacteriol.,169,1651–1655 (1987)
[3] Glaser, L.: Methods Enzymol.,8,240–243 (1966) (Review)

1 NOMENCLATURE

EC number
1.1.1.138

Systematic name
D-Mannitol:NADP$^+$ 2-oxidoreductase

Recommended name
Mannitol 2-dehydrogenase (NADP$^+$)

Synonymes

CAS Reg. No.
37250-68-3

2 REACTION AND SPECIFICITY

Catalysed reaction
D-Mannitol + NADP$^+$ →
→ D-fructose + NADPH

Reaction type
Redox reaction

Natural substrates
D-Mannitol + NADP$^+$ [1–13]

Substrate spectrum
1 D-Mannitol + NADP$^+$ (r) [1–13]
2 D-Arabinitol + NADP$^+$ [6, 11, 12]
3 D-Sorbitol + NADP$^+$ [6, 10–12]
4 D-Glucitol + NADP$^+$ [6]

Product spectrum
1 D-Fructose + NADPH [1–13]
2 D-Xylulose + NADPH [6, 11, 12]
3 L-Sorbose + NADPH [10–12]
4 ?

Inhibitor(s)
Cd^{2+} [5, 6]; Cu^{2+} [5, 6, 11]; Hg^{2+} [5, 12]; Zn^{2+} [5, 6, 11]; Sn^{2+} [5]; Fe^{2+} [5]; 2-Mercaptoethanol [5]; Dithiotreitol [5]; K_2SO_4 [5]; p-Chloromercuribenzoate [10, 11, 13]; KCN [10, 13]

Cofactor(s)/prosthetic group(s)/activating agents

$NADP^+$ [1–13]; NADPH [1–13]

Metal compounds/salts

Mg^{2+} (activation [5, 7], activation of sugar oxidation [10]) [5, 7, 10]; Ca^{2+} (activation of sugar oxidation) [10]; Mn^{2+} (activation of sugar oxidation) [10]; Ba^{2+} (activation of sugar oxidation) [10]

Turnover number (min^{-1})

Specific activity (U/mg)

130–194.4 [1, 6, 11]; 18.1 [5]

K_m-value (mM)

16.2–128 (D-mannitol) [4, 6, 10, 12]; 760 (D-mannitol) [5]; 0.46 (mannitol) [13]; 0.014–0.082 ($NADP^+$) [4–6, 12, 13]; 190–1200 (D-fructose) [4, 10]; 1.4 (D-fructose) [13]; 0.00525–0.0385 (NADPH) [4, 10, 13]; 560 (D-glucitol) [6]; 190 (D-arabinitol) [6]

pH-optimum

10.25 (D-mannitol + $NADP^+$) [4]; 7.9–8.8 (D-mannitol + $NADP^+$) [7, 10, 12]; 9.4 (D-mannitol + $NADP^+$) [11]; 7.0–7.5 (D-fructose + NADPH) [4, 7, 11, 12]; 7.9–8.1 (D-fructose + NADPH) [10]; 7.3 [5]

pH-range

Temperature optimum (°C)

47 (D-fructose + NADPH) [10]

Temperature range (°C)

3 ENZYME STRUCTURE

Molecular weight

160000 (Penicillium chrysogenum, gel filtration, HPLC) [5]
140000 (Aspergillus parasiticus, gel filtration) [6]
115000–130000 (Agaricus bisporus, gel filtration, rate zonal ultracentrifugation) [4]
40000 (Agaricus bisporus, gel filtration) [10]

Subunits

Tetramer (4 × 40000, Penicillium chrysogenum, SDS-PAGE [5], 4 × 32000, Aspergillus parasiticus, SDS-PAGE [6], 4 × 29500, Agaricus bisporus, SDS-PAGE [4]) [4–6]

Glycoprotein/Lipoprotein

–

4 ISOLATION/PREPARATION

Source organism

Lentinus edodes (shiitake mushroom) [1]; Aspergillus nidulans [2]; Aspergillus parasiticus [6]; Aspergillus niger [8]; Cenococcum graniforme [3]; Agaricus bisporus [4, 8, 10]; Agaricus campestris [13]; Penicillium chrysogenum [5]; Penicillium islandicum [8]; Penicillium frequentans [8]; Alternaria alternata [8]; Giberella zeae [8]; Ceratocystis multiannulata [8]; Neurospora crassa [8]; Botrytis cinerea [8]; Cladosporium cladosporioides [8]; Thermomyces lanuginosus [8]; Melampsora lini (flax rust) [9]; Diplodia viticola [12]; Laminaria digitata [7]; Acetobacter melanogenum [11]

Source tissue

Localisation in source

Cytosol [2]

Purification

Agaricus bisporus [4, 10]; Penicillium chrysogenum [5]; Aspergillus parasiticus [6]; Acetobacter melanogenum [11]; Diplodia viticola (partial) [12]; Agaricus campestris [13]

Crystallization

–

Cloned

–

Renaturated

–

5 STABILITY

pH

6.0 (optimum) [5]

Temperature (°C)

45 (not stable above) [10, 12]

Oxidation

Organic solvent

General stability information

Storage

4°C, 1 month [12]; –20°C, 1 month [13]

6 CROSSREFERENCES TO STRUCTURE DATABANKS

PIR/MIPS code

Brookhaven code

7 LITERATURE REFERENCES

[1] Kulkarni, R.K.: Appl. Environ. Microbiol.,56,250–253 (1990)
[2] Singh, M., Scrutton, N.S., Scrutton, M.C.: J. Gen. Microbiol.,134,643–654 (1988)
[3] Martin, F., Canet, D., Marchal, J.P.: Plant Physiol.,77,499–502 (1985)
[4] Morton, N., Dickerson, A.G., Hammond, J.B.W.: J. Gen. Microbiol.,131,2885–2890 (1985)
[5] Boutelje, J., Hult, K., Gatenbeck, S.: Eur. J. Appl. Microbiol. Biotechnol.,17,7–12 (1983)
[6] Niehaus jr., W.G., Dilts jr., R.P.: J. Bacteriol.,151,243–250 (1982)
[7] Grant, C.R., Rees, T.A.: Phytochemistry,20,1505–1511 (1981)
[8] Hult, K., Veide, A., Gatenbeck, S.: Arch. Microbiol.,128,253–255 (1980)
[9] Clancy, F.G., Coffey, M.D.: J. Gen. Microbiol.,120,85–88 (1980)
[10] Ruffner, H.P., Rast, D., Tobler, H., Karesch, H.: Phytochemistry,17,865–868 (1978)
[11] Sasajima, K., Isono, M.: Agric. Biol. Chem.,32,161–169 (1968)
[12] Strobel, G.A., Kosuge, T.: Arch. Biochem. Biophys.,109,622–626 (1965)
[13] Edmundowicz, J.M., Wriston, J.C.: J. Biol. Chem.,238,3539–3541 (1963)

1 NOMENCLATURE

EC number
1.1.1.140

Systematic name
D-Sorbitol-6-phosphate:NAD$^+$ 2-oxidoreductase

Recommended name
Sorbitol-6-phosphate 2-dehydrogenase

Synonymes
Ketosephosphate reductase
Dehydrogenase, D-sorbitol 6-phosphate
D-Sorbitol-6-phosphate dehydrogenase
Sorbitol-6-P-dehydrogenase
D-Glucitol-6-phosphate dehydrogenase [6]
More (a similar enzyme with the name sorbitol-6-phosphate dehydrogenase from loquat (Eriobotrya japonica) catalyzes following reaction: sorbitol-6-phosphate + NADP$^+$ → glucose-6-phosphate + NADPH) [7, 8]

CAS Reg. No.
37250-69-4

2 REACTION AND SPECIFICITY

Catalysed reaction
D-Sorbitol 6-phosphate + NAD$^+$ →
→ D-fructose 6-phosphate + NADH

Reaction type
Redox reaction

Natural substrates
D-Sorbitol 6-phosphate + NAD$^+$ (hexitol catabolic enzyme) [3]

Substrate spectrum
1 D-Sorbitol 6-phosphate + NAD$^+$ (r [1, 4], high specificity for D-sorbitol 6-phosphate [1, 4–6] and fructose 6-phosphate [4], equilibrium far on the side of sorbitol 6-phosphate formation [4]) [1, 4–6]
2 More (slow reaction with mannitol 1-phosphate, at 1–5% of reaction with sorbitol 6-phosphate) [4]

Product spectrum
1 D-Fructose 6-phosphate + NADH [1, 4]
2 ?

Inhibitor(s)

CO_3^{2-} [1]; NADH [4]; Sorbitol 6-phosphate [4]; N-Ethylmaleimide [6]; Diethyl dicarbonate [6]

Cofactor(s)/prosthetic group(s)/activating agents

NAD^+ (specific for [4, 5], inhibited by alteration of the 6-amino position of the adenine moiety of NAD^+ [1, 5]) [1, 4–6]; NADH (specific for) [4]; More (no activity with $NADP^+$) [4]

Metal compounds/salts

Turnover number (min^{-1})

Specific activity (U/mg)

43.2 [4]; 180 [6]; More [1, 5]

K_m-value (mM)

6.25 (sorbitol 6-phosphate) [1]; 0.18 (NADH) [1]; 0.42 (NAD^+) [1]; 10.5 (D-fructose 6-phosphate) [1]; 0.40–0.44 (sorbitol 6-phosphate, glycine-NaOH buffer) [4]; 0.04–0.05 (sorbitol 6-phosphate, Tris-HCl buffer) [4]; 0.19–0.21 (NAD^+ (0.11–0.45 mM), glycine-NaOH buffer) [4]; 0.74–0.84 (NAD^+ (0.45–4.5 mM), glycine-NaOH buffer) [4]; 0.07–0.08 (NAD^+ (0.11–0.45 mM), Tris-HCl buffer) [4]; 0.20–0.25 (NAD^+ (0.45–4.5 mM), Tris-HCl buffer) [4]; 0.02–0.03 (NADH, phosphate buffer) [4]; 2.8–3.0 (fructose 6-phosphate (+ 0.43 mM NADH), phosphate buffer) [4]; 7.5–7.8 (fructose 6-phosphate (+ 1.08 mM NADH), phosphate buffer) [4]; 0.2 (NAD^+) [6]; 3.3 (D-sorbitol 6-phosphate) [6]; More (overview: various buffers, various substrate concentrations) [4]

pH-optimum

6.5–6.7 (reduction of fructose 6-phosphate, phosphate buffer) [4]; 8.5 (oxidation of sorbitol 6-phosphate, Tris-HCl buffer) [4]; 8.7 [6]; 9 (Tris-HCl buffer) [1, 5]; 10 (oxidation of sorbitol 6-phosphate, glycine-NaOH buffer) [4]

pH-range

5.7–7.5 (5.7: about 50% of activity maximum, 7.5: about 35% of activity maximum) [4]; More (in Tris buffer 32times greater activity than in $NaHCO_3$ buffer) [1]

Temperature optimum (°C)

25 (assay at) [4]

Temperature range (°C)

3 ENZYME STRUCTURE

Molecular weight

74000–94000 (Clostridium pasteurianum, gel filtration) [4]
117000 (E. coli, gel filtration) [6]

Subunits

Tetramer (4 × 26000, E. coli, SDS-PAGE) [6]

Glycoprotein/Lipoprotein

–

4 ISOLATION/PREPARATION

Source organism

Clostridium pasteurianum (enzyme induced by growth on sorbitol [2]) [2, 4]; Aerobacter aerogenes (enzyme induced by growth on sorbitol [5]) [1, 5]; Streptococcus mutans [3]; E. coli [6]; Eriobotrya japonica (similar enzyme with the name sorbitol-6-phosphate dehydrogenase from loquat (Eriobotrya japonica) catalyzes following reaction: sorbitol 6-phosphate + $NADP^+ \rightarrow$ glucose 6-phosphate + NADPH) [7, 8]

Source tissue

Cell [4]; Fruit (similar enzyme with the name sorbitol-6-phosphate dehydrogenase from loquat (Eriobotrya japonica) catalyzes the following reaction: sorbitol 6-phosphate + $NADP^+ \rightarrow$ glucose 6-phosphate + NADPH) [7]; Leaf (similar enzyme with the name sorbitol-6-phosphate dehydrogenase from loquat (Eriobotrya japonica) catalyzes following reaction: sorbitol 6-phosphate + $NADP^+ \rightarrow$ glucose 6-phosphate + NADPH) [8]

Localisation in source

Purification

Aerobacter aerogenes [1, 5]; Clostridium pasteurianum [4]; E. coli [6]; Eriobotrya japonica (similar enzyme with the name sorbitol-6-phosphate dehydrogenase from loquat (Eriobotrya japonica) catalyzes following reaction: sorbitol 6-phosphate + $NADP^+ \rightarrow$ glucose 6-phosphate + NADPH) [7, 8]

Crystallization

–

Cloned

–

Renaturated

–

5 STABILITY

pH

Temperature (°C)

40 (20 min, 85% loss of activity) [6]

Oxidation

Organic solvent

General stability information

Storage

Frozen, several weeks [5]; –20°C, more than 4 months [6]

6 CROSSREFERENCES TO STRUCTURE DATABANKS

PIR/MIPS code

PIR1:DEECSP (Escherichia coli)

Brookhaven code

7 LITERATURE REFERENCES

[1] Liss, M., Horwitz, S.B., Kaplan, N.O.: J. Biol. Chem.,237,1342–1350 (1962)
[2] Sadegh Roohi, M., Mitchell, W.J.: J. Gen. Microbiol.,133,2207–2215 (1987)
[3] Dills, S.S., Seno, S.: J. Bacteriol.,153,861–866 (1983)
[4] Du Toit, P.J., Kotze, J.P.: Biochim. Biophys. Acta,206,333–342 (1970)
[5] Horwitz, S.B.: Methods Enzymol.,9,150–155 (1966) (Review)
[6] Novotny, M.J., Reizer, J., Esch, F., Saier, M.: J. Bacteriol.,159,986–990 (1984) (Review)
[7] Hirai, M.: Plant Physiol.,63,715–717 (1979)
[8] Hirai, M.: Plant Physiol.,67,221–224 (1981)

1 NOMENCLATURE

EC number
1.1.1.141

Systematic name
(5Z,13E)-(15S)-11alpha,15-Dihydroxy-9-oxoprost-13-enoate:NAD^+ 15-oxidoreductase

Recommended name
15-Hydroxyprostaglandin dehydrogenase (NAD^+)

Synonymes
NAD^+-dependent 15-hydroxyprostaglandin dehydrogenase (type I) [2]
PGDH [10]
11alpha,15-Dihydroxy-9-oxoprost-13-enoate:NAD^+ 15-oxidoreductase [13]
15-OH^-PGDH [14]
Dehydrogenase, 15-hydroxyprostaglandin
15-Hydroxyprostanoic dehydrogenase
NAD-specific 15-hydroxyprostaglandin dehydrogenase
Prostaglandin dehydrogenase
More (cf. EC 1.1.1.196 and EC 1.1.1.197)

CAS Reg. No.
9030-87-9

2 REACTION AND SPECIFICITY

Catalysed reaction
(5Z,13E)-(15S)-11alpha,15-Dihydroxy-9-oxoprost-13-enoate + NAD^+ →
→ (5Z,13E)-11alpha-hydroxy-9,15-dioxoprost-13-enoate + NADH (single displacement mechanism [5, 12], ordered bi-bi-mechanism [9, 13], NAD^+ added first, followed by prostaglandin E_2, 15-ketoprostaglandin is released followed by NADH [13])

Reaction type
Redox reaction

Natural substrates
15-Hydroxyprostaglandin + NAD^+ (physiological inactivation of prostaglandin by catalyzing the first step in catabolism) [3, 12, 16]

Substrate spectrum

1 (5Z,13E)-(15S)-11alpha,15-Dihydroxy-9-oxoprost-13-enoate + NAD^+ (i.e. prostaglandin E_1, r [4, 12]) [4, 8, 10, 12, 16]
2 Prostaglandin F_{2alpha} + NAD^+ [1, 4, 8, 10, 12]
3 Prostaglandin E_2 + NAD^+ [1, 4, 6, 8, 10, 12, 13, 16]
4 Prostaglandin A_1 + NAD^+ [4, 10, 12, 16]
5 6-Ketoprostaglandin E_1 + NAD^+ [10]
6 6-Ketoprostaglandin E_{1alpha} + NAD^+ [10]
7 Prostaglandin D_2 + NAD^+ [10]
8 Prostaglandin A_2 + NAD^+ [16]
9 More (not: prostaglandin B_1 [10], prostaglandin D_2) [10]

Product spectrum

1 (5Z,13E)-11alpha-Hydroxy-9,15-dioxoprost-13-enoate + NADH (i.e. 15-ketoprostaglandin E_1) [4]
2 15-Ketoprostaglandin F_{2alpha} + NADH
3 15-Ketoprostaglandin E_2 + NADH
4 15-Ketoprostaglandin A_1 + NADH
5 ?
6 ?
7 ?
8 ?
9 ?

Inhibitor(s)

Ethacrynic acid [5]; Epi-7-thiaprostaglandin F_{2alpha} [5]; 7-Thia-13-prostynoic acid [5]; Ent-13-dehydroprostaglandin E_2 [5]; Nat-13-thiaprostaglandin F_{2alpha} [5]; Prostanoic acid [5, 12]; 7-Oxoprostanoic acid [5]; 7-Oxa-13-prostynoic acid [5]; Prostaglandin B_1 [5, 12]; Ent-13-dehydro-15-epiprostaglandin F_{3alpha} [5]; 13-cis-Prostaglandin F_{2alpha} [5]; 15-Epiprostaglandin E_1 [5]; Papaverine [7]; Ethanol (slight) [8]; Glycerol (slight) [8]; ICI 81,008 [9]; p-Chloromercuriphenylsulfonic acid [10]; N-Ethylmaleimide [10, 14]; Adenosine-5'-diphosphoribose [13]; Prostaglandin E_2 (product inhibition) [13]; NAD^+ (product inhibition) [13]; N-Chlorosuccinimide [14]; 2,4,6-Trinitrobenzenesulfonic acid [14]; Diethyldicarbonate [14]; Phenylglyoxal [14]; Indomethacin [15]; Acetylsalicylate [15]; NADH [1, 8, 13]; Tyroxine [1]; 3,3',5-Triiodothyronine [1]; Tetraiodothyroacetic acid [1]; Arachidonic acid [5]; More (not: NADPH, thyronine) [1]

Cofactor(s)/prosthetic group(s)/activating agents

NAD^+ [1, 4, 6, 8, 10, 12, 13, 16]; Thionicotinamide adenine dinucleotide (can replace NAD^+ [4], 5–33% of the activity with NAD^+, depending on substrate [4, 12]) [4, 12]; NADH [4, 12]; More (not: $NADP^+$ [4, 10, 12], acetylpyridine dinucleotide [12]) [4, 10, 12]

Metal compounds/salts

Turnover number (min^{-1})

Specific activity (U/mg)

More (assay procedure using (15S)-[15–^{3}H]prostaglandin E_2 [17]) [1, 11, 12, 17]; 25 [10]; 1791 [3]

K_m-value (mM)

0.0026 (prostaglandin E_2) [9]; 0.0054 (prostaglandin E_1) [9]; 0.0213 (prostaglandin F_{2alpha}) [9]; 0.014 (prostaglandin E_2) [1]; 0.025 (prostaglandin F_{2alpha}) [1]; 0.033 (prostaglandin E_1) [10]; 0.038 (prostaglandin A_1) [10]; 0.0399 (NAD^+ (+ prostaglandin F_{2alpha})) [9]; 0.0405 (NAD^+ (+ prostaglandin E_2)) [9]; 0.0451 (NAD^+ (+ prostaglandin E_1)) [9]; 0.059 (prostaglandin E_2) [10]; 0.066 (6-ketoprostaglandin E_1) [10]; 0.111 (6-ketoprostaglandin E_{1alpha}) [10]; 0.133 (prostaglandin F_{2alpha}) [10]; 0.741 (prostaglandin D_2) [10]; More [5, 12]

pH-optimum

7.2–8.5 [1]; 7.5–8.8 [16]; 9.5 [8]; 10.0–10.4 (prostaglandin E_1, 3-(cyclohexylamine)propanesulfonic acid buffer) [10]

pH-range

Temperature optimum (°C)

25 (assay at) [12]; 37 (assay at) [14]

Temperature range (°C)

3 ENZYME STRUCTURE

Molecular weight

24500 (human, FPLC) [3]
32000 (human, gel filtration in presence of mercaptoethanol) [8]
40000 (bovine, gel filtration) [15]
51500 (human, gel filtration) [4, 12]
54000 (human, gel filtration) [11]
55000 (pig, gel filtration) [14]

Subunits

Monomer (1 × 24000, human, SDS-PAGE [3], 1 × 42000, human, SDS-PAGE [4, 12]) [3, 4, 12]
Dimer (2 × 29000, human [11], pig [14], SDS-PAGE) [11, 14]
? (x × 29000, rabbit, SDS-PAGE) [10]
More (primary structure) [11]

Glycoprotein/Lipoprotein

–

4 ISOLATION/PREPARATION

Source organism

Rat [7]; Rabbit [10]; Monkey [1]; Chicken [1, 7]; Dog [1, 7]; Pig [2, 7, 13, 14, 17]; Human [3–5, 8, 9, 11, 12, 16]; Bovine [6, 15]; Guinea pig [7]

Source tissue

Lung [1, 6, 7, 10, 15]; Heart [1, 7]; Liver [1]; Kidney (medulla, cortex [2]) [1, 2, 7, 13, 14]; Spleen [1]; Placenta [3–5, 8, 9, 11, 12, 16]

Localisation in source

Purification

Rabbit [10]; Chicken (partial) [1]; Pig (partial [2]) [2, 14]; Human [3, 8, 11, 12, 16]; Bovine [6, 15]; Guinea pig (partial) [7]

Crystallization

–

Cloned

–

Renaturated

–

5 STABILITY

pH

Temperature (°C)

Oxidation

Organic solvent

General stability information

Glycerol, 50%, stabilizes [4, 8, 12]; NAD^+ stabilizes [5, 12]; No stabilization by addition of prostaglandins [5]

Storage

–20°C, 50% glycerol, 2 months, 8% loss of activity [4]; –20°C [6, 12]; –20°C, 50% glycerol, 10 mM 2-mercaptoethanol, stable for at least 1 year [8]; –80°C, 3–4 weeks [10]

6 CROSSREFERENCES TO STRUCTURE DATABANKS

PIR/MIPS code

PIR2:A35802 (placental human)

Brookhaven code

7 LITERATURE REFERENCES

[1] Lee, S.-C., Levine, L.: J. Biol. Chem.,250,548–552 (1975)
[2] Lee, S.-C., Pong, S.-S., Katzen, D., Wu, K.-Y., Levine, L.: Biochemistry,14,142–145 (1975)
[3] Tombach, B., Kusseler, R., Schlegel, W.: J. Chromatogr.,521,231–238 (1990)
[4] Braithwaite, S.S., Jabarak, J.: J. Biol. Chem.,250,2315–2318 (1975)
[5] Jabarak, J., Braithwaite, S.S.: Arch. Biochem. Biophys.,177,245–254 (1976)
[6] Saeed, S.A., Roy, A.C.: Biochem. Biophys. Res. Commun.,47,96–102 (1972)
[7] Iijima, Y., Ueno, T., Sasagawa, K., Yamazaki, M.: Biochem. Biophys. Res. Commun.,80,484–489 (1978)
[8] Jung, A., Schlegel, W., Jackisch, R., Friedrich, E.J., Wendel, A., Rückrich, M.F.: Hoppe-Seyler's Z. Physiol. Chem.,356,787–798 (1975)
[9] Rückrich, M.F., Wendel, A., Schlegel, W., Jackisch, R., Jung, A.: Hoppe-Seyler's Z. Physiol. Chem.,356,799–809 (1975)
[10] Bergholte, J.M., Okita, R.T.: Arch. Biochem. Biophys.,245,308–315 (1986)
[11] Krook, M., Marekov, L., Jörnvall, H.: Biochemistry,29,738–743 (1990)
[12] Jabarak, J.: Methods Enzymol.,86,126–130 (1982) (Review)
[13] Kung-Chao, D. T.-Y., Tai, H.-H.: Biochim. Biophys. Acta,614,14–24 (1980)
[14] Mak, O. T., Liu, Y., Tai, H.-H.: Biochim. Biophys. Acta,1035,190–196 (1990)
[15] Hansen, H.S.: Prostaglandins,8,95–105 (1974)
[16] Schlegel, W., Demers, L.M., Hildebrandt-Stark, H.E., Behrman, H.R., Greep, R.O.: Prostaglandins,5,417–433 (1974)
[17] Tai, H.-H.: Methods Enzymol.,86,131–135 (1982) (Review)

1 NOMENCLATURE

EC number
1.1.1.142

Systematic name
5D-5-O-Methyl-chiro-inositol:NADP$^+$ oxidoreductase

Recommended name
D-Pinitol dehydrogenase

Synonymes
Dehydrogenase, D-pinitol
More (no real separation of the activities of EC 1.1.1.142 and EC 1.1.1.143 has been achieved) [1]

CAS Reg. No.
37250-71-8

2 REACTION AND SPECIFICITY

Catalysed reaction
5D-5-O-Methyl-chiro-inositol + NADP$^+$ →
→ 5D-5-O-methyl-2,3,5/4,6-pentahydroxycyclohexanone + NADPH

Reaction type
Redox reaction

Natural substrates
5D-5-O-Methyl-2,3,5/4,6-pentahydroxycyclohexanone + NADPH (final step in biosynthesis of D-pinitol, no real separation of the activities of EC 1.1.1.142 and EC 1.1.1.143 has been achieved) [1]

Substrate spectrum
1 5D-5-O-Methyl-2,3,5/4,6-pentahydroxycyclohexanone + NADPH (r) [1]
2 More (no real separation of the activities EC 1.1.1.142 and EC 1.1.1.14 has been achieved)

Product spectrum
1 5D-5-O-Methyl-chiro-inositol (i.e. pinitol) + NADP$^+$ [1]
2 ?

Inhibitor(s)
SH-compounds (slight) [1]; Cu^{2+} [1]; Cysteine (slight) [1]; Fe^{2+} [1]; Glutathione (slight) [1]; p-Chloromercuribenzoate (slight) [1]; Phenylmercurinitrate (slight) [1]

Cofactor(s)/prosthetic group(s)/activating agents

NADP+ (A-specificity, i.e. transfer of pro-R hydrogen to cofactor) [1]; NADPH (A-specificity, i.e. transfer of pro-R hydrogen from cofactor) [1]

Metal compounds/salts

Turnover number (min^{-1})

Specific activity (U/mg)

K_m-value (mM)

pH-optimum

6–8.5 (broad) [1]

pH-range

4.8–9.5 (4.8: about 45% of activity maximum, 9.5: about 30% of activity maximum) [1]

Temperature optimum (°C)

Temperature range (°C)

---

3 ENZYME STRUCTURE

Molecular weight

34000 (Trifolium incarnatum, gel filtration, no real separation of the activities of EC 1.1.1.142 and EC 1.1.1.143 has been achieved) [1]

Subunits

Glycoprotein/Lipoprotein

–

---

4 ISOLATION/PREPARATION

Source organism

Trifolium incarnatum (no real separation of the activities of EC 1.1.1.142 and EC 1.1.1.143 has been achieved) [1]

Source tissue

Cell [1]

Localisation in source

Purification

Trifolium incarnatum (no real separation of the activities of EC 1.1.1.142 and EC 1.1.1.143 has been achieved) [1]

Crystallization

–

Cloned

–

Renaturated

–

5 STABILITY

pH

More (relatively stable against weak acids) [1]

Temperature (°C)

Oxidation

Organic solvent

General stability information

Purification increases stability [1]; DEAE-cellulose chromatography causes 70% loss of activity [1]

Storage

2°C, 3 weeks, purified state, little loss of activity [1]

6 CROSSREFERENCES TO STRUCTURE DATABANKS

PIR/MIPS code

Brookhaven code

7 LITERATURE REFERENCES

[1] Ruis, H., Hoffmann-Osterhof, O.: Eur. J. Biochem.,7,442–448 (1969)

1 NOMENCLATURE

EC number
1.1.1.143

Systematic name
5-O-Methyl-myo-inositol:NAD^+ oxidoreductase

Recommended name
Sequoyitol dehydrogenase

Synonymes
Dehydrogenase, sequoyitol
More (no real separation of the activities of EC 1.1.1.142 and EC 1.1.1.143 has been achieved) [1]

CAS Reg. No.
37250-72-9

2 REACTION AND SPECIFICITY

Catalysed reaction
5-O-Methyl-myo-inositol + NAD^+ →
→ 5D-5-O-methyl-2,3,5/4,6-pentahydroxycyclohexanone + NADH

Reaction type
Redox reaction

Natural substrates
5-O-Methyl-myo-inositol + NAD^+ (the second but last step in biosynthesis of D-pinitol, no real separation of the activities of EC 1.1.1.142 and EC 1.1.1.143 has been achieved) [1]

Substrate spectrum
1 5-O-Methyl-myo-inositol + NAD^+ (r) [1]

Product spectrum
1 5D-5-O-Methyl-2,3,5/4,6-pentahydroxycyclohexanone + NADH [1]

Inhibitor(s)
SH-reagents [1]; Iodoacetate [1]; p-Chloromercuribenzoate [1]; Phenylmercurinitrate [1]

Cofactor(s)/prosthetic group(s)/activating agents
NADH (B-specificity, i.e. transfer of pro-S hydrogen from cofactor) [1]; NAD^+ (B-specificity, i.e. transfer of pro-S hydrogen to cofactor) [1]

Metal compounds/salts

Turnover number (min^{-1})

Specific activity (U/mg)

K_m-value (mM)

pH-optimum
5.5–8.0 (broad) [1]

pH-range
4.2–10 (4.2: about 20% of activity maximum, 10: about 30% of activity maximum) [1]

Temperature optimum (°C)

Temperature range (°C)

---

3 ENZYME STRUCTURE

Molecular weight
34000 (Trifolium incarnatum, gel filtration, no real separation of the 2 activities of EC 1.1.1.142 and EC 1.1.1.143 has been achieved) [1]

Subunits

Glycoprotein/Lipoprotein
–

---

4 ISOLATION/PREPARATION

Source organism
Trifolium incarnatum (no real separation of the activities of EC 1.1.1.142 and EC 1.1.1.143 has been achieved) [1]

Source tissue
Cell [1]

Localisation in source

Purification

Trifolium incarnatum (no real separation of the activities of EC 1.1.1.142 and EC 1.1.1.143 has been achieved) [1]

Crystallization

–

Cloned

–

Renaturated

–

5 STABILITY

pH

More (relatively stable against weak acids) [1]

Temperature (°C)

Oxidation

Organic solvent

General stability information

DEAE-cellulose chromatography destroys activity completely [1]

Storage

6 CROSSREFERENCES TO STRUCTURE DATABANKS

PIR/MIPS code

Brookhaven code

7 LITERATURE REFERENCES

[1] Ruis, H., Hoffmann-Ostenhof, O.: Eur. J. Biochem.,7,442–448 (1969)

1 NOMENCLATURE

EC number
1.1.1.144

Systematic name
Perillyl-alcohol:NAD^+ oxidoreductase

Recommended name
Perillyl-alcohol dehydrogenase

Synonymes
Dehydrogenase, perillyl alcohol
Perillyl alcohol dehydrogenase

CAS Reg. No.
37250-73-0

2 REACTION AND SPECIFICITY

Catalysed reaction
Perillyl alcohol + NAD^+ →
→ perillyl aldehyde + NADH

Reaction type
Redox reaction

Natural substrates
Perillyl alcohol + NAD^+ (catalyzes dehydrogenation reaction which is a part of the catabolic sequence of alpha-pinenes and beta-pinenes, limonene, $DELTA^1$-p-methene and p-cymene) [1]

Substrate spectrum
1 Perillyl alcohol + NAD^+ (r, rate of the reverse reaction is about 20% of the rate of the forward reaction [1], oxidizes a number of primary alcohols with the alcohol group allylic to an endocyclic double bond and a 6-membered ring, either aromatic or hydroaromatic [1], e.g.: 8-hydroxyphellandrol [1, 2], cumic alcohol [1, 2], p-ethyl benzyl alcohol [1], p-methyl benzyl alcohol [1]) [1, 2]

Product spectrum
1 Perillyl aldehyde + NADH [1, 2]

Inhibitor(s)

o-Phenanthroline (at low concentration of NAD^+) [1]; p-Hydroxymercuribenzoate [1]; Iodoacetate [1]; Hg^{2+} [1]; Ag^{2+} [1]; Cu^{2+} [1]; Ni^{2+} [1]; Zn^{2+} [1]

Cofactor(s)/prosthetic group(s)/activating agents

NAD^+ (also active with some NAD^+ analogues, e.g.: 3-acetylpyridine adenine dinucleotide, 3-pyridinealdehyde adenine dinucleotide, deamino-NAD) [1]; NADH [1]; More (inactive with $NADP^+$) [1]

Metal compounds/salts

Turnover number (min^{-1})

Specific activity (U/mg)

13.0 [1]

K_m-value (mM)

0.077 (NAD^+) [1]; More [1]

pH-optimum

7.0 (NADH + perillyl aldehyde) [1]; 9.4 (NAD^+ + perillyl alcohol) [1]

pH-range

5–8 (5: about 35% of activity maximum, 8: about 70% of activity maximum, NADH + perillyl aldehyde) [1]; 6.3–10.5 (at pH 6.3 and 10.5: about 10% of activity maximum, perillyl alcohol + NAD^+) [1]

Temperature optimum (°C)

Temperature range (°C)

---

3 ENZYME STRUCTURE

Molecular weight

Subunits

Glycoprotein/Lipoprotein

–

---

4 ISOLATION/PREPARATION

Source organism

Pseudomonas sp. (PL-strain [1], from soil [2]) [1, 2]

Source tissue

Cell [1]

Localisation in source

Purification

Pseudomonas sp. (PL-strain [1], from soil [2]) [1, 2]

Crystallization

–

Cloned

–

Renaturated

–

5 STABILITY

pH

Temperature (°C)

Oxidation

Organic solvent

General stability information

Storage

–20°C, for at least 3 days, DEAE eluate [1]

6 CROSSREFERENCES TO STRUCTURE DATABANKS

PIR/MIPS code

Brookhaven code

7 LITERATURE REFERENCES

[1] Ballal, N.R., Bhattacharyya, P.K., Rangachari, P.N.: Indian J. Biochem.,5,1–6 (1968)
[2] Ballal, N.R., Bhattacharyya, P.K., Rangachari, P.N.: Biochem. Biophys. Res. Commun.,23,473–478 (1966)

1 NOMENCLATURE

EC number
1.1.1.145

Systematic name
3beta-Hydroxy-DELTA^5-steroid:NAD$^+$ 3-oxidoreductase

Recommended name
3beta-Hydroxy-DELTA^5-steroid dehydrogenase

Synonymes
Progesterone reductase
Dehydrogenase, 3beta-hydroxy-DELTA^5-steroid
DELTA5–3beta-Hydroxysteroid dehydrogenase
3beta-Hydroxy-5-ene steroid dehydrogenase
3beta-Hydroxy steroid dehydrogenase/isomerase
3beta-Hydroxy-DELTA^5-C_{27}-steroid dehydrogenase/isomerase
3beta-Hydroxy-DELTA^5-C_{27}-steroid oxidoreductase
3beta-Hydroxy-5-ene-steroid oxidoreductase
Steroid-DELTA^5–3beta-ol dehydrogenase
3beta-HSDH [4]
5-Ene-3-beta-hydroxysteroid dehydrogenase [5]
3beta-Hydroxy-5-ene-steroid dehydrogenase [11]

CAS Reg. No.
9044-85-3

2 REACTION AND SPECIFICITY

Catalysed reaction
3beta-Hydroxy-5-ene-teroid + NAD$^+$ →
→ 3-oxo-5-ene-steroid + NADH

Reaction type
Redox reaction

Natural substrates

Substrate spectrum

1 Pregnenolone + $NAD(P)^+$ (i.e. 3beta-hydroxy-5-pregnen-20-one) [1, 3, 9]
2 3beta,17alpha-Dihydroxy-5-pregnen-20-one + $NAD(P)^+$ [1]
3 Dehydroepiandrosterone + NAD^+ (i.e. 3beta-hydroxyandrost-5-en-17-one) [2, 3, 8, 12]
4 3beta-Hydroxyandrostan-17-one + NAD^+ [8]
5 4-Androsten-3beta,17beta-diol + NAD^+ [8]
6 More (specific for 3beta-OH group in steroids with trans configuration of A/B rings and DELTA^5-double bond [13], no reaction with OH in 3alpha-, 17beta-, 11beta-, 21-position [8], no reaction with OH in 17alpha-, 17beta-, 20alpha-, 20beta-position [9]) [8, 9, 13]

Product spectrum

1 Pregn-5-en-3,20-dione + NAD(P)H (or progesterone i.e. pregn-4-ene-3,20-dione [9]) [1, 9]
2 17alpha-Hydroxypregn-5-ene-3,20-dione + NAD(P)H [1]
3 Androst-5-en-3,17-dione + NADH
4 Androstan-3,17-dione + NADH
5 4-Androsten-17beta-ol-3-one + NADH
6 ?

Inhibitor(s)

Diethyldicarbonate [1]; Trilostane (i.e. 4alpha,5-epoxy-3,17beta-dihydroxy-5alpha-androst-2-en-2-carbonitrile) [1]; p-Chloromercuribenzoate [1]; 5,10-Secoestr-4-yne-3,10,17-trione (irreversible) [3]; Mn^{2+} (1mM: weak) [4]; Fe^{2+} (1 mM: weak, 100 mM: strong) [4]; Co^{2+} (1 mM: weak, 100 mM: strong) [4]; Ni^{2+} (1 mM: weak, 100 mM: strong) [4]; Cu^{2+} (1 mM: strong) [4]; Zn^{2+} (1 mM: strong) [4]; Cd^{2+} (1 mM: strong) [4]; Progesterone [5]; 20alpha-Hydroxy-4-pregnen-3-one [5]; 20beta-Hydroxy-4-pregnen-3-one [5]; Androstenedione [5]; Estradiol-17beta [5]; 5'-Fluorosulfonylbenzoyl adenosine [11]; N-(Anilino-naphthyl-4)-maleimide [6]; N-Iodoacetyl-N'-(5-sulfo-1-naphthyl) [6]; 3-Chloroacetyladenine dinucleotide [6]; Norethisterone (Pseudomonas testosteroni) [12]; Norethisterone acetate (Pseudomonas testosteroni) [12]; Ethynylestradiol [12]

Cofactor(s)/prosthetic group(s)/activating agents

NAD^+ (preferred [9], more than 10fold as active as $NADP^+$ [4], cysteinyl and histidyl at NAD^+-binding site [1]) [1–12]; $NADP^+$ (33 times less effective than NAD^+ [9], 10 times less effective than NAD^+ [4], low activity [1], not [8]) [1, 3, 4, 9, 10]; Iodoacetamide (activation) [6]; p-Chloromercuribenzoate (activation) [6]; More (no reaction with 3-acetylpyridine-adenine dinucleotide, nicotinamide-hypoxanthine dinucleotide, 3-acetylpyridine-hypoxanthine dinucleotide) [8]

Metal compounds/salts

More (no effect of Ca^{2+}, Sr^{2+}, Ba^{2+}) [4]

Turnover number (min^{-1})

Specific activity (U/mg)
88.9 [2]; 45 [8]; 0.677 [1]

K_m-value (mM)
0.00007 (pregnenolone) [4]; 0.002 (pregnenolone) [1]; 0.005 (dehydroepiandrosterone, rat) [12]; 0.0053 (3beta,17alpha-dihydroxy-5-pregnen-20-one) [1]; 0.0062–0.0066 (NAD^+) [1]; 0.01 (dehydroepiandrosterone, Pseudomonas testosteroni) [12]; 0.02 (NAD^+) [4]; 0.045 (NAD^+) [9]; 4.9 ($NADP^+$) [1]; 13 ($NADP^+$) [9]; More [5, 6]

pH-optimum
7.2 [8]; 7.5 [13]; 7.8–8.5 [1]; 9.5 [4]

pH-range
6.8–9.0 [8]

Temperature optimum (°C)
40–45 [4]; 45 [13]

Temperature range (°C)
20–55 [4]

---

3 ENZYME STRUCTURE

Molecular weight
300000 (sheep, gel filtration of Triton X-100 solubilized enzyme) [7]

Subunits
? (x × 40000, sheep, SDS-PAGE [7], x × 41000, bovine, SDS-PAGE [1], x × 46500–46800, rat [10, 11]) [1, 7, 10, 11]

Glycoprotein/Lipoprotein
–

---

4 ISOLATION/PREPARATION

Source organism
Bovine [1, 6, 8]; Pseudomonas sp. TB [2]; Human (single enzyme molecule catalyzing 3beta-hydroxy-DELTA^5-steroid dehydrogenation and steroid DELTA-isomerization, EC 1.1.1.145 and EC 5.3.3.1 [3]) [3, 4]; Gallus domesticus [5]; Sheep (single enzyme molecule catalyzing DELTA^5–3beta-hydroxysteroid dehydrogenation and DELTA^5-isomerization) [7]; Rat (single enzyme molecule catalyzing 3beta-hydroxy-5-ene-steroid dehydrogenation and 5-ene-4-ene-isomerization [10, 11]) [9–12]; Pseudomonas testosteroni [12]; Streptomyces griseocarneus [13]

Source tissue

Adrenal gland [1, 6, 8–10]; Placenta [3, 4]; Preovulatory follicular granulosa cells [5]; Ovary [8]; Testis [10–12]

Localisation in source

Microsomes [1, 3–8, 11, 12]

Purification

Bovine [1, 8]; Pseudomonas sp. TB [2]; Sheep [7]; Rat [11]; Streptomyces griseocarneus (partial) [13]

Crystallization

–

Cloned

–

Renaturated

–

5 STABILITY

pH

Temperature (°C)

37 (30 min, 50% loss of activity) [4]; 55 (50 min, in microsomes 40–55% loss of activity, solubilized 80% loss of activity) [5]; 60 (microsomal fraction, inactivation) [5]

Oxidation

Organic solvent

General stability information

Storage

–80°C, 0.1 M potassium phosphate buffer, pH 7.4, 10 mg protein/ml, at least 3 months [1]; –20°C, 2 mM potassium phosphate buffer, pH 7.0, 50% glycerol, 18 months, 50–70% loss of activity [2]

6 CROSSREFERENCES TO STRUCTURE DATABANKS

PIR/MIPS code

PIR3:S30509 (human)

Brookhaven code

7 LITERATURE REFERENCES

[1] Hiwatashi, A., Hamamoto, I., Ichikawa, Y.: J. Biochem.,98,1519–1526 (1985)
[2] Shikita, M., Talalay, P.: Anal. Biochem.,95,286–292 (1979)
[3] Thomas, J.L., Strickler, R.C., Myers, R.P., Covey, D.F: Biochemistry,31,5522–5527 (1992)
[4] Rabe, T., Brandstetter, K., Kellermann, J., Runnebaum, B.: J. Steroid Biochem.,17, 427–433 (1982)
[5] Armstrong, D.G.: J. Steroid Biochem.,17,225–230 (1982)
[6] Vincent, M., Gallay, J., de Paillerets, C., Alfsen, A., Biellmann, J.F.: Biochim. Biophys. Acta,525,1–8 (1978)
[7] Ford, H.C., Engel, L.L.: J. Biol. Chem.,249,1363–1368 (1974)
[8] Cheatum, S.G., Warren, J.C.: Biochim. Biophys. Acta,122,1–13 (1966)
[9] Koritz, S.B.: Biochemistry,3,1098–1102 (1964)
[10] Ishii-Ohba, H., Inano, H., Bunichi, T.: J. Steroid Biochem.,27,775–779 (1987)
[11] Ishii-Ohba, H., Inano, H., Tamaoki, B.: J. Steroid Biochem.,25,555–560 (1986)
[12] Spona, J.: Endocrinol. Exp.,17,107–118 (1983)
[13] Kerenyi, G., Szentirmai, A., Natonek, M.: Acta Microbiol. Acad. Sci. Hung.,22,487–496 (1975)

1 NOMENCLATURE

EC number
1.1.1.146

Systematic name
11beta-Hydroxysteroid:$NADP^+$ 11-oxidoreductase

Recommended name
11beta-Hydroxysteroid dehydrogenase

Synonymes
beta-Hydroxysteroid dehydrogenase [3]
11beta-Hydroxy steroid dehydrogenase
Corticosteroid 11-reductase
Dehydrogenase, 11beta-hydroxy steroid

CAS Reg. No.
9041-46-7

2 REACTION AND SPECIFICITY

Catalysed reaction
11beta-Hydroxysteroid + $NADP^+$ →
→ 11-oxosteroid + NADPH (sequential bi-bi mechanism [2])

Reaction type
Redox reaction

Natural substrates

Substrate spectrum
1 Corticosterone + $NADP^+$ (i.e. 11beta,21-dihydroxypregn-4-en-3,20-dione, r [3, 4], ir [2]) [2–5]
2 Cortisol + $NADP^+$ (i.e. 11beta,17alpha,21-trihydroxy-pregn-4-en-3,20-dione, r [3]) [1, 3]
3 11-Oxoprogesterone + NADPH (i.e. pregn-4-en-3,11,20-trione) [3]
4 Androst-4-en-3,11,17-trione + NADPH [3]
5 Prednisone + NADPH (i.e. 17alpha,21-dihydroxypregna-1,4-diene-3,11,20-trione, r) [3]
6 17alpha,21-Dihydroxy-5beta-pregnan-3,11,20-trione + NADPH [3]
7 17alpha,21-Dihydroxy-5alpha-pregnan-3,11,20-trione + NADPH [3]
8 3alpha-Hydroxy-5beta-androstan-11,17-dione + NADPH [3]
9 3alpha-Hydroxy-5alpha-androstan-11,17-dione + NADPH (r) [3]
10 9alpha-Fluorocortisone + NADPH (ir) [3]
11 9alpha-Fluoro-11-oxoprogesterone + NADPH (ir) [3]

12 12alpha-Fluoro-11-dehydrocorticosterone + NADPH (ir) [3]
13 16beta-Methylcortisol + $NADP^+$ [3]
14 16alpha-Methylcortisol + $NADP^+$ [3]

Product spectrum

1 11-Dehydrocorticosterone + NADPH (i.e. 21-hydroxy-pregn-4-en-3,11,20-trione) [3]
2 Cortisone + NADPH (i.e. 17alpha,21-dihydroxy-pregn-4-en-3,11,20-trione, r [3]) [3]
3 11beta-Hydroxyprogesterone + $NADP^+$ (i.e. 11beta-hydroxy-pregn-4-en-3,20-dione)
4 11beta-Hydroxyandrost-4-en-3,17-dione + $NADP^+$
5 Prednisolone + $NADP^+$ (i.e. 11beta,17alpha,21-trihydroxypregna-1,4-diene-3,20-dione) [3]
6 11beta,17alpha,21-Trihydroxy-5beta-pregnan-3,20-dione + $NADP^+$
7 11beta,17alpha,21-Trihydroxy-5alpha-pregnan-3,20-dione + $NADP^+$
8 3alpha,11beta-Dihydroxy-5beta-androstan-17-one + $NADP^+$ [3]
9 3alpha,11beta-Dihydroxy-5alpha-androstan-17-one + $NADP^+$ [3]
10 9alpha-Fluorocortisol + $NADP^+$
11 9alpha-Fluoro-11beta-hydroxyprogesterone + $NADP^+$
12 12alpha-Fluoro-corticosterone + $NADP^+$
13 16beta-Methylcortisone + NADPH
14 16alpha-Methylcortisone + NADPH

Inhibitor(s)

Glycyrrhetinic acid (i.e. 3beta-hydroxy-11-oxo-18beta-olean-12-en-30-oic acid, inhibition of dehydrogenase and reductase activity [4]) [2, 4]; p-Chloromercuribenzoate [3]; More (substrate inhibition [1], not: iodosobenzoate [3]) [1, 3]

Cofactor(s)/prosthetic group(s)/activating agents

$NADP^+$ [1–5]; NADPH [3, 4]; NAD^+ (low activity) [3]; NADH (low activity) [3]

Metal compounds/salts

Turnover number (min^{-1})

Specific activity (U/mg)

K_m-value (mM)

0.00183 (corticosterone) [7]; 0.0173 (cortisol) [7]; More (overview cortisol, values depend on organism and enzyme preparation) [3]

pH-optimum

7.0 (reduction of 11-ketosteroids) [4]; 8.5 (dehydrogenation of 11beta-hydroxysteroids) [4]; 8.5–9.5 (dehydrogenation of 11beta-hydroxysteroids) [5]; 10 (dehydrogenation of 11beta-hydroxysteroids, latent enzyme form) [5]

pH-range

Temperature optimum (°C)

Temperature range (°C)

---

3 ENZYME STRUCTURE

Molecular weight

Subunits
? (x × 34000, rat, SDS-PAGE) [7]

Glycoprotein/Lipoprotein
Glycoprotein [7]

---

4 ISOLATION/PREPARATION

Source organism
Human [1]; Rat (exposure of microsomes to detergent releases a latent form [5], no 11-oxosteroid reductase activity [7]) [2–5, 7]; Mammals (11beta-hydroxysteroid dehydrogenase and 11-oxosteroid reductase) [6]; Birds (11-oxosteroid reductase) [6]; Dogfish (11-oxosteroid reductase) [6]; More (no 11beta-hydroxysteroid dehydrogenase activity in frog, toad, mud puppy, shark, bird, no 11-oxosteroid reductase activity in amphibians, bony fish) [6]

Source tissue
Placenta [1]; Liver [2–7]

Localisation in source
Microsomes [1–7]

Purification
Rat [3, 7]

Crystallization
–

Cloned
(expression of rat liver cDNA in vaccinia virus) [4]

Renaturated
–

5 STABILITY

pH

Temperature (°C)

Oxidation

Organic solvent

General stability information

Storage

6 CROSSREFERENCES TO STRUCTURE DATABANKS

PIR/MIPS code

PIR1:DXHUBH (human); PIR1:DXRTBH (rat); PIR3:S33117 (sheep)

Brookhaven code

7 LITERATURE REFERENCES

[1] Pearson Murphy, B.E.: J. Steroid Biochem.,14,807–809 (1981)
[2] Monder, C., Lakshmi, V., Miroff, Y.: Biochim. Biophys. Acta,1115,23–29 (1991)
[3] Bush, I.E., Hunter, S.A., Meigs, R.A.: Biochem. J.,107,239–258 (1968)
[4] Agarwal, A.K., Tusie-Luna, M.T., Monder, C., White, P.C.: Mol. Endocrinol.,4, 1827–1832 (1990)
[5] Monder, C., Lakshmi, V.: J. Steroid Biochem.,32,77–83 (1989)
[6] Monder, C., Lakshmi, V.: Steroids,52,515–528 (1988)
[7] Lakshmi, V., Monder, C.: Endocrinology,123,2390–2398 (1988)

1 NOMENCLATURE

EC number
1.1.1.147

Systematic name
16alpha-Hydroxysteroid:NAD(P)$^+$ 16-oxidoreductase

Recommended name
16alpha-Hydroxysteroid dehydrogenase

Synonymes
Dehydrogenase, 16alpha-hydroxy steroid

CAS Reg. No.
37250-74-1

2 REACTION AND SPECIFICITY

Catalysed reaction
16alpha-Hydroxysteroid + NAD(P)$^+$ →
→ 16-oxosteroid + NAD(P)H

Reaction type
Redox reaction

Natural substrates

Substrate spectrum
1 16-Keto-17beta-estradiol + NADH (r) [1, 2]
2 16-Keto-17alpha-estradiol + NADH (r) [1]
3 16-Ketotestosterone + NADH (r) [1]
4 3-Methoxyestra-1,3,5(10)-triene-16alpha,17beta-diol + NAD$^+$ [1]
5 Estra-1,3,5(10)-triene-3,16alpha-diol + NAD$^+$ [1]

Product spectrum
1 16alpha-Hydroxy-17beta-estradiol + NAD$^+$ [1]
2 16alpha-Hydroxy-17alpha-estradiol + NAD$^+$ [1]
3 16alpha,17beta-Dihydroxyandrost-4-en-3-one + NAD$^+$ [1]
4 3-Methoxyestra-1,3,5(10)-triene-17beta-ol-16-one + NADH
5 Estra-1,3,5(19)-triene-3-ol-16-one + NADH

Inhibitor(s)

Cofactor(s)/prosthetic group(s)/activating agents

NADH [1, 2]; NAD^+ [1, 2]; NADPH (can replace NADH under saturating conditions) [1]

Metal compounds/salts

Turnover number (min^{-1})

Specific activity (U/mg)

K_m-value (mM)

0.07 (16-keto-17alpha-estradiol) [1]; 0.089 (16-keto-17beta-estradiol) [1]; 0.28 (16-ketotestosterone) [1]

pH-optimum

6.1–6.2 (reduction of 16-keto-steroids) [1]

pH-range

Temperature optimum (°C)

Temperature range (°C)

---

3 ENZYME STRUCTURE

Molecular weight

Subunits

Glycoprotein/Lipoprotein

–

---

4 ISOLATION/PREPARATION

Source organism

Rat [1]; Human [2]

Source tissue

Kidney [1]; Placenta [2]

Localisation in source

Soluble part of cell [1, 2]

Purification

Rat (partial) [1]

Crystallization

–

Cloned

–

Renaturated

–

5 STABILITY

pH

Temperature (°C)

Oxidation

Organic solvent

General stability information

Storage

6 CROSSREFERENCES TO STRUCTURE DATABANKS

PIR/MIPS code

Brookhaven code

7 LITERATURE REFERENCES

[1] Meigs, R.A., Ryan, K.J.: J. Biol. Chem.,241,4011–4015 (1966)
[2] Preumont, P., Smuk, M.: Acta Endocrinol.,78,760–765 (1975)

1 NOMENCLATURE

EC number
1.1.1.148

Systematic name
17alpha-Hydroxysteroid:NAD(P)$^+$ 17-oxidoreductase

Recommended name
Estradiol 17alpha-dehydrogenase

Synonymes
17alpha-Estradiol dehydrogenase
17alpha-Hydroxy steroid dehydrogenase
17alpha-Hydroxy steroid oxidoreductase
17alpha-Hydroxysteroid oxidoreductase
Estradiol 17alpha-oxidoreductase

CAS Reg. No.
9044-91-1

2 REACTION AND SPECIFICITY

Catalysed reaction
Estradiol-17alpha + NAD(P)$^+$ →
→ estrone + NAD(P)H

Reaction type
Redox reaction

Natural substrates

Substrate spectrum
1 17alpha-Estradiol + NAD(P)$^+$ (r [5]) [1–3, 5, 6]
2 5alpha-Androstan-3beta,17alpha-diol + NADP$^+$ [2]
3 Epitestosterone + NADP$^+$ [2, 3]
4 Testosterone + NADP$^+$ [2]
5 Estradiol-17alpha 3-glucuronide + NADP$^+$ [3, 6]

Product spectrum
1 Estrone + NAD(P)H [5]
2 3beta-Hydroxy-5alpha-androstan-17-one + NADPH
3 ?
4 ?
5 Estrone 3-glucuronide + NADPH

Inhibitor(s)

17beta-(1-Oxoprop-2-ynyl)androst-4-ene-3-one [2]

Cofactor(s)/prosthetic group(s)/activating agents

NAD^+ (preferred [1]) [1–3, 5, 6]; $NADP^+$ (40% of NAD^+ activity [1], preferred [5], activity 20 times greater than with NAD^+ [6]) [1–3, 5, 6]; NADPH [5]

Metal compounds/salts

Turnover number (min^{-1})

Specific activity (U/mg)

0.2 [2]; More [4]

K_m-value (mM)

0.0002–0.0052 (epitestosterone) [3]; 0.001–0.0051 (17alpha-estradiol 3-glucuronide) [3]; 0.0024–0.0088 (17alpha-estradiol) [3]; 0.0017 (17alpha-estradiol) [1]; 0.0038 ($NADP^+$) [5]; 0.0087–0.011 ($NADP^+$) [2]; 0.017–0.025 (17alpha-estradiol) [2]

pH-optimum

9.0–9.5 [1, 2, 5]

pH-range

8–10 [1]

Temperature optimum (°C)

Temperature range (°C)

---

3 ENZYME STRUCTURE

Molecular weight

52500 (horse, gel filtration) [1]

Subunits

? (x × 33000, horse, SDS-PAGE [1], x × 39600, rabbit, SDS-PAGE, all 8 forms [3]) [1, 3]

Glycoprotein/Lipoprotein

–

---

4 ISOLATION/PREPARATION

Source organism

Horse [1]; Chicken [2, 5]; Rabbit [3, 4, 6]

Source tissue

Placenta [1]; Liver [2–6]

Localisation in source
Microsomes [1]; Cytosol [2–5]

Purification
Horse [1]; Chicken (2 forms [2], partial [5]) [2, 5]; Rabbit (8 forms [4]) [4, 6]

Crystallization
–

Cloned
–

Renaturated
–

5 STABILITY

pH

Temperature (°C)

Oxidation

Organic solvent

General stability information

Storage
–20°C, 50% v/v glycerol [2]; 4°C, phosphate buffer, pH 7.0, 1 mM dithiothreitol, at least 1 month [1]; 4°C, 5 mM phosphate buffer, pH 7.2, 20% v/v glycerol, at least 3 days [5]

6 CROSSREFERENCES TO STRUCTURE DATABANKS

PIR/MIPS code

Brookhaven code

7 LITERATURE REFERENCES

[1] LaRhee, L.H., Warren, J.C.: Biochemistry,23,486–491 (1984)
[2] Johnston, J., Renwick, A.G.C.: Biochem. J.,222,761–768 (1984)
[3] Hasnain, S., Williamson, D.G.: Biochem. J.,161,279–283 (1977)
[4] Hasnain, S., Williamson, D.G.: Biochem. J.,147,457–461 (1975)
[5] Renwick, A.G.C., Engel, L.L.: Biochim. Biophys. Acta,146,336–348 (1967)
[6] Hasnain, S., Williamson, D.G.: Can. J. Biochem.,52,120–125 (1974)

1 NOMENCLATURE

EC number
1.1.1.149

Systematic name
20alpha-Hydroxysteroid:NAD(P)$^+$ 20-oxidoreductase

Recommended name
20alpha-Hydroxysteroid dehydrogenase

Synonymes
Dehydrogenase, 20alpha-hydroxy steroid
20alpha-Hydroxy steroid dehydrogenase
20alpha-HSD [2]
20alpha-HSDH [3]

CAS Reg. No.
9040-08-8

2 REACTION AND SPECIFICITY

Catalysed reaction
17alpha,20alpha-Dihydroxypregn-4-en-3-one + NAD(P)$^+$ →
→ 17alpha-hydroxyprogesterone + NAD(P)H

Reaction type
Redox reaction

Natural substrates

Substrate spectrum
1 Progesterone + NAD(P)H (r [5–8]) [1, 5–8, 16]
2 Cortisone + NADH [3]
3 Cortisol + NADH [3]
4 11-Desoxycortisol + NADH (i.e. 17,21-dihydroxy-pregn-4-en-3,20-dione) [3]
5 5beta-Dihydrocortisol + NADH (i.e. 11beta,17,21-trihydroxy-5beta-pregnan-3,20-dione, r) [3]
6 17alpha-Hydroxyprogesterone + NAD(P)H (r [5]) [5, 9–15]
7 More (no 17alpha- or 17beta-hydroxysteroid dehydrogenase activities) [3, 15]

Product spectrum

1 20alpha-Hydroxy-pregn-4-en-3-one + $NAD(P)^+$ [2, 16]
2 17alpha,20alpha,21-Trihydroxy-pregn-4-en-3,11-dione + NAD^+
3 11beta,17alpha,20alpha,21-Tetrahydroxy-pregn-4-en-3-one + NAD^+
4 17,20alpha,21-Trihydroxy-pregn-4-en-3-one + NAD^+
5 11beta,17,20alpha,21-Tetrahydroxy-5beta-pregnan-3-one + NAD^+
6 17alpha,20alpha-Dihydroxy-pregn-4-en-3-one + $NAD(P)^+$ [10, 11, 13–15]
7 ?

Inhibitor(s)

16alpha-Bromoacetoxyprogesterone [1]; 2',5'-Adenosine diphosphate [4]; 5'-Adenylic acid [4]; 2'-Adenylic acid [4]; 5'-Adenosine diphosphate [4]; 3',5'-Cyclic AMP [4]; 5'-Adenosine triphosphate [4]; Adenosine diphosphoribose [4]; N^1-Alkylnicotinamide chlorides [4]; 2'-Phosphoadenosine diphosphoribose [4, 7]; N-Alkylammonium chlorides [4]; Nicotinamide adenine dinucleotide [4]; 3-Aminopyridine adenine dinucleotide phosphate [4, 6]; 3-Aminopyridine 1,N^6-ethenoadenine dinucleotide phosphate [4]; Diazotized 3-aminopyridine adenine dinucleotide phosphate [4]; Diazotized 3-aminopyridine 1,N^6-ethenoadenine dinucleotide phosphate [4]; $NADP^+$ [4]; 5alpha-Dihydrotestosterone [5]; 5beta-Dihydrotestosterone [5]; 2,4-Pentanedione [6]; Fluorescein mercuriacetate [6]; Acetic acid [4]; Propionic acid [4]; Pentanoic acid [4]; Hexanoic acid [4]; Octanoic acid [4]; Nonanoic acid [4]; Decanoic acid [4]; Undecanoic acid [4]; N-Etylmaleimide [6]; N-Butyl-bis-maleimide [6]; N-Octylmaleimide [6]; 2'-Phospho-3-aminopyridine adenine dinucleotide phosphate [6]; Androstendione [8]; Progesterone [8]; 17alpha-Hydroxyprogesterone [8]; Glycerol [10]; p-Chloromercuribenzoate [12]; Iodoacetamide [12]; Ag^+ [12]; Cu^{2+} [12]; Pyridoxal 5'-phoshate [12]

Cofactor(s)/prosthetic group(s)/activating agents

$NADP^+$ (A-specific, i.e. transfer of hydrogen to pro-R position of cofactor [11, 13, 16], preferred [14, 15]) [1, 3, 6–8, 11, 12–16]; NADPH (A-specific, i.e. transfer of hydrogen from pro-R position of cofactor [11, 13, 16], 50% of NADH-activity [3], preferred [14, 15]) [1, 3, 6–8, 11–16]; NADH (not [8]) [3, 12, 14, 15]

Metal compounds/salts

Turnover number (min^{-1})

Specific activity (U/mg)

6.5 [2]; 3.48 [7]; 0.282 [3]; More [12, 15]

K_m-value (mM)

0.0007–0.0008 ($NADP^+$) [7]; 0.00087 ($NADP^+$, isozyme HSD2) [2]; 0.00105 ($NADP^+$, isozyme HSD1) [2]; 0.001–0.0016 (20alpha-hydroxypregn-4-en-3-one, NADPH) [7]; 0.0023 (20alpha-hydroxy-4-pregnen-3-one) [5]; 0.00475 (20alpha-hydroxypregn-4-en-3-one, isozyme HSD1) [2]; 0.00516 (20alpha-hydroxy-pregn-4-en-3-one, isozyme HSD2) [2]; 0.00538 (progesterone) [7]; 0.0081 (NADH) [3]; 0.022–0.032 (cortisone, cortisol [3], $NADP^+$ [8], 17alpha-hydroxyprogesterone, isozyme HSD-I [14]) [3, 8, 14]; 0.041 (11beta,17,20alpha,21-tetrahydroxy-pregn-4-en-3-one) [3]; 0.052 (NADPH) [12]; 0.073 (17-hydroxyprogesterone) [15]; 0.118 (20alpha-hydroxyprogesterone, isozyme HSD-II) [14]; 0.2 (progesterone) [1]; 0.3 (20alpha-hydroxyprogesterone) [8]; 0.526 (NAD^+) [3]; 0.75 (NADH) [12]; 2.5 (20alpha-hydroxyprogesterone) [1]

pH-optimum

5.6 (steroid reduction) [15]; 5.5–7.5 (NADPH) [12]; 6.0–6.5 (steroid reduction) [3]; 6.8–7.2 (NADH) [12]; 7.8–8.2 (oxidation of 20alpha-hydroxyprogesterone) [8]; 8 (steroid oxidation) [3]

pH-range

6.3–9.0 (steroid oxidation) [3]; 6.5–9.0 (steroid oxidation) [8]

Temperature optimum (°C)

50 [12]

Temperature range (°C)

0–55 (10% of maximal reaction rate at 0°C) [12]

3 ENZYME STRUCTURE

Molecular weight

158000–162000 (Clostridium scindens, pore gradient gel electrophoresis, gel filtration HPLC) [3]
75100 (rat, gel filtration) [17]
41000 (rat, size exclusion HPLC, non-reducing conditions) [2]
34000–35000 (rat, non-denaturing gel electrophoresis [7], pig, sucrose and glycerol gradient centrifugation [10], pig, gel filtration [12], bovine, gel filtration [15]) [7, 10, 12, 15]
30000 (pig, gel filtration) [14]

Subunits

Monomer (1 × 33000–40000, rat, SDS-PAGE [2, 7], pig, SDS-PAGE [12, 14], bovine, SDS-PAGE [15]) [2, 7, 12, 14, 15]
Tetramer (4 × 40000, identical, Clostridium scindens, SDS-PAGE, N-terminal amino acid sequence) [3]

Glycoprotein/Lipoprotein

Glycoprotein [14]

---

4 ISOLATION/PREPARATION

Source organism

Human [1, 5]; Rat [2, 4, 6–9, 16, 17]; Clostridium scindens [3]; Pig [10–12, 14]; Bovine [13, 15]

Source tissue

Placenta [1, 5]; Ovary [2, 4, 6–9, 16, 17]; Testis [10, 11, 15]; Adrenal gland [14]

Localisation in source

Cytosol [1, 2, 7, 9, 12]; Microsomes [5]

Purification

Human (copurification with 17beta-estradiol dehydrogenase, EC 1.1.1.62) [1]; Rat (isozyme HSD1, HSD2 [2]) [2, 7, 9]; Clostridium scindens [3]; Pig [12, 14]; Bovine [15]

Crystallization

–

Cloned

–

Renaturated

–

---

5 STABILITY

pH

5.5–10 [12]; 6–7 [3]

Temperature (°C)

25 (half-life 20 min) [7]; 38 (half-life 7.6 min) [7]; 60 (10 min, complete inactivation) [12]; More ($NADP^+$ protects against thermal inactivation) [7]

Oxidation

Organic solvent

General stability information

2-Mercaptoethanol: stabilization [3]; Glycerol: stabilization [3]; Inactivation by freezing/thawing [12]; $NADP^+$ protects against thermal inactivation [7]

Storage

–20°C or 4°C, protein concentration above 0.5 mg/ml, 0.05 M phosphate buffer, pH 6.8, 50% glycerol, 10 mM 2-mercaptoethanol, several days [3]; –15°C, 0.01 M potassium phosphate buffer, pH 8.0, 50% propylene glycol, 0.2 M KCl, protein concentration above 5 microgram/ml, at least 2 months [7]; –20°C, 0.1 M phosphate buffer, pH 7.0, several months [12]; 4°C, glycerol, DTT [15]

6 CROSSREFERENCES TO STRUCTURE DATABANKS

PIR/MIPS code

PIR2:A46379 (bovine (fragment)); PIR2:A44755 (Clostridium scindens (fragment)); PIR2:A46379 (bovine (fragment))

Brookhaven code

7 LITERATURE REFERENCES

[1] Strickler, R.C., Tobias, B., Covey, D.F.: J. Biol. Chem.,256,316–321 (1981)
[2] Noda, K., Shiota, K., Takahashi, M.: Biochim. Biophys. Acta,1079,112–118 (1991)
[3] Krafft, A.E., Hylemon, P.B.: J. Bacteriol.,171,2925–2932 (1989)
[4] Pongsawasdi, P., Anderson, B.M.: Arch. Biochem. Biophys.,238,280–289 (1985)
[5] Blomquist, C.H., Lindemann, N.J., Hakanson, E.Y.: Arch. Biochem. Biophys.,239, 206–215 (1985)
[6] Pongsawasdi, P., Anderson, B.M.: Arch. Biochem. Biophys.,233,481–488 (1984)
[7] Pongsawasdi, P., Anderson, B.M.: Biochim. Biophys. Acta,799,51–58 (1984)
[8] Robertson, W.R., Frost, J., Hoyer, P.E., Weinkove, C.: J. Steroid Biochem.,17, 237–243 (1982)
[9] Mori, M., Wiest, W.G.: J. Steroid Biochem.,11,1443–1449 (1979)
[10] Shikita, M., Tsuneoka, K.: FEBS Lett.,66,4–7 (1976)
[11] Hatano-Sato, F., Takagi, Y., Shikita, M.: J. Biochem.,74,1065–1067 (1973)
[12] Sato, F., Takagi, Y., Shikita, M.: J. Biol. Chem.,247,815–823 (1972)
[13] Pineda, J.A., Murdock, G.L., Watson, R.J., Warren, J.C.: J. Steroid Biochem.,33, 1223–1228 (1989)
[14] Nakajin, S., Kawai, Y., Ohno, S., Shinoda, M.: J. Steroid Biochem.,33,1181–1189 (1989)
[15] Pineda, J.A., Salinas, M.E., Warren, J.C.: J. Steroid Biochem.,23,1001–1006 (1985)
[16] Kersey, W.H., Wilcox, R.B.: Biochemistry,9,1284–1286 (1970)
[17] Kersey, W.H.: Tex. J. Sci.,26,607–611 (1975)